Atmospheric Modeling, Analysis and Applications

Atmospheric Modeling, Analysis and Applications

Editor: Bruce Mullan

R CALLISTO
REFERENCE

www.callistoreference.com

Callisto Reference,
118-35 Queens Blvd., Suite 400,
Forest Hills, NY 11375, USA

Visit us on the World Wide Web at:
www.callistoreference.com

ISBN: 978-1-64116-043-8 (Hardback)

Trademark Notice: Registered trademark of products or corporate names are used only for explanation and identification without intent to infringe.

Cataloging-in-Publication Data

Atmospheric modeling, analysis and applications / edited by Bruce Mullan.
    p. cm.
Includes bibliographical references and index.
ISBN 978-1-64116-043-8
1. Atmospheric models. 2. Air--Analysis . 3. Atmospheric chemistry. I. Mullan, Bruce.
QC866 .A86 2019
551--dc23

Table of Contents

Preface

The atmospheric model is a set of mathematical equations that represent atmospheric processes and motions. Dynamic parameterizations of radiation, heat exchange, soil, vegetation, etc. are used here. The objective of atmosphere modeling is to predict phenomena such as tornadoes, boundary layer eddies, etc. This book delves into innovative techniques of modeling and analysis of the atmosphere along with their applications in diverse fields. This book includes some of the vital pieces of work being conducted across the world, on various topics related to atmospheric modeling. Students, researchers, experts and all associated with atmospheric science will benefit alike from this book.

This book is the end result of constructive efforts and intensive research done by experts in this field. The aim of this book is to enlighten the readers with recent information in this area of research. The information provided in this profound book would serve as a valuable reference to students and researchers in this field.

At the end, I would like to thank all the authors for devoting their precious time and providing their valuable contribution to this book. I would also like to express my gratitude to my fellow colleagues who encouraged me throughout the process.

Editor

Long-term O$_3$–precursor relationships in Hong Kong: field observation and model simulation

Yu Wang[1,*]**, Hao Wang**[1,*]**, Hai Guo**[1]**, Xiaopu Lyu**[1]**, Hairong Cheng**[2]**, Zhenhao Ling**[3]**, Peter K. K. Louie**[4]**,
Isobel J. Simpson**[5]**, Simone Meinardi**[5]**, and Donald R. Blake**[5]

[1]Air Quality Studies, Department of Civil and Environmental Engineering, the Hong Kong Polytechnic
University, Hong Kong SAR, China
[2]Department of Environmental Engineering, School of Resource and Environmental Sciences, Wuhan University,
Wuhan 430079, China
[3]School of Atmospheric Sciences, Sun Yat-sen University, Guangzhou, China
[4]Air Group, Hong Kong Environmental Protection Department, Hong Kong SAR, China
[5]Department of Chemistry, University of California, Irvine, CA, USA
[*]These authors contributed equally to this work.

Correspondence to: Hai Guo (ceguohai@polyu.edu.hk) and Hairong Cheng (chenghr@whu.edu.cn)

Abstract. Over the past 10 years (2005–2014), ground-level O$_3$ in Hong Kong has consistently increased in all seasons except winter, despite the yearly reduction of its precursors, i.e. nitrogen oxides (NO$_x$ = NO + NO$_2$), total volatile organic compounds (TVOCs), and carbon monoxide (CO). To explain the contradictory phenomena, an observation-based box model (OBM) coupled with CB05 mechanism was applied in order to understand the influence of both locally produced O$_3$ and regional transport. The simulation of locally produced O$_3$ showed an increasing trend in spring, a decreasing trend in autumn, and no changes in summer and winter. The O$_3$ increase in spring was caused by the net effect of more rapid decrease in NO titration and unchanged TVOC reactivity despite decreased TVOC mixing ratios, while the decreased local O$_3$ formation in autumn was mainly due to the reduction of aromatic VOC mixing ratios and the TVOC reactivity and much slower decrease in NO titration. However, the decreased in situ O$_3$ formation in autumn was overridden by the regional contribution, resulting in elevated O$_3$ observations. Furthermore, the OBM-derived relative incremental reactivity indicated that the O$_3$ formation was VOC-limited in all seasons, and that the long-term O$_3$ formation was more sensitive to VOCs and less to NO$_x$ and CO in the past 10 years. In addition, the OBM results found that the contributions of aromatics to O$_3$ formation decreased in all seasons of these years, particularly in autumn, probably due to the effective control of solvent-related sources. In contrast, the contributions of alkenes increased, suggesting a continuing need to reduce traffic emissions. The findings provide updated information on photochemical pollution and its impact in Hong Kong.

1 Introduction

Ozone (O$_3$), one of the most important photochemical products influencing atmospheric oxidative capacity, human and vegetation health, and climate change, is formed through a series of photochemical reactions among volatile organic compounds (VOCs) and nitrogen oxides (NO$_x$) in the atmosphere (Seinfeld and Pandis, 2006). Due to the non-linear relationship between O$_3$ and its precursors, the development of appropriate control measures for O$_3$ is still problematic in megacities (Sillman, 1999).

Distinguished from short-term O$_3$ studies, investigation of long-term O$_3$ variations enables us to understand the seasonal and inter-annual characteristics of O$_3$, the influence of meteorological parameters on O$_3$ formation, and the O$_3$–precursor relationships in different years. Subsequently, more effective and sustainable O$_3$ control strategies can be formulated and implemented. Hence, earlier efforts have been made to investigate long-term variations of O_3 in different atmospheric conditions. For example, multi-year data analysis showed that O_3 levels started to decrease around 2000 in Europe (e.g. Jungfraujoch, Zugspitze, Mace Head) and North America excluding western US rural sites (e.g. US Pacific, Lassen Volcanic National Park; Lefohn et al., 2010; Cui et al., 2011; Parrish et al., 2012; Pollack et al., 2013; Lin et al., 2017), due to a decrease in the emissions of O_3 precursors since the early 1990s (Cui et al., 2011; Derwent et al., 2013). In contrast, the O_3 levels in East Asia increased at a rate of 1.0 ppbv yr^{-1} from 1998 to 2006, based on measurements at Mt Happo, Japan (Parrish et al., 2012; Tanimoto, 2009). In China, with rapid economic growth and urbanization over the past three decades, increasing O_3 levels have been found at many locations. For instance, based on the data collected between 1991 and 2006 at Lin'an, a NO_x-limited rural area close to Shanghai, Xu et al. (2008) reported that the maximum mixing ratios of O_3 increased by 2.0, 2.7, 2.4, and 2.0 % yr^{-1} in spring, summer, autumn, and winter, respectively, which were probably related to increased mixing ratios of NO_2. In the North China Plain (NCP), Ding et al. (2008) reported that O_3 in the lower troposphere over Beijing had a positive trend of ~ 2.0 % yr^{-1} from 1995 to 2005, while Zhang et al. (2014) found that the daytime average O_3 in summer in Beijing significantly increased by 2.6 ppbv yr^{-1} from 2005 to 2011, due to decreased NO titration (-1.4 ppbv yr^{-1} of NO_x over the study period) and elevated regional background O_3 levels (~ 0.58–1.0 ppbv yr^{-1}) in the NCP.

Hong Kong, together with the inland Pearl River Delta (PRD) region of southern China, has suffered from high O_3 mixing ratios in recent years (Chan et al., 1998a, b; Wang and Kwok, 2003; Ding et al., 2004; Zhang et al., 2007; Guo et al., 2009, 2013). In 2014, O_3 exceeded the Chinese national air quality standard (80 ppbv, for the daily 8 h maximum average, DMA8) on more than 90 days in some areas of the PRD, with the highest DMA8 value of 165 ppbv (GDEMC and HKEPD, 2015). Based on the observational data at a newly established regional monitoring network, Li et al. (2014) found that O_3 mixing ratios in the inland PRD region increased at a rate of 0.86 ppbv yr^{-1} from 2006 to 2011 because of the rapid reduction of NO in this VOC-limited region. Similarly, Wang et al. (2009) reported a continuous record of increased surface O_3 at Hok Tsui (HT), a regional background site in Hong Kong, with a rate of 0.58 ppbv yr^{-1} based on observations conducted from 1994 to 2007, concluding that the increased NO_2 column concentration in upwind eastern China might significantly contribute to the increased O_3 in Hong Kong. Even so, knowledge gaps still exist regarding the long-term characteristics of O_3, long-term O_3–precursor relationships, and the mechanisms for the varying O_3 trends in the PRD region, because of the lack of long-term observations of VOCs in the region, where photochemical O_3 formation is sensitive to VOCs in

urban areas, and where the levels of VOCs and NO_x have varied significantly due to more stringent control measures since 2005 (Zhong et al., 2013). It is noteworthy that although Xue et al. (2014) reported increasing O_3 trends in the period 2002–2013 in Hong Kong and investigated the roles of VOCs and NO_x in the long-term O_3 variations, only data from autumn were used, which could not provide a consistent and full picture of the long-term variations of O_3, VOCs, NO_x, and their relationships.

In this study, field measurements and model simulations were combined to characterize the long-term variations of O_3 and its precursors, the variations of locally produced O_3, and the impact of regional transport in Hong Kong from 2005 to 2014. In addition, the long-term contribution of different VOC groups to the O_3 formation was explored. The study aims to provide the most up-to-date information on the characteristics of photochemical pollution and its impact in Hong Kong.

2 Methodology

2.1 Site description

Field measurements were carried out at the Tung Chung (TC) Air Quality Monitoring Station managed by the Hong Kong Environmental Protection Department (HKEPD). The sampling site (22.29° N, 113.94° E) is located at about 24 km southwest of downtown Hong Kong and about 3 km south of Hong Kong International Airport (Fig. 1). The elevation of TC is 37.5 m above sea level. It is surrounded by a newly developed residential town on the northern Lantau Island, and is downwind of urban Hong Kong and the inland PRD region when easterly and northeasterly winds are prevailing (Ou et al., 2015). At TC, the prevailing wind varies by season, being from the east in spring and autumn, from the southwest in summer, and from the northeast in winter (see Fig. S1 in the Supplement). The selection of this site for the trend study was due to its downwind location being a good receptor for urban plume, suffering high O_3 pollution, and having the most comprehensive dataset. More detailed description of the TC site can be found in our previous papers (Jiang et al., 2010; Cheng et al., 2010; Ling et al., 2013; Ou et al., 2015).

2.2 Measurement techniques

Hourly observations of O_3, CO, SO_2, NO-NO_2-NO_x, and meteorological parameters at TC from 2005 to 2014 were obtained from the HKEPD (http://epic.epd.gov.hk/ca/uid/airdata). Briefly, O_3 was measured using a commercial UV photometric instrument (Advanced Pollution Instrumentation (API), model 400A) with a detection limit of 0.6 ppbv. CO was measured with a gas filter correlation CO analyser (Thermo Electron Corp. (TECO), model 48C) with a detection limit of 0.04 ppm. SO_2 was measured using a pulsed fluorescence analyser (TECO, model 43A) with a detec-

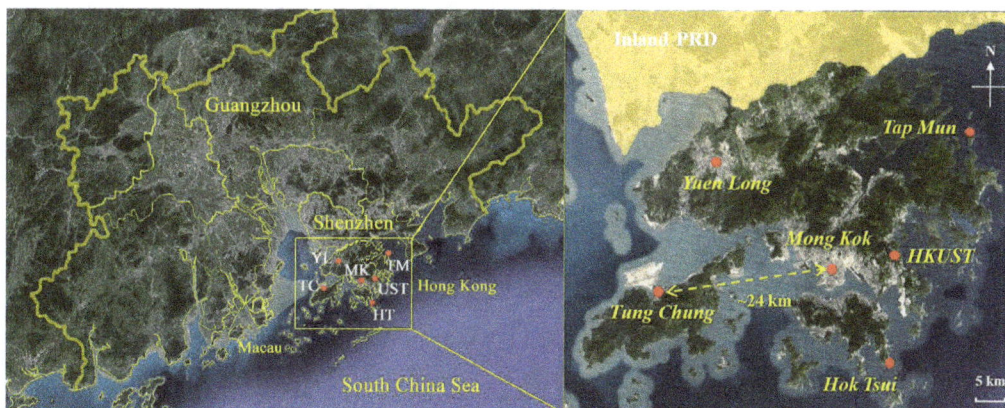

Figure 1. Location of the sampling sites and surrounding environments. Guangzhou and Shenzhen are the two biggest cities in the inland PRD region with a population of over 10 million for each city. Hok Tsui (HT) and Tap Mun (TM) are regional background sites. The Hong Kong University of Science and Technology (HKUST) is an Air Quality Research Supersite located in a suburban area. Yuen Long (YL) is a typical urban site adjacent to main traffic roads and surrounded by residential and industrial blocks. Mong Kok is a typical downtown roadside site with high traffic density.

tion limit of 1.0 ppbv. NO-NO_2-NO_x was detected using a commercial chemiluminescence with an internal molybdenum converter (API, model 200A) and a detection limit of 0.4 ppbv. All the time resolutions for these gas analysers are 1 h. To ensure a high degree of accuracy and precision, the QA/QC procedures for gaseous pollutants were identical to those in the US air quality monitoring programme (http://epic.epd.gov.hk/ca/uid/airdata). The accuracy of the monitoring network is assessed by performance audits, while the precision, a measure of the repeatability, of the measurements is checked in accordance with HKEPD's quality manuals. For the gaseous pollutants, accuracy and precision within the limits of ± 15 and $\pm 20\%$ are adopted, respectively (HKEPD, 2015).

Real-time VOC data at TC were also measured by the HKEPD. An online GC-FID analyser (Synspec GC 955, Series 600/800) was used to collect VOC speciation data continuously with a time resolution of 30 min. The VOC analyser consists of two separate systems for detection of C_2–C_5 and C_6–C_{10} hydrocarbons, respectively. Detailed description about the real-time VOC analyser can be found in Lyu et al. (2016). There were 28 C_3–C_{10} VOC species identified and quantified with this method. In terms of the QA/QC for VOC analysis, built-in computerized programmes of quality control systems such as auto-linearization and auto-calibration were used. Weekly calibrations were conducted by using NPL standard gas (National Physical Laboratory, Teddington, Middlesex, UK). In general, the detection limits of the target VOCs ranged from 2 to 56 pptv. The accuracy of each species measured by online GC-FID was determined by the percentage difference between measured mixing ratio and actual mixing ratio based on weekly span checks and monthly calibrations. The precision was based on the 95 % probability limits for the integrated precision check results. The accu-

racy of the measurements was about 1–7 %, depending on the species, and the measurement precision was about 1–10 % (Table S1 in the Supplement). In addition, the quality of the real-time data was assured by regular comparison with whole-air canister samples collected and analysed by the University of California at Irvine (UCI). More details can be found in reports of previous studies in Hong Kong (Xue et al., 2014; Ou et al., 2015; Lyu et al., 2016).

For data analysis, linear regression and error bars representing 95 % confidence intervals were used. Trends of O_3 and its precursors with a p value < 0.05 were considered significant (Guo et al., 2009).

2.3 Observation-based model

In this study, an observation-based box model (OBM) coupled with a carbon bond mechanism (CB05) was used to simulate photochemical O_3 formation and to evaluate the sensitivity of O_3 formation to its precursors. The CB05 mechanism is a condensed mechanism with high computational efficiency and reliable simulation, and has been successfully applied in many emission-based modelling systems such as Weather Research and Forecasting with Chemistry (WRF/Chem) and the Community Multiscale Air Quality (CMAQ; Yarwood et al., 2005; Coates and Butler, 2015). Unlike emission-based models, the OBM in this study is based on the real-time observations at the TC site in Hong Kong. The simulation was constrained by observed hourly data of meteorological parameters (temperature, relative humidity, and pressure) and air pollutants (NO, NO_2, CO, SO_2, and 22 C_3–C_{10} VOCs). In the CB05 module, VOCs are grouped according to carbon bond type and the reactions of individual VOCs were condensed using a lumped structure technique (conversions from measured VOCs to CB05 grouped species are shown in Table S2). To better describe the photochemical reactions in Hong Kong, the photolysis rates of different species in the OBM were determined using the output of the Tropospheric Ultraviolet and Visible Radiation model (TUV v5; Madronich and Flocke, 1999) based on the actual conditions of Hong Kong, i.e. meteorological parameters, location, and time period of the field campaign. However, it is noteworthy that the atmospheric physical processes (i.e. vertical and horizontal transport), the deposition of species, and the radical loss to aerosol (George et al., 2013; Lakey et al., 2015) were not considered in the OBM. In addition, a "spin-up" time was not applied in the model to get the radical intermediates steady which might have caused a slight underestimation of the simulated O_3 production (Fig. S2) and its sensitivity to precursors (Fig. S3). In this study, we performed day-by-day OBM simulations for 2688 days during 2005–2014, where the missing days were due to lack of real-time VOC data (see Table S3). For each daily simulation, the model was run for a 24 h period with 00:00 (local time, LT) as the initial time. The model output simulated mixing ratios of O_3, radicals (i.e. OH, HO_2, RO, and RO_2), and intermediates. The model performance was evaluated using the index of agreement (IOA; Huang et al., 2005; Wang et al., 2015, 2013; Lyu et al., 2015a):

$$\text{IOA} = 1 - \frac{\sum_{i=1}^{n}(O_i - S_i)^2}{\sum_{i=1}^{n}(|O_i - \overline{O}| + |S_i - \overline{O}|)^2}, \quad (1)$$

where S_i and O_i represent simulated and observed values, respectively, $\overline{O}$ represents the mean of observed values, and n is the number of samples. The IOA value lies between 0 and 1. The better the agreement between simulated results and observed data, the higher the IOA (Huang et al., 2005).

Apart from the OBM (CB05), which is mainly for condensed VOC groups, a Master Chemical Mechanism (MCM, v3.2) was applied to inter-compare the modelling performance of the OBM (CB05; shown in Sect. 3.2). Since the MCM utilizes the near-explicit mechanism describing the degradation of 143 primary VOCs and contains around 16 500 reactions involving 5900 chemical species, it has a better performance in calculating the contribution of individual VOCs to O_3 production (Jenkin et al., 1997, 2003; Saunders et al., 2003). The hourly input data of meteorological parameters, air pollutants, and the photolysis rates in MCM were the same as in CB05. It was assumed that the measured VOCs contributed a dominant fraction to O_3 production, and that the initial concentrations of those VOCs not measured but needed by MCM were zero. We acknowledge that the use of a limited number of VOCs causes some photochemical reactivity to be overlooked. For a more detailed description of the MCM, see Jenkin et al. (1997, 2003) and Saunders et al. (2003). Some developments on localization of the MCM for Hong Kong and the addition of chemical reaction pathways of more biogenic VOC species and alkyl nitrates are

given in our previous papers (Lam et al., 2013; Cheng et al., 2013; Ling et al., 2014; Lyu et al., 2015b).

The measured precursors (i.e. VOCs, NO, and NO_2) at TC are a mixture of regional background values augmented by local source influences, and the two parts are very difficult to fully separate. It is worth noting that the regional background values are those observed at locations where there is little influence from urban sources of pollution, while the baseline values mentioned in Sect. 3.2 are observations made at a site not influenced by recent, locally emitted (or produced) pollution (TF HTAP, 2010). To minimize the influence of regional transport from the inland PRD region, the real-time regional background values in this study were simply subtracted from the observations at TC. Previous studies have reported that Tap Mun (TM; 22.47° N, 114.36° E) and Hok Tsui (HT; 22.217° N, 114.25° E) are two background sites of Hong Kong (Lyu et al., 2016; So and Wang, 2003, 2004; Wang et al., 2005, 2009). TM is a rural site that is upwind of Hong Kong in autumn/winter seasons and HT is a background site at the southeastern tip of Hong Kong. Good trace gas correlations have been found between the two sites (Lyu et al., 2016). Since not all the data during the entire 10-year period were available at one background site, the hourly measured VOCs at HT and NO_2 at TM were treated as background values. The background data were excluded using the following equations:

$$[\text{VOC}]_{\text{local}} = [\text{VOC}]_{\text{observed}} - [\text{VOC}]_{\text{background}}, \quad (2)$$

$$[\text{NO}]_{\text{local}} = [\text{NO}]_{\text{observed}} - [\text{NO}]_{\text{background}}, \quad (3)$$

$$[\text{NO}_2]_{\text{local}} = [\text{NO}_2]_{\text{observed}} - [\text{NO}_2]_{\text{background}}, \quad (4)$$

where $[xx]_{\text{local}}$, $[xx]_{\text{observed}}$, and $[xx]_{\text{background}}$ represent the local, observed, and background values, respectively. In this study, mixing ratios of 21 anthropogenic VOC species with relatively long lifetimes (5 h–14 days) at HT were selected as the background values to deduct from the observed data at TC. The lifetimes of these VOCs were estimated based on the reactions with OH radicals (Simpson et al., 2010). The rate constants used were from Atkinson and Arey (2003) – assuming a 12 h daytime average OH radical concentration of 2.0×10^6 molecules cm^{-3}. Isoprene was considered as not having a regional impact due to its short lifetime (1–2 h; Ling et al., 2011). Furthermore, the lifetime of NO_2 is determined by the main sinks of $OH+NO_2$ reaction and the hydrolysis of N_2O_5 at the surface of wet aerosols, which highly depends on meteorological conditions, such as temperature and humidity (Dils, 2008; Evans and Jacob, 2005). Previous experimental studies showed an exponential relationship between the NO_2 lifetime and temperature (Dils, 2008; Merlaud et al., 2011; Rivera et al., 2013), which was used to estimate the lifetime of NO_2 in this study. The lifetime of NO_2 was calculated to be approximately 3.4 ± 0.3, 2.2 ± 0.1, 2.8 ± 0.2, and 5.2 ± 0.3 h in spring, summer, autumn, and winter, respectively, consistent with the lifetimes of NO_2 in different seasons in the PRD region (Beirle et al., 2011). Considering the shortest distance between the inland PRD and TC (i.e. from the centre of Shenzhen to the TC site, ~ 30 km) and the average wind speed in different seasons (Ou et al., 2015), it would take approximately 3.4 ± 0.3, 4.3 ± 0.3, 4.0 ± 0.5, and 3.7 ± 0.4 h in spring, summer, autumn, and winter, respectively, for NO_2 originating in the inland PRD to arrive at TC. Hence, although NO_2 emitted from the inland PRD is slightly more likely to arrive at TC in winter and spring than in summer and autumn, the differences in travel time between the seasons are relatively small and it is difficult to be precise about seasonal average estimates of NO_2 lifetime and travel time. We have excluded background NO_2 values in spring and winter in this study during model simulations, but we recognize the limitations in these calculations.

In addition to the simulation of O_3 formation, the precursor sensitivity of O_3 formation was assessed by the OBM using the relative incremental reactivity (RIR; Cardelino and Chameides, 1995; Lu et al., 2010; Cheng et al., 2010; Ling et al., 2011; Xue et al., 2014). A higher positive RIR of a given precursor means a greater probability that reducing emissions of this precursor will significantly reduce O_3 production. The RIR is defined as the percentage change in daytime O_3 production per percent change in precursor. The RIR for precursor X is given by

$$\mathrm{RIR}(X) = \frac{[\mathrm{P}^{\mathrm{S}}_{O_3-NO}(X) - \mathrm{P}^{\mathrm{S}}_{O_3-NO}(X-\Delta X)]/\mathrm{P}^{\mathrm{S}}_{O_3-NO}(X)}{\frac{\Delta S(X)}{S(X)}}, \quad (5)$$

where X represents a specific precursor (i.e. VOCs, NO_x, or CO); the superscript "S" is used to denote the specific site where the measurements were made; $S(X)$ is the measured mixing ratio of species X (ppbv); $\Delta S(X)$ is the hypothetical change in the mixing ratio of X; and $\mathrm{P}^{\mathrm{S}}_{O_3-NO}(X)$ and $\mathrm{P}^{\mathrm{S}}_{O_3-NO}(X-\Delta X)$ represent net O_3 production in a base run with original mixing ratios, and in a run with a hypothetical change ($\Delta S(X)$; 10 % $S(X)$ in this study) in species X. In both runs, O_3 production modulated by NO titration is considered during the evaluation period. The O_3 production $\mathrm{P}^{\mathrm{S}}_{O_3-NO}$ was calculated by the output parameters of the OBM.

$\mathrm{P}^{\mathrm{S}}_{O_3-NO}$ is derived from the difference between O_3 gross production rate $\mathrm{G}^{\mathrm{S}}_{O_3-NO}$ and O_3 destruction rate $D^{\mathrm{S}}_{O_3-NO}$ (Eq. 6). $\mathrm{G}^{\mathrm{S}}_{O_3-NO}$ is calculated by the oxidation of NO by HO_2 and RO_2 (Eq. 7), while $D^{\mathrm{S}}_{O_3-NO}$ is calculated by O_3 photolysis, reactions of O_3 with OH, HO_2 and alkenes, and reaction of NO_2 with OH (Eq. 8):

$$P^{\mathrm{S}}_{O_3-NO} = G^{\mathrm{S}}_{O_3-NO} - D^{\mathrm{S}}_{O_3-NO}, \quad (6)$$

$$G^{\mathrm{S}}_{O_3-NO} = k_{HO_2+NO}[HO_2][NO]$$
$$+ \sum k_{RO_2i+NO}[RO_2i][NO], \quad (7)$$

$$D^{\mathrm{S}}_{O_3-NO} = k_{HO_2+O_3}[HO_2][O_3] + k_{OH+O_3}[OH][O_3]$$
$$+ k_{o(^1D)+H_2O}[O(^1D)][H_2O]$$
$$+ k_{OH+NO_2}[OH][NO_2] + k_{alkenes+O_3}[alkenes][O_3]. \quad (8)$$

In Eqs. (7) and (8), k constants are the rate coefficients of their subscript reactions. Values of radicals and intermediates are simulated by the OBM. Details of the calculation can be found in Ling et al. (2014) and Xue et al. (2014).

Furthermore, the sensitivities in the OBM to the uncertainties in initial concentrations of O_3 precursors have been examined by running the model with varying NO_2 or VOCs initial concentrations in the range of ± 95 % confidence intervals, respectively. The results demonstrate that the modelled O_3 production was more sensitive to NO_2 than VOCs, with a percentage variation about ± 13 and ± 3.9 %, respectively (see Table S4, Figs. S4 and S5). In addition, the uncertainties associated with removing the background concentrations are also evaluated, suggesting a similar trend for simulated local O_3 production for both approaches (see Fig. S6 and Tables S5–S7).

3 Results and discussion

3.1 Long-term trends of O_3 and its precursors

Figure 2 shows trends of monthly averaged mixing ratios of O_3 and its precursors, namely NO_x, total VOCs (TVOCs), and CO measured at TC in the past 10 years. Please note that arithmetic means were used here in order to compare with other studies. TVOC was defined as the sum of the 22 VOC species listed in Text S1. Note that not all detected VOCs were included in this study because of high rates of missing data. The limited number of VOC precursors causes a reduced reactivity which was estimated at < 30 % for total hydrocarbons based on our previous study (Guo et al., 2004). The missing reactivity would be larger if carbonyls were considered (Cheng et al., 2010). It was found that both monthly averaged O_3 and monthly maximum O_3 increased at a rate of 0.56 ± 0.01 ppbv yr^{-1} ($p < 0.01$) and 1.92 ± 0.15 ($p < 0.05$), respectively. The monthly maximum O_3 level, which was defined as the maximum of DMA8 O_3 in 1 month, increased from about 68 ppbv in 2005 to 86 ppbv in 2014, exceeding the ambient air quality standard in Hong Kong (i.e. 80 ppbv). The number of days per year (days yr^{-1}) that DMA8 exceeded 80 ppbv also increased during the period 2005–2014 (1.16 ± 0.26 days yr^{-1}, $p < 0.05$; see Fig. S7), indicating increasing O_3 pollution in Hong Kong. This finding is consistent with other big cities and regions in the world, such as Beijing (Tang et al., 2009), the West Plains of Taiwan (Chou et al., 2006), and Osaka (Itano et al., 2007). The annual average O_3 concentration in Hong Kong increased by 0.56 ppbv yr^{-1} in the period 2004–2015, which is close to that reported for Osaka (0.6 ppbv yr^{-1}) in the period 1985–2002, and in agreement with Lin et al. (2017), who found that the annual mean O_3 over Hong Kong increased by about 0.5 ppbv yr^{-1} in the period 2000–2014. In contrast, NO_x and CO significantly decreased at an average rate of -0.71 ± 0.01 ($p < 0.01$) and -29.4 ± 0.05 ppbv yr^{-1} ($p < 0.01$), respectively, while TVOC remained unchanged ($p = 0.71$) for these years. The decreasing trends of NO_x and CO, also observed in many other high-population industrial urban areas (Geddes et al., 2009; Tang et al., 2009), suggest effective reduction of local emissions from transportation, power plants, and other industrial activities (HKEPD, 2016). Unlike O_3 and NO_x, the trend of TVOC varied across different areas – for example, increasing in Beijing (Tang et al., 2009), decreasing in Toronto (Geddes et al., 2009) and Taiwan (Chou et al., 2006), while remaining almost unchanged ($p > 0.05$) in Hong Kong (Fig. 2). Although the 10-year TVOC trend did not change, their levels showed clear inter-annual variations in spring and autumn (Fig. 3). Moreover, the long-term trends of individual VOCs, except for BVOC, were different from that of TVOCs (see Fig. S8) because many control measures were taken in the last decade, which altered the composition of VOCs in the atmosphere, such as the reduction of toluene by solvent usage control and the increase in alkanes in liquefied petroleum gas (LPG) in 2005–2013 (Ou et al., 2015) and the decrease in LPG alkanes in 2013–2014 (Lyu et al., 2016).

Figure 3 displays variations of measured O_3 and its precursors (NO_x, TVOC, and CO) in four seasons in 2005–2014. Here we defined December–February, March–May, June–August, and September–November as winter, spring, summer, and autumn, respectively. Generally, all precursors showed low values in summer and high levels in winter, mainly due to typical Asian monsoon circulations, which brought in clean marine air in summer and delivered pollutant-laden air from mainland China in winter (Wang et al., 2009). A similar seasonal variation was observed for the TVOC averages at different locations over Hong Kong (see Table S8). With lower (diluted) precursor concentrations, together with a high frequency of rainy days, it is not uncommon for Hong Kong to see the lowest O_3 values in summertime (see Fig. 3).

The long-term trends of CO in all seasons (slopes from spring to winter: -39.2 ± 0.20, -16.8 ± 0.12, -14.4 ± 0.16, and -50.3 ± 0.23 ppbv yr^{-1}, respectively) and NO_x and TVOCs in spring (NO_x: -0.69 ± 0.01; TVOCs: -0.26 ± 0.01 ppbv yr^{-1}) and autumn (NO_x: -0.50 ± 0.02; TVOCs: -0.32 ± 0.01 ppbv yr^{-1}) showed significant decreases ($p < 0.05$), whereas NO_x and TVOCs did not have statistical variations in summer and winter during the 10 years ($p > 0.05$). The different inter-annual trends of NO_x and TVOCs in spring/autumn from those in summer/winter were probably because marine air significantly diluted air pollution in summer, while continental air masses remarkably burdened air pollution in winter, which concealed the decreased local emissions of NO_x and TVOCs in summer and winter (Wang et al., 2009). In

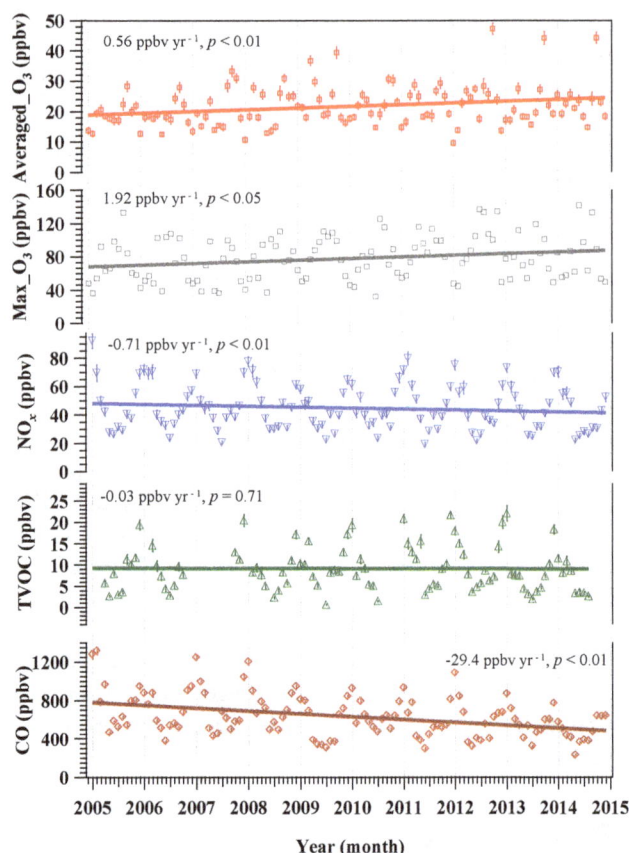

Figure 2. Trends of monthly averages of O_3 and its precursors, i.e. NO_x, TVOC, and CO at TC during 2005–2014. Error bars represent 95 % confidence intervals of monthly averages.

contrast, the measured O_3 trends significantly increased in spring, summer, and autumn, at the rate of 0.51 ± 0.05, 0.50 ± 0.04, and 0.67 ± 0.07 ppbv yr^{-1}, respectively ($p < 0.05$), while winter O_3 levels showed no significant trend (0.23 ± 0.05 ppbv yr^{-1}, $p = 0.11$). It is noteworthy that there were three extremely high O_3 data points in October months in the period 2012–2014 (Fig. 3), which seemed to bias the O_3 trend in autumn. However, it can be seen from Fig. S9, which shows the daily average O_3 values in autumn (i.e. more data points than in Fig. 3), that the values varied much more significantly in the period 2012–2014 than in previous years. It is worth emphasizing that the overall O_3 trend was determined by all the measured data points including both extremely high and low values in all the study years. Apart from the impact of regional transport, the increased spring and autumn O_3 in these years was probably due to the reduction of NO titration overriding the O_3 decrease owing to the reduction of TVOCs, leading to a net O_3 increase. Here the NO titration refers to the "titration reaction" between NO and O_3. Although NO–NO_2–O_3 reaction cycling (including the effects of NO titration; see Reactions R1–R3) can be theoretically regarded as a null

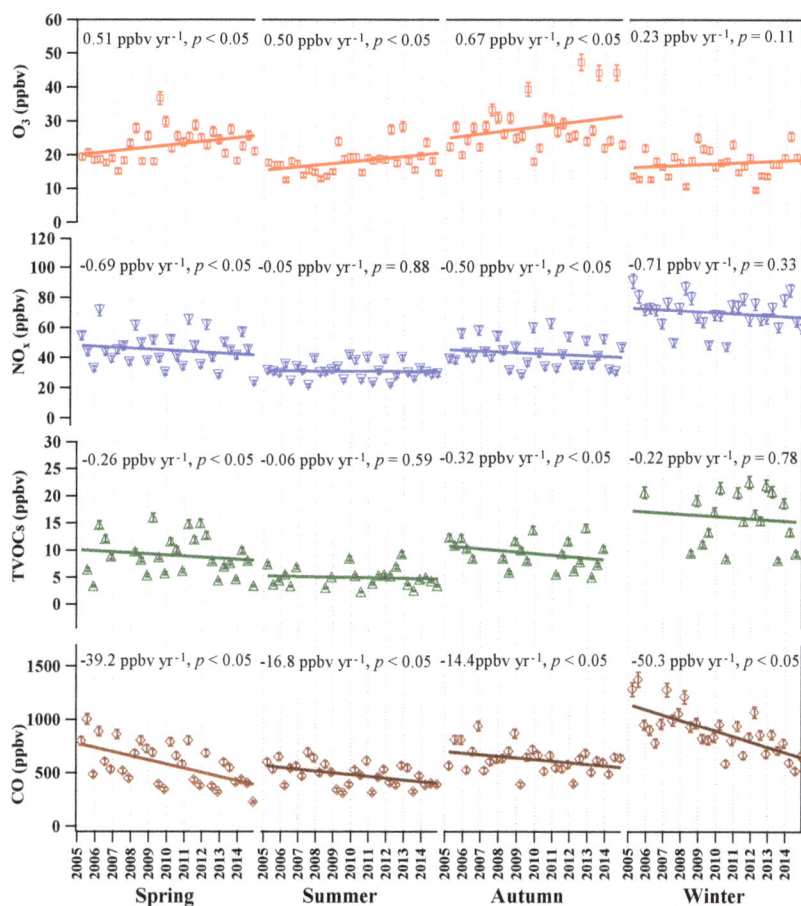

Figure 3. Variations of O_3 and its precursors in four seasons at TC during 2005–2014. Each data point in the figure is obtained by averaging hourly values into a monthly value. Error bars represent 95 % confidence intervals of the averages. In the sub-plot for O_3 trend in autumn, the three extremely high values are not considered as outliers as each of them represents 1 month of data with relatively small uncertainty (see Fig. S9).

cycle and provides rapid cycling between NO and NO_2, the NO titration effect can retard the accumulation of O_3 in an urban environment by means of substantial NO emissions (Chou et al., 2006). Indeed, the observed NO at TC site decreased significantly during the 2005–2010 period (shown in Fig. S10), which mitigated the effects of NO titration and led to the increase in O_3:

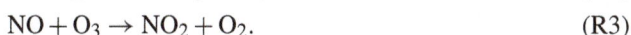

$$NO_2 + h\upsilon \rightarrow NO + O, \qquad (R1)$$

$$O + O_2 + M \rightarrow O_3 + M, \qquad (R2)$$

$$NO + O_3 \rightarrow NO_2 + O_2. \qquad (R3)$$

Interestingly, summer O_3 significantly increased although NO_x and TVOCs showed no differences in these years. Further investigation found that temperature and solar radiation in summer increased ($p < 0.05$) in these years (see Fig. S11), whereas they had no significant change in other seasons (the reasons remain unclear), consistent with the fact that an increase in temperature and solar radiation would enhance the photochemical reaction rates, resulting in O_3 increase in

summer (Lee et al., 2014). On the other hand, the unchanged winter O_3 trend was in line with the unchanged NO_x and TVOC values in winters of the past 10 years. Again, the impact of regional transport could not be ignored. To better understand the mechanisms of long-term trends of O_3 in different seasons in this study, the source origins of O_3, i.e. whether it was locally formed or transported from other regions, were explored.

3.2 Long-term trends of locally produced O_3 and regional contribution

In this study, the OBM (CB05) was used to simulate the long-term trends of O_3 produced by in situ photochemical reactions – hereinafter locally produced O_3 (simulated). The comparisons of simulated and observed O_3 at TC during the period 2005–2014 are shown in Figs. S12 (by year) and S13 (by month); Table S9 lists the IOA values between observed and locally produced O_3 (simulated) at TC site in each year. As shown, the IOA values range from 0.71 to 0.89, indicating that the performance of the OBM in the O_3 simulation was acceptable.

It is noteworthy that MCM has better simulation performance than CB05 due to its near-explicit rather than condensed chemical mechanism. Indeed, the overall IOA of MCM modelling (0.89) was higher than that of CB05 (0.81), according to our test on the same sampling days of 2005–2014 (shown in Fig. S14; rainy days were excluded). Despite this, the high computational efficiency OBM (CB05) was used for the 10-year day-by-day O_3 simulations to investigate the long-term trend of O_3 in this study, because the simulated results of both CB05 and MCM models followed similar temporal patterns ($p > 0.05$), and the difference of simulated values between the two models was reasonable (IOA value: 0.89 vs. 0.81), revealing that the condensed mechanism of CB05 would not significantly affect the long-term trends of O_3 in this study (shown in Fig. S15).

Previous studies suggest that a wind speed of $2\,m\,s^{-1}$ could be used as a threshold to classify regional and local air masses in Hong Kong (Guo et al., 2013; Ou et al., 2015; Cheung et al., 2014). That is, the O_3 values measured with $<2\,m\,s^{-1}$ in this study were considered as locally produced O_3 – hereinafter locally produced O_3 (filtered). Figure 4 presents the long-term trends of observed daytime O_3 (07:00–19:00 LT) and locally produced O_3 (filtered/simulated) in four seasons during the period 2005–2014. Although the actual duration of daytime in Hong Kong varies by 1–2 h according to season, the expected uncertainty from it would be limited when considering weak photochemical reactions in the early morning and in the late afternoon. Also note that the trends of observed daytime O_3 in the four seasons were consistent with those of 24 h observed O_3 (see Fig. S16). It can be seen that the locally produced O_3 (simulated) increased in spring (0.28 ± 0.01 ppbv yr^{-1}, $p < 0.05$), decreased in autumn (-0.39 ± 0.02 ppbv yr^{-1}, $p<0.05$), and showed no change in summer and winter ($p > 0.05$). Interestingly, the long-term trend of locally produced O_3 (simulated) in autumn was opposite to that in spring although both NO_x and TVOCs decreased in the two seasons. The reasons were because (1) NO_x decreased faster while TVOCs decreased more slowly (Fig. 3) in spring than in autumn, leading to a net increase in O_3 formation in spring and a decrease in autumn, and (2) TVOC reactivity (described in Text S2 and Table S10) decreased in autumn ($-0.03\,s^{-1}$ yr^{-1}, $p < 0.05$) but showed insignificant variations in other seasons ($p > 0.1$; Fig. S17), resulting in the reduction of O_3 production in autumn. The simulated springtime O_3 increase and unchanged winter values were consistent with the observed trends, whereas the simulated autumn O_3 decrease was opposite to the observed trend for the overall observations. However, locally produced O_3 (filtered) values clearly showed similar trends to locally produced O_3 (simulated) in spring, autumn, and winter (see Fig. 4), confirming that locally produced O_3 indeed increased in spring and decreased in autumn in these years.

In comparison, a significant difference ($\Delta O_3 = O_3$ overall$_{observed}$ $-$ $O_{3\,simulated}$, $p < 0.01$) was found between measured and simulated O_3 in spring, implying the contribution of regional transport to the measured O_3. The 10-year average ΔO_3 was 8.26 ± 1.77 ppbv and the long-term trend of ΔO_3 showed no significant change ($p = 0.91$), suggesting that the contribution of regional transport in spring has been stable during the last decade. The spring pattern of O_3 in this study is consistent with the findings of Li et al. (2014), who reported the increasing O_3 trend (2.0 ppbv yr^{-1}) in spring at urban clusters of PRD from 2006 to 2011. In conclusion, the increasing O_3 trend in spring at TC was caused by the increased local O_3 production, and the contribution of regional transport was steady in the 2005–2014 period.

Unlike in spring, though the observed and locally produced O_3 (filtered) displayed increasing trends in summer (0.67 ± 0.34 and 0.61 ± 0.41 ppbv yr^{-1}, respectively; $p < 0.05$), locally produced O_3 (simulated) showed no significant change ($p = 0.18$), consistent with the unchanged trends of precursors (NO_x and TVOCs) in summer (Fig. 3). Note that the influence of annual variation in solar radiation over the 10 years was not considered while the TUV model was used to calculate the photolysis rates, which could mask the actual trends of O_3 mixing ratios. Indeed, the total solar radiation ($0.24 \pm 0.16\,MJ\,m^{-2}$ yr^{-1}, $p < 0.01$) and temperature ($0.095 \pm 0.034\,K$ yr^{-1}, $p < 0.05$) in summer significantly increased during the past 10 years (see Fig. S11), subsequently resulting in the enhanced in situ photochemical reactivity of VOCs, although their quantitative contributions remain unknown and require further investigation. The increase in solar radiation might be due to the decreasing haze as the air quality has been getting better in Hong Kong and the PRD (Louie et al., 2013). Moreover, the summertime wind speeds significantly decreased at a rate of $-0.062 \pm 0.041\,m\,s^{-1}$ yr^{-1} ($p < 0.05$), which might contribute to accumulation of the locally produced O_3. Lastly, the locally produced O_3 (filtered) trend was comparable to observed O_3 ($p = 0.12$) and locally produced O_3 (simulated; $p = 0.32$), indicating a negligible impact of regional transport on the summer O_3 trend in the period 2005–2014 (and thereby ΔO_3 in summer, not shown in Fig. 4). As such, the increasing trend of summer O_3 was partly attributed to the increase in solar radiation and temperature from 2005 to 2014.

Consistent with the decreasing trends of NO_x and TVOCs in autumn, both locally produced O_3 (simulated; $p < 0.05$) and locally produced O_3 (filtered; $p < 0.1$) remarkably decreased, suggesting the dominant impact of VOC reduction over the reduction of NO titration. The decreased locally produced O_3 in autumn was consistent with the results of Xue et al. (2014), who found that local O_3 production decreased in autumn from 2002 to 2013. Furthermore, the 10-year average ΔO_3 was 7.35 ± 3.16 ppbv and the long-term trend of ΔO_3 increased at a rate of 1.09 ± 0.21 ppbv yr^{-1} ($p < 0.05$), suggesting an increased contribution of regional transport in autumn during the last decade, in line with the fact that au-

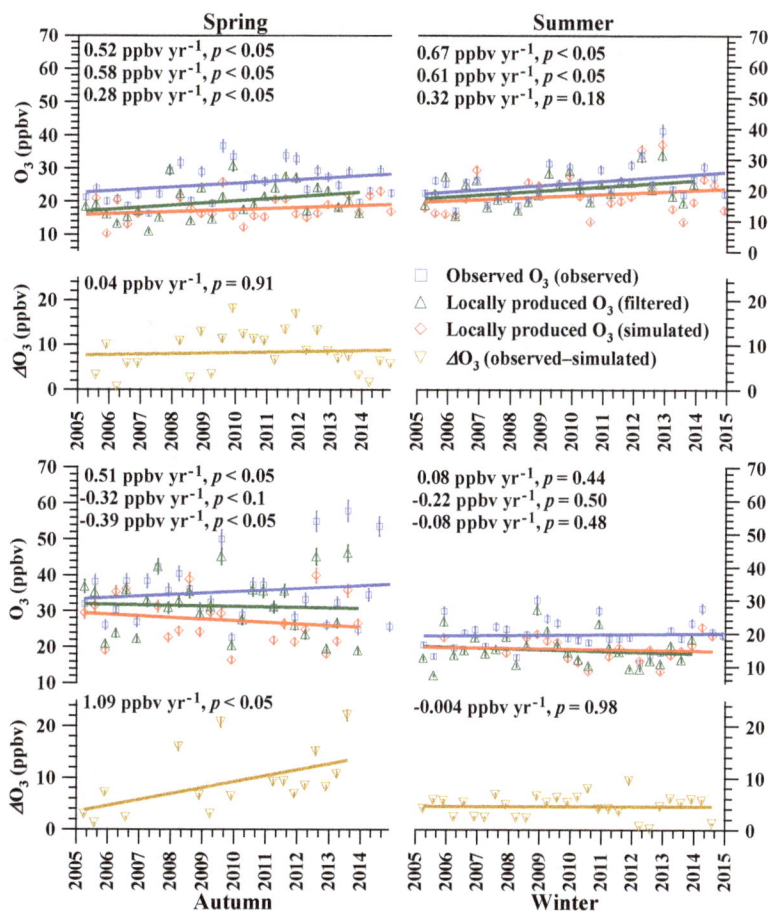

Figure 4. Trends of locally produced O_3 simulated by OBM (red line), observed O_3 (blue line: overall observed O_3; green line: locally produced O_3 (filtered), i.e. observed O_3 with hourly wind speed $< 2\,\mathrm{m\,s^{-1}}$), and regional O_3 (gold line, $\Delta O_3 = O_3$ overall observed $-\ O_3$ simulated) in four seasons at TC during the period 2005–2014. Note: all the data are based on daytime hours (07:00–19:00 LT). The regional O_3 in summer was negligible and is not shown in the graph. Error bars represent 95 % confidence intervals of the averages.

tumn O_3 level in inland PRD was higher and increased more rapidly than in Hong Kong (Fig. 5), and high O_3 mixing ratios were frequently observed in this season due to stronger solar radiation, lower wind speeds, and less vertical dilution of air pollution than in other seasons in this region. In summary, locally produced O_3 in autumn decreased due to the reduction of dominant VOC precursors, while an increased contribution of regional transport negated the local reduction, leading to an elevated O_3 observation.

In winter, locally produced O_3 (filtered and simulated) had similar trends ($p = 0.93$) and the trends showed no significant changes ($p > 0.05$), confirming similar locally produced O_3 in these years, due to insignificant variations of NO_x and TVOCs levels (Fig. 3). Also, locally produced O_3 (both filtered and simulated) presented significant differences from the observed O_3 ($p < 0.01$), implying a regional contribution in winter. The 10-year average ΔO_3 was 4.56 ± 0.78 ppbv and the long-term trend of ΔO_3 was not significant ($p = 0.98$).

To further investigate the regional transport from the PRD region to Hong Kong, the observed O_3 values at PRD sites and at TC site in four seasons between 2006 and 2014 are compared (Fig. 5). Generally, the observed O_3 levels in PRD were all higher than those at TC in the four seasons ($p < 0.05$). Considering that the PRD is upwind of Hong Kong in spring/autumn/winter (Ou et al., 2015), high O_3-laden air in the PRD region could transport to Hong Kong in these three seasons. Moreover, comparable long-term trends were found between the sites in PRD and the TC site in spring (PRD: 0.49 ± 0.06; TC: 0.51 ± 0.05 ppbv yr^{-1}) and winter (PRD: 0.30 ± 0.06; TC: 0.23 ± 0.05 ppbv yr^{-1}), indicating that regional transport in spring and winter was stable in these years. In comparison, autumn O_3 level in inland PRD increased (0.84 ± 0.08 ppbv yr^{-1}) more rapidly than in Hong Kong (0.67 ± 0.07 ppbv yr^{-1}), implying an elevated regional contribution to Hong Kong. Therefore, the differences of observed O_3 between PRD and TC in spring/autumn/winter were consistent with the above calculations of average ΔO_3, confirming regional contribution to the observed O_3 in Hong

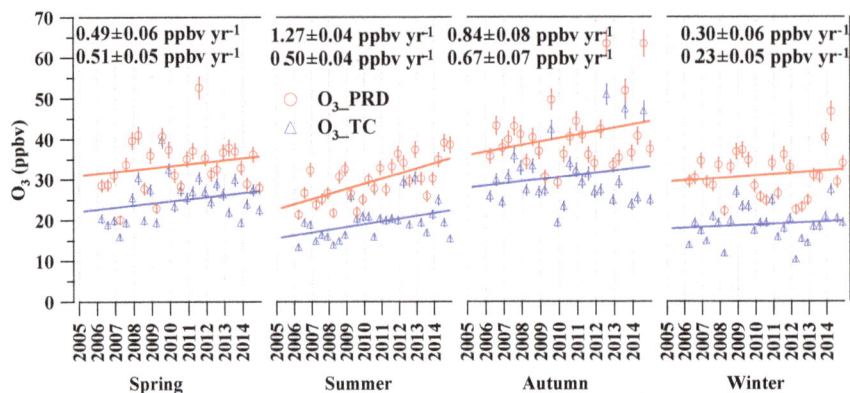

Figure 5. Trends of observed O_3 in inland PRD (red line) and at TC (blue line) in four seasons during the period 2006–2014. Each data point in the figure is obtained by averaging hourly values into a monthly value. Note: due to the location of sites which might have O_3 transport to TC (Guo et al., 2009), three regional background sites (i.e. Wanqingsha, Jinguowan, and Tianhu) and an urban site (i.e. Haogang) in Dongguan are used for comparison. Error bars represent 95 % confidence interval of monthly averages.

Kong. In contrast, though the increasing rate of O_3 level in PRD was much faster than at TC in summer, the impact of regional transport from the PRD region was insignificant due to the dominance of southerly and southwesterly winds from the South China Sea.

Overall, locally produced O_3 (simulated) in Hong Kong varied by season, showing an increase in spring, a decrease in autumn, and no change in summer and winter. The elevated observed O_3 in spring/summer/autumn was mainly attributed to the increase in locally produced springtime O_3 and constant regional contribution, increased summertime in situ photochemical reactivity, and regional contribution in autumn. Moreover, since the NO_x and NO levels significantly decreased during the last decade (Fig. S10), the reduced effect of NO titration, to a certain extent, made contribution to local O_3 levels. The effect of NO titration has also been reported in other areas (i.e. Beijing, Taiwan, Guangdong Province in China, and Osaka in Japan; Chou et al., 2006; Itano et al., 2007; Tang et al., 2009; Li et al., 2014). To confirm the reduction of NO titration in this study, the variation of O_x, the total oxidant estimated by O_3+NO_2 was investigated. According to the reaction of NO titration ($NO+O_3 \rightarrow NO_2+O_2$), the sum of O_3 and NO_2 (i.e. total oxidant O_x) remained essentially constant regardless the variation of NO (Chou et al., 2006). Indeed, the mixing ratio of local O_x (filtered by wind speed $< 2\,\mathrm{m\,s^{-1}}$) showed no significant change ($p = 0.42$) at TC site, during the period 2005–2013 (Fig. 6), suggesting that the increase in O_3 was a result of the reduced NO titration. The reduction of NO titration was also confirmed by the increasing NO_2/NO_x ratio at a roadside site (Mong Kok) in Hong Kong. The NO_2/NO_x emission ratio is a parameter that can be used to examine the variation of NO titration (Carslaw, 2005; Yao et al., 2005; Dallmann et al., 2011; Tian et al., 2011; Ning et al., 2012; Lau et al., 2015). Generally, higher ratios of NO_2/NO_x mean a lower potential of O_3 titration by NO, resulting in

higher O_3. Indeed, the NO_2/NO_x ratio at Mong Kok significantly increased, with enhanced traffic-related NO_2/NO_x ratios observed at night, from 2005 to 2014 ($p < 0.01$), leading to increased local O_3 levels (Fig. 7). This finding was supported by the Hong Kong emission inventory, which indicated that the NO_x emission decreased from 1997 to 2014 in Hong Kong (HKEPD, 2016), and studies conducted by Tian et al. (2011) and Lau et al. (2015), who found an increasing trend of primary NO_2 emission in Hong Kong due to several diesel retrofit programmes in the period 1998–2008.

Apart from the regional and local impact on O_3 trends, the impact of variations of baseline O_3 was also considered. Oltmans et al. (2013) reported that O_3 at mid-latitudes of the Northern Hemisphere was flat or declining during the period 1996–2010 and the limited data in the subtropical Pacific suggested very little change during the same period. Thus, the O_3 trend in Hong Kong might be unaffected or underestimated given the flatness or decline of baseline O_3. Therefore, the increasing trend of O_3 in Hong Kong over the last decade was the integrated influence of its precursors, meteorological parameters, and regional transport.

3.3 Ozone–precursor relationships

Ozone–precursor relationships are critical to determine the reduction plan of precursors for future O_3 control. In this study, the RIR of major O_3 precursors was calculated by the OBM, to directly reflect the O_3 alteration in response to the percentage changes of its precursors (see Sect. 2.3). Furthermore, the long-term trend of RIR was used to evaluate the variation of the sensitivity of O_3 formation to each individual precursor. Figure 8a shows the average RIR values of O_3 precursors in the four seasons during the last decade. The RIR values of TVOCs (AVOC + BVOC, where AVOC means anthropogenic VOC and BVOC means biogenic VOC; see Sect. 3.4 for the definition of BVOC) ranged from

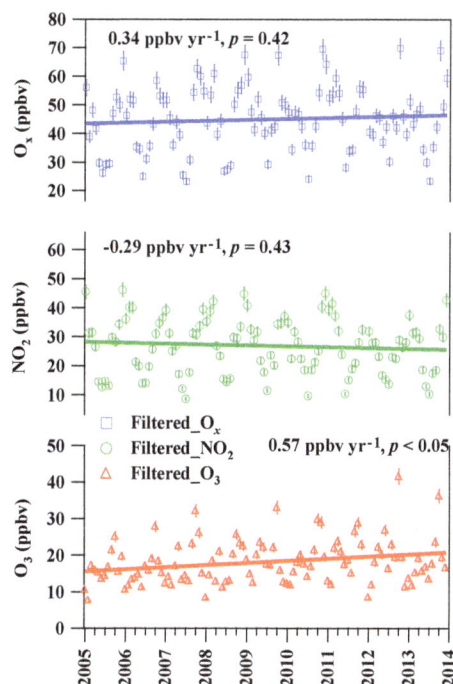

Figure 6. Annual trend of O_x, O_3, and NO_2 (filtered) at TC in the period 2005–2013. Error bars represent the 95 % confidence intervals of the averages.

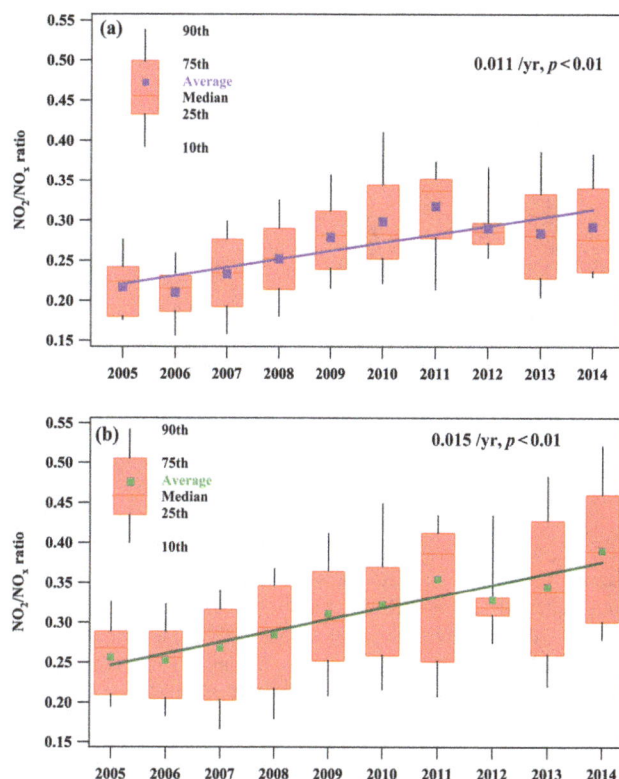

Figure 7. Annual trend of monthly average NO_2 / NO_x ratio at MK site in daytime **(a)** and night-time **(b)** in Hong Kong in the period 2005–2014. Hourly observations of NO_2 and NO_x are obtained at MK from the HKEPD (http://epic.epd.gov.hk/ca/uid/airdata). Note that data of October in 2014 are excluded due to the impact of Occupy Central event in Hong Kong.

0.78 ± 0.04 to 0.89 ± 0.04, followed by CO (0.32 ± 0.01 to 0.37 ± 0.01) in all seasons, suggesting the dominant role of TVOCs in photochemical O_3 formation. Among TVOCs, AVOCs had their highest RIR value in winter (0.80 ± 0.03, $p < 0.05$) and lowest in summer (0.39 ± 0.02, $p < 0.05$). Since RIR values are highly dependent on precursor mixing ratios (Eq. 5), the difference of RIR values of AVOCs in the four seasons was mainly caused by seasonal variations of observed AVOC levels (Fig. 3). In contrast, BVOCs had the highest RIR in summer (0.47 ± 0.02, $p < 0.05$), followed by autumn, spring, and winter (0.30 ± 0.02, 0.28 ± 0.02, and 0.09 ± 0.01, respectively). The higher RIR of BVOCs in summer was mainly due to the higher biogenic emissions in summer. In addition, higher photochemical reactivity of BVOCs also contributed to higher RIR of BVOCs (Atkinson and Arey, 2003; Tsui et al., 2009). The RIR values of NO_x, in contrast, were negative in all seasons, indicating that reducing NO_x would lead to an increase in photochemical O_3 formation. The RIR values of NO_x were lower in spring (-1.15 ± 0.02) and summer (-1.22 ± 0.02) than in autumn (-1.05 ± 0.02) and winter (-1.00 ± 0.01, $p < 0.05$), suggesting that reducing NO_x would increase more O_3 in spring and summer. The aforementioned findings were consistent with the results of previous studies conducted in autumn in Hong Kong, which were based on modelled and observed VOC / NO_x ratios (Zhang et al., 2007; Cheng et al., 2010; Ling et al., 2013; Guo et al., 2013). The relationship

analyses suggest that the O_3 formation in Hong Kong was VOC-limited in all seasons in these years; that is, the O_3 formation was dominated by AVOCs in winter and was sensitive to both AVOCs and BVOCs in the other three seasons, whereas reducing NO_x emissions enhanced O_3 formation – the more so in spring and summer. The findings suggest that a simultaneous cut of AVOCs and NO_x (which is often the case in real situations) would be most effective in O_3 pollution control in winter, but least efficient in summer.

Figure 8b presents the long-term trends of RIR values of O_3 precursors from 2005 to 2014. The RIR values of TVOCs and NO_x increased at an average rate of 0.014 ± 0.012 ($p < 0.05$) and $0.009 \pm 0.01 \, \mathrm{yr}^{-1}$ ($p < 0.05$), respectively, while the RIR of CO decreased at an average rate of $-0.014 \pm 0.007 \, \mathrm{yr}^{-1}$ ($p < 0.01$). The evolution of RIR values suggested that the O_3 formation was more sensitive to TVOCs and less to CO and NO_x, indicating that VOCs reduction strategies would be more effective at O_3 control. The decreasing sensitivities of O_3 formation to both CO and NO_x were consistent with the decrease in their mixing ratios during the last decade. The sensitivity of O_3 formation to TVOCs increased although the 10-year levels of TVOCs

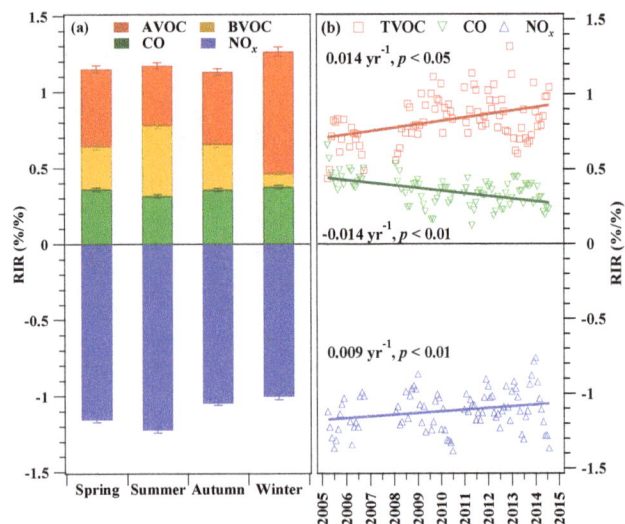

Figure 8. (a) Average RIR values of O_3 precursors in the four seasons and **(b)** trends of RIR values of O_3 precursors at TC from 2005 to 2014. AVOC represents anthropogenic VOCs, including 20 VOC species. BVOC means biogenic VOCs, including isoprene. Note: all the data are based on daytime hours (07:00–19:00 LT).

showed no significant trend, which might be attributed to the variations of speciated VOC levels and the VOC / NO_x ratios in these years (Ou et al., 2015). Furthermore, the monthly variation of TVOC / NO_x ratios showed a statistically significant decreasing trend at a rate of $-0.02 \, \text{yr}^{-1}$ ($p < 0.05$; see Fig. S18). The weak declining trend moderately supports the view that VOC reduction has become more effective in reducing O_3 in the past 10 years, which is consistent with the conclusions from the above modelling results.

3.4 Contribution of VOC groups to O_3 formation

Since the local O_3 production was VOC-limited in Hong Kong, it is important to study the contribution of VOC species to the O_3 formation. To facilitate analysis and interpretation, AVOC species were categorized into three groups – namely, AVOC (aromatics; including benzene, toluene, m/o-xylenes, ethylbenzene, and three trimethylbenzene isomers), AVOC (alkenes; including propene, three butene isomers, and 1,3-butadiene), and AVOC (alkanes; including propane, n/i-butanes, n/i-pentanes, n/i-hexanes, and n-heptane). It is noteworthy that C_2 hydrocarbons were not included in the groups due to high missing rates of the C_2 data. The variations of the daytime averaged contributions of the four VOC groups (i.e. AVOC (alkanes/alkenes/aromatics) and BVOC) to O_3 mixing ratios were calculated by OBM and are shown in Fig. 9. Two scenarios were selected for data analysis. The first scenario was "origin", which used all originally measured data as input. The second scenario was "AVOC (aromatics/alkenes/alkanes) or BVOC group", which excluded each of the four VOC groups in turn from the input data in the

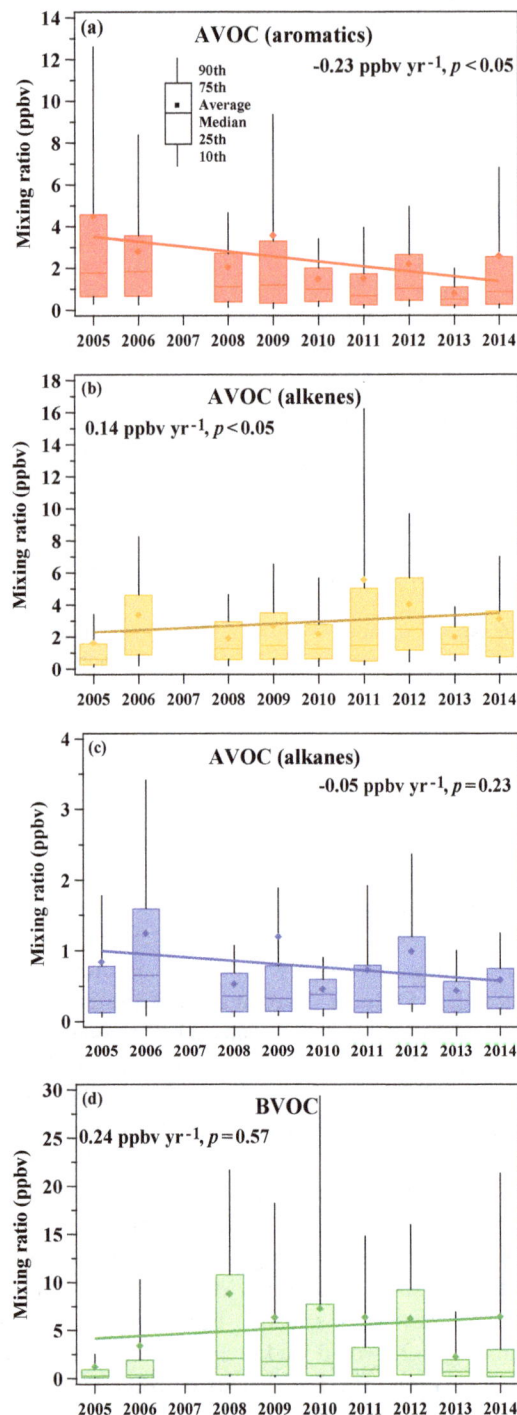

Figure 9. Trends of the daytime averaged contribution of four VOC groups to O_3 mixing ratio: **(a)** AVOC (aromatics), **(b)** AVOC (alkenes), **(c)** AVOC (alkanes), and **(d)** BVOC at TC during the period 2005–2014.

"origin" scenario. Hence, the contributions of VOC groups ((AVOC (aromatics/alkenes/alkanes) and BVOC) were obtained from the difference of simulated O_3 between the scenario "origin" and the related "VOCs group (AVOC (aromatics/alkenes/alkanes) or BVOC))". Clearly, the contribution of AVOC (aromatics) to O_3 mixing ratios significantly decreased at an average rate of -0.23 ± 0.01 ppbv yr^{-1} ($p < 0.05$), while AVOC (alkenes) made increasing contribution to O_3 mixing ratio ($p < 0.05$). BVOC and AVOC (alkanes) showed no significant changes ($p > 0.05$). The decreased contribution of AVOC (aromatics) to the O_3 mixing ratio was probably due to the decrease in C_6–C_8 aromatics, consistent with previous studies which found that aromatic levels decreased during the period 2005–2013 in Hong Kong (Ou et al., 2015). In fact, the Hong Kong Government has implemented a series of VOC-control measures since 2007 (HKEPD, 2016). In April 2007, the Air Pollution Control (VOCs) Regulation was implemented to control VOC emissions from regulated products, including architectural paints/coatings, printing inks, and six selected categories of consumer products. In January 2010, the regulation was extended to other high VOC-containing products, namely vehicle refinishing paints/coatings, vessel and pleasure craft paints/coatings, adhesives and sealants. The reduced contribution of AVOC (aromatics) to O_3 formation in these years also agreed well with the decreasing O_3 production rate of aromatics in autumn at TC from 2002 to 2013 reported by Xue et al. (2014). Furthermore, source apportionment results from Lyu et al. (2017) showed that solvent-related VOCs decreased at a rate of 204.7 ± 39.7 pptv yr^{-1} at the TC site, confirming that the reduction of solvent usage in these years was effective in decreasing the contribution of aromatics to O_3 production.

In contrast, the contributions of AVOC (alkenes) to O_3 production in these years showed a significant increasing trend with a rate of 0.14 ± 0.01 ppbv yr^{-1} ($p < 0.05$), perhaps attributed to the increased emissions of alkenes from changes in the composition of the traffic fleet and from increased traffic volume. During the period 2005–2014, the Hong Kong government launched a series of measures to reduce vehicular emissions, including diesel, LPG, and gasoline vehicles (http://www.epd.gov.hk/epd/english/environmentinhk/air/prob_solutions/air_problems.html). Among the measures, gasoline and LPG vehicular emissions caused ambient alkenes to increase during the same period due to the increasing number of LPG/gasoline vehicles and some short-term/non-mandatory measures (Lyu et al., 2017). The diesel commercial vehicle (DCV) programme (2007–2013) was shown to be effective in reducing the emission of alkenes from diesel vehicles (Lyu et al., 2017). However, these vehicles emit significantly lower level of VOCs than gasoline-propelled vehicles. In consequence, the overall emissions of alkenes from traffic-related sources increased during the period 2005–2014, leading to the increased contribution of AVOC (alkenes) to O_3 formation (Lyu et al., 2017).

Unlike AVOC (aromatics/alkenes), the contribution of AVOC (alkanes) to O_3 formation during the period 2005–2014 showed no significant change ($p = 0.23$) despite the increase in total alkane levels in the atmosphere in the 2005–2013 period (Ou et al., 2015). This is because alkanes include a number of compounds which have different but generally low reactivity with OH radicals. Hence, although the level of total alkanes increased over the years, it did not cause an increase in its contribution to O_3 formation. For example, one possible case is that some alkanes with relatively high reactivity decreased, offset by an increase in some low-reactivity alkanes. In addition, the seasonal variation of O_3 formation, of which the reaction rate of alkanes with OH radicals was high in summer and low in winter, would also blur the trend.

Furthermore, BVOC showed no evident change in its contribution to O_3 mixing ratios during the last decade ($p = 0.57$), which is probably attributable to the lack of significant change of isoprene levels at the TC site during the 2005–2014 period (shown in Fig. S8). In this study, isoprene is defined as a biogenic VOC. The main known sources of isoprene are biogenic and anthropogenic (Borbon et al., 2001; Barletta et al., 2002; Reiman et al., 2000). It is noteworthy that according to the tunnel study in Hong Kong (Ho et al., 2009), vehicular emissions of isoprene are not significant in this city. Another tunnel study in the PRD region (Tsai et al., 2006) found that isoprene was not present in diesel-fuelled vehicular emissions in Hong Kong, probably related to variations in fuel type and vehicular engines used in different countries (Ho et al., 2009). In addition, in low-latitude areas like Hong Kong, with a high level of plant coverage (more than 70 %), isoprene is mainly produced by biogenic emissions. The source of isoprene at the TC site has been also investigated and confirmed by previous long-term source apportionment studies, which reported that during the period 2005–2013 about 90 % of isoprene was from biogenic emissions, with minor contribution from traffic emissions, consumer products, and printing processes (Ou et al., 2015). Therefore, in this study the traffic source of isoprene in Hong Kong was disregarded and isoprene was defined as a biogenic VOC.

4   Conclusions

In this study, the long-term trends of O_3 and its precursors (NO_x, TVOCs and CO) were analysed at TC from 2005 to 2014. It was found that NO_x and CO decreased while TVOCs remained unchanged, suggesting the effective reduction of some emissions in Hong Kong. However, ambient O_3 levels increased in these years and the locally produced O_3 showed different variations in the four seasons, reflecting the complexity of photochemical pollution in Hong Kong. To effectively control locally produced O_3, VOC control plays a vital role, since O_3 formation in Hong Kong was shown to be VOC-limited in these years. Moreover, trend studies found that the sensitivity of O_3 formation gradually increased with VOCs and decreased with NO_x and CO, indicating that controlling VOCs will be increasingly effective for O_3 control in the future. Among the VOCs, the contribution of aromatics to O_3 formation decreased in the period 2005–2014, consistent with their declining abundance over this period and implying effective control measures of solvent-related sources. In contrast, the contribution of anthropogenic alkenes increased, suggesting a continuing need for the control of traffic-related sources. In addition, of the four seasons, the highest sensitivity of O_3 formation to AVOC and the relatively low sensitivity to NO_x concurrently appeared in winter, suggesting that winter is the best time for O_3 control. Lastly, in addition to locally produced O_3, regional transport of O_3 from the PRD region made a substantial contribution to ambient O_3 in Hong Kong and even increased in autumn. In the future, the Hong Kong government should collaborate closely with Guangdong Province to mitigate O_3 pollution in this region.

Competing interests. The authors declare that they have no conflict of interest.

Disclaimer. The opinions expressed in this paper are those of the authors and do not necessarily reflect the views or policies of the government of the Hong Kong Special Administrative Region, nor does the mention of trade names or commercial products constitute an endorsement or recommendation of their use.

Acknowledgements. We thank the Hong Kong Environmental Protection Department for providing us with the data. This work was supported by the Natural Science Foundation of China (41275122), the Research Grants Council (RGC) of the Hong Kong Government of Special Administrative Region (PolyU5154/13E, PolyU152052/14E, PolyU152052/16E and CRF/C5004-15E), the Innovation and Technology Commission of the HKSAR to the Hong Kong Branch of National Rail Transit Electrification and Automation Engineering Technology Research Center, and the Hong Kong Polytechnic University PhD scholarships (project #RTTA). This study is partly supported by the Hong Kong PolyU internal grant (1-BBW4 and 1-BBYD).

Edited by: Andreas Hofzumahaus

References

Atkinson, R. and Arey, J.: Atmospheric degradation of volatile organic compounds, Chem. Rev., 103, 4605–4638, 2003.

Barletta, B., Meinardi, S., Simpson, I. J., Khwaja, H. A., Blake, D. R., and Rowland, F. S.: Mixing ratios of volatile organic compounds (VOCs) in the atmosphere of Karachi, Pakistan, Atmos. Environ., 36, 3429–3443, 2002.

Beirle, S., Boersma, K. F., Platt, U., Lawrence, M. G., and Wagner, T.: Megacity Emissions and Lifetimes of Nitrogen Oxides Probed from Space, Science, 333, 1737–1739, 2011.

Borbon, A., Fontaine, H., Veillerot, M., Locoge, N., Galloo, J. C., and Guillermo, R.: An investigation into the traffic-related fraction of isoprene at an urban location, Atmos. Environ., 35, 3749–3760, 2001.

Cardelino, C. A. and Chameides, W. L.: An Observation-Based Model for Analyzing Ozone Precursor Relationships in the Urban Atmosphere, J. Air Waste Manage., 45, 161–180, https://doi.org/10.1080/10473289.1995.10467356, 1995.

Carslaw, D. C.: Evidence of an increasing NO_2 / NO_x emissions ratio from road traffic emissions, Atmos. Environ., 39, 4793–4802, 2005.

Chan, L. Y., Chan, C. Y., and Qin, Y.: Surface ozone pattern in Hong Kong, J. Appl. Meteorol., 37, 1153–1165, 1998a.

Chan, L. Y., Liu, H. Y., Lam, K. S., Wang, T., Oltmans, S. J., and Harris, J. M.: Analysis of the seasonal behavior of tropospheric ozone at Hong Kong, Atmos. Environ., 32, 159–168, 1998b.

Cheng, H. R., Guo, H., Wang, X. M., Saunders, S. M., Lam, S. H., Jiang, F., Wang, T. J., Ding, A. J., Lee, S. C., and Ho, K. F.: On the relationship between ozone and its precursors in the Pearl River Delta: application of an observation-based model (OBM), Environ. Sci. Pollut. R., 17, 547–560, 2010.

Cheng, H. R., Saunders, S. M., Guo, H., Louie, P. K., and Jiang, F.: Photochemical trajectory modeling of ozone concentrations in Hong Kong, Environ. Pollut., 180, 101–110, 2013.

Cheung, K., Guo, H., Ou, J. M., Simpson, I. J., Barletta, B., Meinardi, S., and Blake, D. R.: Diurnal profiles of isoprene, methacrolein and methyl vinyl ketone at an urban site in Hong Kong, Atmos. Environ., 84, 323–331, 2014.

Chou, C. C. K., Liu, S. C., Lin, C. Y., Shiu, C. J., and Chang, K. H.: The trend of surface ozone in Taipei, Taiwan, and its causes: Implications for ozone control strategies, Atmos. Environ., 40, 3898–3908, 2006.

Coates, J. and Butler, T. M.: A comparison of chemical mechanisms using tagged ozone production potential (TOPP) analysis, Atmos. Chem. Phys., 15, 8795–8808, https://doi.org/10.5194/acp-15-8795-2015, 2015.

Cui, J., Pandey Deolal, S., Sprenger, M., Henne, S., Staehelin, J., Steinbacher, M., and Nédélec, P.: Free tropospheric ozone changes over Europe as observed at Jungfraujoch (1990–2008): An analysis based on backward trajectories, J. Geophys. Res., 116, D10304, https://doi.org/10.1029/2010JD015154, 2011.

Dallmann, T. R., Harley, R. A., and Kirchstetter, T. W.: Effects of Diesel Particle Filter Retrofits and Accelerated Fleet Turnover on Drayage Truck Emissions at the Port of Oakland, Environ. Sci. Technol., 45, 10773–10779, 2011.

Derwent, R. G., Manning, A. J., Simmonds, P. G., Spain, T. G., and O'Doherty, S.: Analysis and interpretation of 25 years of ozone observations at the Mace Head Atmospheric Research Station on the Atlantic Ocean coast of Ireland from 1987 to 2012, Atmos. Environ., 80, 361–368, 2013.

Dils, B.: Long Range Transport of Tropospheric NO_2 as simulated by FLEXPART, Product Specification document TEM/LRT2/001, TEMIS, De Bilt, The Netherlands, 2008.

Ding, A. J., Wang, T., Zhao, M., Wang, T. J., and Li, Z. K.: Simulation of sea-land breezes and a discussion of their implications on the transport of air pollution during a multi-day ozone episode in the Pearl River Delta of China, Atmos. Environ., 38, 6737–6750, 2004.

Ding, A. J., Wang, T., Thouret, V., Cammas, J.-P., and Nédélec, P.: Tropospheric ozone climatology over Beijing: analysis of aircraft data from the MOZAIC program, Atmos. Chem. Phys., 8, 1–13, https://doi.org/10.5194/acp-8-1-2008, 2008.

Evans, M. J. and Jacob, D. J.: Impact of new laboratory studies of N2O5 hydrolysis on global model budgets of tropospheric nitrogen oxides, ozone, and OH, Geophys. Res. Lett., 32, L09813, https://doi.org/10.1029/2005GL022469, 2005.

GDEMC and HKEPD (Guangdong Environmental Monitoring Centre and Hong Kong Environmental Protection Department): Pearl River Delta Regional Air Quality Monitoring Network – A Report of Monitoring Results in 2014, available at: http://www.epd.gov.hk/epd/english/resources_pub/publications/m_report.html (last access: 10 September 2017), 2015.

Geddes, J. A., Murphy, J. G., and Wang, D. K.: Long term changes in nitrogen oxides and volatile organic compounds in Toronto and the challenges facing local ozone control, Atmos. Environ., 43, 3407–3415, 2009.

George, I. J., Matthews, P. S. J., Whalley, L. K., Brooks, B., Goddard, A., Baeza-Romero, M. T., and Heard, D. E.: Measurements of uptake coefficients for heterogeneous loss of HO2 onto submicron inorganic salt aerosols, Phys. Chem. Chem. Phys., 15, 12829–12845, 2013.

Guo, H., Lee, S. C., Louie, P. K., and Ho, K. F.: Characterization of hydrocarbons, halocarbons and carbonyls in the atmosphere of Hong Kong, Chemosphere, 57, 1363–1372, 2004.

Guo, H., Jiang, F., Cheng, H. R., Simpson, I. J., Wang, X. M., Ding, A. J., Wang, T. J., Saunders, S. M., Wang, T., Lam, S. H. M., Blake, D. R., Zhang, Y. L., and Xie, M.: Concurrent observations of air pollutants at two sites in the Pearl River Delta and the implication of regional transport, Atmos. Chem. Phys., 9, 7343–7360, https://doi.org/10.5194/acp-9-7343-2009, 2009.

Guo, H., Ling, Z. H., Cheung, K., Jiang, F., Wang, D. W., Simpson, I. J., Barletta, B., Meinardi, S., Wang, T. J., Wang, X. M., Saunders, S. M., and Blake, D. R.: Characterization of photochemical pollution at different elevations in mountainous areas in Hong Kong, Atmos. Chem. Phys., 13, 3881–3898, https://doi.org/10.5194/acp-13-3881-2013, 2013.

HKEPD (Hong Kong Environmental Protection Department): Air Quality in Hong Kong 2014, available at: http://www.aqhi.gov.hk/api_history/english/report/files/AQR2014e_Update0616.pdf (last access: 10 September 2017), 2015.

HKEPD (Hong Kong Environmental Protection Department): 2014 Hong Kong Emission Inventory Report, available at: http://www.epd.gov.hk/epd/sites/default/files/epd/2014Summary_of_Updates_eng_2.pdf (last access: 10 September 2017), 2016.

Ho, K. F., Lee, S. C., Ho, W. K., Blake, D. R., Cheng, Y., Li, Y. S., Ho, S. S. H., Fung, K., Louie, P. K. K., and Park, D.: Vehicular emission of volatile organic compounds (VOCs) from a tunnel study in Hong Kong, Atmos. Chem. Phys., 9, 7491–7504, https://doi.org/10.5194/acp-9-7491-2009, 2009.

Huang, J. P., Fung, J. C. H., Lau, A. K. H., and Qin, Y.: Numerical simulation and process analysis of typhoon-related ozone episodes in Hong Kong, J. Geophys. Res., 110, D05301,

https://doi.org/10.1029/2004JD004914, 2005.

Itano, Y., Bandow, H., Takenaka, N., Saitoh, Y., Asayama, A., and Fukuyama, J.: Impact of NOx reduction on long-term ozone trends in an urban atmosphere, Sci. Total Environ., 379, 46–55, 2007.

Jenkin, M. E., Saunders, S. M., and Pilling, M. J.: The tropospheric degradation of volatile organic compounds: a protocol for mechanism development, Atmos. Environ., 31, 81–104, 1997.

Jenkin, M. E., Saunders, S. M., Wagner, V., and Pilling, M. J.: Protocol for the development of the Master Chemical Mechanism, MCM v3 (Part B): tropospheric degradation of aromatic volatile organic compounds, Atmos. Chem. Phys., 3, 181–193, https://doi.org/10.5194/acp-3-181-2003, 2003.

Jiang, F., Guo, H., Wang, T. J., Cheng, H. R., Wang, X. M., Simpson, I. J., Ding, A. J., Saunders, S. M., Lam, S. H. M., and Blake, D. R.: An ozone episode in the Pearl River Delta: Field observation and model simulation, J. Geophys. Res., 115, D22305, https://doi.org/10.1029/2009JD013583, 2010.

Lakey, P. S. J., George, I. J., Whalley, L. K., Baeza-Romero, M. T., and Heard, D. E.: Measurements of the HO2 Uptake Coefficients onto Single Component Organic Aerosols, Environ. Sci. Technol., 49, 4878–4885, 2015.

Lam, S. H. M., Saunders, S. M., Guo, H., Ling, Z. H., Jiang, F., Wang, X. M., and Wang, T. J.: Modelling VOC source impacts on high ozone episode days observed at a mountain summit in Hong Kong under the influence of mountain-valley breezes, Atmos. Environ., 81, 166–176, 2013.

Lau, C. F., Rakowska, A., Townsend, T., Brimblecombe, P., Chan, T. L., Yam, Y. S., Mocnik, G., and Ning, Z.: Evaluation of diesel fleet emissions and control policies from plume chasing measurements of on-road vehicles, Atmos. Environ., 122, 171–182, 2015.

Lee, Y. C., Shindell, D. T., Faluvegi, G., Wenig, M., Lam, Y. F., Ning, Z., Hao, S., and Lai, C. S.: Increase of ozone concentrations, its temperature sensitivity and the precursor factor in South China, Tellus B, 66, 23455, https://doi.org/10.3402/tellusb.v66.23455, 2014.

Lefohn, A. S., Shadwick, D., and Oltmans, S. J.: Characterizing changes in surface ozone levels in metropolitan and rural areas in the United States for 1980–2008 and 1994–2008, Atmos. Environ., 44, 5199–5210, 2010.

Li, J. F., Lu, K. D., Lv, W., Li, J., Zhong, L. J., Ou, Y. B., Chen, D. H., Huang, X., and Zhang, Y. H.: Fast increasing of surface ozone concentrations in Pearl River Delta characterized by a regional air quality monitoring network during 2006–2011, J. Environ. Sci., 26, 23–36, 2014.

Lin, M., Horowitz, L. W., Payton, R., Fiore, A. M., and Tonnesen, G.: US surface ozone trends and extremes from 1980 to 2014: quantifying the roles of rising Asian emissions, domestic controls, wildfires, and climate, Atmos. Chem. Phys., 17, 2943–2970, https://doi.org/10.5194/acp-17-2943-2017, 2017.

Ling, Z. H., Guo, H., Cheng, H. R., and Yu, Y. F.: Sources of ambient volatile organic compounds and their contributions to photochemical ozone formation at a site in the Pearl River Delta, southern China, Environ. Pollut., 159, 2310–2319, 2011.

Ling, Z. H., Guo, H., Zheng, J. Y., Louie, P. K. K., Cheng, H. R., Jiang, F., Cheung, K., Wong, L. C., and Feng, X. Q.: Establishing a conceptual model for photochemical ozone pollution in subtropical Hong Kong, Atmos. Environ., 76, 208–220, 2013.

Ling, Z. H., Guo, H., Lam, S. H. M., Saunders, S. M., and Wang, T.: Atmospheric photochemical reactivity and ozone production at two sites in Hong Kong: Application of a Master Chemical Mechanism-photochemical box model, J. Geophys. Res., 119, 10567–10582, 2014.

Louie, P. K. K., Zhong, L. J., Zheng, J. Y. A., and Lau, A. K. H.: A Special Issue of Atmospheric Environment on "Improving Regional Air Quality over the Pearl River Delta and Hong Kong: From Science to Policy", Atmos. Environ., 76, 1–2, 2013.

Lu, K. D., Zhang, Y. H., Su, H., Brauers, T., Chou, C. C., Hofzumahaus, A., Liu, S. C., Kita, K., Kondo, Y., Shao, M., Wahner, A., Wang, J. L., Wang, X. S., and Zhu, T.: Oxidant (O_3+NO_2) production processes and formation regimes in Beijing, J. Geophys. Res., 115, D07303, https://doi.org/10.1029/2009JD012714, 2010.

Lyu, X., Guo, H., Simpson, I. J., Meinardi, S., Louie, P. K. K., Ling, Z., Wang, Y., Liu, M., Luk, C. W. Y., Wang, N., and Blake, D. R.: Effectiveness of replacing catalytic converters in LPG-fueled vehicles in Hong Kong, Atmos. Chem. Phys., 16, 6609–6626, https://doi.org/10.5194/acp-16-6609-2016, 2016.

Lyu, X. P., Chen, N., Guo, H., Zhang, W. H., Wang, N., Wang, Y., and Liu, M.: Ambient volatile organic compounds and their effect on ozone production in Wuhan, central China, Sci. Total Environ., 541, 200–209, 2015a.

Lyu, X. P., Ling, Z. H., Guo, H., Saunders, S. M., Lam, S. H. M., Wang, N., Wang, Y., Liu, M., and Wang, T.: Re-examination of C1–C5 alkyl nitrates in Hong Kong using an observation-based model, Atmos. Environ., 120, 28–37, 2015b.

Lyu, X. P., Zeng, L. W., Guo, H., Simpson, I. J., Ling, Z. H., Wang, Y., Murray, F., Louie, P. K. K., Saunders, S. M., Lam, S. H. M., and Blake, D. R.: Evaluation of the effectiveness of air pollution control measures in Hong Kong, Environ. Pollut., 220, 87–94, 2017.

Madronich, S. and Flocke, S.: The Role of Solar Radiation in Atmospheric Chemistry, Handbook of Environmental Chemistry, Vol. 2 Part L, Springer-Verlag, Berlin, German, 1999.

Merlaud, A., Van Roozendael, M., Theys, N., Fayt, C., Hermans, C., Quennehen, B., Schwarzenboeck, A., Ancellet, G., Pommier, M., Pelon, J., Burkhart, J., Stohl, A., and De Mazière, M.: Airborne DOAS measurements in Arctic: vertical distributions of aerosol extinction coefficient and NO2 concentration, Atmos. Chem. Phys., 11, 9219–9236, https://doi.org/10.5194/acp-11-9219-2011, 2011.

Ning, Z., Wubulihairen, M., and Yang, F.: PM, NO_x and butane emissions from on-road vehicle fleets in Hong Kong and their implications on emission control policy, Atmos. Environ., 61, 265–274, 2012.

Oltmans, S. J., Lefohn, A. S., Shadwick, D., Harris, J. M., Scheel, H. E., Galbally, I., Tarasick, D. W., Johnson, B. J., Brunke, E. G., Claude, H., Zeng, G., Nichol, S., Schmidlin, F., Davies, J., Cuevas, E., Redondas, A., Naoe, H., Nakano, T., and Kawasato, T.: Recent tropospheric ozone changes – A pattern dominated by slow or no growth, Atmos. Environ., 67, 331–351, 2013.

Ou, J. M., Guo, H., Zheng, J. Y., Cheung, K., Louie, P. K. K., Ling, Z. H., and Wang, D. W.: Concentrations and sources of non-methane hydrocarbons (NMHCs) from 2005 to 2013 in Hong Kong: A multi-year real-time data analysis, Atmos. Environ., 103, 196–206, 2015.

Parrish, D. D., Law, K. S., Staehelin, J., Derwent, R., Cooper, O. R., Tanimoto, H., Volz-Thomas, A., Gilge, S., Scheel, H.-E., Steinbacher, M., and Chan, E.: Long-term changes in lower tropospheric baseline ozone concentrations at northern mid-latitudes, Atmos. Chem. Phys., 12, 11485–11504, https://doi.org/10.5194/acp-12-11485-2012, 2012.

Pollack, I. B., Ryerson, T. B., Trainer, M., Neuman, J. A., Roberts, J. M., and Parrish, D. D.: Trends in ozone, its precursors, and related secondary oxidation products in Los Angeles, California: A synthesis of measurements from 1960 to 2010, J. Geophys. Res., 118, 5893–5911, 2013.

Reimann, S., Calanca, P., and Hofer, P.: The anthropogenic contribution to isoprene concentrations in a rural atmosphere, Atmos. Environ., 34, 109–115, 2000.

Rivera, C., Stremme, W., and Grutter, M.: Nitrogen dioxide DOAS measurements from ground and space: comparison of zenith scattered sunlight ground-based measurements and OMI data in Central Mexico, Atmósfera, 26, 401–414, 2013.

Saunders, S. M., Jenkin, M. E., Derwent, R. G., and Pilling, M. J.: Protocol for the development of the Master Chemical Mechanism, MCM v3 (Part A): tropospheric degradation of non-aromatic volatile organic compounds, Atmos. Chem. Phys., 3, 161–180, https://doi.org/10.5194/acp-3-161-2003, 2003.

Seinfeld, J. H. and Pandis, S. N.: Atmos. Chem. Phys.: from air pollution to climate change, 2nd Edn., Wiley Publisher, New Jersey, USA, 2006.

Sillman, S.: The relation between ozone, NO_x and hydrocarbons in urban and polluted rural environments, Atmos. Environ., 33, 1821–1845, 1999.

Simpson, I. J., Blake, N. J., Barletta, B., Diskin, G. S., Fuelberg, H. E., Gorham, K., Huey, L. G., Meinardi, S., Rowland, F. S., Vay, S. A., Weinheimer, A. J., Yang, M., and Blake, D. R.: Characterization of trace gases measured over Alberta oil sands mining operations: 76 speciated C_2–C_{10} volatile organic compounds (VOCs), CO_2, CH_4, CO, NO, NO_2, NO_y, O_3 and SO_2, Atmos. Chem. Phys., 10, 11931–11954, https://doi.org/10.5194/acp-10-11931-2010, , 2010.

So, K. L. and Wang, T.: On the local and regional influence on ground-level ozone concentrations in Hong Kong, Environ. Pollut., 123, 307–317, 2003.

So, K. L. and Wang, T.: C3-C12 non-methane hydrocarbons in subtropical Hong Kong: spatial-temporal variations, source-receptor relationships and photochemical reactivity, Sci. Total Environ., 328, 161–174, 2004.

Tang, G., Li, X., Wang, Y., Xin, J., and Ren, X.: Surface ozone trend details and interpretations in Beijing, 2001–2006, Atmos. Chem. Phys., 9, 8813–8823, https://doi.org/10.5194/acp-9-8813-2009, 2009.

Tanimoto, H.: Increase in springtime tropospheric ozone at a mountainous site in Japan for the period 1998–2006, Atmos. Environ., 43, 1358–1363, 2009.

TF HTAP (Task Force on Hemispheric Transport of Air Pollution): Hemispheric Transport of Air Pollution 2010, Part A. Ozone and Particulate Matter, United Nations Economic Commission for Europe, Geneva, 304 pp., 2010.

Tian, L. W., Hossain, S. R., Lin, H. L., Ho, K. F., Lee, S. C., and Yu, I. T. S.: Increasing trend of primary NO2 exhaust emission fraction in Hong Kong, Environ. Geochem. Hlth., 33, 623–630, 2011.

Tsai, W. Y., Chan, L. Y., Blake, D. R., and Chu, K. W.: Vehicular fuel composition and atmospheric emissions in South China: Hong Kong, Macau, Guangzhou, and Zhuhai, Atmos. Chem. Phys., 6, 3281–3288, https://doi.org/10.5194/acp-6-3281-2006, 2006.

Tsui, J. K.-Y., Guenther, A., Yip, W.-K., and Chen, F.: A biogenic volatile organic compound emission inventory for Hong Kong, Atmos. Environ., 43, 6442–6448, 2009.

Wang, N., Guo, H., Jiang, F., Ling, Z. H., and Wang, T.: Simulation of ozone formation at different elevations in mountainous area of Hong Kong using WRF-CMAQ model, Sci. Total Environ., 505, 939–951, 2015.

Wang, T. and Kwok, J. Y. H.: Measurement and analysis of a multiday photochemical smog episode in the Pearl River delta of China, J. Appl. Meteorol., 42, 404–416, 2003.

Wang, T., Guo, H., Blake, D. R., Kwok, Y. H., Simpson, I. J., and Li, Y. S.: Measurements of Trace Gases in the Inflow of South China Sea Background Air and Outflow of Regional Pollution at Tai O, Southern China, J. Atmos. Chem., 52, 295–317, 2005.

Wang, T., Wei, X. L., Ding, A. J., Poon, C. N., Lam, K. S., Li, Y. S., Chan, L. Y., and Anson, M.: Increasing surface ozone concentrations in the background atmosphere of Southern China, 1994–2007, Atmos. Chem. Phys., 9, 6217–6227, https://doi.org/10.5194/acp-9-6217-2009, 2009.

Wang, X. M., Liu, H., Pang, J. M., Carmichael, G., He, K. B., Fan, Q., Zhong, L. J., Wu, Z. Y., and Zhang, J. P.: Reductions in sulfur pollution in the Pearl River Delta region, China: Assessing the effectiveness of emission controls, Atmos. Environ., 76, 113–124, 2013.

Xu, X., Lin, W., Wang, T., Yan, P., Tang, J., Meng, Z., and Wang, Y.: Long-term trend of surface ozone at a regional background station in eastern China 1991–2006: enhanced variability, Atmos. Chem. Phys., 8, 2595–2607, https://doi.org/10.5194/acp-8-2595-2008, 2008.

Xue, L. K., Wang, T., Louie, P. K., Luk, C. W., Blake, D. R., and Xu, Z.: Increasing external effects negate local efforts to control ozone air pollution: a case study of Hong Kong and implications for other chinese cities, Environ. Sci. Technol., 48, 10769–10775, 2014.

Yao, X., Lau, N. T., Chan, C. K., and Fang, M.: The use of tunnel concentration profile data to determine the ratio of NO_2 / NO_x directly emitted from vehicles, Atmos. Chem. Phys. Discuss., https://doi.org/10.5194/acpd-5-12723-2005, 2005.

Yarwood, G., Rao, S., Yocke, M., and Whitten, G. Z.: Updates to the Carbon Bond Chemical Mechanism: CB05, Tech. rep., US Environmental Protection Agency, Novato, California, USA, 2005.

Zhang, J., Wang, T., Chameides, W. L., Cardelino, C., Kwok, J., Blake, D. R., Ding, A., and So, K. L.: Ozone production and hydrocarbon reactivity in Hong Kong, Southern China, Atmos. Chem. Phys., 7, 557–573, https://doi.org/10.5194/acp-7-557-2007, 2007.

Zhang, Q., Yuan, B., Shao, M., Wang, X., Lu, S., Lu, K., Wang, M., Chen, L., Chang, C.-C., and Liu, S. C.: Variations of ground-level O_3 and its precursors in Beijing in summertime between 2005 and 2011, Atmos. Chem. Phys., 14, 6089–6101, https://doi.org/10.5194/acp-14-6089-2014, 2014.

Zhong, L. J., Louie, P. K. K., Zheng, J. Y., Yuan, Z. B., Yue, D. L., Ho, J. W. K., and Lau, A. K. H.: Science-policy interplay: Air quality management in the Pearl River Delta region and Hong Kong, Atmos. Environ., 76, 3–10, 2013.

Determination of enhancement ratios of HCOOH relative to CO in biomass burning plumes by the Infrared Atmospheric Sounding Interferometer (IASI)

Matthieu Pommier[1,a]**, Cathy Clerbaux**[1,2]**, and Pierre-Francois Coheur**[2]

[1]LATMOS/IPSL, UPMC Univ. Paris 06 Sorbonne Universités, UVSQ, CNRS, Paris, France
[2]Spectroscopie de l'Atmosphère, Chimie Quantique et Photophysique, Université Libre de Bruxelles (ULB), Brussels, Belgium
[a]now at: Norwegian Meteorological Institute, Oslo, Norway

Correspondence to: Matthieu Pommier (matthieup@met.no)

Abstract. Formic acid (HCOOH) concentrations are often underestimated by models, and its chemistry is highly uncertain. HCOOH is, however, among the most abundant atmospheric volatile organic compounds, and it is potentially responsible for rain acidity in remote areas. HCOOH data from the Infrared Atmospheric Sounding Interferometer (IASI) are analyzed from 2008 to 2014 to estimate enhancement ratios from biomass burning emissions over seven regions. Fire-affected HCOOH and CO total columns are defined by combining total columns from IASI, geographic location of the fires from Moderate Resolution Imaging Spectroradiometer (MODIS), and the surface wind speed field from the European Centre for Medium-Range Weather Forecasts (ECMWF). Robust correlations are found between these fire-affected HCOOH and CO total columns over the selected biomass burning regions, allowing the calculation of enhancement ratios equal to $7.30 \times 10^{-3} \pm 0.08 \times 10^{-3}\,\mathrm{mol\,mol^{-1}}$ over Amazonia (AMA), $11.10 \times 10^{-3} \pm 1.37 \times 10^{-3}\,\mathrm{mol\,mol^{-1}}$ over Australia (AUS), $6.80 \times 10^{-3} \pm 0.44 \times 10^{-3}\,\mathrm{mol\,mol^{-1}}$ over India (IND), $5.80 \times 10^{-3} \pm 0.15 \times 10^{-3}\,\mathrm{mol\,mol^{-1}}$ over Southeast Asia (SEA), $4.00 \times 10^{-3} \pm 0.19 \times 10^{-3}\,\mathrm{mol\,mol^{-1}}$ over northern Africa (NAF), $5.00 \times 10^{-3} \pm 0.13 \times 10^{-3}\,\mathrm{mol\,mol^{-1}}$ over southern Africa (SAF), and $4.40 \times 10^{-3} \pm 0.09 \times 10^{-3}\,\mathrm{mol\,mol^{-1}}$ over Siberia (SIB), in a fair agreement with previous studies. In comparison with referenced emission ratios, it is also shown that the selected agricultural burning plumes captured by IASI over India and Southeast Asia correspond to recent plumes where the chemistry or the sink does not occur. An additional classification of the enhancement ratios by type of fuel burned is also provided, showing a diverse origin of the plumes sampled by IASI, especially over Amazonia and Siberia. The variability in the enhancement ratios by biome over the different regions show that the levels of HCOOH and CO do not only depend on the fuel types.

1 Introduction

Formic acid (HCOOH) is one of the most abundant carboxylic acids present in the atmosphere. HCOOH is mainly removed from the troposphere through wet and dry deposition, and to a lesser extent by the OH radical. It is a relatively short-lived species with an average lifetime in the troposphere of 3–4 days (Paulot et al., 2011; Stavrakou et al., 2012). HCOOH contributes a large fraction to acidity in precipitation in remote areas (e.g., Andreae et al., 1988).

HCOOH is mainly a secondary product from other organic precursors. The largest global source of HCOOH is biogenic and follows the emissions of isoprene, monoterpenes, other terminal alkenes (e.g., Neeb et al., 1997; Lee et al., 2006; Paulot et al., 2011), alkynes (Hatakeyama et al., 1986; Bohn et al., 1996), and acetaldehyde (Andrews et al., 2012; Clubb et al., 2012). There are also small direct emissions by vegetation (Keene and Galloway, 1984, 1988; Gabriel et al., 1999) and biomass burning (e.g., Goode et al., 2000). Other studies highlighted the existence of other sources, such as from ants (Graedel and Eisner, 1988), dry savanna soils (Sanhueza and Andreae, 1991), motor vehicles (Kawamura et al., 1985; Grosjean, 1989), abiological formation on rock surfaces (Ohta et al., 2000), and cloud processing (Chameides and Davis, 1983). Their contributions are very uncertain and most are probably minor.

More generally there are still large uncertainties on the sources and sinks of HCOOH, and on the relative contribution of anthropogenic and natural sources, despite the fact that recent progress has been made possible by using the synergy between atmospheric models and satellite data (e.g., Stavrakou et al., 2012; Chaliyakunnel et al., 2016). These uncertainties have an impact on our understanding of the HCOOH tropospheric chemistry, as on the oxidizing capacity of the atmosphere (i.e., the chemistry of OH in cloud water – Jacob, 1986; the heterogeneous oxidation of organic aerosols – Paulot et al., 2011) or the origin of the acid rains. One of the large uncertainties in the HCOOH tropospheric budget seems to be the underestimation of the emissions from forest fires, as recently suggested by Stavrakou et al. (2012), Cady-Pereira et al. (2014), and Chaliyakunnel et al. (2016).

One way to estimate the atmospheric emissions of pyrogenic species is the use of emission factors. The emission factors are often obtained from ground and airborne measurements or from small fires burned under controlled laboratory conditions. The emission factors can also be derived from enhancement ratios of the target species relative to a reference species, which is often carbon monoxide (CO) or carbon dioxide (CO_2) due to their long lifetime (e.g., Hurst et al., 1994) and are based on the characteristic of the combustible and hence depend on the type of biomass burning. However, the difference between an emission ratio and an enhancement ratio is that emission ratios are calculated from measurements at the time of emission and enhancement ratios are related to the ongoing chemistry. To correctly convert these enhancement ratios to emission ratios, the decay of the chemical species needs to be taken into account or assumptions need to be made, suggesting that the enhancement ratios are equivalent to emission ratios, hence measured at the source and not impacted by chemistry.

Compilations of numerous enhancement ratios, emission ratios, and emission factors for several trace gases from measurements at various locations worldwide are published regularly (e.g., Akagi et al., 2011) in order to facilitate their use in chemistry transport models.

There has been a recent interest in calculating enhancement ratios and emission factors from satellite data (e.g., Rinsland et al., 2007; Coheur et al., 2009; Tereszchuk et al., 2011). The above difficulty of inferring emission factors using the satellite observations comes from the fact that these observations are indeed typically made in the free–upper troposphere and further downwind of the fires. The fact that satellites mainly probe transported plumes where chemistry

modifies the original composition explains why the use of the enhancement ratio is more relevant than emission ratio.

Only a few papers have reported on the use of satellite retrievals to study tropospheric HCOOH, including the nadir-viewing Infrared Atmospheric Sounding Interferometer (IASI; e.g., Razavi et al., 2011; Stavrakou et al., 2012; R'Honi et al., 2013; Pommier et al., 2016) and Tropospheric Emission Spectrometer (TES; e.g., Cady-Pereira et al., 2014; Chaliyakunnel et al., 2016). Other studies have used the solar occultation Atmospheric Chemistry Experiment–Fourier Transform Spectrometer (ACE-FTS), which measures the atmospheric composition in the upper troposphere (e.g., Rinsland et al., 2006; Gonzalez Abad et al., 2009; Tereszchuk et al., 2011, 2013), and the Michelson Interferometer for Passive Atmospheric Sounding (MIPAS) limb instrument, which is sensitive to around 10 km (Grutter et al., 2010).

These infrared (IR) sounders have limited vertical sensitivity as compared to ground-based or airborne measurements, but their spatial coverage represents a major advantage, which allows observation of remote regions which are sparsely studied by field measurements, like the biomass burning regions.

This work aims to provide a list of enhancement ratios of HCOOH relative to CO over several biomass burning regions. For this, we analyzed 7 years of IASI measurements, between 2008 and 2014. Section 2 describes the IASI satellite mission and the retrieval characteristics for the CO and the HCOOH total columns. Section 3 presents the fire product used from the Moderate Resolution Imaging Spectroradiometer (MODIS) to identify the fire locations. Section 4 details the methodology used to identify of the IASI fire-affected observations. In Sect. 5 we describe and analyze the enhancement ratios obtained from the IASI measurements, including an analysis of these ratios by type of fuel burned, and we compare these values to those available in the literature. Finally, the conclusions are presented in Sect. 6.

2 HCOOH and CO columns from IASI

2.1 The IASI mission

IASI is a nadir-viewing Fourier transform spectrometer. Two models are currently in orbit. The first model (IASI-A) was launched onboard the Meteorological Operational Satellite (METOP)-A platform in October 2006. The second instrument was launched in September 2012 onboard METOP-B. Owing to its wide swath, IASI delivers near-global coverage twice a day with observation at around 09:30 and 21:30 local time. Each atmospheric view is composed of 2×2 circular pixels with a 12 km footprint diameter, spaced out by 50 km at nadir. IASI measures in the thermal infrared part of the spectrum, between 645 and 2760 cm^{-1}. It records radiance from the Earth's surface and the atmosphere with an apodized spectral resolution of 0.5 cm^{-1}, spectrally sampled at $0.25 \, \text{cm}^{-1}$. IASI has a wavenumber-dependent radiometric noise ranging from 0.1 to 0.4 K for a reference blackbody at 280 K (Clerbaux et al., 2009), and more specifically around 0.15 K for HCOOH and 0.20 K for the CO spectral ranges ($\sim 1105 \, \text{cm}^{-1}$ and $\sim 2150 \, \text{cm}^{-1}$, respectively).

The HCOOH and CO columns from IASI are used hereafter to determine the enhancement ratios of HCOOH. CO is chosen as reference due to its longer tropospheric lifetime (a few weeks to a few months, depending on latitude and time of year) as compared to HCOOH. In our study we use CO as the reference and not CO_2, since variations in CO_2 concentration are difficult to measure with sufficient accuracy from IASI (Crevoisier et al., 2009).

2.2 The CO retrieval characteristics

The CO concentrations are retrieved from IASI using the FORLI-CO software (Hurtmans et al., 2012), which uses an optimal estimation method based on Rodgers (2000). The spectral range used for the retrieval is between 2143 and $2181.25 \, \text{cm}^{-1}$. The CO total columns have been validated for different locations and atmospheric conditions (e.g., De Wachter et al., 2012; Kerzenmacher et al., 2012), and the comparisons with other data have shown good overall agreement, even if some discrepancies were found within CO-enriched plumes (reaching 12 % over the Arctic in summer, see Pommier et al., 2010; and reaching 17 % in comparison with other IR sounders, see George et al., 2009). These data were also used previously to study biomass burning plumes (e.g., Turquety et al., 2009; Pommier et al., 2010; Krol et al., 2013; Whitburn et al., 2015).

In order to keep only the most reliable retrievals, the selected data used have a root-mean-square error lower than $2.7 \times 10^{-9} \, \text{W} \, (\text{cm}^2 \, \text{cm}^{-1} \, \text{sr})^{-1}$ and a bias ranging between -0.15 and 0.25×10^{-9} as recommended in Hurtmans et al. (2012).

2.3 The HCOOH retrieval characteristics

The retrieval is based on the determination of the brightness temperature difference (ΔT_b) between spectral channels with and without the signature of HCOOH. The reference channels used for the calculation of ΔT_b were chosen on both sides of the HCOOH Q-branch ($1105 \, \text{cm}^{-1}$), i.e., at 1103.0 and $1109.0 \, \text{cm}^{-1}$. These ΔT_b were converted into total columns of HCOOH using conversion factors compiled in lookup tables. This simple and efficient retrieval method is described in more detail in Pommier et al. (2016).

As shown in Pommier et al. (2016), the vertical sensitivity of the IASI HCOOH total column ranges between 1 and 6 km. That study also showed that large HCOOH total columns were detected over biomass burning regions (e.g., Africa, Siberia) even if the largest values were found to be underestimated. This underestimation, which is less than 35 % for the columns smaller than $2.5 \times 10^{16} \, \text{molec} \, \text{cm}^{-2}$

(Pommier et al., 2016), will affect the enhancement ratios calculated in this work.

On the other hand, a large overestimation of the IASI HCOOH columns was shown in comparison with ground-based FTIR (Fourier transform infrared) measurements. This overestimation was larger for background columns (expected to reach 80 % for a column close to $0.3 \times 10^{16} \, \text{molec} \, \text{cm}^{-2}$), which can also impact our enhancement ratios.

3 MODIS

To identify the fire locations (hotspots), the fire product from MODIS onboard the polar-orbiting sun-synchronous NASA Terra and Aqua satellites (Justice et al., 2002; Giglio et al., 2006) are used. The Terra and Aqua satellites' equatorial overpass times are $\sim$ 10:30 and 22:30 and $\sim$ 01:30 and 13:30 local time, respectively. Fire pixels are $1 \, \text{km} \times 1 \, \text{km}$ in size at nadir.

For this work, we more specifically use the Global Monthly Fire Location Product (MCD14ML, Level 2, Collection 5) developed by the University of Maryland (https://earthdata.nasa.gov/data/near-real-time-data/firms/active-fire-data#tab-content-6) which, for each detected fire pixel, includes the geographic location of the fire, the fire radiative power (FRP), the confidence in detection, and the acquisition date and time. The FRP provides a measure of fire intensity that is linked to the fire fuel consumption rate (e.g., Wooster et al., 2005). Only data presenting a high confidence percentage are used, i.e., higher than or equal to 80 % as recommended in the MODIS user's guide (Giglio, 2013).

To characterize each MODIS hotspot by the type of fuel burned, the Global Mosaics of the standard MODIS land-cover-type data product (MCD12Q1) in the IGBP land-cover-type classification (Friedl et al., 2010; Channan et al., 2014) with a $0.5° \times 0.5°$ horizontal resolution has also been used (http://glcf.umd.edu/data/lc/). As the annual variability in this product is limited (not shown) and since the period available (from 2001 to 2012) does not fully match the period of the IASI mission, only the data for 2012 have been used. Whitburn et al. (2017) have also used this MCD12Q1 product to determine their IASI-derived NH_3 enhancement ratios by vegetation types.

4 Identifying fire-affected IASI observations

4.1 The selected areas

The determination of the biomass burning regions is based on the MODIS fire product. Figure 1 highlights the main areas that contributed to the biomass burning for the period between 2008 and 2014. Seven regions were selected for this work: Amazonia (AMA; 5–15° S, 40–60° W), corresponding mainly to the Brazilian Cerrado; Australia (AUS;

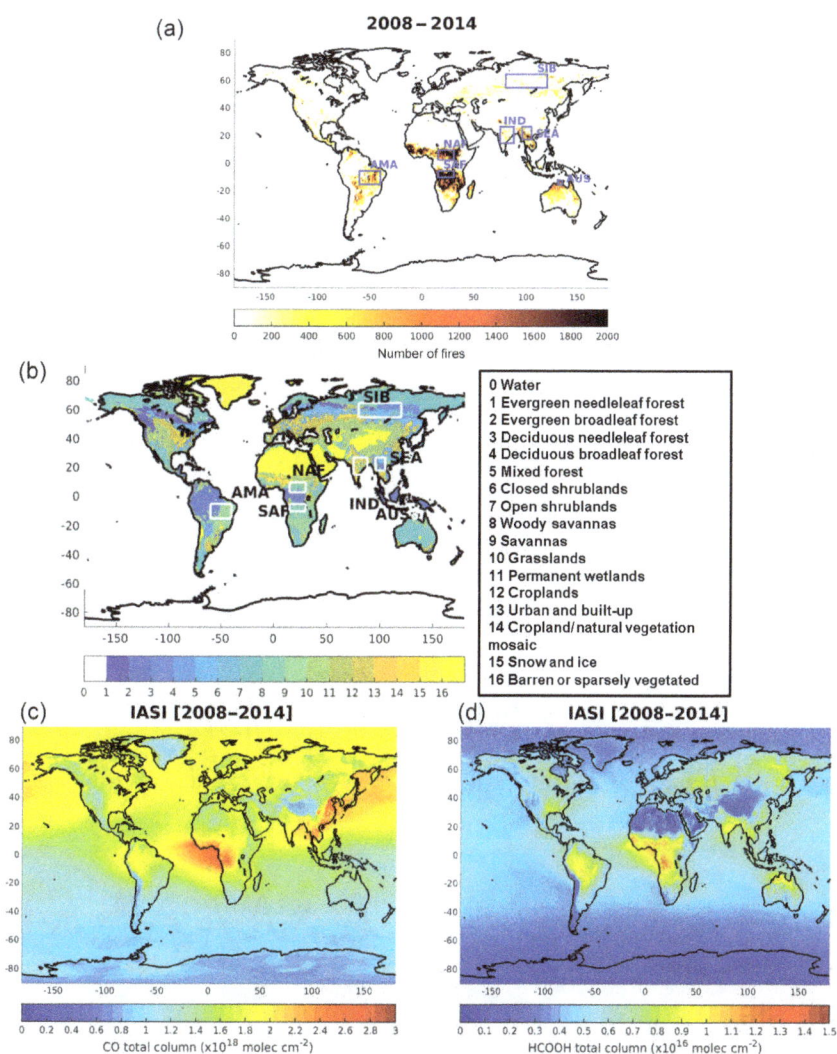

Figure 1. (a) Number of MODIS fire hotspots with a confidence percentage higher than or equal to 80 %, averaged on a $0.5° \times 0.5°$ grid, for the period between 2008 and 2014. The blue boxes are the regions studied in this work. **(b)** Classification of the land cover type from MODIS on the same grid and highlighting the studied regions in white. Each number corresponds to the type of vegetation. Only the data between $64°$ S and $84°$ N are available. The IASI CO total column distribution **(c)** and the IASI HCOOH total column distribution **(d)**, averaged between 2008 and 2014 and on the same grid.

12–15° S, 131–135° E); northern Africa (NAF; 3–10° N, 15–30° E); southern Africa (SAF; 5–10° S, 15–30° E); Southeast Asia (SEA; 18–27° N, 96–105° E); India (IND; 15–27° N, 75–88° E); and Siberia (SIB; 55–65° N, 80–120° E). Among these regions, India and Siberia do not represent the most active regions in terms of number of fires. It seemed, however, important to also investigate them. One first reason for this is that Pommier et al. (2016) showed a misrepresentation of the fire emissions of HCOOH over India. Secondly, India also encounters excess acidity in rainwater, which could be partly attributed to biomass burning (e.g., Bisht et al., 2014). Concerning Siberia, this region and the surrounding areas experienced intense fires over some years, such as during the summer 2010 (Pommier et al., 2016; and R'Honi et al., 2013, for

the region close to Moscow). The classification of the vegetation from the MODIS product has also been used for a detailed analysis of the enhancement ratios for these regions (Fig. 1).

4.2 The IASI data used

For this work, both the daytime and nighttime IASI data were used. We have verified that using only the daytime retrievals did not change the results. Figure 2 presents the time series of the monthly mean for the HCOOH and CO total columns over the seven selected regions. The number of fires and their FRP are also indicated. The variation in the total columns of HCOOH and CO matches relatively well with the variation

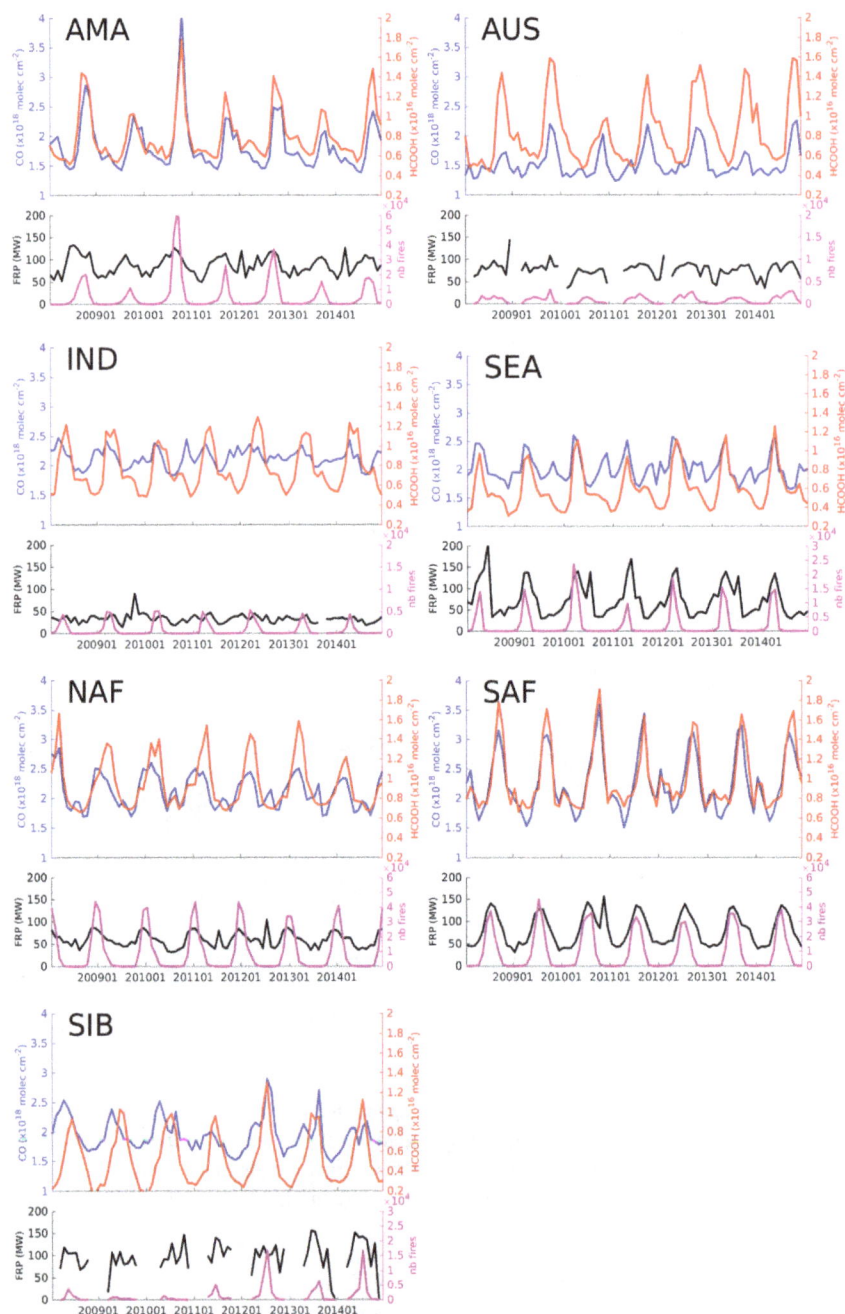

Figure 2. Time series from 2008 to 2014 of the monthly means of IASI CO (blue) and HCOOH (red) total columns in 10^{18} molec cm^{-2} and in 10^{16} molec cm^{-2}, respectively; FRP (black) in megawatts (MW); and the number of fires (nb fires, magenta) from MODIS over the seven regions (AMA, Amazonia; AUS, Australia; IND, India; SEA, Southeast Asia; NAF, northern Africa; SAF, southern Africa; SIB, Siberia).

of the number of fires. It is also worth noting that these variations in the total columns do not depend on the intensity of the fires as shown by Fig. 2 and by the scatterplots with the values characterizing each fire as described below (not shown).

The monthly HCOOH and CO total columns are found to be highly correlated over the selected biomass burning regions (correlation coefficient, r, from 0.75 to 0.91), ex-

cept over India ($r = 0.34$) and Siberia ($r = 0.58$). Over both regions, the impact of sources other than biomass burning is thus not negligible. Over India, the CO budget is influenced by long-range transport (e.g., Srinivas et al., 2016) and the anthropogenic emissions also have a large impact (e.g., Ohara et al., 2007). This could explain why the variation in CO does not perfectly follow the variation in the number of fires and why the difference between the background level and the CO peaks is less marked than for the HCOOH. Over Siberia, a temporal shift between the highest peaks for CO and for HCOOH is noticed for some years, such as for 2009, 2010, and 2011. For these years, the variation in CO does not follow the variation in the number of fires. The large region selected over Siberia is known to also be impacted by CO-enriched plumes transported from other regions, such as polluted air masses from China (e.g., Paris et al., 2008) or from Europe (e.g., Pochanart et al., 2003). These external influences interfere with the CO plumes originating from forest fires measured over this region.

Despite the overall good match between the number of fires and the variation in HCOOH and CO, we are not certain that HCOOH and the CO were emitted solely by fires, and the discrimination between a natural and an anthropogenic origin for each compound is challenging. This assessment is particularly obvious for IND and SIB. To isolate the HCOOH and CO signals measured by IASI, potentially emitted by a fire, we decided to only use the data in the vicinity of each MODIS hotspot. To do so, we co-located the IASI data at 50 km around each MODIS pixel and between 0 and 5 h from the time registered by MODIS for each detected fire, so that each MODIS pixel is associated with a mean value of HCOOH and CO total columns from IASI.

With these co-location criteria, good correlation coefficients, calculated by linear-least-squares fitting, are found between the HCOOH and CO total columns as shown in Table 1 (upper row). The smaller correlation coefficients, i.e., less than 0.7, are found for India, Australia, Siberia, and northern Africa. It is also important to note that the HCOOH and CO columns are better correlated for India and Siberia compared to the monthly time series shown in Fig. 2. The three other regions present a large correlation, around 0.8. The high correlation suggests that IASI sampled the same biomass burning air mass for these compounds.

4.3 Importance of the meteorological conditions

As shown in earlier studies, the wind speed can have a large influence on the detection of tropospheric plumes of trace gases from space (e.g., NO_2: Beirle et al., 2011; CO: Pommier et al., 2013; SO_2: Fioletov et al., 2015). We have chosen to assign a surface wind speed value for each MODIS hotspot. These meteorological fields were taken from the ECMWF (European Centre for Medium-Range Weather Forecasts) reanalysis data (http://apps.ecmwf.int/datasets/data/interim-full-daily/levtype=sfc/; Dee et al., 2011). The horizontal resolution of these fields is 0.125° on longitude and latitude with a 6 h time step. As shown in Fig. 3, the three regions where the HCOOH : CO correlations are found to be high (r close to 0.8) correspond to the regions where the surface wind speed was lower, i.e., for AMA, SEA, and SAF. IND also has a low mean and median surface wind speed, but the distribution of this surface wind speed over IND is more spread out than for AMA, SEA, and SAF. It is

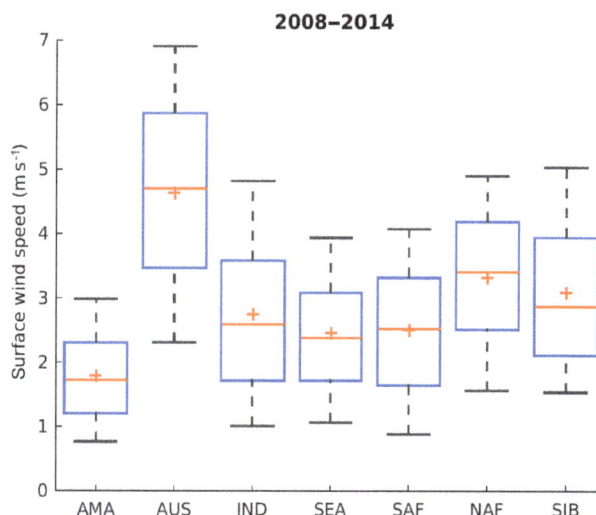

Figure 3. Box and whisker plots showing the mean (red central cross), median (red central line), and 25th and 75th percentiles (blue box edges) of surface wind speed for each MODIS hotspot over the studied regions (AMA = Amazonia, AUS = Australia, IND = India, SEA = Southeast Asia, NAF = northern Africa, SAF = southern Africa, SIB = Siberia). The whiskers encompass values from the $25th - 1.5 \times (75th - 25th)$ to the $75th + 1.5 \times (75th - 25th)$ percentiles. This range of values corresponds to approximately 99.3 % coverage if the data are normally distributed.

also noteworthy that the IND and SEA regions are both characterized by higher wind speed at higher altitudes, i.e., for the pressure levels 650 and 450 hPa (not shown). This shows that the wind speed at higher altitudes has a lower influence on our correlations than the surface wind. When filtering out the data associated with a large surface wind (higher than $1.44 \, \mathrm{m \, s^{-1}}$), new correlations between the HCOOH and the CO total columns from IASI are calculated (Table 1 – lower row). This value of $1.44 \, \mathrm{m \, s^{-1}}$ for the surface wind speed corresponds to the 25th percentile of the distribution of the three regions characterized by the lowest surface wind speed (Fig. 3).

The correlation coefficients, shown on the scatterplots in Fig. 4 and summarized in Table 1 (lower row), increase for all regions except over NAF, where the coefficient is found to be slightly lower than the previous correlation (Table 1 – upper row). The correlation coefficient is significantly improved over IND and SIB (Table 1 – lower row). These results confirm a robust correlation between the HCOOH and the CO total columns measured by IASI in the vicinity of each MODIS fire location.

Table 1. Upper row: correlation coefficients between the HCOOH total columns and the CO total columns measured by IASI for the period between 2008 and 2014 over the seven studied regions. Lower row: as upper row but with only MODIS fire hotspot having a surface wind speed lower than $1.44 \, \mathrm{m \, s^{-1}}$. Each IASI datapoint is selected in an area of $50 \, \mathrm{km}$ around the MODIS fire hotspot and up to $5 \, \mathrm{h}$ after the time recorded for each fire. The number of fires characterized by HCOOH and CO total columns is given in parenthesis.

	AMA	AUS	IND	SEA	SAF	NAF	SIB
r	0.78 (13 342)	0.63 (1525)	0.53 (1641)	0.84 (1865)	0.78 (12 227)	0.58 (21 139)	0.65 (22 353)
	0.79 (4580)	0.65 (93)	0.65 (340)	0.86 (528)	0.80 (895)	0.53 (1095)	0.72 (2097)

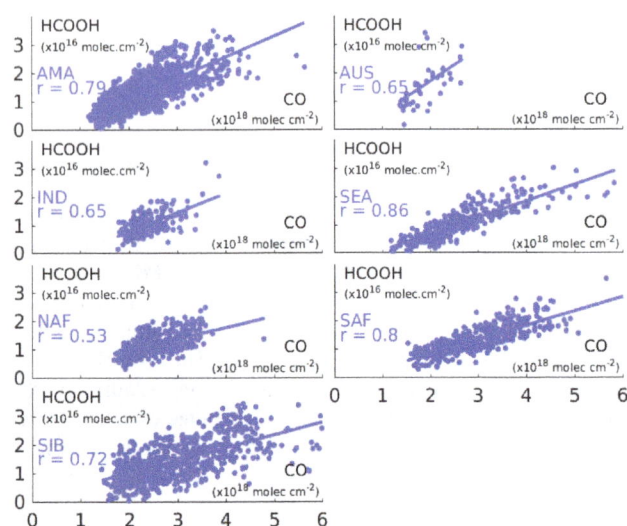

Figure 4. Scatterplots between the IASI fire-affected HCOOH total columns (in $10^{16} \, \mathrm{molec \, cm^{-2}}$) and the CO total columns (in $10^{18} \, \mathrm{molec \, cm^{-2}}$) over the seven regions (AMA = Amazonia, AUS = Australia, IND = India, SEA = Southeast Asia, NAF = northern Africa, SAF = southern Africa, SIB = Siberia). The linear regression is represented by the blue line and the correlation coefficient is also provided for each region.

5 Analysis of the data over the fire regions

5.1 Determination of the enhancement ratios

5.1.1 General analysis

Based on scatterplots in Fig. 4, an enhancement ratio can be calculated for each region. These enhancement ratios, defined as $\mathrm{ER_{(HCOOH/CO)}}$, correspond to the value of the slope $\partial[\mathrm{HCOOH}] / \partial[\mathrm{CO}]$ found in Fig. 4. This technique to determine the $\mathrm{ER_{(HCOOH/CO)}}$ is more reliable than using only the columns themselves, i.e., by estimating an $\mathrm{ER_{(HCOOH/CO)}}$ for each measurement pair (HCOOH, CO). Indeed, performing scatterplots helps to identify a common origin for HCOOH and CO. The values of the $\mathrm{ER_{(HCOOH/CO)}}$ over each region are summarized in Table 2.

It is known that trace gas concentrations within smoke plumes can vary rapidly with time and are very sensitive to chemistry, so a comparison with previous work is always challenging, especially if these studies were performed over another altitude range, at a different location, or at a different period of the year.

A good agreement is, however, generally found with previous studies, even if it is important to keep in mind that an underestimation of our $\mathrm{ER_{(HCOOH/CO)}}$ is possible due to the underestimation in the highest values of HCOOH as over the forest fires (see Sect. 2.3). On the other hand, the overestimation in the background column can also impact the calculation of our $\mathrm{ER_{(HCOOH/CO)}}$. The effects of both biases are, however, limited since most of the HCOOH total columns used in our analysis over the selected regions are higher than $0.3 \times 10^{16} \, \mathrm{molec \, cm^{-2}}$ and lower than $2.5 \times 10^{16} \, \mathrm{molec \, cm^{-2}}$, as explained in Sect. 2.3.

Nevertheless, in order to investigate the possible impact of the overestimation in the lower columns and the underestimation in the higher columns on the calculated ratios, a test was performed, by using only HCOOH columns with a thermal contrast (TC) larger than $10 \, \mathrm{K}$. Indeed, the increase in the thermal contrast (i.e., the temperature difference between the surface and the first layer in the retrieved profile) leads to reducing the detection limit as shown in Pommier al. (2016). This enhancement of the detection level helps to minimize the bias in the retrieved total columns as explained in Crevoisier et al. (2014). For the analysis performed here, similar slopes and correlation coefficients were generally calculated, suggesting a negligible effect of this parameter on the biases. The only exception is an increase in $\mathrm{ER_{(HCOOH/CO)}}$ over Siberia ($6.5 \times 10^{-3} \pm 0.19 \times 10^{-3} \, \mathrm{mol \, mol^{-1}}$ when using only IASI measurements with TC above $10 \, \mathrm{K}$ against $4.4 \times 10^{-3} \, \mathrm{mol \, mol^{-1}} \pm 0.09 \times 10^{-3}$ in Table 2). It is worth noting that only 48 % of the selected scenes remain over Siberia when applying this filter on thermal contrast (60 % for SEA, 77 % for AMA, 80 % for SAF, 83 % for AUS and NAF, and 89 % for IND). This implies that the statistics on the fire emissions in the higher latitudes of Siberia are dominated by measurements with a low thermal contrast and thus with HCOOH total columns with higher uncertainties. However, the limited changes in slopes and correlation coefficients give us confidence that the results presented in Table 2 are representative.

Table 2. Enhancement ratio of HCOOH relative to CO (mol mol^{-1}) with its standard deviation compared to enhancement ratios of HCOOH relative to CO and emissions ratios of HCOOH reported in the literature for the seven studied regions.

Region	Enhancement ratio to CO (mol mol^{-1}), this work	Enhancement ratio to CO (mol mol^{-1}) found in literature	Emission ratio to CO (mol mol^{-1}) found in literature	Instrument used
AMA	7.3×10^{-3} $\pm 0.08 \times 10^{-3}$	$5.1 \times 10^{-3} \pm 1.5 \times 10^{-3}$ (González Abad et al., 2009)* $6.7 \times 10^{-3} \pm 2.1 \times 10^{-3}$ (Chaliyakunnel et al., 2016)		ACE-FTS TES
AUS	11.1×10^{-3} $\pm 1.37 \times 10^{-3}$	$4.5 \times 10^{-3} \pm 5.1 \times 10^{-3}$ (Chaliyakunnel et al., 2016) $21.0 \times 10^{-3} \pm 10.0 \times 10^{-3}$ (Paton-Walsh et al., 2005)*		TES ground-based FTIR
IND	6.8×10^{-3} $\pm 0.44 \times 10^{-3}$	None		–
SEA	5.8×10^{-3} $\pm 0.15 \times 10^{-3}$	None		–
NAF	4.0×10^{-3} $\pm 0.19 \times 10^{-3}$	$2.8 \times 10^{-3} \pm 0.4 \times 10^{-3}$ (Chaliyakunnel et al., 2016)		TES
SAF	5.0×10^{-3} $\pm 0.13 \times 10^{-3}$	$2.6 \times 10^{-3} \pm 0.3 \times 10^{-3}$ (Chaliyakunnel et al., 2016) $4.6 \times 10^{-3} \pm 0.3 \times 10^{-3}$ (Vigouroux et al., 2012) 5.1×10^{-3} (Coheur et al., 2007) $11.3 \times 10^{-3} \pm 7.6 \times 10^{-3}$ (Rinsland et al., 2006)*	$5.9 \times 10^{-3} \pm 2.2 \times 10^{-3}$ (Yokelson et al., 2003) 5.1–8.7×10^{-3} (Sinha al., 2003)	TES ground-based FTIR ACE-FTS ACE-FTS airborne FTIR airborne FTIR
SIB	4.4×10^{-3} $\pm 0.09 \times 10^{-3}$	0.77–6.41×10^{-3} (Tereszchuk et al., 2013) 2.69–15.93×10^{-3} (Viatte et al., 2015) 10.0–32.0×10^{-3} (R'Honi et al., 2013)		ACE-FTS ground-based FTIR IASI

* Their "emission ratios" are re-qualified as enhancement ratios in this study since their ratios were not measured at the origin the fire emission but at high altitudes and/or further downwind of the fires.

5.1.2 Analysis over each region

A few backward trajectories (along 5 days, not shown) have been calculated for our hotspots with the online version of the HYSPLIT atmospheric transport and dispersion modeling system (Rolph, 2017). These trajectories, initialized at different altitudes, confirm a main origin close to the surface of our IASI fire-affected columns. However, it is impossible to properly compare the origin of the air masses with previous studies as our studied period (2008–2014) or our studied fires do not necessarily match with plumes described in other publications. It is also difficult to estimate the age of our stud-

ied air masses by gathering the plumes during a 7-year period and without an accurate knowledge of the altitude of the plumes.

When compared with other studies, the best agreement for the values presented in Table 2 is found over southern Africa where the $\mathrm{ER}_{(HCOOH/CO)}$ ($5 \times 10^{-3} \pm 0.13 \times 10^{-3}$ mol mol^{-1}) is similar to the value calculated by Vigouroux et al. (2012) and Coheur et al. (2007). It also agrees with the broad range of values of emission ratios ($\mathrm{EmR}_{(HCOOH/CO)}$) referenced by Sinha et al. (2003). This result corroborates the relevance of the methodology used in this work over this region for the identification of fire-affected IASI columns close to the source. Vigouroux et al. (2012) sampled biomass burning outflow of southern Africa and Coheur et al. (2007) calculated their $ER_{(HCOOH/CO)}$ in plumes observed over Tanzania in the upper troposphere, while Sinha et al. (2003) did it within plumes over Zambia at the origin of the fire.

A few assumptions are needed in order to interpret our $ER_{(HCOOH/CO)}$, but the analysis given hereafter is only indicative since these previous studies did not measure the same plumes as those presented in this work. Our $ER_{(HCOOH/CO)}$ is also calculated without making any distinction on the seasonal variation or on the type of biomass burning plumes sampled (e.g., emitted by a savanna fire or by a forest fire). The analysis by biome is presented in Sect. 5.2. Since these $ER_{(HCOOH/CO)}$ values from previous studies and the $EmR_{(HCOOH/CO)}$ from Sinha et al. (2003) agree with our $ER_{(HCOOH/CO)}$, and since HCOOH has a short lifetime, this may suggest that the selected plumes measured by IASI from 2008 to 2014 and those sampled by Vigouroux et al. (2012) and Coheur et al. (2007) encountered a limited secondary production or a low sink (deposition or reaction with OH in the troposphere during their transport). To quantify the role of the chemistry or of the deposition within the plumes, a modeling work should be performed. However, this is beyond the scope of this paper.

Another important point is that the decay of HCOOH is faster than that of CO. As our $ER_{(HCOOH/CO)}$ is similar to the $ER_{(HCOOH/CO)}$ from the other studies and to the $EmR_{(HCOOH/CO)}$ given in Sinha et al. (2003), this could suggest that all these plumes (from our study, from Vigouroux et al., 2012, and from Coheur et al., 2007) are rapidly advected in the troposphere. Our $ER_{(HCOOH/CO)}$ differs from the value in Rinsland et al. (2006) ($11.3 \times 10^{-3} \pm 7.6 \times 10^{-3}$ mol mol^{-1}), since our ratio is 55 % lower. One possible explanation is the multi-origin of the plumes studied by Rinsland et al. (2006), since, based on their backward trajectories, their plumes could be influenced by biomass burning originating from southern Africa and/or from South America. The travel during the few days across the Atlantic Ocean may explain the change in their $ER_{(HCOOH/CO)}$.

It is worth noting that the ACE-FTS instrument used in their study works in a limb solar occultation mode. This means that the atmospheric density sampled by the instrument is larger than the one measured by the nadir geometry with IASI. However, the difference in geometry cannot explain why we find an agreement with the ACE-FTS measurement values reported by Coheur et al. (2007) and a disagreement with those from Rinsland et al. (2006). Part of the difference could be associated with the difference in the assumptions used in both retrievals (e.g., the a priori profile).

The $ER_{(HCOOH/CO)}$ from our work is also 15 % lower than the $EmR_{(HCOOH/CO)}$ in Yokelson et al. (2003) ($5.9 \times 10^{-3} \pm 2.2 \times 10^{-3}$ mol mol^{-1}), who calculated

their value within plumes over Zambia, Zimbabwe, and South Africa. With this difference we can also suggest the presence of a sink of HCOOH within the plumes detected by IASI or that this slight underestimation is simply related to the faster decay of HCOOH than that of CO. Conversely, the $ER_{(HCOOH/CO)}$ retrieved from IASI is twice that of Chaliyakunnel et al. (2016) ($2.6 \times 10^{-3} \pm 0.3 \times 10^{-3}$ mol mol^{-1}). Chaliyakunnel et al. (2016) developed an approach allowing the determination of pyrogenic $ER_{(HCOOH/CO)}$ by reducing the impact of the mix with the ambient air. To do so, they calculated the $ER_{(HCOOH/CO)}$ in the vicinity of the fire count from MODIS (averaged in a cell having the resolution of the GEOS-Chem model, i.e., $2° \times 2.5°$) and they differentiated this value with a background $ER_{(HCOOH/CO)}$ defined by the concentrations distant from these fires. They concluded that their most reliable value on the amount of HCOOH produced from fire emissions was obtained from African fires.

Over northern Africa, the calculated $ER_{(HCOOH/CO)}$ ($4 \times 10^{-3} \pm 0.19 \times 10^{-3}$ mol mol^{-1}) is 42 % higher than the $ER_{(HCOOH/CO)}$ calculated in Chaliyakunnel et al. (2016) ($2.8 \times 10^{-3} \pm 0.4 \times 10^{-3}$ mol mol^{-1}). It is worth noting that NAF is the region characterized by a scatterplot with the lowest correlation coefficient (Fig. 4).

A larger difference is found over Australia where the $ER_{(HCOOH/CO)}$ is $11.1 \times 10^{-3} \pm 1.37 \times 10^{-3}$ mol mol^{-1}. This $ER_{(HCOOH/CO)}$ is roughly the mean of both values reported by Paton-Walsh et al. (2005) and Chaliyakunnel et al. (2016). The difference between our work and that of Paton-Walsh et al. (2005) may be explained by the different origin of the probed plume. In our case, the studied area corresponds to the northern part of the Northern Territory with savanna-type vegetation (as shown in Sect. 5.2), while Paton-Walsh et al. (2005) sampled bush fire plumes coming from the eastern coast of Australia, representative of Australian temperate forest. In the work done by Chaliyakunnel et al. (2016), a quite uncertain value is reported ($4.5 \times 10^{-3} \pm 5.1 \times 10^{-3}$ mol mol^{-1}), with an error larger than their $ER_{(HCOOH/CO)}$.

Over Amazonia, our $ER_{(HCOOH/CO)}$ ($7.3 \times 10^{-3} \pm 0.08 \times 10^{-3}$ mol mol^{-1}) is similar to the value given in Chaliyakunnel et al. (2016), who report a larger bias over Amazonia. Over this region, our $ER_{(HCOOH/CO)}$ is higher than the one obtained by González Abad et al. (2009) with ACE-FTS in the upper troposphere ($5.1 \times 10^{-3} \pm 1.5 \times 10^{-3}$ mol mol^{-1}). This difference with the study done by González Abad et al. (2009) may be explained by the difference in the altitude of the detection of the forest fire plume between IASI (mid-troposphere) and ACE-FTS (upper troposphere) and thus by a difference in the ongoing chemistry within their respective sampled plumes. The geometry of the sampling (nadir vs. limb) or the difference in the retrieval may also have an impact in the retrieved HCOOH.

The Siberian $ER_{(HCOOH/CO)}$ $(4.4 \times 10^{-3}\,\mathrm{mol\,mol^{-1}}$ $\pm 0.09 \times 10^{-3})$ is found to be in good agreement with the wide range of values obtained by Tereszchuk et al. (2013) and Viatte et al. (2015). This $ER_{(HCOOH/CO)}$ is, however, lower than the ratios calculated by R'Honi et al. (2013), who focused on the extreme fire event that occurred in 2010.

For India and Southeast Asia, a comparison is not possible since no previous studies were reported. The comparison is performed next, based on the emission factors.

5.2 Analysis based on the type of vegetation

We have complemented our comparison of the enhancement ratios by comparing our ratios to emissions ratios calculated from emission factors found in literature. The main argument for performing such comparison is the lack of measurements of enhancement ratios over IND and SEA. Furthermore, such comparison from emission factors facilitates an analysis based on hypotheses about the type of vegetation burned.

Even if our methodology attempts to characterize the HCOOH emitted by biomass burning close to the source, our columns are probably not representative of the emission at the origin of the fire. The altitude of the sampling (mid-troposphere), even if an influence from the surface is shown, and the age of the plumes (at least a few hours) have a large impact on our enhancement ratios.

To perform a proper comparison with emission ratios, our enhancement ratios should be converted to emission ratios. To do so, it would be essential to take into account the decay of the compounds during the transport of the plume. However, due to the methodology used, i.e., averaging the data collected during a few hours (between 0 and 5 h from the time registered by MODIS for each detected fire), the calculation of the decay of each compound is not possible. We therefore have compared our enhancement ratios to emission ratios, and the comparison presented hereafter is mostly illustrative.

For both IND and SEA regions, the emission ratios have been calculated from the emission factors provided in Akagi et al. (2011). For the other regions, in addition to the values from Akagi et al. (2011), emission ratios were similarly calculated from emission factors given in other studies (listed in Table 3).

Based on the emission ratios, the emission factors are usually derived by this following equation:

$$EF_{HCOOH} = EF_{CO} \times MW_{HCOOH}/MW_{CO}$$
$$\times EmR_{(HCOOH/CO)}. \tag{1}$$

EF_{HCOOH} is the emission factor for HCOOH; $EmR_{(HCOOH/CO)}$ is the molar emission ratio of HCOOH with respect to CO; MW_{HCOOH} is the molecular weight of HCOOH; MW_{CO} is the molecular weight of CO; and EF_{CO}

is the emission factor for CO for dry matter, set to the value taken from Akagi et al. (2011).

Thus, based on Eq. (1), $EmR_{(HCOOH/CO)}$ values were calculated and compared with our $ER_{(HCOOH/CO)}$ (Table 3). In this calculation, the vegetation type characterizing each region is important. Some regions are composed of a mix of vegetation types as shown in Fig. 1. This is, for example, the case for AMA and SAF (e.g., White, 1981). Thus, following the classification from Akagi et al. (2011), AMA and SAF are composed of tropical forest and savanna, characterized by an EF_{CO} of 93 ± 27 and $63 \pm 17\,\mathrm{g\,kg^{-1}}$, respectively (Akagi et al., 2011). AUS and NAF correspond to a savanna fuel type. SIB is a boreal forest area with an EF_{CO} of $127 \pm 45\,\mathrm{g\,kg^{-1}}$. Based also on the maps shown by Fig. 9 in Schreier et al. (2014) and Fig. 13 in van der Werf et al. (2010), the soil for IND is supposed to be mainly composed of cropland (agriculture), which is associated with an EF_{CO} of $102 \pm 33\,\mathrm{g\,kg^{-1}}$, and probably also by extratropical forest, which is characterized by an EF_{CO} equal to $122 \pm 44\,\mathrm{g\,kg^{-1}}$, and savanna, with an EF_{CO} of $63 \pm 17\,\mathrm{g\,kg^{-1}}$. The fuel type for SEA is supposed to be a mix of extratropical forest and savanna, with an EF_{CO} of 122 ± 44 and $63 \pm 17\,\mathrm{g\,kg^{-1}}$, respectively. Cropland fuel type was also used, since large agricultural biomass burning is occurring in this region (e.g., Duc et al., 2016).

In addition to the $EmR_{(HCOOH/CO)}$ calculated from the EF_{HCOOH} given in the literature, a classification for our $ER_{(HCOOH/CO)}$ has also been done, based on the data from the MCD12Q1 product. As each hotspot is associated with a land cover value defined by the MCD12Q1 product, enhancement ratios by biome have been calculated. The limitations of this dataset are its coarse resolution ($0.5° \times 0.5°$) and the lack of seasonal variation. It gives supplementary information on the type of fuel burned identified by MODIS. The corresponding $ER_{(HCOOH/CO)}$ values are provided in Table 3. Only the values calculated from a scatterplot with a correlation coefficient higher than 0.4 are reported.

Despite the assumptions made, a fair agreement is found over southern Africa. Our $ER_{(HCOOH/CO)}$ $(5 \times 10^{-3} \pm 0.13 \times 10^{-3}\,\mathrm{mol\,mol^{-1}})$ is indeed similar to the $EmR_{(HCOOH/CO)}$ calculated from Sinha et al. (2004) by using savanna fuel type, and the $ER_{(HCOOH/CO)}$ is between both values calculated from Yokelson et al. (2003). This agreement is consistent since both previous studies sampled plumes emitted by savanna fires. Yokelson et al. (2003) and Sinha et al. (2004) both used the same sampling strategy. They sampled fire plumes by penetrating several minutes-old plumes at relatively low altitude (up to 1.3 km for Sinha et al., 2004, and just above the flame front for Yokelson et al., 2003). This agreement shows, as already described in the previous section, that our $ER_{(HCOOH/CO)}$ over southern Africa is similar to their $EmR_{(HCOOH/CO)}$. It is also noteworthy, based on the MODIS land-cover-type product, that all the studied hotspots are defined as savanna fires. On other hand, our $ER_{(HCOOH/CO)}$ is also similar to the

Table 3. Enhancement ratio of HCOOH relative to CO (mol mol^{-1}) with its standard deviation and enhancement ratio of HCOOH relative to CO (mol mol^{-1}) by biome with its standard deviation calculated in this work. For each enhancement ratio by biome, the correlation coefficient and the number of MODIS hotspots are provided. The enhancement ratios are compared to emission ratios calculated from emission factors given in the literature for the seven studied regions. For the calculation of these emission ratios, the emission factors of CO for the corresponding fuel type given in Akagi et al. (2011) are used. Emission ratios of HCOOH relative to CO (mol mol^{-1}) calculated from the emission factors of HCOOH given in Akagi et al. (2011) for the corresponding fuel type are also provided.

Region	Enhancement ratio to CO (mol mol^{-1}), this work	Enhancement ratio to CO (mol mol^{-1})[a] by biome[b], this work	Emission ratio to CO (mol mol^{-1}) calculated from EF$_{HCOOH}$ given in literature and using EF$_{CO}$ from Akagi et al. (2011)	Instrument used
AMA	7.3×10^{-3} $\pm 0.08 \times 10^{-3}$	$6.3 \times 10^{-3} \pm 0.22 \times 10^{-3}$ (evergreen broadleaf forest, $r = 0.81, n = 454$) $3.0 \times 10^{-3} \pm 0.81 \times 10^{-3}$ (open shrubland, $r = 0.91, n = 5$) $7.0 \times 10^{-3} \pm 2.47 \times 10^{-3}$ (woody savanna, $r = 0.63, n = 14$) $7.6 \times 10^{-3} \pm 0.09 \times 10^{-3}$ (savanna, $r = 0.79, n = 3909$) $8.4 \times 10^{-3} \pm 0.39 \times 10^{-3}$ (grassland, $r = 0.88, n = 143$) $4.6 \times 10^{-3} \pm 0.35 \times 10^{-3}$ (cropland, $r = 0.88, n = 54$)	1.8×10^{-3} – tropical forest (Yokelson et al., 2007, 2008)[c] 2.7×10^{-3} – savanna (Yokelson et al., 2007, 2008)[c] 2.0×10^{-3} – savanna (Akagi et al., 2011) 5.2×10^{-3} – tropical forest (Akagi et al., 2011)	airborne FTIR (Yokelson et al., 2007); laboratory (Yokelson et al., 2008); catalog
AUS	11.1×10^{-3} $\pm 1.37 \times 10^{-3}$	$5.7 \times 10^{-3} \pm 2.55 \times 10^{-3}$ (woody savanna, $r = 0.6, n = 11$) $11.2 \times 10^{-3} \pm 1.49 \times 10^{-3}$ (savanna, $r = 0.65, n = 80$)	2.0×10^{-3} – savanna (Akagi et al., 2011)	catalog
IND	6.8×10^{-3} $\pm 0.44 \times 10^{-3}$	$6.6 \times 10^{-3} \pm 0.77 \times 10^{-3}$ (woody savanna, $r = 0.65$, $n = 103$) $6.2 \times 10^{-3} \pm 0.62 \times 10^{-3}$ (cropland, $r = 0.58, n = 198$) $8.8 \times 10^{-3} \pm 1.19 \times 10^{-3}$ (cropland/natural vegetation mosaic, $r = 0.85, n = 23$)	2.0×10^{-3} – savanna (Akagi et al., 2011) 2.7×10^{-3} – extratropical forest (Akagi et al., 2011) 6.0×10^{-3} – cropland (Akagi et al., 2011)	catalog
SEA	5.8×10^{-3} $\pm 0.15 \times 10^{-3}$	$5.6 \times 10^{-3} \pm 0.20 \times 10^{-3}$ (evergreen broadleaf forest, $r = 0.83, n = 334$) $6.3 \times 10^{-3} \pm 0.66 \times 10^{-3}$ (mixed forest, $r = 0.76, n = 70$) $6.2 \times 10^{-3} \pm 0.38 \times 10^{-3}$ (woody savanna, $r = 0.86, n = 99$) $7.1 \times 10^{-3} \pm 0.99 \times 10^{-3}$ (cropland/natural vegetation mosaic, $r = 0.84, n = 23$)	2.0×10^{-3} – savanna (Akagi et al., 2011) 2.7×10^{-3} – extratropical forest (Akagi et al., 2011) 6.0×10^{-3} – cropland (Akagi et al., 2011)	catalog
NAF	4.0×10^{-3} $\pm 0.19 \times 10^{-3}$	$3.4 \times 10^{-3} \pm 0.63 \times 10^{-3}$ (evergreen broadleaf forest, $r = 0.52, n = 78$) $3.3 \times 10^{-3} \pm 0.28 \times 10^{-3}$ (woody savanna, $r = 0.44$, $n = 569$) $4.4 \times 10^{-3} \pm 0.29 \times 10^{-3}$ (savanna, $r = 0.59, n = 441$) $22.6 \times 10^{-3} \pm 11.06 \times 10^{-3}$ (cropland/natural vegetation mosaic, $r = 0.67, n = 7$)	2.0×10^{-3} – savanna (Akagi et al., 2011)	catalog

Table 3. Continued.

Region	Enhancement ratio to CO $(mol\,mol^{-1})$, this work	Enhancement ratio to CO $(mol\,mol^{-1})^a$ by biome[b], this work	Emission ratio to CO $(mol\,mol^{-1})$ calculated from EF_{HCOOH} given in literature and using EF_{CO} from Akagi et al. (2011)	Instrument used
SAF	5.0×10^{-3} $\pm 0.13 \times 10^{-3}$	all hotspots are woody savanna	3.3×10^{-3} – tropical forest (Sinha et al., 2004)[d]	airborne FTIR
			4.8×10^{-3} – savanna (Sinha et al., 2004)[d]	
			4.1×10^{-3} – tropical forest (Yokelson et al., 2003)	airborne FTIR
			6.0×10^{-3} – savanna (Yokelson et al., 2003)	
			13×10^{-3} – tropical forest (Rinsland et al., 2006)	ACE-FTS
			19.2×10^{-3} – savanna (Rinsland et al., 2006)	
			2.0×10^{-3} – savanna (Akagi et al., 2011)	catalog
			5.2×10^{-3} – tropical forest (Akagi et al., 2011)	
SIB	4.4×10^{-3} $\pm 0.09 \times 10^{-3}$	$4.0 \times 10^{-3} \pm 0.31 \times 10^{-3}$ (evergreen needleleaf forest, $r = 0.63, n = 245$) $3.6 \times 10^{-3} \pm 0.16 \times 10^{-3}$ (deciduous needleleaf forest, $r = 0.66, n = 659$) $3.4 \times 10^{-3} \pm 0.18 \times 10^{-3}$ (mixed forest, $r = 0.57, n = 759$) $6.6 \times 10^{-3} \pm 0.48 \times 10^{-3}$ (open shrubland, $r = 0.76$, $n = 143$) $6.0 \times 10^{-3} \pm 0.41 \times 10^{-3}$ (woody savanna, $r = 0.76$, $n = 155$) $3.8 \times 10^{-3} \pm 0.65 \times 10^{-3}$ (permanent wetland, $r = 0.6$, $n = 63$)	2.7×10^{-3} – boreal forest (Akagi et al., 2011)	catalog

[a] Only the enhancement ratios to CO calculated from a scatterplot with a correlation coefficient higher than 0.4 are reported. [b] The type of vegetation is defined by the land-cover-type data product (MCD12Q1). [c] The EF_{HCOOH} values were corrected based on the comment from Yokelson et al. (2013) (EF_{HCOOH} used: 0.281 for Yokelson et al., 2007; 0.2767 for Yokelson et al., 2008). [d] The means of both EF_{HCOOH} values provided in Sinha et al. (2004) were used for our $EmR_{HCOOH/CO}$ calculation.

$EmR_{(HCOOH/CO)}$ from Akagi et al. (2011) but for the tropical forest. A large underestimation compared to Rinsland et al. (2006) is found. This underestimation confirms the disagreement with their study already shown in Table 2.

Over northern Africa, our $ER_{(HCOOH/CO)}$ is twice as large as the $EmR_{(HCOOH/CO)}$ provided by Akagi et al. (2011), probably due to the lower correlation found in our scatterplot. It is highly probable that our presumed fire-affected IASI columns are indeed impacted by other air masses. The land classification based on the MODIS product also shows a diverse origin of the hotspots.

For Amazonia, the calculated $ER_{(HCOOH/CO)}$ $(7.3 \times 10^{-3} \pm 0.08 \times 10^{-3}\,mol\,mol^{-1})$ is close to the $EmR_{(HCOOH/CO)}$ given in Akagi et al. (2011) for the tropical

forest $(5.2 \times 10^{-3}\,mol\,mol^{-1})$, but it is 3 times higher than the values derived from Yokelson et al. (2007, 2008) for the same vegetation type. For the latter, it is worth noting that their factors have been corrected a posteriori (scaled down by a factor of 2.1), as described in their comment following the paper done by R'Honi et al. (2013) (see Yokelson et al., 2013). As Yokelson et al. (2007, 2008) sampled the forest fire plumes by penetrating recent columns of smoke 200–1000 m above the flame front, our $ER_{(HCOOH/CO)}$ may reflect a secondary production of HCOOH. This assumed secondary production is less substantial if we compare with the $EmR_{(HCOOH/CO)}$ from Akagi et al. (2011). The classification based on the type of fuel burned shows a diverse origin of the fire plumes over Amazonia. Six biomes have been identified following the classification from the MCD12Q1 product.

Over Australia and over Siberia, the calculated $ER_{(HCOOH/CO)}$ is overestimated compared to the $EmR_{(HCOOH/CO)}$ given in Akagi et al. (2011) for a savanna fire and for a boreal forest, respectively. If our value for near-source estimation is correct, this would probably mean that the direct emission is underestimated (by 450 % over Australia and by 60 % over Siberia) or that a large secondary production of HCOOH from Australian and Siberian fires occurred. These hypotheses in biased emissions and/or secondary production need to be verified with modeling studies. Over Australia, the difference is very large even though the comparison done by Pommier al. (2016) with FTIR measurements showed that the lowest bias was found for the Australian site (−2 % at Wollongong). Over Siberia, we also note that the region is characterized by fires emitted from six types of biome based on the classification from MODIS.

Finally, in this comparison, the studied plumes over India and Southeast Asia are certainly related to agricultural fires, even if the evergreen broadleaf forest seems to dominate in the MODIS land-cover-type product. This is strongly possible as agricultural residue burning is prevalent in these regions (e.g., Kaskaoutis et al., 2014; Vadrevu et al., 2015). Over India and over Southeast Asia, our $ER_{(HCOOH/CO)}$ values ($6.8 \times 10^{-3} \pm 0.44 \times 10^{-3}$ mol mol^{-1} for India and $5.8 \times 10^{-3} \pm 0.15 \times 10^{-3}$ mol mol^{-1} for Southeast Asia) are close to the value referenced by Akagi et al. (2011) for cropland fires (6×10^{-3} mol mol^{-1}). Since our $ER_{(HCOOH/CO)}$ values are close to the $EmR_{(HCOOH/CO)}$ derived from the EF_{HCOOH} in Akagi et al. (2011), this may suggests that the plumes studied over the 7-year period correspond to fresh plumes where the chemistry or the physical sink is small. This is further supported by the fact that among the seven regions, IND and SEA have larger vertical velocity means close to the surface, indicating a larger rising motion of the air masses (not shown).

In general, the $ER_{(HCOOH/CO)}$ calculated for a specific biome varies with the regions. This shows that the type of vegetation is not the only factor influencing the $ER_{(HCOOH/CO)}$. The ongoing chemistry within a plume is important, and the age of the air masses impact the level of HCOOH and CO in the plumes.

6 Conclusions

A total of 7 years of HCOOH data measured by IASI over seven different fire regions around the world were analyzed (AMA, Amazonia; AUS, Australia; IND, India; SEA, Southeast Asia; NAF, northern Africa; SAF, southern Africa; SIB, Siberia). By taking into account the surface wind speed and by characterizing each MODIS fire hotspot with a value of HCOOH and CO total columns, this work established en-

hancement ratios for the seven biomass burning areas and compared them to previously reported values found in literature.

The difficulties in performing such a comparison are associated with the difference in locations, altitude of the sampling, and age of each fire plume studied in these previous publications. However, a fair agreement was found for the enhancement ratios calculated in this work, in comparison with other studies, using satellite, airborne, or FTIR measurements.

In agreement with previous studies, the plumes from southern African savanna fires may reflect a limited secondary production or a limited sink occurring in the upper layers of the troposphere during their transport. Such assumptions, however, are difficult to verify by comparing individual plumes (from previous studies) with plumes gathered during a 7-year period (from IASI) and remain speculative without a detailed modeling study. Plumes from agricultural fires over India and Southeast Asia probably correspond to fresh plumes as our $ER_{(HCOOH/CO)}$ values based on the 7-year IASI measurements are similar to the $EmR_{(HCOOH/CO)}$ values calculated from emission factors provided by Akagi et al. (2011).

A very good agreement in $ER_{(HCOOH/CO)}$ was found over Amazonia, especially in comparison with the work done by Chaliyakunnel et al. (2016), who determined pyrogenic $ER_{(HCOOH/CO)}$.

Fires over Australia and over Siberia are probably underestimated in terms of direct emission or secondary production of HCOOH. The analysis over Australia is, however, complicated as our $ER_{(HCOOH/CO)}$ approximately corresponds to the mean of the values reported in Paton-Walsh et al. (2005) and in Chaliyakunnel et al. (2016), and it is also 450 % higher than the $EmR_{(HCOOH/CO)}$ derived from Akagi et al. (2011). The underestimation by 60 % over Siberia is consistent with conclusions given in R'Honi et al. (2013). The calculation of the $ER_{(HCOOH/CO)}$ by biome shows that Siberian plumes are related to the burning of six different vegetation classes. The underestimation reported is thus difficult to confirm without the use of a chemical transport model.

The values found over northern Africa were more difficult to interpret as this region is characterized by a poorer correlation between our fire-affected HCOOH and CO total columns.

Finally, the estimation of the $ER_{(HCOOH/CO)}$ calculated by the type of vegetation burned, as referenced in the MODIS product, varies with the regions. This shows that other parameters than the type of fuel burned also influence the $ER_{(HCOOH/CO)}$.

With these findings and by updating the enhancement ratios, an interesting modeling study could be performed to estimate a new tropospheric budget for HCOOH. This IASI dataset may also be used in the future to study a single plume at different times to inform on the loss during transport. Further insight into the transport and chemistry may be gained by using IASI's capability to measure several fire species simultaneously, such as HCN or C_2H_2 (e.g., Duflot et al., 2015). This would be useful for the characterization of the chemistry ongoing in a fire plume outflow.

An intercomparison with other space-borne instruments such as TES and ACE-FTS will be helpful to interpret the difference and the biases between the retrieved HCOOH columns and thus between their respective $ER_{(HCOOH/CO)}$ values.

Competing interests. The authors declare that they have no conflict of interest.

Acknowledgements. The IASI mission is a joint mission of EUMETSAT and the Centre National d'Etudes Spatiales (CNES, France). The IASI L1 data are distributed in near real time by EUMETSAT through the EUMETCast system distribution. We thank the MODIS team for providing public access to fire products MCD14ML and the land-cover-type data product MCD12Q1. This MCD14ML MODIS dataset was provided by the University of Maryland and NASA FIRMS operated by NASA/GSFC/ESDIS with funding provided by NASA/HQ. The authors thank Simon Whitburn (ULB) for his help on the MODIS files. They also thank Juliette Hadji-Lazaro (LATMOS) and Lieven Clarisse (ULB) for preparing the IASI ΔT_b dataset. The authors also acknowledge ECMWF for free access to the meteorological data.

Edited by: Paul Monks

References

Akagi, S. K., Yokelson, R. J., Wiedinmyer, C., Alvarado, M. J., Reid, J. S., Karl, T., Crounse, J. D., and Wennberg, P. O.: Emission factors for open and domestic biomass burning for use in atmospheric models, Atmos. Chem. Phys., 11, 4039–4072, https://doi.org/10.5194/acp-11-4039-2011, 2011.

Andreae, M. O., Andreae, T. W., Talbot, R. W., and Harriss, R. C.: Formic and acetic acid over the central Amazon region, Brazil, I. Dry season, J. Geophys. Res., 93, 1616–1624, https://doi.org/10.1029/JD093iD02p01616, 1988.

Andrews, D. U., Heazlewood, B. R., Maccarone, A. T., Conroy, T., Payne, R. J., Jordan, M. J. T., and Kable, S. H.: Photo-tautomerization of acetaldehyde to vinyl alcohol: a potential route to tropospheric acids, Science, 337, 1203–1206, https://doi.org/10.1126/science.1220712, 2012.

Beirle, S., Boersma, K. F., Platt, U., Lawrence, M. G., and Wagner, T.: Megacity emissions and lifetimes of nitrogen oxides probed from space, Science, 333, 1737–1739, https://doi.org/10.1126/science.1207824, 2011.

Bisht, D. S., Tiwari, S., Srivastava, A. K., Singh, J. V., Singh, B. P., and Srivastava, M. K.: High concentration of acidic species in rainwater at Varanasi in the Indo-Gangetic Plains, India, Nat. Hazards, 75, 2985–3003, https://doi.org/10.1007/s11069-014-1473-0, 2014.

Bohn, B., Siese, M., and Zetzschn, C.: Kinetics of the $OH + C_2H_2$ reaction in the presence of O_2, J. Chem. Soc. Faraday T., 92, 1459–1466, 1996.

Cady-Pereira, K. E., Chaliyakunnel, S., Shephard, M. W., Millet, D. B., Luo, M., and Wells, K. C.: HCOOH measurements from space: TES retrieval algorithm and observed global distribution, Atmos. Meas. Tech., 7, 2297–2311, https://doi.org/10.5194/amt-7-2297-2014, 2014.

Chaliyakunnel, S., Millet, D. B., Wells, K. C., Cady-Pereira, K. E., and Shephard, M. W.: A Large Underestimate of Formic Acid from Tropical Fires: Constraints from Space-Borne Measurements, Environ. Sci. Technol., 50, 5631–5640, https://doi.org/10.1021/acs.est.5b06385, 2016.

Chameides, W. L. and Davis, D. D.: Aqueous-phase source of formic acid in clouds, Nature, 304, 427–429, 1983.

Channan, S., Collins, K., and Emanuel, W. R.: Global mosaics of the standard MODIS land cover type data. University of Maryland and the Pacific Northwest National Laboratory, College Park, Maryland, USA, 2014.

Clerbaux, C., Boynard, A., Clarisse, L., George, M., Hadji-Lazaro, J., Herbin, H., Hurtmans, D., Pommier, M., Razavi, A., Turquety, S., Wespes, C., and Coheur, P.-F.: Monitoring of atmospheric composition using the thermal infrared IASI/MetOp sounder, Atmos. Chem. Phys., 9, 6041–6054, https://doi.org/10.5194/acp-9-6041-2009, 2009.

Clubb, A. E., Jordan, M. J. T., Kable, S. H., and Osborn, D. L.: Phototautomerization of Acetaldehyde to vinyl alcohol: a primary process in UV-irradiated acetaldehyde from 295 to 335 nm, J. Phys. Chem. Lett., 3, 3522–3526, 2012.

Coheur, P.-F., Herbin, H., Clerbaux, C., Hurtmans, D., Wespes, C., Carleer, M., Turquety, S., Rinsland, C. P., Remedios, J., Hauglustaine, D., Boone, C. D., and Bernath, P. F.: ACE-FTS observation of a young biomass burning plume: first reported measurements of C_2H_4, C_3H_6O, H_2CO and PAN by infrared occultation from space, Atmos. Chem. Phys., 7, 5437–5446, https://doi.org/10.5194/acp-7-5437-2007, 2007.

Coheur, P.-F., Clarisse, L., Turquety, S., Hurtmans, D., and Clerbaux, C.: IASI measurements of reactive trace species in biomass burning plumes, Atmos. Chem. Phys., 9, 5655–5667, https://doi.org/10.5194/acp-9-5655-2009, 2009.

Crevoisier, C., Chédin, A., Matsueda, H., Machida, T., Armante, R., and Scott, N. A.: First year of upper tropospheric integrated content of CO_2 from IASI hyperspectral infrared observations, Atmos. Chem. Phys., 9, 4797–4810, https://doi.org/10.5194/acp-9-4797-2009, 2009.

Crevoisier, C., Clerbaux, C., Guidard, V., Phulpin, T., Armante, R., Barret, B., Camy-Peyret, C., Chaboureau, J.-P., Coheur, P.-F., Crépeau, L., Dufour, G., Labonnote, L., Lavanant, L., Hadji-Lazaro, J., Herbin, H., Jacquinet-Husson, N., Payan, S., Péquignot, E., Pierangelo, C., Sellitto, P., and Stubenrauch, C.: Towards IASI-New Generation (IASI-NG): impact of improved spectral resolution and radiometric noise on the retrieval of thermodynamic, chemistry and climate variables, Atmos. Meas. Tech., 7, 4367–4385, https://doi.org/10.5194/amt-7-4367-2014, 2014.

Dee, D. P., Uppala, S. M., Simmons, A. J., Berrisford, P., Poli, P., Kobayashi, S., Andrae, U., Balmaseda, M. A., Balsamo, G., Bauer, P., Bechtold, P., Beljaars, A. C. M., van de Berg, L., Bidlot, J., Bormann, N., Delsol, C., Dragani, R., Fuentes, M., Geer, A. J., Haimberger, L., Healy, S. B., Hersbach, H., Hólm, E. V.,

Isaksen, L., Kållberg, P., Köhler, M., Matricardi, M. , McNally, A. P., Monge-Sanz, B. M., Morcrette, J.-J., Park, B.-K., Peubey, C., de Rosnay, P., Tavolato, C., Thépaut, J.-N., and Vitart, F.: The ERA-Interim reanalysis: configuration and performance of the data assimilation system, Q. J. Roy. Meteor. Soc., 137, 553–597, https://doi.org/10.1002/qj.828, 2011.

Duflot, V., Wespes, C., Clarisse, L., Hurtmans, D., Ngadi, Y., Jones, N., Paton-Walsh, C., Hadji-Lazaro, J., Vigouroux, C., De Mazière, M., Metzger, J.-M., Mahieu, E., Servais, C., Hase, F., Schneider, M., Clerbaux, C., and Coheur, P.-F.: Acetylene (C_2H_2) and hydrogen cyanide (HCN) from IASI satellite observations: global distributions, validation, and comparison with model, Atmos. Chem. Phys., 15, 10509–10527, https://doi.org/10.5194/acp-15-10509-2015, 2015.

De Wachter, E., Barret, B., Le Flochmoën, E., Pavelin, E., Matricardi, M., Clerbaux, C., Hadji-Lazaro, J., George, M., Hurtmans, D., Coheur, P.-F., Nedelec, P., and Cammas, J. P.: Retrieval of MetOp-A/IASI CO profiles and validation with MOZAIC data, Atmos. Meas. Tech., 5, 2843–2857, https://doi.org/10.5194/amt-5-2843-2012, 2012.

Duc, H. N., Bang, H. Q., and Quang, N. X.: Modelling and prediction of air pollutant transport during the 2014 biomass burning and forest fires in peninsular Southeast Asia, Environ. Monit. Assess, 188, 106, https://doi.org/10.1007/s10661-016-5106-9, 2016.

Fioletov, V. E., McLinden, C. A., Krotkov, N., and Li, C.: Lifetimes and emissions of SO_2 from point sources estimated from OMI, Geophys. Res. Lett., 42, 1969–1976, https://doi.org/10.1002/2015GL063148, 2015.

Friedl, M. A., Sulla-Menashe, D., Tan, B., Schneider, A., Ramankutty, N., Sibley, A., and Huang, X., MODIS Collection 5 global land cover: Algorithm refinements and characterization of new datasets, 2001–2012, Collection 5.1 IGBP Land Cover, Remote Sens. Environ., 114, 168–182, https://doi.org/10.1016/j.rse.2009.08.016, 2010.

Gabriel, R., Schäfer, L., Gerlach, C., Rausch, T., and Kesselmeier, J.: Factors controlling the emissions of volatile organic acids from leaves of *Quercus ilex* L. (Holm oak), Atmos. Environ., 33, 1347–1355, 1999.

George, M., Clerbaux, C., Hurtmans, D., Turquety, S., Coheur, P.-F., Pommier, M., Hadji-Lazaro, J., Edwards, D. P., Worden, H., Luo, M., Rinsland, C., and McMillan, W.: Carbon monoxide distributions from the IASI/METOP mission: evaluation with other space-borne remote sensors, Atmos. Chem. Phys., 9, 8317–8330, https://doi.org/10.5194/acp-9-8317-2009, 2009.

Giglio, L.: MODIS Collection 5 Active Fire Product User's Guide, v2.5, available at: http://modis-fire.umd.edu/pages/manuals.php or http://modis-fire.umd.edu/files/MODIS_Fire_Users_Guide_2.5.pdf (last access: 31 March 2013), 2013.

Giglio, L., van der Werf, G. R., Randerson, J. T., Collatz, G. J., and Kasibhatla, P.: Global estimation of burned area using MODIS active fire observations, Atmos. Chem. Phys., 6, 957–974, https://doi.org/10.5194/acp-6-957-2006, 2006.

González Abad, G., Bernath, P. F., Boone, C. D., McLeod, S. D., Manney, G. L., and Toon, G. C.: Global distribution of upper tropospheric formic acid from the ACE-FTS, Atmos. Chem. Phys., 9, 8039–8047, https://doi.org/10.5194/acp-9-8039-2009, 2009.

Goode, J., Yokelson, R., Ward, D., Susott, R., Babbitt, R., Davies, M., and Hao, W.: Measurements of excess O_3, CO_2, CO, CH_4, C_2H_4, C_2H_2, HCN, NO, NH_3, HCOOH, CH_3COOH, HCHO, and CH_3OH in 1997 Alaskan biomass burning plumes by airborne Fourier transform infrared spectroscopy (AFTIR), J. Geophys. Res., 105, 22147, https://doi.org/10.1029/2000JD900287, 2000.

Graedel, T. and Eisner, T.: Atmospheric formic acid from formicine ants: a preliminary assessment, Tellus B, 40, 335–339, 1988.

Grosjean, D.: Organic acids in southern California air: ambient concentrations, mobile source emissions, in situ formation and removal processes, Environ. Sci. Technol., 23, 1506–1514, 1989.

Grutter, M., Glatthor, N., Stiller, G. P., Fischer, H., Grabowski, U., Höpfner, M., Kellmann, S., Linden, A., and von Clarmann, T.: Global distribution and variability of formic acid as observed by MIPAS-ENVISAT, J. Geophys. Res., 115, D10303, https://doi.org/10.1029/2009JD012980, 2010.

Hatakeyama, S., Washida, N., and Akimoto, H.: Rate constants and mechanisms for the reaction of hydroxyl (OD) radicals with acetylene, propyne, and 2-butyne in air at 297 ± 2 K, J. Phys. Chem., 6, 90, 173–178, 1986.

Hurtmans, D., Coheur, P.-F., Wespes, C., Clarisse, L., Scharf, O., Clerbaux, C., Hadji-Lazaro, J., George, M., and Turquety, S.: FORLI radiative transfer and retrieval code for IASI, J. Quant. Spectrosc. Ra., 113, 1391–1408, https://doi.org/10.1016/j.jqsrt.2012.02.036, 2012.

Hurst, D. F., Griffith, D. W. T., and Cook, G. D.: Trace gas emissions from biomass burning in tropical Australian savannas, J. Geophys. Res., 99, 16441–16456, https://doi.org/10.1029/94JD00670, 1994.

Jacob, D.: Chemistry of OH in remote clouds and its role in the production of formic acid and peroxymonosulfate, J. Geophys. Res., 91, 9807–9826, 1986.

Justice, C. O., Giglio, L., Korontzi, S., Owens, J., Morisette, J. T., Roy, D., Descloitres, J., Alleaume, S., Petitcolin, F., and Kaufman, Y.: The MODIS fire products, Remote Sens. Environ. 83, 244–262, 2002.

Kaskaoutis, D. G., Kumar, S., Sharma, D., Singh, R. P., Kharol, S. K., Sharma, M., Singh, A. K., Singh, S., Singh, A., and Singh, D.: Effects of crop residue burning on aerosol properties, plume characteristics, and long-range transport over northern India, J. Geophys. Res.-Atmos., 119, 5424–5444, https://doi.org/10.1002/2013JD021357, 2014.

Kawamura, K., Ng, L.-L., and Kaplan, I.: Determination of organic acids (C_1–C_{10}) in the atmosphere, motor exhausts, and engine oils, Environ. Sci. Technol., 19, 1082–1086, 1985.

Keene, W. and Galloway, J.: Organic acidity in precipitation of North America, Atmos. Environ., 18, 2491–2497, 1984.

Keene, W. and Galloway, J.: The biogeochemical cycling of formic and acetic acids through the troposphere: An overview of current understanding, Tellus B, 40, 322–334, 1988.

Kerzenmacher, T., Dils, B., Kumps, N., Blumenstock, T., Clerbaux, C., Coheur, P.-F., Demoulin, P., García, O., George, M., Griffith, D. W. T., Hase, F., Hadji-Lazaro, J., Hurtmans, D., Jones, N., Mahieu, E., Notholt, J., Paton-Walsh, C., Raffalski, U., Ridder, T., Schneider, M., Servais, C., and De Mazière, M.: Validation of IASI FORLI carbon monoxide retrievals using FTIR data from NDACC, Atmos. Meas. Tech., 5, 2751–2761, https://doi.org/10.5194/amt-5-2751-2012, 2012.

Krol, M., Peters, W., Hooghiemstra, P., George, M., Clerbaux, C., Hurtmans, D., McInerney, D., Sedano, F., Bergamaschi,

P., El Hajj, M., Kaiser, J. W., Fisher, D., Yershov, V., and Muller, J.-P.: How much CO was emitted by the 2010 fires around Moscow?, Atmos. Chem. Phys., 13, 4737–4747, https://doi.org/10.5194/acp-13-4737-2013, 2013.

Lee, A., Goldstein, A. H., Kroll, J. H., Ng, N. L., Varutbangkul, V., Flagan, R. C., and Seinfeld, J. H.: Gas-phase products and secondary aerosol yields from the photooxidation of different terpenes, J. Geophys. Res., 111, D17305, https://doi.org/10.1029/2006JD007050, 2006.

Neeb, P., Sauer, F., Horie, O., and Moortgat, G. K.: Formation of hydroxymethyl hydroperoxide and formic acid in alkene ozonolysis in the presence of water vapour, Atmos. Environ., 31, 1417–1423, 1997.

Ohara, T., Akimoto, H., Kurokawa, J., Horii, N., Yamaji, K., Yan, X., and Hayasaka, T.: An Asian emission inventory of anthropogenic emission sources for the period 1980–2020, Atmos. Chem. Phys., 7, 4419–4444, https://doi.org/10.5194/acp-7-4419-2007, 2007.

Ohta, K., Ogawa, H., and Mizuno, T.: Abiological formation of formic acid on rocks in nature, Appl. Geochem., 15, 91–95, 2000.

Paris, J.-D., Ciais, P., Nedelec, P., Ramonet, M., Belan, B. D., Arshinov, M. Yu., Golitsyn, G. S., Granberg, I., Stohl, A., Cayez, G., Athier, G., Boumard, F., and Cousin, J.-M.: The YAK-AEROSIB transcontinental aircraft campaigns: new insights on the transport of CO_2, CO and O_3 across Siberia, Tellus B, 60, 551–568, https://doi.org/10.1111/j.1600-0889.2008.00369.x, 2008.

Paton-Walsh, C., Jones, N. B., Wilson, S. R., Haverd, V., Meier, A., Griffith, D. W. T., and Rinsland, C. P.: Measurements of trace gas emissions from Australian forest fires and correlations with coincident measurements of aerosol optical depth, J. Geophys. Res., 110, D24305, https://doi.org/10.1029/2005JD006202, 2005.

Paulot, F., Wunch, D., Crounse, J. D., Toon, G. C., Millet, D. B., DeCarlo, P. F., Vigouroux, C., Deutscher, N. M., González Abad, G., Notholt, J., Warneke, T., Hannigan, J. W., Warneke, C., de Gouw, J. A., Dunlea, E. J., De Mazière, M., Griffith, D. W. T., Bernath, P., Jimenez, J. L., and Wennberg, P. O.: Importance of secondary sources in the atmospheric budgets of formic and acetic acids, Atmos. Chem. Phys., 11, 1989–2013, https://doi.org/10.5194/acp-11-1989-2011, 2011.

Pochanart, P., Akimoto, H., Kajii, Y., Potemkin, V. M., and Khodzher, T. V.: Regional background ozone and carbon monoxide variations in remote Siberia/East Asia, J. Geophys. Res., 108, 4028, https://doi.org/10.1029/2001JD001412, 2003.

Pommier, M., Law, K. S., Clerbaux, C., Turquety, S., Hurtmans, D., Hadji-Lazaro, J., Coheur, P.-F., Schlager, H., Ancellet, G., Paris, J.-D., Nédélec, P., Diskin, G. S., Podolske, J. R., Holloway, J. S., and Bernath, P.: IASI carbon monoxide validation over the Arctic during POLARCAT spring and summer campaigns, Atmos. Chem. Phys., 10, 10655–10678, https://doi.org/10.5194/acp-10-10655-2010, 2010.

Pommier, M., McLinden, C. A., and Deeter, M.: Relative changes in CO emissions over megacities based on observations from space, Geophys. Res. Lett., 40, 3766–3771, https://doi.org/10.1002/grl.50704, 2013.

Pommier, M., Clerbaux, C., Coheur, P.-F., Mahieu, E., Müller, J.-F., Paton-Walsh, C., Stavrakou, T., and Vigouroux, C.: HCOOH distributions from IASI for 2008–2014: comparison with ground-based FTIR measurements and a global

chemistry-transport model, Atmos. Chem. Phys., 16, 8963–8981, https://doi.org/10.5194/acp-16-8963-2016, 2016.

Razavi, A., Karagulian, F., Clarisse, L., Hurtmans, D., Coheur, P. F., Clerbaux, C., M"uller, J. F., and Stavrakou, T.: Global distributions of methanol and formic acid retrieved for the first time from the IASI/MetOp thermal infrared sounder, Atmos. Chem. Phys., 11, 857–872, https://doi.org/10.5194/acp-11-857-2011, 2011.

R'Honi, Y., Clarisse, L., Clerbaux, C., Hurtmans, D., Duflot, V., Turquety, S., Ngadi, Y., and Coheur, P.-F.: Exceptional emissions of NH_3 and HCOOH in the 2010 Russian wildfires, Atmos. Chem. Phys., 13, 4171–4181, https://doi.org/10.5194/acp-13-4171-2013, 2013.

Rinsland, C. P., Boone, C. D., Bernath, P. F., Mahieu, E., Zander, R., Dufour, G., Clerbaux, C., Turquety, S., Chiou, L., McConnell, J. C., Neary, L., and Kaminski, J. W.: First space-based observations of formic acid (HCOOH): Atmospheric Chemistry Experiment austral spring 2004 and 2005 Southern Hemisphere tropical-mid-latitude upper tropospheric measurements, Geophys. Res. Lett., 33, L23804, https://doi.org/10.1029/2006GL027128, 2006.

Rinsland, C. P., Dufour, G., Boone, C. D., Bernath, P. F., Chiou, L., Coheur, P.-F., Turquety, S., and Clerbaux, C.: Satellite boreal measurements over Alaska and Canada during June–July 2004: Simultaneous measurements of upper tropospheric CO, C_2H_6, HCN, CH_3Cl, CH_4, C_2H_2, CH_3OH, HCOOH, OCS, and SF_6 mixing ratios, Global Biogeochem. Cy., 21, GB3008, https://doi.org/10.1029/2006GB002795, 2007.

Rolph, G. D.: Real-time Environmental Applications and Display sYstem (READY) Website, NOAA Air Resources Laboratory, College Park, MD, available at: http://www.ready.noaa.gov (last access: 18 September 2017), 2017.

Rodgers, C. D.: Inverse methods for atmospheric sounding: theory and practice, Ser. Atmos. Ocean. Planet. Phys. 2, World Sci., Hackensack, NJ, 2000.

Sanhueza, E. and Andreae, M.: Emission of formic and acetic acids from tropical savanna soils, Geophys. Res. Lett., 18, 1707–1710, 1991.

Schreier, S. F., Richter, A., Kaiser, J. W., and Burrows, J. P.: The empirical relationship between satellite-derived tropospheric NO_2 and fire radiative power and possible implications for fire emission rates of NO_x, Atmos. Chem. Phys., 14, 2447–2466, https://doi.org/10.5194/acp-14-2447-2014, 2014.

Sinha, P. P., Hobbs, V., Yokelson, R. J., Bertschi, I. T., Blake, D. R., Simpson, I. J., Gao, S., Kirchstetter, T. W., and Novakov, T.: Emissions of trace gases and particles from savanna fires in southern Africa, J. Geophys. Res., 108, 8487, https://doi.org/10.1029/2002JD002325, 2003.

Sinha, P. P., Hobbs, V., Yokelson, R. J., Blake, D. R., Gao, S., and Kirchstetter, T. W.: Emissions from miombo woodland and dambo grassland savanna fires, J. Geophys. Res., 109, D11305, https://doi.org/10.1029/2004JD004521, 2004.

Srinivas, R., Beig, G., and Peshin S. K.: Role of transport in elevated CO levels over Delhi during onset phase of monsoon, Atmos. Environ., 140, 234–241, https://doi.org/10.1016/j.atmosenv.2016.06.003, 2016.

Stavrakou, T., Müller, J.-F., Peeters, J., Razavi, A., Clarisse, L., Clerbaux, C., Coheur, P.-F., Hurtmans, D., and De Mazière, M.: Satellite evidence for a large source of formic acid from boreal and tropical forests, Nat. Geosci., 5, 26–30, https://doi.org/10.1038/ngeo1354, 2012.

Tereszchuk, K. A., González Abad, G., Clerbaux, C., Hurtmans, D., Coheur, P.-F., and Bernath, P. F.: ACE-FTS measurements of trace species in the characterization of biomass burning plumes, Atmos. Chem. Phys., 11, 12169–12179, https://doi.org/10.5194/acp-11-12169-2011, 2011.

Tereszchuk, K. A., González Abad, G., Clerbaux, C., Hadji-Lazaro, J., Hurtmans, D., Coheur, P.-F., and Bernath, P. F.: ACE-FTS observations of pyrogenic trace species in boreal biomass burning plumes during BORTAS, Atmos. Chem. Phys., 13, 4529–4541, https://doi.org/10.5194/acp-13-4529-2013, 2013.

Turquety, S., Hurtmans, D., Hadji-Lazaro, J., Coheur, P.-F., Clerbaux, C., Josset, D., and Tsamalis, C.: Tracking the emission and transport of pollution from wildfires using the IASI CO retrievals: analysis of the summer 2007 Greek fires, Atmos. Chem. Phys., 9, 4897–4913, https://doi.org/10.5194/acp-9-4897-2009, 2009.

Vadrevu, K. P., Lasko, K., Giglio, L., and Justice, C.: Vegetation fires, absorbing aerosols and smoke plume characteristics in diverse biomass burning regions of Asia, Environ. Res. Lett., 10, 2371–2379, https://doi.org/10.1039/c4em00307a, 2015.

van der Werf, G. R., Randerson, J. T., Giglio, L., Collatz, G. J., Mu, M., Kasibhatla, P. S., Morton, D. C., DeFries, R. S., Jin, Y., and van Leeuwen, T. T.: Global fire emissions and the contribution of deforestation, savanna, forest, agricultural, and peat fires (1997–2009), Atmos. Chem. Phys., 10, 11707–11735, https://doi.org/10.5194/acp-10-11707-2010, 2010.

Viatte, C., Strong, K., Hannigan, J., Nussbaumer, E., Emmons, L. K., Conway, S., Paton-Walsh, C., Hartley, J., Benmergui, J., and Lin, J.: Identifying fire plumes in the Arctic with tropospheric FTIR measurements and transport models, Atmos. Chem. Phys., 15, 2227–2246, https://doi.org/10.5194/acp-15-2227-2015, 2015.

Vigouroux, C., Stavrakou, T., Whaley, C., Dils, B., Duflot, V., Hermans, C., Kumps, N., Metzger, J.-M., Scolas, F., Vanhaelewyn, G., Müller, J.-F., Jones, D. B. A., Li, Q., and De Mazière, M.: FTIR time-series of biomass burning products (HCN, C_2H_6, C_2H_2, CH_3OH, and HCOOH) at Reunion Island (21° S, 55° E) and comparisons with model data, Atmos. Chem. Phys., 12, 10367–10385, https://doi.org/10.5194/acp-12-10367-2012, 2012.

Whitburn, S., Van Damme, M., Kaiser, J. W., van der Werf, G. R., Turquety, S., Hurtmans, D., Clarisse, L., Clerbaux, C., and Coheur, P.-F.: Ammonia emissions in tropical biomass burning regions: Comparison between satellite-derived emissions and bottom-up fire inventories, Atmos. Environ, 121, 42–54, https://doi.org/10.1016/j.atmosenv.2015.03.015, 2015.

Whitburn, S., Van Damme, M., Clarisse, L., Hurtmans, D., Clerbaux, C., and Coheur, P.-F.: IASI-derived NH_3 enhancement ratios relative to CO for the tropical biomass burning regions, Atmos. Chem. Phys. Discuss., https://doi.org/10.5194/acp-2017-331, in review, 2017.

White, F.: UNESCO/AETFAT/UNSO vegetation map of Africa, scale 1 : 5,000,000, UNESCO, Paris, 1981.

Wooster, M. J., Roberts, G., Perry, G. L. W., and Kaufman, Y. J.: Retrieval of biomass combustion rates and totals from fire radiative power observations: FRP derivation and calibration relationships between biomass consumption and fire radiative energy release, J. Geophys. Res., 110, D24311, https://doi.org/10.1029/2005JD006318, 2005.

Yokelson, R. J., Bertschi, I. T., Christian, T. J., Hobbs, P. V., Ward, D. E., and Hao, W. M.: Trace gas measurements in nascent, aged, and cloud-processed smoke from African savanna fires by airborne Fourier transform infrared spectroscopy (AFTIR), J. Geophys. Res., 108, 8478, https://doi.org/10.1029/2002JD002322, 2003.

Yokelson, R. J., Karl, T., Artaxo, P., Blake, D. R., Christian, T. J., Griffith, D. W. T., Guenther, A., and Hao, W. M.: The Tropical Forest and Fire Emissions Experiment: overview and airborne fire emission factor measurements, Atmos. Chem. Phys., 7, 5175–5196, https://doi.org/10.5194/acp-7-5175-2007, 2007.

Yokelson, R. J., Christian, T. J., Karl, T. G., and Guenther, A.: The tropical forest and fire emissions experiment: laboratory fire measurements and synthesis of campaign data, Atmos. Chem. Phys., 8, 3509–3527, https://doi.org/10.5194/acp-8-3509-2008, 2008.

Yokelson, R. J., Akagi, S. K., Griffith, D. W. T., and Johnson, T. J.: Interactive comment on "Exceptional emissions of NH_3 and HCOOH in the 2010 Russian wildfires" by Y. R'Honi et al., Atmos. Chem. Phys. Discuss., 12, C11864–C11868, 2013.

3

Exploring sources of biogenic secondary organic aerosol compounds using chemical analysis and the FLEXPART model

Johan Martinsson[1,2], **Guillaume Monteil**[3], **Moa K. Sporre**[4], **Anne Maria Kaldal Hansen**[5], **Adam Kristensson**[1], **Kristina Eriksson Stenström**[1], **Erik Swietlicki**[1], and **Marianne Glasius**[5]

[1]Division of Nuclear Physics, Lund University, P.O. Box 118, 22100, Lund, Sweden

[2]Centre for Environmental and Climate Research, Lund University, Ecology Building, 22362, Lund, Sweden

[3]Department of Physical Geography, Lund University, Lund, P.O. Box 118, 22100, Lund, Sweden

[4]Department of Geosciences, University of Oslo, P.O. Box 1022, Blindern, 0315, Oslo, Norway

[5]Department of Chemistry and iNANO, Aarhus University, Langelandsgade 140, 8000, Aarhus C, Denmark

Correspondence to: Johan Martinsson (johan.martinsson@nuclear.lu.se)

Abstract. Molecular tracers in secondary organic aerosols (SOAs) can provide information on origin of SOA, as well as regional scale processes involved in their formation. In this study 9 carboxylic acids, 11 organosulfates (OSs) and 2 nitrooxy organosulfates (NOSs) were determined in daily aerosol particle filter samples from Vavihill measurement station in southern Sweden during June and July 2012. Several of the observed compounds are photo-oxidation products from biogenic volatile organic compounds (BVOCs). Highest average mass concentrations were observed for carboxylic acids derived from fatty acids and monoterpenes (12.3 ± 15.6 and 13.8 ± 11.6 ng m^{-3}, respectively). The FLEXPART model was used to link nine specific surface types to single measured compounds. It was found that the surface category "sea and ocean" was dominating the air mass exposure (56 %) but contributed to low mass concentration of observed chemical compounds. A principal component (PC) analysis identified four components, where the one with highest explanatory power (49 %) displayed clear impact of coniferous forest on measured mass concentration of a majority of the compounds. The three remaining PCs were more difficult to interpret, although azelaic, suberic, and pimelic acid were closely related to each other but not to any clear surface category. Hence, future studies should aim to deduce the biogenic sources and surface category of these compounds. This study bridges micro-level chemical speciation to air mass surface exposure at the macro level.

1 Introduction

Carbonaceous aerosols are abundant in ambient air around the world and account for 40 % of the European PM$_{2.5}$ mass (Putaud et al., 2010). The carbonaceous aerosol fraction has severe effects on human health as well as a profound effect on the Earth climate system (Dockery et al., 1993; Pope et al., 1995). During summer, carbonaceous aerosols are mainly of biogenic origin, emitted either through primary emissions or gas-phase oxidation products from biogenic volatile organic compounds (BVOCs) (Genberg et al., 2011; Yttri et al., 2011). BVOCs are primarily emitted from plants as a tool for communication and to handle biotic and abiotic stress (Laothawornkitkul et al., 2009; Monson et al., 2013; Penuelas and Llusia, 2003; Sharkey et al., 2008). The emissions of BVOCs tend to increase with increasing temperature and photosynthetically active radiation (PAR) (Guenther et al., 1995, 1993; Hakola et al., 2003). Global BVOC emissions are dominated by isoprene (C$_5$H$_8$) and monoterpenes (C$_{10}$H$_{16}$) (Laothawornkitkul et al., 2009). Isoprene is emitted from a variety of plants, but mainly from deciduous forests and shrubs, which may account for more than 70 % of the emissions (Guenther et al., 2006). Monoterpenes are largely emitted from coniferous trees like pine and spruce, but also from some deciduous trees, such as birch (Mentel et al., 2009). The most abundant monoterpenes in the boreal forests include α-pinene, β-pinene, Δ^3-carene and limonene (Hakola et al., 2012; Räisänen et al., 2008).

Biogenic secondary organic aerosols (BSOAs) are formed by photo-oxidation of BVOCs, a process which tends to lower the saturation vapor pressure of the oxidation products relative to that of the BVOCs, thus forcing the gas-phase products to partition in the aerosol phase. BSOA has been shown to dominate over combustion source aerosols during summer (Genberg et al., 2011; Yttri et al., 2011). Yttri et al. (2011) performed source apportionment at four sites in Scandinavia during August 2009 and found that the biogenic contribution to the carbonaceous aerosol dominated (69–86 %) at all four sites. Genberg et al. (2011) performed a 1-year source apportionment at one site in southern Sweden where they apportioned 80 % of the summertime carbonaceous aerosol to biogenic sources. Gelencser et al. (2007) also reported biogenic source dominance (63–76 %) of the carbonaceous aerosol at six sites in south-central Europe during summer. Castro et al. (1999) observed a maximum and minimum in SOA in Europe during summer and winter, respectively. The relative SOA contribution was higher in rural forest and ocean measurement sites compared to urban sites (Castro et al., 1999).

BSOA consists of a myriad of organic compounds. Small (carbon number: C_3–C_6) and larger (C_7–C_9) dicarboxylic acids are highly hydrophilic and hygroscopic, which have shown to result in potential strong climate effect due to their cloud condensation properties (Cruz and Pandis, 1998; Kerminen, 2001). Dicarboxylic acid contribution to carbon mass has been estimated to 1–3 % in urban and semi-urban areas and up to 10 % in remote marine areas (Kawamura and Ikushima, 1993; Kawamura and Sakaguchi, 1999). Primary aerosol sources of dicarboxylic acids in atmospheric aerosols include ocean emissions, engine exhausts and biomass burning (Kawamura and Kaplan, 1987; Kundu et al., 2010; Mochida et al., 2003). However, the main source of dicarboxylic acids are oxidation/photo-oxidation processes of VOCs (Zhang et al., 2010). These VOC precursors may originate from both anthropogenic and biogenic sources (Mochida et al., 2003). However, BVOCs constitute more than 50 % of all atmospheric VOCs, which is approximately equal to 1150 Tg carbon yr^{-1} (Guenther et al., 1995; Hallquist et al., 2009).

Organosulfates (OSs) and nitrooxy organosulfates (NOSs) are low-volatility SOA products that in recent years have gained increased attention due to their potential properties as tracers for atmospheric ageing of aerosols in polluted air masses (Hansen et al., 2015, 2014; Kristensen, 2014; Kristensen and Glasius, 2011; Nguyen et al., 2014). Many of these compounds are formed from isoprene and monoterpene oxidation products that react with sulfuric acid in the aerosol phase (Iinuma et al., 2007; Surratt et al., 2010, 2007b). Since atmospheric sulfuric acid is mainly of anthropogenic origin (Zhang et al., 2009), presence of OSs from biogenic organic precursors thus indicates an effect of anthropogenic influence on BSOA (Hansen et al., 2014). Recently, OSs from anthropogenic organic precursors such as alkanes and poly-

cyclic aromatic hydrocarbons (PAHs) have also been discovered (Riva et al., 2016, 2015). Tolocka and Turpin (2012) estimated that OSs could comprise up to 10 % of the total organic aerosol mass in the US.

Many carboxylic acids and OSs originate from biogenic sources, however, the exact vegetation types emitting the precursor are poorly explored (Mochida et al., 2003; Tolocka and Turpin, 2012). Coniferous forests, deciduous forests, arable land, pastures etc. are all examples of potential BVOCs sources. Information on specific land surface type BVOCs and BSOA emissions is potentially crucial if an increased understanding should be reached on how land-use changes will affect organic aerosol levels and composition. Van Pinxteren et al. (2010) demonstrated how air mass exposure to land cover affected the measured size-resolved organic carbon (OC), elemental carbon (EC) and inorganic compounds at a receptor site in Germany by using the HYS-PLIT model. Yttri et al. (2011) measured one dicarboxylic acid (pinic acid), four OSs and two NOSs at four locations in Scandinavia and connected this measurement data to the FLEXPART model (Stohl et al., 2005) footprint of specific surface landscape types. They used 13 types of surface landscapes and found that the two NOSs (MW 295 and MW 297, both formed from monoterpenes) correlated with air mass exposure to mixed forest (Yttri et al., 2011).

In this study, a comprehensive measurement campaign was conducted in order to investigate sources and levels of BSOA. Thirty-eight sequential 24 h filter samples were analysed for 9 species of carboxylic acids, 11 species of OSs and 2 species of NOSs at a rural background station in southern Sweden. FLEXPART model simulations at the time and location of the observations were then used to estimate the potential origin of the aerosols sampled.

2 Methods

2.1 Location and sampling

The Vavihill measurement station is a rural background station in southern Sweden (56°01′ N, 13°09′ E; 172 m a.s.l.) within ACTRIS (Aerosols, Clouds and Trace gases Research Infrastructure) and EMEP (European Monitoring and Evaluation Programme). The surrounding landscape consists of pastures, mixed forest and arable land. The largest nearby cities are Helsingborg (140 000 inhabitants), Malmö (270 000 inhabitants) and Copenhagen (1 990 000 inhabitants) at a distance of 25, 45 and 50 km, respectively. These cities are in the west and southwest direction from the measurement station. Previous observations have shown that air masses from continental Europe are usually more polluted than air masses from the north and westerly direction, i.e. Norwegian Sea and Atlantic Ocean (Kristensson et al., 2008).

Thirty-eight filter samples of aerosols were collected at the Vavihill field station in southern Sweden from 10 June to 18 July 2012. Aerosols were collected on 150 mm quartz fibre filters (Advantec) using a high-volume sampler (Digitel, DHA-80) with a PM_1 inlet. The filters were heated to 900 °C for 4 h prior to sampling, with the purpose of removing adsorbed organic compounds from the filters. The sampling air flow was $530 \, L \, min^{-1}$ and total sampling time per filter was 24 h. Sampled filters were wrapped in aluminium foil and stored at -18 °C until extraction.

2.2 BSOA analysis

The method for extraction and analysis is based on previous studies (Hansen et al., 2014; Kristensen and Glasius, 2011; Nguyen et al., 2014) and thus only described briefly here. For extraction each filter was placed in a beaker and spiked with 15 µL of a $100 \, \mu g \, mL^{-1}$ recovery standard (camphoric acid). The filter was covered with 90 % acetonitrile with 10 % Milli-Q water and extracted in a cooled ultrasound bath for 30 min. The extract was filtered through a Teflon filter (0.45 µm pore size, Chromafil) and evaporated until dryness using a rotary evaporator. The sample was then redissolved twice in 0.5 mL 3 % acetonitrile, 0.1 % acetic acid, and stored in a refrigerator (3–5 °C) until analysis. The samples were analysed with an ultra-high-performance liquid chromatograph (UHPLC, Dionex) coupled to a quadrupole time-of-flight mass spectrometer (q-TOF-MS, Bruker Daltonics) through an electro-spray ionization (ESI) inlet. The UHPLC stationary phase was an Acquity T3 1.8 µm (2.1 × 100 mm) column from Waters, and the mobile phase consisted of eluent A (0.1 % acetic acid in Milli-Q water) and eluent B (acetonitrile with 0.1 % acetic acid). The operational eluent flow was $0.3 \, mL \, min^{-1}$ and an 18 min multistep gradient was applied: from 1 to 10 min eluent B increased from 3 to 30 %, then eluent B increased to 90 % during 1 min, where it was held for 1 min, before eluent B was increased further to 95 % (during 0.5 min) kept here for 3.5 min before reduction to 3 % (during 0.5 min) for the remaining 0.5 min of the analysis. The ESI-q-TOF-MS instrument was operated in negative ionization mode with a nebulizer pressure of 3.0 bar and a dry gas flow of $8 \, L \, min^{-1}$. All data were acquired and processed using Bruker Compass software. Analysed dicarboxylic acids are summarized in Table 1 and OSs and NOSs are summarized in Table 2. Authentic standards were used for identification and quantification of all carboxylic acids, while OSs and NOSs were identified based on their MS/MS loss of HSO_4^- ($m/z = 97$) and an additional neutral loss of HNO_3 ($u = 63$) in the case of NOSs. This work focused on identification of OSs from biogenic organic precursors, since OSs from alkanes and PAHs had not been discovered at the time of the analysis. OSs and NOSs were quantified using surrogate standards of OS 250 derived from β-pinene (synthesized in-house), octyl sulfate sodium salt ($\geq$ 95 % Sigma-Aldrich) or D-mannose-6-sulfate

sodium salt ($\geq$ 90 % Sigma-Aldrich) based on their retention times in the UHPLC-q-TOF-MS system (Table 2). A linear or quadratic relationship between peak area and concentration was demonstrated for all standards and surrogates, and the correlation coefficients, R^2, of all calibration curves were better than 0.98 ($n = 7$ data points).

The analytical uncertainty was estimated to be < 20 % for carboxylic acids and < 25 % for OSs and NOSs. The uncertainty of the absolute concentrations of OSs and NOSs are higher than carboxylic acids due to lack of authentic standards.

2.3 Auxiliary measurements and analysis

$PM_{2.5}$ was measured with 1 h time resolution using a tapered element oscillating microbalance (TEOM, Thermo, 8500 FDMS), and estimated uncertainty was less than 25 %. Geographical air mass origin was analysed with the Hybrid Single Particle Lagrangian Integrated Trajectory (HYSPLIT) model (Draxler and Hess, 1998; Stein et al., 2015). Gridded meteorological data from the Center for Environmental Prediction (NCEP) Global Data Assimilation System (GDAS) were used as input by the trajectory model. Back-trajectories were calculated at an hourly frequency 120 h backward in time and the trajectories started 100 m above ground at the Vavihill measurement site. For each filter sample, 24 trajectories were used since the sampling time was 24 h.

2.4 Source apportionment

The concentration and chemical composition of an aerosol sample depends on the trajectory of the sampled air mass in the days preceding the observation (whether or not it comes in contact with a source of aerosols or of aerosol precursors), but also on other meteorological factors such as the temperature and the amount of solar radiation (which control the chemical reactions that lead to production, destruction and transformation of aerosols), and the occurrence of precipitation, which can lead to a rapid scavenging of aerosol particles.

A formal source apportionment would typically involve using a complex chemistry-transport model, able to account for the most important of these factors, and comparing this model results with the observations to validate or refute hypotheses on the origin of the aerosols. The size of our observation dataset is unfortunately too limited for such an exercise to provide meaningful results. Instead, we opted for a much simpler approach: we first used the FLEXPART model to compute back-trajectories corresponding to the air masses sampled. We then used these back-trajectories to estimate the exposure of each sample to various land surface types. Finally, we analysed the relations between the surface type exposures and the aerosols chemical composition of the samples to deduce information about the origin of the sampled aerosols.

Table 1. Analysed organic acids in the Vavihill aerosol samples. Measured m/z, molecular formula, possible molecular structure, suggested precursor and assigned precursor class.

Precursor class	Name	Measured m/z	Molecular formula	Possible structure	Suggested precursor
Anthropogenic	Adipic acid	145.050	$C_6H_{10}O_4$		Cyclohexene[a]
	Pimelic acid	159.065	$C_7H_{12}O_4$		Cycloheptene[a]
Fatty-acid-derived	Suberic acid	173.081	$C_8H_{14}O_4$		Unsaturated fatty acid[b,c]
	Azelaic acid	187.097	$C_9H_{16}O_4$		Unsaturated fatty acid[b,c]
First-generation monoterpene	Pinic acid	185.081	$C_9H_{14}O_4$		$\alpha\text{-}/\beta$-Pinene[d,e]
	Pinonic acid	183.102	$C_{10}H_{16}O_3$		$\alpha\text{-}/\beta$-Pinene[d,e]
	Terpenylic acid	171.065	$C_8H_{12}O_4$		α-Pinene[f]
Second-generation monoterpene	3-Methyl-1,2,3-butane-tricarboxylic acid (MBTCA)	203.055	$C_8H_{12}O_6$		α-Pinene[d]
	Diaterpenylic acid acetate (DTAA)	231.086	$C_{10}H_{16}O_6$		α-Pinene[f]

[a] Hatakeyama et al. (1987). [b] Stephanou and Stratigakis (1993). [c] Kawamura and Gagosian (1987). [d] Szmigielski et al. (2007). [e] Ma et al. (2007). [f] Claeys et al. (2009).

2.4.1 Footprint computations

For each observation, 7-day footprints (i.e. sensitivity of the observations to surface processes) are computed, using the FLEXPART Lagrangian particle dispersion model in its version 10.0 (Seibert and Frank, 2004; Stohl et al., 2005). The response functions are computed hourly, 7 days backward, on a $0.2° \times 0.2°$ grid ranging from 30 to 65° N and from 2° W to 32° E. Only one (surface) layer is used, ranging from the surface to 400 m altitude. This choice of a relatively thick surface layer is a compromise between the necessity to account for a maximum of the aerosol production, which does not occurs only at the earth (or canopy) surface, and the fact that the higher the altitude, the more mixed the air. This setting also means that we do not compute the sensitivity of the observations to aerosol production/destruction above 400 m. Even though aerosol formation occurs throughout the whole troposphere (de Reus et al., 2000), it would be impossible, with our simple model approach, to distinguish in situ aerosol production from long-range transport.

Each footprint was computed based on the dispersion 7 days backward in time of 100 000 particles. An average particle size of 250 nm was used, with a size distribution parameter ("dsigma") of 12.5, meaning that 68 % of the total particles mass is in a 250/12.5 to 250×12.5 nm range. Previous particle-size measurements at Vavihill measurement station have shown a distribution around a mean of ~ 100 nm (Kristensson et al., 2008). The particles density was set to $1500 \, \mathrm{kg\,m^{-3}}$. We briefly discuss the impact of these selected parameters in Sect. 3.4. FLEXPART configuration files are provided in the Supplement.

Table 2. Analysed organosulfates (OSs) and nitrooxy organosulfates (NOSs) in the Vavihill aerosol samples. Measured m/z, molecular formula, possible molecular structure, suggested precursor and assigned precursor class.

Precursor class	Name	Measured m/z	Molecular formula	Possible structure	Suggested precursor
Isoprene/ anthropogenic	OS 140[1]	138.970	$C_2H_4O_5S$		Glycolaldehyde[a]
	OS 154[1]	152.985	$C_3H_6O_5S$		Hydroxyacetone[a]/methacrolein[b]/ methyl vinyl ketone[b]
	OS 156[1]	154.961	$C_2H_4O_6S$		Glycolic acid[c,d]/methyl vinyl ketone[b]
	OS 170[1]	168.979	$C_3H_6O_6S$		Methylglycolic acid[c,d]
	OS 200[1]	198.991	$C_4H_8O_7S$		2-Methylglyceric acid[a,e]
Isoprene	OS 212[1]	210.991	$C_5H_8O_7S$		Isoprene[f,g]
	OS 214[1]	213.007	$C_5H_{10}O_7S$	More isomers	Isoprene[f]
	OS 216[1]	215.021	$C_5H_{12}O_7S$		C_5-epoxydiols from isoprene (IEPOX)[h]
Monoterpene	OS 250[2]	249.080	$C_{10}H_{18}O_5S$	More isomers	α-/β-Pinene and limonene[f]
	OS 268[2]	267.053	$C_9H_{16}O_7S$		Limonene[f]
	OS 280[2]	279.054	$C_{10}H_{16}O_7S$		α-/β-Pinene[f]
Monoterpene NOS	NOS 295[3]	294.062	$C_{10}H_{17}O_7NS$	More isomers	α-/β-Pinene, Limonene[a,f]
	NOS 297[2]	296.044	$C_9H_{15}O_8NS$	More isomers	Limonene[f]

[a] Surratt et al. (2007a). [b] Schindelka et al. (2013). [c] Olson et al. (2011). [d] Shalamzari et al. (2013). [e] Gomez-Gonzalez et al. (2008). [f] Surratt et al. (2008). [g] Hettiyadura et al. (2015). [h] Surratt et al. (2010). The OSs and NOSs were quantified with D-mannose 6-sulfate (1), β-pinene OS 250 (2) or octyl sulfate (3).

2.4.2 Land surface type exposures

To compute the exposure of each sample to different land surface types, we coupled the information from the footprints to the CORINE 2012 land cover map (Copernicus, 2012). CORINE 2012 is a high-resolution (250 m × 250 m) map of the land surface types in the European Union (44 land surface categories, to which we added a "sea and ocean" category). The exposure E_i of one observation to the land type i is given by $E_i = \sum_j f_j^i R_j$, where j is one pixel of the domain, f_j^i is the fraction of the land surface type i in that pixel, and R_j is the sensitivity of the observation to that pixel (i.e. the value of the footprint at that location), divided by the height of the surface layer (400 m) and by the size of the grid cell.

It is important to remember that since aerosol formation/destruction along the particles trajectories is not accounted for in the FLEXPART simulations (except for deposition processes), these land surface exposures are not a proper source apportionment, only a tool to interpret the observations.

2.4.3 Principal component analysis (PCA)

In order to deduce potential sources of measured BSOA compounds a PCA was performed on measured chemical compounds together with air mass exposure to the landscape surface types derived from the FLEXPART model. The principle of PCA is that if measured parameters from the same source are strongly correlated they are treated as one principal component (PC), i.e. PCA identifies variables that have a prominent role by analysis of correlation and variance. PCA has been an extensively used tool in order to reduce the complexity of atmospheric data and has been applied in several studies on aerosol chemical composition (Almeida et al., 2006; Chan and Mozurkewich, 2007; Ito et al., 2004; Nyanganyura et al., 2007; van Pinxteren et al., 2010, 2014; Viana et al., 2006; Wehner and Wiedensohler, 2003). PCA with VARIMAX rotation was performed by using the software SPSS (version 23, IBM). VARIMAX rotation was chosen due to its property of producing uncorrelated PCs, which aids interpretation of the data. In PCA, it is of good practice to transform all variables into a standardized format (i.e. Z score); however, the PCA solution from the standardized variables did not differ from the unstandardized one. Hence, unstandardized variables were used in the analysis. Extracted factors were varied from 2 to 6 in order to achieve the best logical and physical interpretation of the derived factors. The most interpretable result was found using four extracted factors.

Table 3. Ranges of concentrations, means and standard deviation (SD) of the analysed compounds in aerosol samples collected at the Vavihill measurement station 10 June to 18 July 2012.

Compound	N	Minimum	Maximum	Mean	±SD
		\multicolumn{4}{c}{($ng\,m^{-3}$)}			
Adipic acid	36	0.03	19.27	1.76	3.87
Pimelic acid	36	0.02	1.21	0.38	0.28
Suberic acid	31	0.05	9.03	2.45	2.42
Azelaic acid	35	0.03	55.27	10.52	13.83
Pinic acid	38	0.28	4.71	1.31	1.04
Pinonic acid	38	0.82	10.66	2.89	2.00
Terpenylic acid	38	0.72	8.86	2.57	1.87
DTAA	38	0.04	5.67	0.84	1.23
MBTCA	38	0.38	29.42	6.18	7.00
OS 140	38	0.02	0.28	0.11	0.07
OS 154	38	0.15	2.95	0.76	0.64
OS 156	32	0.02	2.35	0.65	0.61
OS 170	38	0.08	0.78	0.33	0.17
OS 200	38	0.06	2.02	0.41	0.40
OS 212	38	0.16	4.63	0.91	0.95
OS 214	38	0.06	3.08	0.50	0.58
OS 216	38	0.06	5.83	0.63	1.07
OS 250	38	0.02	3.48	0.51	0.64
OS 268	38	0.01	0.48	0.13	0.12
OS 280	32	0.01	0.70	0.09	0.17
NOS 295	38	0.02	0.53	0.12	0.11
NOS 297	37	0.01	0.18	0.05	0.03

3 Results and discussion

3.1 Variations and features in BSOA compounds

A total of 9 organic acids, 11 OSs and 2 NOSs of anthropogenic and biogenic origin were determined in the samples (Tables 1 and 2). All organic acids were quantified with authentic standards, whereas the other compounds were quantified with surrogates (see experimental section). On average, the total mass of the organic chemical species from filters contributed to 0.3 % (±0.2 %, standard deviation) to $PM_{2.5}$. However, it is worth noting that the particles were sampled through a PM_1 inlet, which may have excluded a considerable portion of the mass collected on filters compared to the $PM_{2.5}$ mass measured by the TEOM. On the other hand, it has been shown that PM_1 can comprise up to 90 % of $PM_{2.5}$ in rural locations during summertime (Gomiscek et al., 2004). Since no gravimetric analysis of filters was performed, no information on the total mass loading of PM_1 is available.

In Table 3 and Fig. 1a concentrations of observed compounds during the sampling period are given. The compounds have been merged into groups based on their likely precursors in Fig. 1a (see Tables 1 and 2). It should be noted that pimelic acid, in Table 1 listed as having cycloheptene as a suggested precursor (i.e. to be of anthropogenic origin), can also be synthesized from salicylic acid

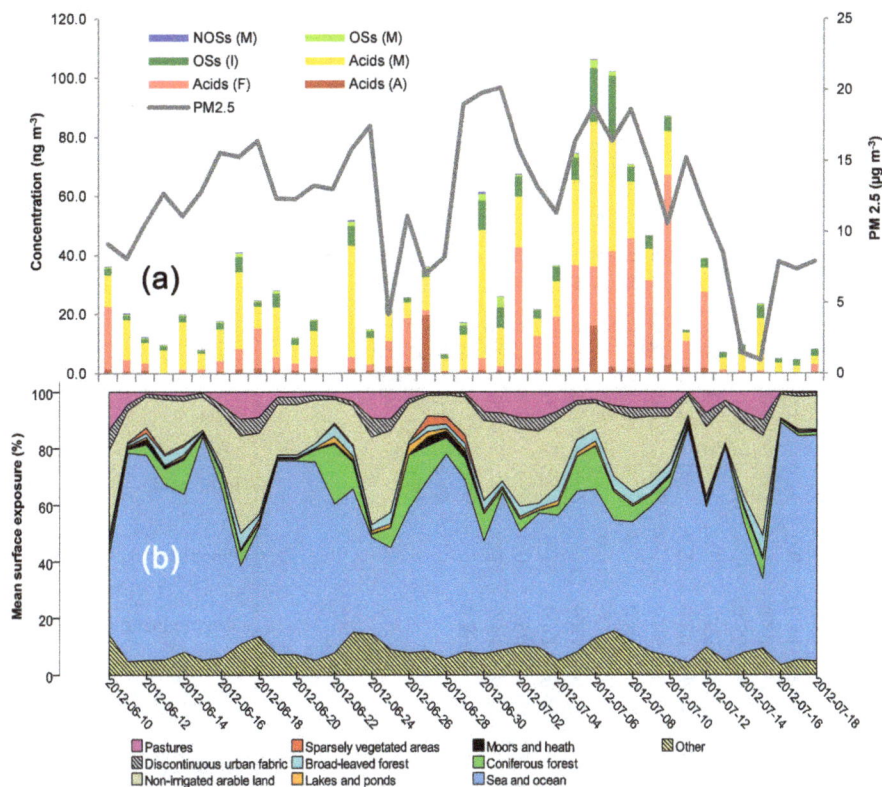

Figure 1. (a) Total concentration of all measured carboxylic acids, organosulfates (OSs) and nitrooxy organosulfates (NOSs) in PM_1 collected at the Vavihill measurement station. The thick grey line displays the $PM_{2.5}$ concentration. Capital letters in parentheses in the legend are the precursor class given in Tables 1 and 2. A: anthropogenic; F: fatty acid; I: isoprene; and M: monoterpenes. **(b)** FLEXPART generated mean exposure from the nine mean largest surface categories. The exposure is a mean of 3-, 5- and 7-day back-trajectories. The category "Other" represents the remaining 34 surface categories. More detailed information on the surface categories can be found in the Supplement.

(Müller, 1931), which is a compound naturally found in plants. Hence, whether the main formation route of pimelic acid is anthropogenic or natural is unclear. On the other hand, adipic acid is rarely found naturally and is originally synthesized from benzene (Tuttle Musser, 2000). Table 3 summarizes concentration ranges, means and standard deviations (SDs) for individual dicarboxylic acids, OSs and NOSs. In general the organic acids from monoterpenes and fatty acids dominate the total concentration over the entire period, where the concentration of acids from monoterpenes range from 1.7 to 49.0 $ng\,m^{-3}$ and the concentration of organic acids from fatty acids range from 0.03 to 64.1 $ng\,m^{-3}$. The concentration of isoprene-derived OSs ranges from 0.34 to 21.6 $ng\,m^{-3}$ over the sampling period and dominates over the monoterpene-derived OSs. This pattern has also been observed in other studies in the Nordic countries (Yttri et al., 2011), and is in line with high emissions of isoprene during summer. The NOSs are low in average concentration (NOS 295 = 0.12 ± 0.11 $ng\,m^{-3}$, NOS 297 = 0.05 ± 0.03 $ng\,m^{-3}$), and are lower than the observed mean concentration by Yttri et al. (2011) from the summer of 2011 (NOS 295 = 0.74 $ng\,m^{-3}$, NOS 297 = 1.2 $ng\,m^{-3}$). This could be

due to differences in aerosol sources and surrogate standards for quantification between the two studies.

The fatty-acid-derived azelaic acid was found to be the most abundant dicarboxylic acid with a concentration range from 0.03 to 55.3 $ng\,m^{-3}$ (mean = 10.5 ± 13.8 $ng\,m^{-3}$). Hyder et al. (2012), who measured nine dicarboxylic acids in aerosol samples obtained at the Vavihill measurement station 2008–2009, also found azelaic acid to be the most prominent with peak concentration during summer (16.2 $ng\,m^{-3}$). The concentration of the anthropogenic acids is low (mean ≈ 2 $ng\,m^{-3}$) except during 27 June and 6 July, when the concentration reaches 19.6 and 16.0 $ng\,m^{-3}$, respectively. The spike in concentration of anthropogenic acids during these 2 days is caused by an increase in the concentration of adipic acid.

Correlations between the different compounds was investigated by Pearson correlation. All Pearson r coefficients are given in Table 4. In general, the biogenic compounds (derived from isoprene and monoterpenes) correlated well ($r \geq 0.8$) with each other. The only exception was OS 250, which showed low to medium correlation with the other compounds.

Table 4. Correlation matrix displaying the Pearson product-moment coefficient (r) for measured chemical species. Typefaces represent degree of correlation – italic: |0.7–0.8|; bold: |0.8–0.9|; bold and italic: |0.9–1.0|.

	Adipic acid	Pimelic acid	Suberic acid	Azelaic acid	Pinic acid	Pinonic acid	Terpenylic acid	DTAA	MBTCA	OS 140	OS 154	OS 156	OS 170	OS 200	OS 212	OS 214	OS 216	OS 250	OS 268	OS 280	NOS 295	NOS 297
Adipic acid																						
Pimelic acid	0.16																					
Suberic acid	0.02	***0.95***																				
Azelaic acid	0.01	**0.87**	***0.95***																			
Pinic acid	0.25	0.20	0.01	0.20																		
Pinonic acid	0.05	0.02	0.32	0.00	**0.81**																	
Terpenylic acid	0.33	0.35	0.18	0.40	**0.80**	0.39																
DTAA	0.35	0.29	0.18	0.37	0.66	0.20	**0.89**															
MBTCA	0.32	0.22	0.06	0.18	*0.71*	0.29	***0.94***	***0.92***														
OS 140	0.13	0.41	0.27	0.26	0.47	0.06	***0.90***	***0.94***	*0.70*													
OS 154	0.33	0.36	0.22	0.50	0.67	0.20	***0.92***	***0.93***	**0.83**	**0.82**												
OS 156	0.22	0.36	0.26	0.43	0.62	0.21	**0.83**	**0.87**	**0.84**	**0.84**	***0.92***											
OS 170	0.24	0.24	0.00	0.34	0.67	0.21	*0.77*	*0.73*	*0.73*	*0.76*	**0.86**	**0.83**										
OS 200	0.27	0.32	0.19	0.31	0.58	0.10	**0.81**	***0.93***	*0.76*	**0.80**	***0.96***	***0.93***	**0.84**									
OS 212	0.34	0.35	0.22	0.58	0.58	0.17	**0.86**	***0.97***	**0.86**	**0.84**	***0.97***	***0.92***	**0.81**	***0.98***								
OS 214	0.33	0.30	0.19	0.41	0.65	0.15	**0.80**	***0.96***	***0.93***	**0.88**	**0.89**	***0.97***	*0.74*	***0.97***	***0.98***							
OS 216	0.33	0.26	0.00	0.43	0.61	0.06	0.65	*0.70*	***0.92***	*0.76*	**0.80**	**0.89**	0.57	**0.89**	***0.91***	***0.96***						
OS 250	0.20	0.00	0.21	0.38	0.50	0.26	**0.89**	***0.96***	*0.76*	**0.80**	0.56	0.68	*0.74*	*0.79*	0.50	***0.91***	0.31					
OS 268	0.19	0.12	0.18	0.33	0.06	0.36	**0.84**	**0.83**	**0.82**	**0.84**	0.63	0.65	*0.72*	0.53	0.69	0.50	0.45	0.55				
OS 280	0.38	0.24	0.15	0.08	0.26	0.08	0.53	0.38	0.62	0.65	0.08	0.44	*0.76*	0.32	*0.78*	*0.75*	0.45	0.66	*0.75*			
NOS 295	0.00	0.00	0.00	0.14	0.56	0.56	0.53	0.38	0.35	0.50	0.56	0.44	0.50	0.33	0.33	0.27	0.09	0.33	*0.70*	**0.85**		
NOS 297	0.01	0.14	0.00	0.14	0.53	0.35	0.67	0.57	0.77	0.57	0.68	0.61	0.68	0.59	0.57	0.50	0.31	0.42	*0.70*	0.55	**0.88**	

Table 5. Ranges, means and standard deviations (SD) of the FLEXPART surface type exposure of incoming air masses during 10 June to 18 July 2012.

Surface type	N	Minimum	Maximum	Mean	±SD
			(%)		
Pasture	38	0	13	4.4	3.6
Discontinuous urban fabric	38	1	7	2.6	1.7
Non-irrigated arable land	38	7	35	18.8	8.3
Sparsely vegetated areas	38	0	3	0.4	0.9
Broad-leaved forest	38	0	8	2.6	1.7
Lakes and ponds	38	0	3	0.9	0.6
Moors and heath	38	0	3	0.5	0.7
Coniferous forest	38	0	22	5.5	5.2
Sea and ocean	38	24.6	86	56.0	16.3
Other	38	3	15	8.3	3.2

Table 6. Correlation matrix displaying the Pearson product-moment coefficient (r) for surface types. Typefaces represent degree of correlation: italic: $|\,0.7\text{--}0.8\,|$; bold: $|\,0.8\text{--}0.9\,|$; bolditalic: $|\,0.9\text{--}1.0\,|$.

	Pasture	Discontinuous urban fabric	Non-irrigated arable land	Sparsely vegetated areas	Broad-leaved forest	Lakes and ponds	Moors and heath	Coniferous forest	Sea and ocean	Other
Pasture										
Discontinuous urban fabric	*0.92*									
Non-irrigated arable land	**0.89**	*0.9*								
Sparsely vegetated areas	−0.47	−0.42	−0.49							
Broad-leaved forest	0.48	0.32	0.53	−0.13						
Lakes and ponds	0	−0.12	−0.13	0.18	0.2					
Moors and heath	−0.46	−0.4	−0.47	***0.98***	−0.17	0.14				
Coniferous forest	−0.17	−0.31	−0.22	0.23	0.43	**−0.8**	0.17			
Sea and ocean	**−0.84**	*−0.78*	**−0.84**	0.27	*−0.73*	−0.31	0.28	−0.28		
Other	0.59	0.57	0.53	−0.16	0.42	0.29	−0.18	0.23	*−0.77*	

Three dicarboxylic acids (azelaic, pimelic and suberic acid) correlated well with each other ($r > 0.87$). It is likely that the fatty-acid-derived dicarboxylic acids have a different origin than isoprene- and monoterpene-generated acids, a conclusion that also was reached in a previous study (Hyder et al., 2012). It was expected that adipic acid would show good agreement with pimelic acid since they are both suggested to be of anthropogenic origin. However, this correlation was poor ($r = 0.16$) and is believed to be explained by two strong concentration peaks in adipic acid (27 June and 6 July, Fig. 1a) with no corresponding peak in pimelic acid. Removing these two concentration peaks led to a better agreement between the two acids ($r = 0.67$).

3.2 Air mass surface exposure

Figure 1b displays the exposures of the samples to the nine largest surface categories as percentage contribution and Tables 5 and 6 present the mean exposures and a correlation matrix for the investigated surface types. These surface categories are explained in more detail in the Supplement. The "sea and ocean" category is dominating the exposure with an average of 56 % (± 16 %). This is hardly surprising since a majority of the incoming air mass is from the westerly region where the North Atlantic Ocean, North Sea and Norwegian Sea are situated. The second most common surface exposure is from "non-irrigated arable land" (mean $= 19 \pm 8$ %). This is a common land type in continental Europe which is anti-correlated ($r = -0.84$) to the "sea and ocean" surface category. The fact that several land-based surface categories anti-correlated to the "sea and ocean" category may be an indicator of the model working properly. The category "other" has a significant contribution to the total exposure (mean $= 8 \pm 3$ %), but it groups 34 surface categories and is therefore difficult to interpret beyond the common fact that all these categories are land masses. It is important to remember that these exposures should not be read as a representation of the contribution of the land surface types to the production of the aerosols measured. For that, an estimation of the aerosol production (or transformation) associated with each surface category would be required. However, correlating the land surface exposures to the measured aerosol time series can provide an indication on the origin of the aerosols.

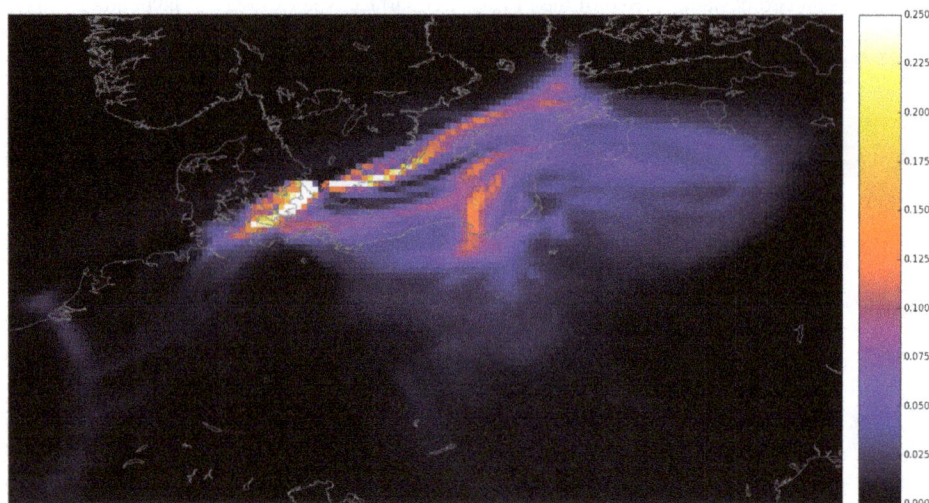

Figure 2. A 120 h back-trajectory air mass covering the concentration peak dates, 6–8 July. The FLEXPART model back-trajectories are shown in shaded colours. The colour bar displays the FLEXPART footprint, normalized to 1 (the colour range has been limited to 0–0.3 to highlight grid points with low but a non-zero contribution). Together, the grid points with a value larger than 0.1 contribute 17 % of the total sensitivity, while grid boxes with a value larger than 0.01 contribute 81 % of the total sensitivity. The 120 h back-trajectory was chosen for easier interpretation of the illustration.

During a period of increased concentrations of molecular BSOA compounds (6–8 July) the air mass was more exposed to land surface categories such as "non-irrigated arable land", "coniferous forest", "broad-leaved forest" and "pastures" on the expense of "sea and ocean" (Fig. 1a, b). Further, the category "other" is also increased during this particular period. Within the "other" category, "mixed forest", "complex cultivation patterns", "land principally occupied by agriculture, with significant areas of natural vegetation" and "transitional woodland/shrub" are dominant (more information about the surface categories can be found on the CORINE database website) (EEA, 2016). This particular concentration increase is caused by the fatty-acid-derived organic acids, monoterpene-derived organic acids and isoprene-derived OSs (Fig. 1a). The concentration of $PM_{2.5}$ does not provide any explanation of the cause of the high concentrations, since $PM_{2.5}$ is in general high during the entire campaign period. Both the HYSPLIT and FLEXPART model revealed that arriving air masses during this period mainly had an origin from continental Europe (Fig. 2). As stated earlier, it has been observed that air masses arriving from this direction usually carry more PM and OSs than from other directions (Nguyen et al., 2014; Kristensson et al., 2008).

The period of increased concentrations of molecular BSOA compounds (6–8 July) is in large contrast to the "clean periods" observed during 12–16 June and 16–18 July (Fig. 1a, b). In particular, the latter period shows very low values of molecular BSOA compounds and a corresponding "sea and ocean" exposure of 79–86 %. Hence, "sea and ocean" exposure does not seem to contribute to the mea-

sured mass of molecular BSOA compounds. Similarly, the "non-irrigated arable land" contributes to a significant fraction during 16–18 July (8–12 %) and most probably does not contribute to the mass of measured BSOA species either.

3.3 Connection between surface type and measured species

To further investigate the impact of surface types on measured BSOA species a PCA was conducted as described in Sect. 2. A four-PC VARIMAX-rotated solution was chosen. This solution explained 80.3 % of the total variance. Table 7 shows the individual parameter contribution to the respective PC. PC1 accounts for 49.1 % of the total variance and has strong positive contributions from several of the monoterpene-derived dicarboxylic acids and both monoterpene- and isoprene-derived OSs and NOSs. The strongest positive surface category in PC1 is "coniferous forest", suggesting that the species with a bold number in PC1 within Table 7 are originating, or that their mass concentration have a positive response, from coniferous forest. Coniferous forests are mainly known as large-scale emitters of monoterpenes. Despite this, the PCA illustrates that isoprene oxidation products are positively correlated to this surface category. Steinbrecher et al. (1999) observed negligible emissions of isoprene from common conifers as Scots pine (*Pinus sylvestris*) and common juniper (*Juniperus communis*). However, they found significant emissions from Norway spruce (*Picea abies*) which may explain some of the isoprene-derived compounds in this study. Although the less strong positive contribution of 0.53, isoprene-

Table 7. Principal component (PC) loadings. The loadings display the variation (between −1 and 1) explained by the PC. Numbers in bold indicates absolute number > 0.6. PC1 explained 49.1 %, PC2 14.9 %, PC3 9.3 % and PC4 6.9 %.

	Principal component			
	1	2	3	4
Adipic acid	0.37	−0.25	0.08	0.59
Pimelic acid	0.24	0.20	**0.75**	−0.21
Suberic acid	0.20	0.26	**0.82**	−0.19
Azelaic acid	0.21	0.39	**0.74**	−0.17
Pinic acid	**0.70**	−0.04	−0.25	0.14
Pinonic acid	0.19	−0.15	−0.37	0.16
Terpenylic acid	**0.88**	0.29	−0.11	0.04
DTAA	**0.93**	0.24	0.04	0.09
MBTCA	**0.89**	0.28	−0.26	−0.02
OS 140	**0.76**	0.30	0.12	−0.41
OS 154	**0.96**	0.22	0.04	−0.10
OS 156	**0.93**	0.06	0.04	−0.14
OS 170	**0.79**	0.20	−0.17	−0.28
OS 200	**0.92**	0.18	0.10	−0.12
OS 212	**0.95**	0.18	0.13	−0.01
OS 214	**0.92**	0.13	0.15	0.04
OS 216	**0.87**	0.03	0.26	0.11
OS 250	0.48	−0.06	−0.38	−0.06
OS 268	**0.67**	0.24	−0.51	−0.18
OS 280	**0.87**	0.13	−0.20	−0.05
NOS 295	0.43	0.16	**−0.69**	−0.25
NOS 297	0.59	0.28	−0.48	−0.35
Pastures	0.22	**0.85**	0.15	−0.37
Discontinuous urban fabric	−0.02	**0.92**	0.12	−0.29
Non-irrigated arable land	0.20	**0.94**	0.10	−0.14
Broad-leaved forest	0.53	**0.77**	0.21	0.11
Sparsely vegetated areas	−0.11	−0.10	−0.18	**0.86**
Lakes and ponds	**0.76**	0.34	0.02	0.42
Moors and heath	−0.16	−0.04	−0.23	**0.85**
Coniferous forest	**0.79**	0.35	−0.04	0.39
Sea and ocean	0.37	**0.62**	0.27	0.34
Other	**0.60**	**0.65**	0.19	0.19

emitting "broad-leaved forest" may also have contributed to the above-described pattern in PC1.

PC2 accounts for 14.9 % of the total variation and can roughly be classified as surface categories with low contribution to measured BSOA compounds. Six of the 10 investigated surface categories show strong positive contribution to PC2 while many of the measured compounds show low and in some cases negative contribution to PC2. The observed pattern of high "sea and ocean" and "non-irrigated arable land" exposure when the mass concentration of BSOA compounds was low, further strengthening the explanation of PC2.

PC3 accounts for 9.3 % of the total variance. The main contributors are suberic acid, azelaic acid and pimelic acid. They are all similar in chemical structure, although suberic and azelaic acid probably originate from fatty acids, while pimelic acid likely is of anthropogenic origin (Table 1). Further, azelaic acid has been found to be involved in the trigering of the plant immune system (Jung et al., 2009). Hyder et al. (2012), who also found these three acids to be highly correlated in ambient aerosol, inferred that pimelic acid was either produced from the same source as suberic and azelaic acid or that pimelic acid is produced by continued oxidation of suberic and azelaic acid down to acids of lower carbon number. None of the land surface categories displayed a high contribution to PC3: "broad-leaved forest" had the highest contribution of 0.21, while the other forest category, "conifer forest", had a 1 order of magnitude lower contribution of −0.04.

PC4 accounted for 6.9 % of the total variance and is harder to interpret than the previous three PCs. The anthropogenically derived adipic acid has a positive PC contribution (0.59) as well as the surface categories "sparsely vegetated areas" (0.86) and "moors and heath" (0.85). The used land cover maps reveals that both "sparsely vegetated areas" and "moors and heath" are mainly found in Norway and northern Sweden, i.e. in the north and northwesterly direction of Vavihill measurement station. The overall interpretation of PC4 is difficult since adipic acid is thought to be of anthropogenic origin but, in this case, seems to correlate with landscape surface types that are sparsely populated and are associated with low human activity (i.e. "sparsely vegetated areas" and "moors and heath").

The complexity in PC4 may be caused by the concentration peaks in adipic acid that occurred 27 June and 6 July (Fig. 1a). During 27 June, the air mass mainly arrived from the Atlantic Ocean and southern Norway, while the air mass during 6 July mainly originated from the Baltic countries and central Europe (partially illustrated in Fig. 2). Removing the two concentration peaks in adipic acid gave a different PCA solution. Adipic acid now falls into the same PCA as pimelic, suberic and azelaic acid with PC contributions of 0.52, 0.66, 0.70 and 0.73, respectively. Further, the new PC solution show that the aforementioned acids are associated with "pastures" (PC contribution = 0.82), "discontinuous urban fabric" (0.84), "non-irrigated arable land" (0.82),"broad-leaved forest" (0.81), "sea and ocean" (0.69) and the "other" category (0.66). Hence, the nature of adipic acid remains unclear since it shows good agreement with the other acids when concentration peaks are removed, implying that adipic is derived from fatty acids or salicylic acid. On the other hand, including the concentration peaks, neither this study nor the study by Hyder et al. (2012) found any strong correlation between adipic and pimelic acid. It can be speculated whether the observed concentration peaks in adipic acid have their explanation in local emission sources of benzene or cyclohexene, followed by a fast oxidation into adipic acid. Future studies should repeat the presented methodology to focus on heavily anthropogenically influenced surface categories (i.e. cities, industries etc.) and their impact on anthropogenic acids and newly discovered anthropogenic OSs (Riva et al., 2016, 2015).

3.4 Uncertainties and limits

In this study, our analysis approach relies on two steps: first the calculation of the exposures, using FLEXPART, and then the estimation of land type contributions using a PCA. Both steps suffer from uncertainties which limit the robustness of our results.

The longer the back-trajectories used in FLEXPART, the larger the error is likely to be. On the other hand, shorter back-trajectories lead to neglecting a larger proportion of "older" aerosols. We tested the impact of the footprint length choice on the exposure time series by repeating the analysis with footprints of 3 and 5 days (instead of 7 days in our default setup). Overall, the exposures are not significantly affected, except for the exposure to the "sea and ocean" surface type during the 8–10 July peak, which show an uncertainty of 6 % (Fig. S1 in the Supplement).

Besides the length of the simulations, a number of FLEXPART settings can impact the results. The size of the aerosols particles has a strong impact on the lifetime of the aerosols in the atmosphere and therefore on the footprints. We have repeated the experiment with mean aerosol sizes of 50 nm and 1 µm, and the results of the PCAs remained reasonably similar (Table S1 and S2 in the Supplement). This is mainly because the PCA is sensitive to correlations, and not to absolute values.

The calculation of the observation exposures is based on the assumption that the measured aerosol compositions scale linearly with the aerosol production within the backplume of the observation. This is not the case in reality: processes such as coagulation, nucleation, chemical reactions between aerosols and surrounding reactive gas species, photo-dissociation and wet and dry deposition (removal of aerosols from the atmosphere by the rain and by gravitational settling) alter the aerosol composition and concentration all along the air mass trajectory. Our approach also ignores the influence of aerosol particles (or precursors) older than 7 days on the observations. Accounting adequately for all these processes would require a comprehensive aerosol model, which is out of the scope of this study. This mainly means that our approach cannot be used to quantify the aerosol production associated with, for example, a specific forest type.

The main limit to the PCA is the shortness of the time series. In particular, there is only one strong event during the campaign (6–8 July), which is not enough for drawing strong conclusions. Our study can, however, be regarded as a proof of concept: computing FLEXPART footprints is relatively easy and lightweight, and could be performed routinely. The conclusions of a PCA are likely to be a lot more robust with longer time series with more observations included, and/or multi-site observation campaigns (provided that the footprints of the different sites overlap sufficiently).

4 Conclusions

Nine carboxylic acids along with 11 organosulfates (OSs) and 2 nitrooxy organosulfates (NOSs) were analysed from 38 daily aerosol samples sampled at Vavihill measurement station in southern Sweden during June and July 2012. Most of the measured compounds can be considered as photo-oxidation products from biogenic volatile organic compounds (BVOCs), hence derived from terrestrial plants. The FLEXPART model was used to identify exposure of the aerosol samples to several different surface categories. For easier interpretation, the study was focused on four potential source-specific components using 22 chemical species and the 9 largest surface categories. The "sea and ocean" category was found to dominate the exposure, and other important categories were "non-irrigated arable land" and "pastures". A principal component analysis (PCA) of four principal components (PCs) was used to explore the impact and connection of surface categories on mass concentration of measured biogenic secondary organic aerosol compounds. It was found that coniferous forest had a positive effect on several of the measured monoterpene-derived compounds. The remaining three PCs were harder to interpret; however, future studies should aim to investigate the sources of azelaic, suberic and pimelic acids which dominate in mass concentration but showed no clear correlation to surface categories.

This study demonstrates the interest of using an atmospheric transport model in aerosol source apportionment on specific chemical compounds. With the presented methodology it is possible to connect single chemical tracer compounds to potential local and long-range aerosol sources, i.e. surface categories. More advanced applications may include particle age estimation and its relation to surface categories; this could be achieved by measuring first- and second-generation BVOC oxidation products and relating these to its measurable gas-phase precursor.

Author contributions. JM designed the study, compiled all data, performed the PCA and wrote most of the paper. GM ran the FLEXPART simulations. MKS ran the HYSPLIT simulations. AMKH and MG ran the chemical analysis. AK, ES and KES assisted in the writing process.

Competing interests. The authors declare that they have no conflict of interest.

Acknowledgements. This work was supported by the Swedish Research Council FORMAS (project 2011-743).

Edited by: Jason Surratt

References

Almeida, S. M., Pio, C. A., Freitas, M. C., Reis, M. A., and Trancoso, M. A.: Source apportionment of atmospheric urban aerosol based on weekdays/weekend variability: evaluation of road resuspended dust contribution, Atmos. Environ., 40, 2058–2067, https://doi.org/10.1016/j.atmosenv.2005.11.046, 2006.

Castro, L. M., Pio, C. A., Harrison, R. M., and Smith, D. J. T.: Carbonaceous aerosol in urban and rural European atmospheres: estimation of secondary organic carbon concentrations, Atmos. Environ., 33, 2771–2781, https://doi.org/10.1016/S1352-2310(98)00331-8, 1999.

Chan, T. W. and Mozurkewich, M.: Application of absolute principal component analysis to size distribution data: identification of particle origins, Atmos. Chem. Phys., 7, 887–897, https://doi.org/10.5194/acp-7-887-2007, 2007.

Claeys, M., Iinuma, Y., Szmigielski, R., Surratt, J. D., Blockhuys, F., Van Alsenoy, C., Boge, O., Sierau, B., Gomez-Gonzalez, Y., Vermeylen, R., Van der Veken, P., Shahgholi, M., Chan, A. W. H., Herrmann, H., Seinfeld, J. H., and Maenhaut, W.: Terpenylic Acid and Related Compounds from the Oxidation of alpha-Pinene: Implications for New Particle Formation and Growth above Forests, Environ. Sci. Technol., 43, 6976–6982, https://doi.org/10.1021/es9007596, 2009.

Copernicus: Land Monitoring Services, http://land.copernicus.eu/pan-european/corine-land-cover/clc-2012 (last acess: 20 October 2016), 2012.

Cruz, C. N. and Pandis, S. N.: The effect of organic coatings on the cloud condensation nuclei activation of inorganic atmospheric aerosol, J. Geophys. Res., 103, 13111–13123, https://doi.org/10.1029/98JD00979, 1998.

de Reus, M., Ström, J., Curtius, J., Pirjola, L., Vignati, E., Arnold, F., Hansson, H. C., Kulmala, M., Lelieveld, J., and Raes, F.: Aerosol production and growth in the upper free troposphere, J. Geophys. Res., 105, 24751–24762, https://doi.org/10.1029/2000JD900382, 2000.

Dockery, D. W., Pope, C. A., Xu, X. P., Spengler, J. D., Ware, J. H., Fay, M. E., Ferris, B. G., and Speizer, F. E.: An Association between Air-Pollution and Mortality in 6 United-States Cities, New Engl. J. Med., 329, 1753–1759, https://doi.org/10.1056/Nejm199312093292401, 1993.

Draxler, R. R. and Hess, G. D.: An overview of the HYSPLIT_4 modelling system for trajectories, dispersion and deposition, Aust. Meteorol. Mag., 47, 295–308, 1998.

EEA: Corine Reports, http://www.eea.europa.eu/publications/COR0-part2/page001.html (last acess: 20 October 2016), 2016.

Gelencsér, A., May, B., Simpson, D., Sanchez-Ochoa, A., Kasper-Giebl, A., Puxbaum, H., Caseiro, A., Pio, C., and Legrand, M.: Source apportionment of $PM_{2.5}$ organic aerosol over Europe: Primary/secondary, natural/anthropogenic, and fossil/biogenic origin, J. Geophys. Res., 112, D23S04, https://doi.org/10.1029/2006JD008094, 2007.

Genberg, J., Hyder, M., Stenström, K., Bergström, R., Simpson, D., Fors, E. O., Jönsson, J. Å., and Swietlicki, E.: Source apportionment of carbonaceous aerosol in southern Sweden, Atmos. Chem. Phys., 11, 11387–11400, https://doi.org/10.5194/acp-11-11387-2011, 2011.

Gomez-Gonzalez, Y., Surratt, J. D., Cuyckens, F., Szmigielski, R., Vermeylen, R., Jaoui, M., Lewandowski, M., Offenberg, J. H., Kleindienst, T. E., Edney, E. O., Blockhuys, F., Van Alsenoy, C., Maenhaut, W., and Claeys, M.: Characterization of organosulfates from the photooxidation of isoprene and unsaturated fatty acids in ambient aerosol using liquid chromatography/(-) electrospray ionization mass spectrometry, J. Mass. Spectrom., 43, 371–382, https://doi.org/10.1002/jms.1329, 2008.

Gomiscek, B., Hauck, H., Stopper, S., and Preining, O.: Spatial and temporal variations Of PM_1, $PM_{2.5}$, PM_{10} and particle number concentration during the AUPHEP-project, Atmos. Environ., 38, 3917–3934, https://doi.org/10.1016/j.atmosenv.2004.03.056, 2004.

Guenther, A., Hewitt, C. N., Erickson, D., Fall, R., Geron, C., Graedel, T., Harley, P., Klinger, L., Lerdau, M., Mckay, W. A., Pierce, T., Scholes, B., Steinbrecher, R., Tallamraju, R., Taylor, J., and Zimmerman, P.: A Global-Model of Natural Volatile Organic-Compound Emissions, J. Geophys. Res., 100, 8873–8892, https://doi.org/10.1029/94JD02950, 1995.

Guenther, A., Karl, T., Harley, P., Wiedinmyer, C., Palmer, P. I., and Geron, C.: Estimates of global terrestrial isoprene emissions using MEGAN (Model of Emissions of Gases and Aerosols from Nature), Atmos. Chem. Phys., 6, 3181–3210, https://doi.org/10.5194/acp-6-3181-2006, 2006.

Guenther, A. B., Zimmerman, P. R., Harley, P. C., Monson, R. K., and Fall, R.: Isoprene and monoterpene emission rate variability: Model evaluations and sensitivity analyses, J. Geophys. Res., 98, 12609–12617, https://doi.org/10.1029/93JD00527, 1993.

Hakola, H., Tarvainen, V., Laurila, T., Hiltunen, V., Hellen, H., and Keronen, P.: Seasonal variation of VOC concentrations above a boreal coniferous forest, Atmos. Environ., 37, 1623–1634, https://doi.org/10.1016/S1352-2310(03)00014-1, 2003.

Hakola, H., Hellén, H., Hemmilä, M., Rinne, J., and Kulmala, M.: In situ measurements of volatile organic compounds in a boreal forest, Atmos. Chem. Phys., 12, 11665–11678, https://doi.org/10.5194/acp-12-11665-2012, 2012.

Hallquist, M., Wenger, J. C., Baltensperger, U., Rudich, Y., Simpson, D., Claeys, M., Dommen, J., Donahue, N. M., George, C., Goldstein, A. H., Hamilton, J. F., Herrmann, H., Hoffmann, T., Iinuma, Y., Jang, M., Jenkin, M. E., Jimenez, J. L., Kiendler-Scharr, A., Maenhaut, W., McFiggans, G., Mentel, Th. F., Monod, A., Prévôt, A. S. H., Seinfeld, J. H., Surratt, J. D., Szmigielski, R., and Wildt, J.: The formation, properties and impact of secondary organic aerosol: current and emerging issues, Atmos. Chem. Phys., 9, 5155–5236, https://doi.org/10.5194/acp-9-5155-2009, 2009.

Hansen, A. M. K., Kristensen, K., Nguyen, Q. T., Zare, A., Cozzi, F., Nøjgaard, J. K., Skov, H., Brandt, J., Christensen, J. H., Ström, J., Tunved, P., Krejci, R., and Glasius, M.: Organosulfates and organic acids in Arctic aerosols: speciation, annual variation and concentration levels, Atmos. Chem. Phys., 14, 7807–7823, https://doi.org/10.5194/acp-14-7807-2014, 2014.

Hansen, A. M. K., Hong, J., Raatikainen, T., Kristensen, K., Ylisirniö, A., Virtanen, A., Petäjä, T., Glasius, M., and Prisle,

N. L.: Hygroscopic properties and cloud condensation nuclei activation of limonene–derived organosulfates and their mixtures with ammonium sulfate, Atmos. Chem. Phys., 15, 14071–14089, https://doi.org/10.5194/acp-15-14071-2015, 2015.

Hatakeyama, S., Ohno, M., Weng, J. H., Takagi, H., and Akimoto, H.: Mechanism for the Formation of Gaseous and Particulate Products from Ozone–Cycloalkene Reactions in Air, Environ. Sci. Technol., 21, 52–57, https://doi.org/10.1021/Es00155a005, 1987.

Hettiyadura, A. P. S., Stone, E. A., Kundu, S., Baker, Z., Geddes, E., Richards, K., and Humphry, T.: Determination of atmospheric organosulfates using HILIC chromatography with MS detection, Atmos. Meas. Tech., 8, 2347–2358, https://doi.org/10.5194/amt-8-2347-2015, 2015.

Hyder, M., Genberg, J., Sandahl, M., Swietlicki, E., and Jönsson, J. A.: Yearly trend of dicarboxylic acids in organic aerosols from south of Sweden and source attribution, Atmos. Environ., 57, 197–204, https://doi.org/10.1016/j.atmosenv.2012.04.027, 2012.

Iinuma, Y., Muller, C., Berndt, T., Boge, O., Claeys, M., and Herrmann, H.: Evidence for the existence of organosulfates from beta-pinene ozonolysis in ambient secondary organic aerosol, Environ. Sci. Technol., 41, 6678–6683, https://doi.org/10.1021/es070938t, 2007.

Ito, K., Xue, N., and Thurston, G.: Spatial variation of PM2.5 chemical species and source-apportioned mass concentrations in New York City, Atmos. Environ., 38, 5269–5282, https://doi.org/10.1016/j.atmosenv.2004.02.063, 2004.

Jung, H. W., Tschaplinski, T. J., Wang, L., Glazebrook, J., and Greenberg, J. T.: Priming in Systemic Plant Immunity, Science, 324, 89–91, https://doi.org/10.1126/science.1170025, 2009.

Kawamura, K. and Gagosian, R. B.: Implications of Omega-Oxocarboxylic Acids in the Remote Marine Atmosphere for Photooxidation of Unsaturated Fatty-Acids, Nature, 325, 330–332, https://doi.org/10.1038/325330a0, 1987.

Kawamura, K. and Ikushima, K.: Seasonal-Changes in the Distribution of Dicarboxylic-Acids in the Urban Atmosphere, Environ. Sci. Technol., 27, 2227–2235, doi10.1021/Es00047a033, 1993.

Kawamura, K. and Kaplan, I. R.: Motor Exhaust Emissions as a Primary Source for Dicarboxylic-Acids in Los-Angeles Ambient Air, Environ. Sci. Technol., 21, 105–110, https://doi.org/10.1021/Es00155a014, 1987.

Kawamura, K. and Sakaguchi, F.: Molecular distributions of water soluble dicarboxylic acids in marine aerosols over the Pacific Ocean including tropics, J. Geophys. Res., 104, 3501–3509, https://doi.org/10.1029/1998jd100041, 1999.

Kerminen, V. M.: Relative roles of secondary sulfate and organics in atmospheric cloud condensation nuclei production, J. Geophys. Res., 106, 17321–17333, https://doi.org/10.1029/2001jd900204, 2001.

Kristensen, K.: Anthropogenic Enhancement of Biogenic Secondary Organic Aerosols – Investigation of Organosulfates and Dimers of Monoterpene Oxidation Products, PhD thesis, Department of Chemistry and iNano, Aarhus University, Aarhus, Denmark, 2014.

Kristensen, K. and Glasius, M.: Organosulfates and oxidation products from biogenic hydrocarbons in fine aerosols from a forest in North West Europe during spring, Atmos. Environ., 45, 4546–4556, https://doi.org/10.1016/j.atmosenv.2011.05.063, 2011.

Kristensson, A., Dal Maso, M., Swietlicki, E., Hussein, T., Zhou, J., Kerminen, V. M., and Kulmala, M.: Characterization of new particle formation events at a background site in Southern Sweden: relation to air mass history, Tellus B, 60, 330–344, https://doi.org/10.1111/j.1600-0889.2008.00345.x, 2008.

Kundu, S., Kawamura, K., Andreae, T. W., Hoffer, A., and Andreae, M. O.: Molecular distributions of dicarboxylic acids, ketocarboxylic acids and α-dicarbonyls in biomass burning aerosols: implications for photochemical production and degradation in smoke layers, Atmos. Chem. Phys., 10, 2209–2225, https://doi.org/10.5194/acp-10-2209-2010, 2010.

Laothawornkitkul, J., Taylor, J. E., Paul, N. D., and Hewitt, C. N.: Biogenic volatile organic compounds in the Earth system, New Phytol., 183, 27–51, https://doi.org/10.1111/j.1469-8137.2009.02859.x, 2009.

Ma, Y., Willcox, T. R., Russell, A. T., and Marston, G.: Pinic and pinonic acid formation in the reaction of ozone with alpha-pinene, Chem. Commun., 1328–1330, https://doi.org/10.1039/B617130C, 2007.

Mentel, Th. F., Wildt, J., Kiendler-Scharr, A., Kleist, E., Tillmann, R., Dal Maso, M., Fisseha, R., Hohaus, Th., Spahn, H., Uerlings, R., Wegener, R., Griffiths, P. T., Dinar, E., Rudich, Y., and Wahner, A.: Photochemical production of aerosols from real plant emissions, Atmos. Chem. Phys., 9, 4387–4406, https://doi.org/10.5194/acp-9-4387-2009, 2009.

Mochida, M., Kawabata, A., Kawamura, K., Hatsushika, H., and Yamazaki, K.: Seasonal variation and origins of dicarboxylic acids in the marine atmosphere over the western North Pacific, J. Geophys. Res., 108, 4193, https://doi.org/10.1029/2002JD002355, 2003.

Monson, R. K., Jones, R. T., Rosenstiel, T. N., and Schnitzler, J. P.: Why only some plants emit isoprene, Plant Cell Environ., 36, 503–516, https://doi.org/10.1111/pce.12015, 2013.

Müller, A.: Pimelic acid from salicylic acid, Org. Synth., 11, 42, https://doi.org/10.15227/orgsyn.011.0042, 1931.

Musser, M. T: Adipic Acid, Ullman's Encyclopedia of Industrial Chemistry, https://doi.org/10.1002/14356007.a01_269, 2000.

Nguyen, Q. T., Christensen, M. K., Cozzi, F., Zare, A., Hansen, A. M. K., Kristensen, K., Tulinius, T. E., Madsen, H. H., Christensen, J. H., Brandt, J., Massling, A., Nøjgaard, J. K., and Glasius, M.: Understanding the anthropogenic influence on formation of biogenic secondary organic aerosols in Denmark via analysis of organosulfates and related oxidation products, Atmos. Chem. Phys., 14, 8961–8981, https://doi.org/10.5194/acp-14-8961-2014, 2014.

Nyanganyura, D., Maenhaut, W., Mathuthua, M., Makarau, A., and Meixner, F. X.: The chemical composition of tropospheric aerosols and their contributing sources to a continental background site in northern Zimbabwe from 1994 to 2000, Atmos. Environ., 41, 2644–2659, https://doi.org/10.1016/j.atmosenv.2006.11.015, 2007.

Olson, C. N., Galloway, M. M., Yu, G., Hedman, C. J., Lockett, M. R., Yoon, T., Stone, E. A., Smith, L. M., and Keutsch, F. N.: Hydroxycarboxylic Acid-Derived Organosulfates: Synthesis, Stability, and Quantification in Ambient Aerosol, Environ. Sci. Technol., 45, 6468–6474, https://doi.org/10.1021/es201039p, 2011.

Penuelas, J. and Llusia, J.: BVOCs: plant defense against climate warming?, Trends Plant Sci., 8, 105–109, https://doi.org/10.1016/S1360-1385(03)00008-6, 2003.

Pope, C. A., Thun, M. J., Namboodiri, M. M., Dockery, D. W., Evans, J. S., Speizer, F. E., and Heath, C. W.: Particulate Air-Pollution as a Predictor of Mortality in a Prospective-Study of US Adults, Am. J. Resp. Crit. Care, 151, 669–674, 1995.

Putaud, J. P., Van Dingenen, R., Alastuey, A., Bauer, H., Birmili, W., Cyrys, J., Flentje, H., Fuzzi, S., Gehrig, R., Hansson, H. C., Harrison, R. M., Herrmann, H., Hitzenberger, R., Huglin, C., Jones, A. M., Kasper-Giebl, A., Kiss, G., Kousa, A., Kuhlbusch, T. A. J., Loschau, G., Maenhaut, W., Molnar, A., Moreno, T., Pekkanen, J., Perrino, C., Pitz, M., Puxbaum, H., Querol, X., Rodriguez, S., Salma, I., Schwarz, J., Smolik, J., Schneider, J., Spindler, G., ten Brink, H., Tursic, J., Viana, M., Wiedensohler, A., and Raes, F.: A European aerosol phenomenology-3: Physical and chemical characteristics of particulate matter from 60 rural, urban, and kerbside sites across Europe, Atmos. Environ., 44, 1308–1320, https://doi.org/10.1016/j.atmosenv.2009.12.011, 2010.

Räisänen, T., Ryyppö, A., and Kellomäki, S.: Effects of elevated CO_2 and temperature on monoterpene emission of Scots pine (*Pinus sylvestris* L.), Atmos. Environ., 42, 4160–4171, https://doi.org/10.1016/j.atmosenv.2008.01.023, 2008.

Riva, M., Tomaz, S., Cui, T. Q., Lin, Y. H., Perraudin, E., Gold, A., Stone, E. A., Villenave, E., and Surratt, J. D.: Evidence for an Unrecognized Secondary Anthropogenic Source of Organosulfates and Sulfonates: Gas-Phase Oxidation of Polycyclic Aromatic Hydrocarbons in the Presence of Sulfate Aerosol, Environ. Sci. Technol., 49, 6654–6664, https://doi.org/10.1021/acs.est.5b00836, 2015.

Riva, M., Da Silva Barbosa, T., Lin, Y.-H., Stone, E. A., Gold, A., and Surratt, J. D.: Chemical characterization of organosulfates in secondary organic aerosol derived from the photooxidation of alkanes, Atmos. Chem. Phys., 16, 11001–11018, https://doi.org/10.5194/acp-16-11001-2016, 2016.

Schindelka, J., Iinuma, Y., Hoffmann, D., and Herrmann, H.: Sulfate radical-initiated formation of isoprene-derived organosulfates in atmospheric aerosols, Faraday Discuss., 165, 237–259, https://doi.org/10.1039/c3fd00042g, 2013.

Seibert, P. and Frank, A.: Source-receptor matrix calculation with a Lagrangian particle dispersion model in backward mode, Atmos. Chem. Phys., 4, 51–63, https://doi.org/10.5194/acp-4-51-2004, 2004.

Shalamzari, M. S., Ryabtsova, O., Kahnt, A., Vermeylen, R., Herent, M. F., Quetin-Leclercq, J., Van der Veken, P., Maenhaut, W., and Claeys, M.: Mass spectrometric characterization of organosulfates related to secondary organic aerosol from isoprene, Rapid Commun. Mass Sp., 27, 784–794, https://doi.org/10.1002/rcm.6511, 2013.

Sharkey, T. D., Wiberley, A. E., and Donohue, A. R.: Isoprene emission from plants: Why and how, Ann. Bot.-London, 101, 5–18, https://doi.org/10.1093/aob/mcm240, 2008.

Stein, A. F., Draxler, R. R., Rolph, G. D., Stunder, B. J. B., Cohen, M. D., and Ngan, F.: Noaa's Hysplit Atmospheric Transport and Dispersion Modeling System, B. Am. Meteorol. Soc., 96, 2059–2077, https://doi.org/10.1175/Bams-D-14-00110.1, 2015.

Steinbrecher, R., Hauff, K., Hakola, H., and Rössler, J.: A revised parameterisation for emission modelling of isoprenoids for boreal plants, The European Commission, Luxembourg, 29–43, 1999.

Stephanou, E. G. and Stratigakis, N.: Oxocarboxylic and Alpha,Omega-Dicarboxylic Acids – Photooxidation Products of Biogenic Unsaturated Fatty-Acids Present in Urban Aerosols, Environ. Sci. Technol., 27, 1403–1407, https://doi.org/10.1021/Es00044a016, 1993.

Stohl, A., Forster, C., Frank, A., Seibert, P., and Wotawa, G.: Technical note: The Lagrangian particle dispersion model FLEXPART version 6.2, Atmos. Chem. Phys., 5, 2461–2474, https://doi.org/10.5194/acp-5-2461-2005, 2005.

Surratt, J. D., Kroll, J. H., Kleindienst, T. E., Edney, E. O., Claeys, M., Sorooshian, A., Ng, N. L., Offenberg, J. H., Lewandowski, M., Jaoui, M., Flagan, R. C., and Seinfeld, J. H.: Evidence for organosulfates in secondary organic aerosol, Environ. Sci. Technol., 41, 517–527, https://doi.org/10.1021/es062081q, 2007a.

Surratt, J. D., Lewandowski, M., Offenberg, J. H., Jaoui, M., Kleindienst, T. E., Edney, E. O., and Seinfeld, J. H.: Effect of acidity on secondary organic aerosol formation from isoprene, Environ. Sci. Technol., 41, 5363–5369, https://doi.org/10.1021/es0704176, 2007b.

Surratt, J. D., Gomez-Gonzalez, Y., Chan, A. W. H., Vermeylen, R., Shahgholi, M., Kleindienst, T. E., Edney, E. O., Offenberg, J. H., Lewandowski, M., Jaoui, M., Maenhaut, W., Claeys, M., Flagan, R. C., and Seinfeld, J. H.: Organosulfate formation in biogenic secondary organic aerosol, J. Phys. Chem. A, 112, 8345–8378, https://doi.org/10.1021/jp802310p, 2008.

Surratt, J. D., Chan, A. W. H., Eddingsaas, N. C., Chan, M. N., Loza, C. L., Kwan, A. J., Hersey, S. P., Flagan, R. C., Wennberg, P. O., and Seinfeld, J. H.: Reactive intermediates revealed in secondary organic aerosol formation from isoprene, P. Natl. Acad. Sci. USA, 107, 6640–6645, https://doi.org/10.1073/pnas.0911114107, 2010.

Szmigielski, R., Surratt, J. D., Gomez-Gonzalez, Y., Van der Veken, P., Kourtchev, I., Vermeylen, R., Blockhuys, F., Jaoui, M., Kleindienst, T. E., Lewandowski, M., Offenberg, J. H., Edney, E. O., Seinfeld, J. H., Maenhaut, W., and Claeys, M.: 3-methyl-1,2,3-butanetricarboxylic acid: An atmospheric tracer for terpene secondary organic aerosol, Geophys. Res. Lett., 34, L24811, https://doi.org/10.1029/2007GL031338, 2007.

Tolocka, M. P. and Turpin, B.: Contribution of Organosulfur Compounds to Organic Aerosol Mass, Environ. Sci. Technol., 46, 7978–7983, https://doi.org/10.1021/es300651v, 2012.

van Pinxteren, D., Brüggemann, E., Gnauk, T., Müller, K., Thiel, C., and Herrmann, H.: A GIS based approach to back trajectory analysis for the source apportionment of aerosol constituents and its first application, J. Atmos. Chem., 67, 1–28, https://doi.org/10.1007/s10874-011-9199-9, 2010.

van Pinxteren, D., Neusüß, C., and Herrmann, H.: On the abundance and source contributions of dicarboxylic acids in size-resolved aerosol particles at continental sites in central Europe, Atmos. Chem. Phys., 14, 3913–3928, https://doi.org/10.5194/acp-14-3913-2014, 2014.

Wehner, B. and Wiedensohler, A.: Long term measurements of submicrometer urban aerosols: statistical analysis for correlations with meteorological conditions and trace gases, Atmos. Chem. Phys., 3, 867–879, https://doi.org/10.5194/acp-3-867-2003, 2003.

Viana, M., Querol, X., Alastuey, A., Gil, J. I., and Menendez, M.: Identification of PM sources by principal component analysis (PCA) coupled with wind direction data, Chemosphere, 65, 2411–2418, https://doi.org/10.1016/j.chemosphere.2006.04.060, 2006.

Yttri, K. E., Simpson, D., Nøjgaard, J. K., Kristensen, K., Genberg, J., Stenström, K., Swietlicki, E., Hillamo, R., Aurela, M., Bauer, H., Offenberg, J. H., Jaoui, M., Dye, C., Eckhardt, S., Burkhart, J. F., Stohl, A., and Glasius, M.: Source apportionment of the summer time carbonaceous aerosol at Nordic rural background sites, Atmos. Chem. Phys., 11, 13339–13357, https://doi.org/10.5194/acp-11-13339-2011, 2011.

Zhang, R. Y., Wang, L., Khalizov, A. F., Zhao, J., Zheng, J., McGraw, R. L., and Molina, L. T.: Formation of nanoparticles of blue haze enhanced by anthropogenic pollution, P. Natl. Acad. Sci. USA, 106, 17650–17654, https://doi.org/10.1073/pnas.0910125106, 2009.

Zhang, Y. Y., Müller, L., Winterhalter, R., Moortgat, G. K., Hoffmann, T., and Pöschl, U.: Seasonal cycle and temperature dependence of pinene oxidation products, dicarboxylic acids and nitrophenols in fine and coarse air particulate matter, Atmos. Chem. Phys., 10, 7859–7873, https://doi.org/10.5194/acp-10-7859-2010, 2010.

4

Brominated VSLS and their influence on ozone under a changing climate

Stefanie Falk[1], **Björn-Martin Sinnhuber**[1], **Gisèle Krysztofiak**[2], **Patrick Jöckel**[3], **Phoebe Graf**[3], and **Sinikka T. Lennartz**[4]

[1]Institute of Meteorology and Climate Research, Karlsruhe Institute of Technology, Karlsruhe, Germany
[2]LPC2E, Université d'Orléans, CNRS, UMR7328, Orléans, France
[3]Deutsches Zentrum für Luft- und Raumfahrt e.V., Oberpfaffenhofen, Germany
[4]Geomar, Helmholtz Centre for Ocean Research Kiel, Kiel, Germany

Correspondence to: Stefanie Falk (stefanie.falk@kit.edu)

Abstract. Very short-lived substances (VSLS) contribute as source gases significantly to the tropospheric and stratospheric bromine loading. At present, an estimated 25 % of stratospheric bromine is of oceanic origin. In this study, we investigate how climate change may impact the ocean–atmosphere flux of brominated VSLS, their atmospheric transport, and chemical transformations and evaluate how these changes will affect stratospheric ozone over the 21st century.

Under the assumption of fixed ocean water concentrations and RCP6.0 scenario, we find an increase of the ocean–atmosphere flux of brominated VSLS of about 8–10 % by the end of the 21st century compared to present day. A decrease in the tropospheric mixing ratios of VSLS and an increase in the lower stratosphere are attributed to changes in atmospheric chemistry and transport. Our model simulations reveal that this increase is counteracted by a corresponding reduction of inorganic bromine. Therefore the total amount of bromine from VSLS in the stratosphere will not be changed by an increase in upwelling. Part of the increase of VSLS in the tropical lower stratosphere results from an increase in the corresponding tropopause height. As the depletion of stratospheric ozone due to bromine depends also on the availability of chlorine, we find the impact of bromine on stratospheric ozone at the end of the 21st century reduced compared to present day. Thus, these studies highlight the different factors influencing the role of brominated VSLS in a future climate.

1 Introduction

Ozone is an important trace gas in the Earth's atmosphere. The stratospheric layer with its highest abundance, the ozone layer, absorbs harmful ultraviolet radiation threatening all life forms on the Earth's surface and acts as a potent greenhouse gas (GHG). In the troposphere, ozone is considered a harmful pollutant. Catalytic cycles involving bromine and mixed halogen reactions, namely with chlorine, efficiently deplete ozone (e.g., Sinnhuber et al., 2009). The ozone depletion efficiency of bromine is strongly related to the available amount of activated chlorine in the atmosphere (Yang et al., 2014; Sinnhuber and Meul, 2015; Oman et al., 2016). Long-lived, anthropogenically emitted, halogenated source gases (SG), e.g., CH_3Br and halons, have been restricted by the Montreal Protocol and its amendments. Their atmospheric concentrations have started to decline globally (see Global Ozone Research and Monitoring Project, 2011, Chap. 1). Still, they contribute about 75 % to the overall bromine loading in the stratosphere. The remainder is provided by organic SG of oceanic origin of which methyl bromide (CH_3Br), bromoform ($CHBr_3$), and dibromomethane (CH_2Br_2) are the most abundant. Minor brominated very short-lived substances (VSLS) include the mixed bromochlorocarbons $CHCl_2Br$, $CHClBr_2$, and CH_2ClBr. The tropospheric lifetime of these gases lies between several days to weeks. They are produced by plankton and macroalgae and are predominantly produced in coastal

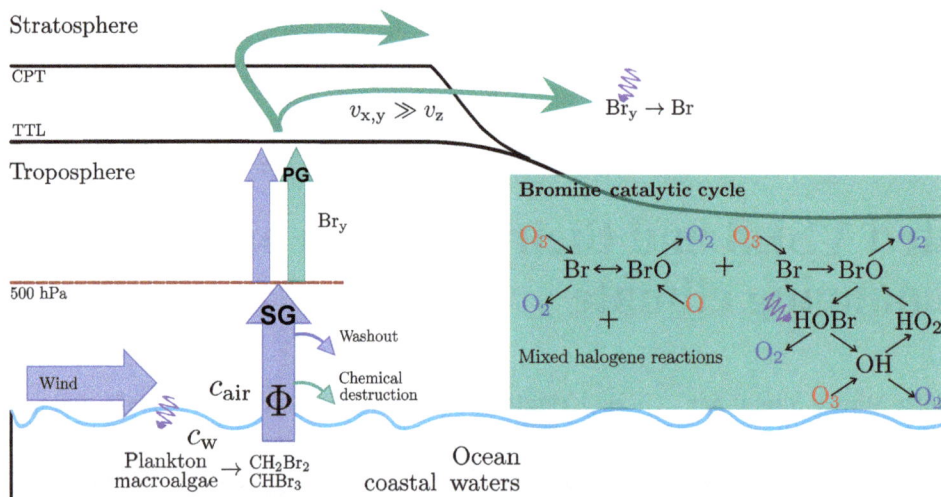

Figure 1. Scheme of VSLS emission and catalytic cycle of ozone depletion involving bromine. (left) VSLS are produced by plankton and macroalgae predominantly in coastal waters. They are emitted through gas exchange between ocean and atmosphere. These organic source gases (SG) undergo chemical transformation into inorganic product gases (PG). Both are convectively transported through the tropical tropopause layer (TTL). Through photochemical decomposition, reactive bromine Br_y is provided to the stratosphere. (right) Two examples of catalytic cycles of ozone depletion involving bromine. A + indicates increasing order of catalytic complexity. Reactants are shown in red, catalysts in black, and products in blue. Photochemical reactions are indicated by a violet wave.

waters (Moore et al., 1996; Lin and Manley, 2012; Hughes et al., 2013; Stemmler et al., 2015). Through gas exchange governed by the concentration gradient between ocean water and atmosphere, solubility, and wind stress, VSLS are emitted into the atmosphere. Transport to the stratosphere, as shown by different model studies (Aschmann et al., 2009; Hossaini et al., 2012; Liang et al., 2014), occurs in tropical regions of deep convection, most importantly the western Pacific and Maritime Continent, in Southeast Asia, and over the Gulf of Mexico. Organic SG are transported through the tropical tropopause layer (TTL) together with their inorganic product gases (PG). PG are produced through photochemical decomposition of VSLS and provide reactive bromine (Br_y, from Br, Br_2, HBr, BrO, $BrONO_2$, $BrNO_2$, BrCl, and HOBr) to the stratosphere. This is schematically shown in Fig. 1. In recent years, several approaches have been taken to describe the stratospheric or regional abundance of bromine from VSLS. Top-down scenarios (Warwick et al., 2006; Liang et al., 2010; Ordonez et al., 2012) match atmospheric observations by setting constant fluxes or boundary concentrations. Bottom-up scenarios (e.g., Ziska et al., 2013) developed emission climatologies by extrapolating measurements in the surface ocean and marine boundary layer and calculate emissions accordingly. As shown by Lennartz et al. (2015), the bottom-up fluxes based on the oceanic water concentrations of Ziska ct al. (2013) are in good agreement with available atmospheric VSLS observations. Recently, Ziska et al. (2017) have investigated the future evolution of the ocean–atmosphere fluxes of VSLS through the 21st century based on Coupled Model Intercomparison Project (CMIP) 5

model output and fixed atmospheric VSLS concentrations. They found fluxes of CH_2Br_2 and $CHBr_3$ increasing by 6.4 % (23.3 %) and 9.0 % (29.4 %), respectively, dependent on the Representative Concentration Pathways (RCP) 2.6 (RCP8.0) scenario.

In this study, we will address the open questions on how these oceanic emissions of VSLS evolve in response to a changing climate and changing atmospheric concentrations (Sect. 3), how transport and tropospheric chemistry influence the stratospheric bromine abundance in a changing climate (Sect. 4), and how stratospheric ozone will be affected by the assumed changes in VSLS abundance (Sect. 5). Details about the model and simulations used in this study will be given in Sect. 2.

2   Model and experiments

All model experiments have been performed using the ECHAM/MESSy Atmospheric Chemistry (EMAC) model (Jöckel et al., 2010). Table 1 gives an overview over the key factors of the simulations.

Future changes in fluxes of brominated VSLS from the ocean are studied with a free-running long-term simulation (SC_free, 1979–2100) using a simplified chemistry (Sect. 3), augmented by a similar simulation, but nudged towards the European Centre for Medium-Range Weather Forecasts (ECMWF) ERA-Interim reanalysis over the period 1979–2012. Therein, VSLS emission fluxes are computed online from prescribed seawater concentrations. The simplified chemistry simulations use EMAC version 2.50 with sub-

Table 1. EMAC model experiments used in this study. All experiments follow the RCP6.0 scenario of GHG emissions and have accordingly prescribed SST and SIC from HadGEM2.

Experiment	Model version	Resolution	Timespan	Chemistry	VSLS emission	Interactive aerosols
SC_nudged	2.50	T42L39MA	1979–2012	simplified bromine	airsea	no
SC_free	2.50	T42L39MA	1979–2100	simplified bromine	airsea	no
RC2-base-05	2.51	T42L47MA	1950–2100	full	Warwick et al. (2006)	no
RT1a	2.52	T42L90MA	2075–2100	full + sulfur	airsea	yes
RT1b	2.52	T42L90MA	2075–2100	full + sulfur	none	yes

models airsea (Pozzer et al., 2006) (with the water-side transfer velocity parameterization $k_w = 0.222\,u^2 + 0.333\,u$ with respect to wind speed according to Nightingale et al., 2000), cloud, cloudopt, convect (with operational ECMWF convection scheme), cvtrans, ddep (Kerkweg et al., 2006a), ptrac (Jöckel et al., 2008), rad, scav (Tost et al., 2006), surface, and tnudge (Kerkweg et al., 2006b). The setup is as in Lennartz et al. (2015); Hossaini et al. (2016). Chemical reactions are not computed interactively, e.g., via the EMAC submodel mecca (Sander et al., 2011a). The VSLS lifetime due to reaction with OH has been fixed to monthly mean values from the National Centre for Meteorological Research (CNRM) (Michou et al., 2011; Morgenstern et al., 2016) model calculations, while photolysis rates are computed within the EMAC submodel jval (Sander et al., 2014). Only in these simulations with simplified chemistry, OH concentrations have been set to zero in the lower troposphere (700–1000 hPa) to reduce the variability of ground level volume mixing ratio (VMR) of VSLS. The chemical lifetime of VSLS in the lower troposphere is therefore overestimated. Due to the longer lifetime, VSLS are more abundant in the lower troposphere leading to a flux suppression. Water concentrations of CH_2Br_2 and $CHBr_3$ have been held constant using the climatology of Ziska et al. (2013). For mixed bromochlorocarbons ($CHBrCl_2$, $CHBr_2Cl$, CH_2BrCl), water concentrations have been estimated by scaling $CHBr_3$ concentrations to obtain a better agreement between model simulation and tropical mean profile and surface observations of these VSLS. Based on the lifetime estimate, VSLS are decomposed and converted to Br_y. The partitioning of Br_y into Br, Br_2, HBr, BrO, $BrONO_2$, $BrNO_2$, BrCl, and HOBr in these simplified chemistry simulations has been computed offline from a full-chemistry EMAC simulation of 1-year duration with 6-hourly output. Scavenging is applied to Br, Br_2, HBr, $BrNO_2$, $BrONO_2$, BrCl, and HOBr. Concentrations of CO_2, CH_4, CFC, and N_2O in SC_free are taken from a CNRM CM5 model (Voldoire et al., 2013) simulation with RCP6.0 scenario (Fujino et al., 2006; Hijioka et al., 2008).

Data of a full-chemistry long-term simulation (RC2-base-05, Jöckel et al., 2016) over a timespan of 150 years (1950–2100) and performed as part of a Chemistry-Climate Model Initiative (CCMI) recommended set of simulations by the Earth System Chemistry integrated Modelling (ESCiMo) Consortium will be used for studying changes in transport and photochemical transformation of bromine species (Sect. 4). In this simulation, VSLS fluxes have been held constant following scenario five of Warwick et al. (2006).

An intermediate-term experiment, consisting of a set of two simulations and spanning the years 2075–2100, has been performed for assessing implications on ozone depletion in a future climate with significantly lower chlorine loading in the atmosphere (Sect. 5). The simulations named RT1a and RT1b both include online computation of aerosol formation. Fluxes of CH_2Br_2 and $CHBr_3$ are computed online from sea-water concentrations of Ziska et al. (2013) using the EMAC submodel airsea as in RC2-base-05 with the k_w parameterization according to Wanninkhof (1992), which is strictly quadratic with respect to wind speed ($k_w = 0.31\,u^2$). For assessing the impact of VSLS on ozone, all VSLS emissions have been switched off in RT1b. The impact of various k_w parameterizations on VSLS emission has been previously studied. The differences on the global level are < 15 % (see Table 4 in Lennartz et al., 2015, when comparing Nightingale et al., 2000, and Wanninkhof and McGillis, 1999). For wind speeds exceeding $10\,\mathrm{ms^{-1}}$ the Wanninkhof (1992) k_w parameterization diverges slightly stronger towards higher transfer velocities compared to the Nightingale et al. (2000) parameterization (cf. Fig. 1 in Wanninkhof and McGillis, 1999, and Fig. 2 in Lennartz et al., 2015). Regarding integrated global emissions of VSLS, both parameterizations result in similar fluxes given that the mean global wind speed lies in a range where these parameterizations do not differ drastically. However, the Nightingale et al. (2000) parameterization reacts more sensitively to changes in wind speed, which introduces a further uncertainty when assessing changes over time in a changing climate. The full-chemistry experiments use EMAC version 2.51 (RC2-base-05) and 2.52 (RT1a/b), respectively. The dynamics have not been specified except for a weak nudging of the equatorial wind quasi-biennial oscillation (QBO). RC2-base-05 combines hindcast with future projections. The setup of RT1a and RT1b is almost identical to RC2-base-05, and so we refer to the corresponding paper by the ESCiMo Consortium (Jöckel et al., 2016) for general information. The major difference lies in the aforementioned treatment of VSLS emission, which is handled analogous to SC_free except for mixed bromochlorocarbons emissions taken from Warwick et al. (2006). Since heterogeneous reaction and chlorine activation are important for the depletion process of ozone, tropospheric and stratospheric aerosol formation is computed online using the submodel GMXe (Pringle et al., 2010) of EMAC. The setup has been adapted from RC1-aero-07 (Jöckel et al., 2016) with modifications as described by Brühl et al. (2012, 2015). Radiation coupling had been activated in GMXe, but cloud coupling had not been activated. In this regard, an additional oceanic sulfur source, carbonyl sulfide (COS), which is a major source of stratospheric sulfur, has been included in addition to dimethyl sulfide (DMS). Whereas the emission of the latter is computed from prescribed ocean concentrations, constant fluxes of COS have been adopted from Kettle et al. (2002). Additional reaction pathways of sulfur have been enabled accordingly. RT1a and RT1b have been initialized with available monthly mean values from RC2-base-05. COS has been initialized from a simulation whose results have been published recently (Glatthor et al., 2015), including an artificially increased oceanic source to close the atmospheric budget.

The model's spatial resolution is T42L39MA for the simplified chemistry experiments, T42L47MA for RC2-base-05, and T42L90MA for RT1a/RT1b, respectively, corresponding to a $2.8° \times 2.8°$ grid, with a top level at 0.01 hPa, and 39, 47, or 90 vertical hybrid-pressure levels. The mean tropical troposphere (below 100 hPa) is discretized into 16, 26, or 27 levels, and the mean tropical stratosphere between 100 and 1 hPa consists of 15, 15, or 48 levels, respectively. Emissions of GHG follow the RCP6.0 scenario and sea surface temperature (SST) and sea ice cover (SIC) are prescribed from Hadley Centre Global Environment Model version 2 (HadGEM2) forced with the RCP6.0 scenario for all simulations accordingly.

3 Long-term trends in oceanic emission fluxes

In this section, we investigate how a changing climate may influence emission fluxes of VSLS from the ocean. We will assess the impact of changing physical factors (e.g., SST, SIC, and wind speed) on ocean–atmosphere gas exchange driven by the RCP6.0 scenario. Here we assume constant oceanic concentrations of VSLS over the course of the century (following Ziska et al., 2013; Lennartz et al., 2015). This specific assumption might not hold since the effects of climate change, e.g., increase of ocean temperature, acidification, change of salinity, and nutrient input, on marine organisms and thus the production of CH_2Br_2 and $CHBr_3$ is not yet fully understood. Recent combined marine ecosystem model studies imply a global decrease of net primary production by plankton over the course of the 21st century (Laufkötter et al., 2015, 2016). However, the impact on bromocar-

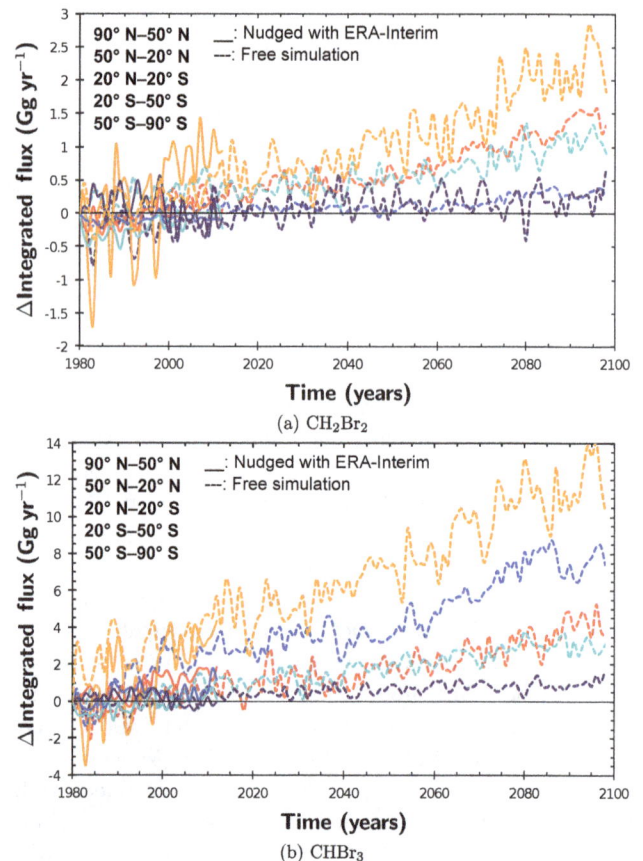

(a) CH_2Br_2

(b) $CHBr_3$

Figure 2. Difference of integrated flux separated in different zonal bands. Simplified chemistry EMAC simulation (SC_free and SC_nudged) with airsea gas exchange and water concentrations held constant. Solid lines represent ERA-Interim nudged (1979–2012) and dashed lines free-running (1979–2100).

bon concentration, predominantly produced by macroalgae in coastal regions, remains unclear.

As implemented in the EMAC submodel airsea (Pozzer et al., 2006), the flux of a gas dissolved in ocean water to the atmosphere is governed by its concentration gradient Δc and transfer velocity k:

$$\Phi = k \cdot \Delta c$$
$$= k \cdot (c_w - H \cdot c_{air}), \tag{1}$$

with $k = (1/k_w + R \cdot H \cdot T_{air}/k_{air})^{-1}$, wherein R is the universal gas constant and H is the Henry coefficient for a specific gas. The transfer velocity depends largely on air temperature T_{air} and surface wind speed, which is taken into account by distinguishing between water- and air-side transport velocities (k_w, k_{air}). k_w is a polynomial function of wind speed depending on the chosen parameterization as mentioned in Sect. 2. The corresponding water and atmospheric concentrations are named c_w and c_{air}.

In Fig. 2, the difference of VSLS fluxes with respect to the start of the simulation in 1979 is shown for the free-

Table 2. Average absolute flux for the year 2000 in $Gg\,yr^{-1}$ and percentage of relative increase in VSLS flux between 2000 and 2100 from SC_free. The numbers have been obtained by linear regression of the data shown in Fig. 2 and evaluated at the given years.

Region	CH_2Br_2		$CHBr_3$	
	($Gg\,yr^{-1}$)	(%)	($Gg\,yr^{-1}$)	(%)
90–50° N	0.6	54.6	23.5	25.0
50–20° N	8.2	14.6	41.7	8.7
20° N–0	19.2	6.6	52.4	8.6
0–20° S	5.5	11.9	44.4	10.0
20–50° S	4.3	18.0	33.2	8.2
50–90° S	6.9	6.8	12.7	8.9

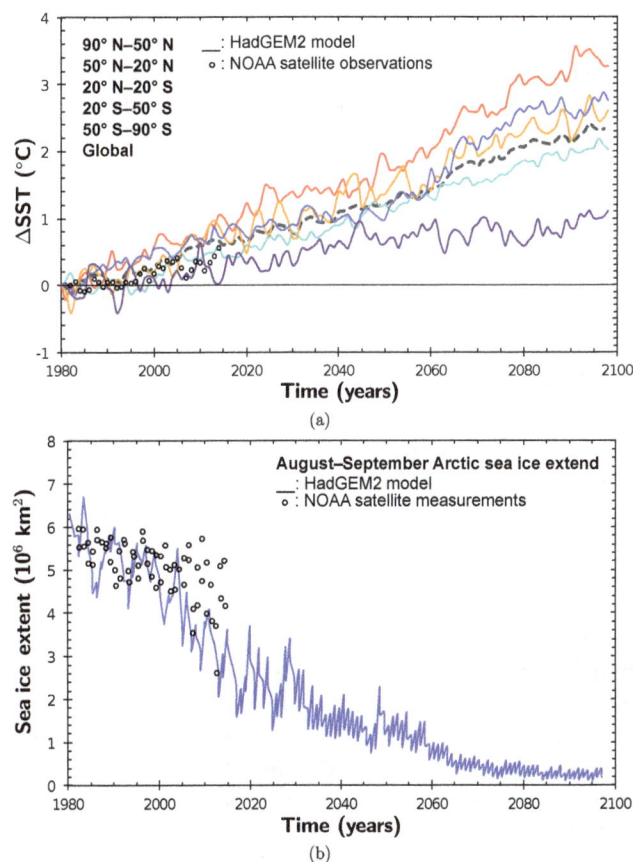

Figure 3. HadGEM2 prescribed ocean properties in the simplified chemistry simulations compared to National Oceanic and NOAA Optimum Interpolation (OI) V2 fields (Reynolds et al., 2002). **(a)** Change in sea surface temperature for different latitude bands. Global average is shown as dashed gray line. **(b)** Arctic sea ice extend in August and September.

running and nudged simplified chemistry simulation. For both CH_2Br_2 and $CHBr_3$, all zonal bands display linearly rising fluxes. The strongest increase with respect to 1979 values is found in the tropical zone (20° N–20° S) with roughly 2.5 and 13 $Gg\,yr^{-1}$ for $CHBr_3$ and CH_2Br_2 respectively. Rela-

tive to the absolute value of the zonally averaged fluxes, this yields an increase of about 10 % over the course of the century (Table 2). The increase is slightly stronger in the southern tropics. The strongest relative increase in flux is found in the northern hemispheric polar region (90–50° N), with 25 % and roughly 55 % for $CHBr_3$ and CH_2Br_2, respectively.

Regarding the changing physical factors, the HadGEM2 prescribed SSTs are increasing almost linearly over the course of the century (Fig. 3a). Under the RCP6.0 scenario, this increase in SST ranges between 1 and 3.5 °C. The weak rise in Antarctic SST is accompanied by a weakly increasing Antarctic flux of VSLS. The corresponding HadGEM2 prognosticated retreat of Arctic sea ice is shown in Fig. 3b. Sea ice is not regarded as a source of VSLS in our study and therefore only acts as a lid blocking the ocean-to-atmosphere flux. Since the water concentrations from Ziska et al. (2013) used in our simulations do not take SIC into account, water concentrations have been extrapolated for regions typically covered by ice at present. In the `airsea` submodel, if SIC (fraction of grid box) is larger than 0.5, the transfer velocity (k_w) is equal to zero, in other cases k_w is scaled depending on the fraction of SIC. Hence, a polar sea which is to a large extent free of sea ice has increased fluxes of VSLS in our future simulations. However, there are large uncertainties regarding the VSLS water concentrations in the future polar sea, for the polar ecosystem as a whole is undergoing a drastic change. In accordance to the general increase in flux, the Arctic August–September maximum of flux is expected to be more pronounced. In both hemispheres, seasonal cycles in zonally averaged VSLS fluxes peak in the summer months and show minima in late winter. There is a slightly stronger increase of fluxes in the future during the time periods of maxima but no change in phase. Negative emissions representing a net sink of atmospheric CH_2Br_2 are found during winter at high latitudes in the Northern Hemisphere. In the northern tropics, $CHBr_3$ shows a distinct maximum in northern hemispheric summer, while the southern tropics do not display any seasonal cycle. Despite increased ocean–atmosphere fluxes in the future, only taking the changes of physical factors into account, seasonal cycles remain largely the same. In our simulations, zonally averaged absolute wind speed at 10 m is only slightly changing over the course of the 21st century and with varying sign (−4–2 %). Thus, it is indicated by our simulations but not explicitly shown that the important factor regarding an increase of ocean–atmosphere flux of VSLS is the change in SSTs.

In Fig. 4, resulting VMR profiles of organic (Br_{org}) and inorganic bromine (Br_y) from VSLS are shown. VSLS data have been averaged over a time period 1990–2000 and 2090–2100. The VMR profile of Br_{org} displays a steep decline in the lower troposphere (400–1000 hPa) by more than 50 % of ground level VMR and stays almost constant before entering the stratosphere, where VSLS are quickly dissociated. Comparing present-day and future values (keeping in mind that OH concentrations are nudged towards monthly mean val-

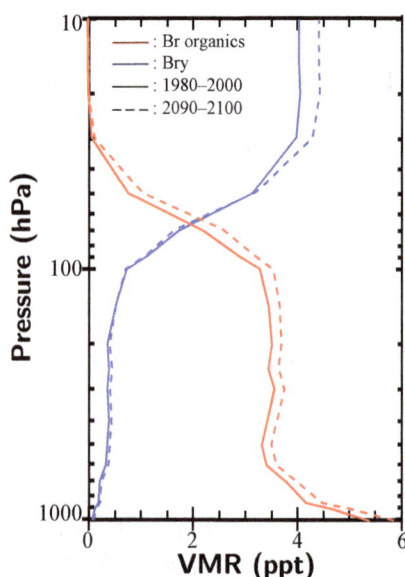

Figure 4. Tropical zonal mean (20° N–20° S), temporally averaged, vertical profiles of organic (Br_{org}) and inorganic bromine (Br_y) for the time periods 1990–2000 and 2090–2100 from SC_free. In consistency with increasing VSLS fluxes by 10 % in the tropics, an increase of roughly 10 % in Br_y from VSLS in the stratosphere is found, while the increase in Br_{org} amounts to 8 %.

ues of OH and photolysis rates are fixed in SC_free), Br_{org} is found to have increased throughout the atmosphere by about 0.1–0.4 ppt. Surface values of VSLS increase by 0.47 ppt (9 %), while in the lower stratosphere the increase amounts to 0.3 ppt (8 %). VMR of Br_y is increased from the lower stratosphere upwards by roughly 0.4 ppt (10 %). These changes in the vertical profiles can be attributed to enhanced emissions in a future climate which, as shown, are of the order of 10 % in the tropics.

As can be inferred from Eq. (1), ocean–atmosphere fluxes are sensitive to the abundance of VSLS in the atmosphere and differing wind speed parameterization. An increased chemical dissociation of VSLS in the lowermost troposphere (e.g., due to a probable future increase in OH) would reduce the atmospheric concentration and therefore increase the flux from the ocean to the atmosphere without necessarily increasing the actual amount of bromine which is transported to the stratosphere. The total amount of bromine from VSLS transported through the UTLS strongly depends on the washout of inorganic PG (Br_y^{VSLS}) and hence on the partitioning and heterogeneous reactions converting Br_y^{VSLS} between soluble, e.g., HBr and HOBr, and insoluble, e.g., BrO, species (e.g., Aschmann et al., 2009; Liang et al., 2014). Since OH concentrations in the lower troposphere have been set to zero in SC_free, the atmospheric lifetime and the resulting abundance of VSLS in the lower troposphere is enhanced. Therefore, the total ocean–atmosphere flux is suppressed. In this regard, fluxes from RT1a at the end of the

21st century have been compared to SC_free within the same time period. Much stronger fluxes (1.3–1.5 times) have been found in RT1a in comparison to SC_free. Particularly, no net sink for CH_2Br_2 occurs at high latitudes in RT1a. This partly explains the smaller increase in comparison to results recently published by Ziska et al. (2017). Ziska et al. (2017) diagnosed the flux from parameters such as SST and wind speed for a fixed VSLS concentration gradient and for different CMIP5 model simulations. They found an increase in flux of $CHBr_3/CH_2Br_2$ of 29.4/23.3 % for the RCP8.0 scenario and 9.0/6.4 % for RCP2.6, respectively. In addition to the smaller absolute fluxes due to the artificial suppression caused by setting OH to zero in the lower troposphere, we expect a smaller increase in flux from a theoretical point, taking Eq. (1) in to account, since we allow atmospheric concentrations to respond to changing flux. Underlying changes in photochemical dissociation and tracer transport due to a changing climate have not been disentangled at this point and will be studied in detail in the following section.

4   Stratospheric bromine loading

In addition to the possible increase in oceanic VSLS emissions due to climate change, discussed in the previous section, atmospheric transport and chemical transformation processes are also sensitive to climate change and may contribute to a change in the future stratospheric bromine loading from VSLS. These aspects will be studied in this section, based on the RC2-base-05 ESCiMo simulation, spanning 150 years from 1950 to 2100, assuming constant VSLS fluxes. Hence the fluxes of VSLS do not response to changes in the ground level abundances of VSLS.

In Fig. 5, profiles of brominated substances are shown for the tropics. The profiles are weighted by the amount of bromine atoms per molecule. The whole 150-year data set has been smoothed using a moving average with a box window size of 11 years to account for, e.g., seasonal variations and the solar cycle. From these smoothed data, three reference years have been chosen for the analysis: 1980, 2016, and 2100. Therefore, 2100 is referring to June of the last valid year of the smoothed data (2094). To guide the eye, the corresponding mean tropical tropopause heights from the model output are shown together with the profiles. There is an upward shift of the tropopause height of about 8 hPa between present day and future. An upward shift of VSLS VMR profiles in 2100 in comparison to past/present-day profiles is also visible. In the RCP6.0 scenario, ground level VMR of CH_3Br and VSLS are constant from 2016 onward. In case of CH_3Br, this roughly amounts to 1980 values. For all years, we find a fast decrease of VSLS of 5 ppt with a standard deviation of 0.25 ppt (or about 50 % compared to ground level VMR) between the surface and the mid-troposphere at about 500 hPa. The comparison of the difference of profiles between future and past/present (Fig. 5a, lower panel) reveals

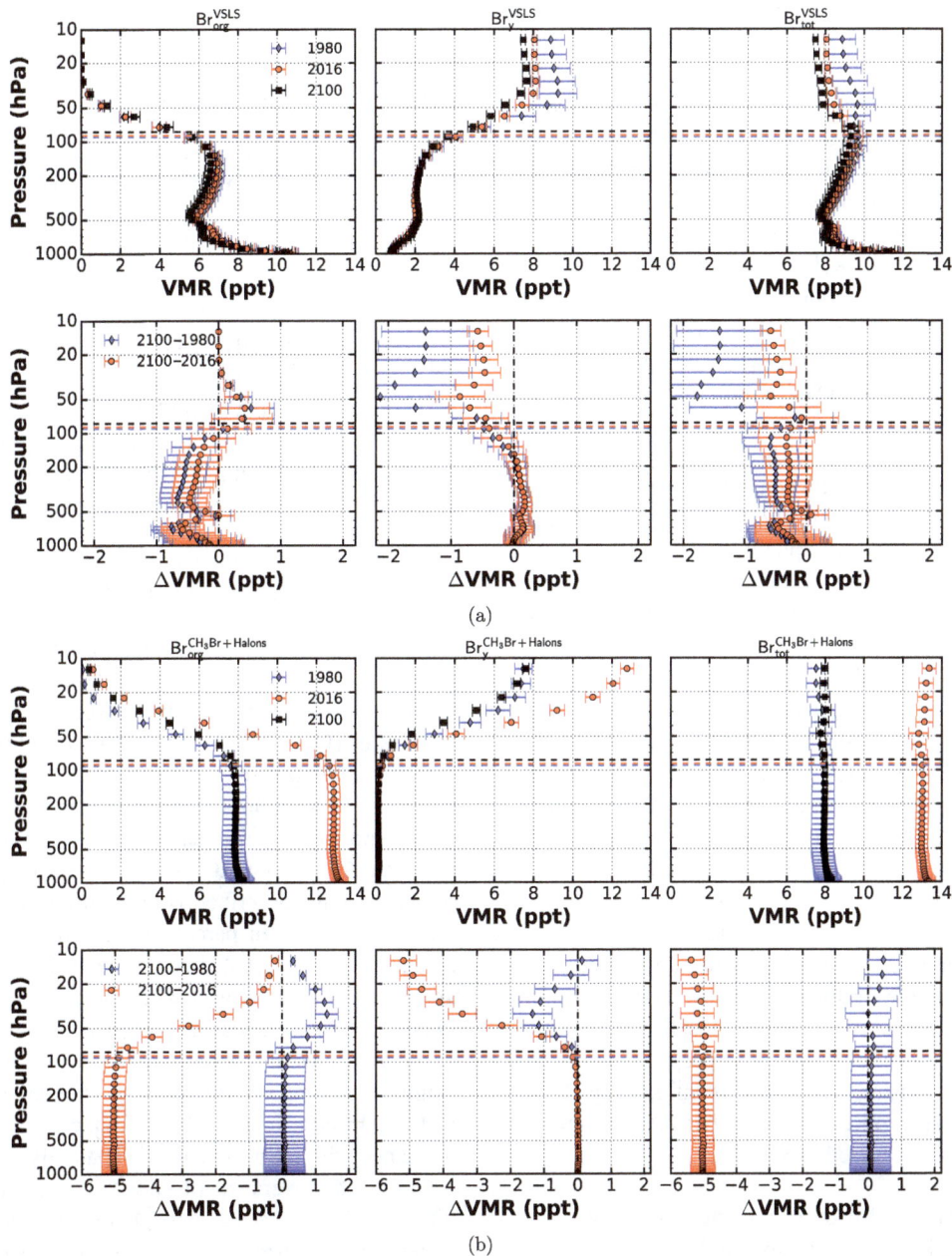

Figure 5. Vertical profiles of brominated substances divided into SG (Br_{org}), PG (Br_y), and SG + PG (Br_{tot}) in the tropics (20° N–20° S). Data from ESCiMo RC2-base-05 simulation (Jöckel et al., 2016). Absolute values of VMR in upper panel, difference ΔVMR with respect to 2100 values in lower panel. **(a)** Bromine from VSLS; **(b)** bromine from CH_3Br and halons.

decreasing bromine values from VSLS by about 0.1–0.8 ppt throughout the troposphere, while there is an increase of the same order of magnitude in the lower stratosphere. Similar results have been published for RCP4.5 and RCP8.5 scenarios, attributing these to changes in the tropospheric circulation and to the primary oxidant OH (Hossaini et al., 2012). The amount of inorganic PG from VSLS (Br_y^{VSLS}) in the UTLS is decreasing by the same order of magnitude due to the enhanced upwelling in the tropics. As air in the UTLS be-

comes younger in a future climate, less SG (Br_{org}^{VSLS}) will be dissociated into PG (Br_y^{VSLS}) compared to present day. For 2016, this decline is compatible with a decreasing amount of VSLS in the troposphere. A slight excess of Br_y^{VSLS} in the stratosphere is found for 1980 in comparison to 2016 and 2100. This excess and the strong variability (denoted by the shown standard deviation) can be attributed to the hindcast period (1950–2005) of the simulation including volcanic eruptions. Large volcanic eruptions can influence the transport of bromine from VSLS into the stratosphere which may be related to a similar effect as seen in stratospheric water vapor (Löffler et al., 2016). Since volcanic activity has not been included in the future scenario, there is no such impact on Br_y^{VSLS} from 2005 onward.

The largest change between present and future stems from the estimated decrease of long-lived SG, in particular halons and CH_3Br. At present, halons contribute about 6–7 ppt to the total bromine loading of the lower stratosphere (~ 23 ppt), which is about the same amount as VSLS and CH_3Br in RC2-base-05, whereas by the end of the century their contribution is reduced significantly to 1–2 ppt of total bromine (~ 17 ppt). This decline in long-lived, anthropogenically emitted SG is altering the amount of bromine released in the stratosphere on longer timescales. VSLS are already reduced due to photochemical dissociation when entering the TTL, while halons are dissociating more slowly, providing a long lasting source of bromine to the stratosphere (Fig. 5b, lower panel). It is important to note that although there is an increase of Br_{org}^{VSLS} of 0.5 ppt in the stratosphere assuming constant ocean–atmosphere fluxes, the overall amount of bromine in the stratosphere due to VSLS (Br_{tot}^{VSLS}) might be decreasing in the future. This depends on whether PG (Br_y^{VSLS}) are transported alongside the VSLS into the UTLS or removed through washout in the troposphere. The model representations of underlying processes, e.g., conversion between soluble and insoluble inorganic bromine species through heterogeneous chemical reactions, are still uncertain.

In the following, we will derive a semi-analytic model to separate various aspects affecting the future VSLS distribution in the atmosphere (Sect. 4.1). Since the atmospheric window for air entering the stratosphere is located in the tropics, we will focus on averaged tropical atmospheric quantities. Subsequently, the transition between troposphere and stratosphere caused by a rising tropopause is influencing the interpretation of VMR profile differences between present and future. This will be discussed in Sect. 4.2.

4.1 Quantification of future atmospheric changes affecting VSLS mixing ratio profiles

The increase of VSLS in the stratosphere in the future can be attributed to changes in chemical and photolytical dissociation rates and alternating transport from source regions through the TTL caused by a speed-up of the Brewer–Dobson circulation (BDC) (Hossaini et al., 2012). All of these factors influence the lifetime of VSLS in the atmosphere. A volume of air in a certain height (or rather pressure coordinate) shall have an associated mean temperature T, OH concentration [OH], photolysis frequency J, and age of air (AOA). In the model, VSLS are dissociated photochemically via

$$CH_2Br_2 + OH \rightarrow 2Br + H_2O, \tag{R1}$$
$$CHBr_3 + OH \rightarrow 3Br + H_2O, \tag{R2}$$

and

$$CH_2Br_2 + h\nu \rightarrow 2Br + products, \tag{R3}$$
$$CHBr_3 + h\nu \rightarrow 3Br + products. \tag{R4}$$

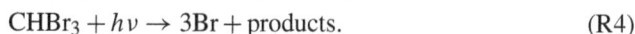

The simplification in these equations compared to reality is justified since the intermediate reaction products CBr_2O and $CHBrO$ insignificantly amount to the total bromine PG (Hossaini et al., 2010). From the resulting first-order differential equation

$$\frac{d[A]}{dt} = -k_A(T) \cdot [OH] \cdot [A] - J_A \cdot [A], \tag{2}$$

with [A] any of the VSLS species concentration, $k_A(T)$ the temperature dependent rate coefficient, J_A the photolysis frequency (Sander et al., 2011b), and assuming [OH] is unchanged by the reaction, a simple solution of Eq. (2) is derived:

$$[A](t) = \exp\left(-(k_A(T) \cdot [OH] + J_A) \cdot (t - t_0)\right) \cdot [A]_0. \tag{3}$$

Based on the above equation, the influence of [OH], temperature, transport, and photolysis rate can be studied. For inferring the change in chemical dissociation, 10-year average profiles of [OH] and 1-year average profiles of temperature have been computed from RC2-base-05 data for the present day (2016) and future (2100). The idea is to assess the effect of transport timescales ($t_0 \rightarrow t$) by using 10-year averages of mean AOA from RC2-base-05 data (neglecting the age spectrum in the described volume of air). AOA shall refer to the time since an air parcel has been in contact with ground level. It has been evaluated in the EMAC simulation from an artificial, passive tracer with linearly increasing emission. Photolysis frequencies have been computed from averaged tropical profiles of temperature, humidity, and ozone column using the column version of `jval` (Sander et al., 2014) from EMAC. In case of photolysis frequencies, temperature dependence will not be discussed separately.

In Fig. 6a, averaged profiles of temperature, AOA, and [OH] are shown. Mean tropospheric temperatures are higher, while stratospheric temperatures are lower in the future, which is in line with other studies (e.g., IPCC – Intergovernmental Panel on Climate Change, 2013, Chap. 12, Global Ozone Research and Monitoring Project, 2014, Chap. 2). The concentration of OH is increased throughout the atmosphere, apart from the lowermost levels. AOA will become notably younger within the stratosphere by the end of the century, as shown in various other studies (Austin et al., 2007, 2013; Butchart et al., 2006; Li et al., 2008; Muthers et al., 2016) and become slightly older (by a few days) in the troposphere. Vertical profiles of the $CHBr_3$ lifetime for present and future are shown in Fig. 6b. Because of an increase in photolysis rates due to increasing temperatures, the $CHBr_3$ lifetime is decreasing. CH_2Br_2 is not shown, since its lifetime with respect to photolysis is almost infinite in the troposphere and thus determined by reaction with OH.

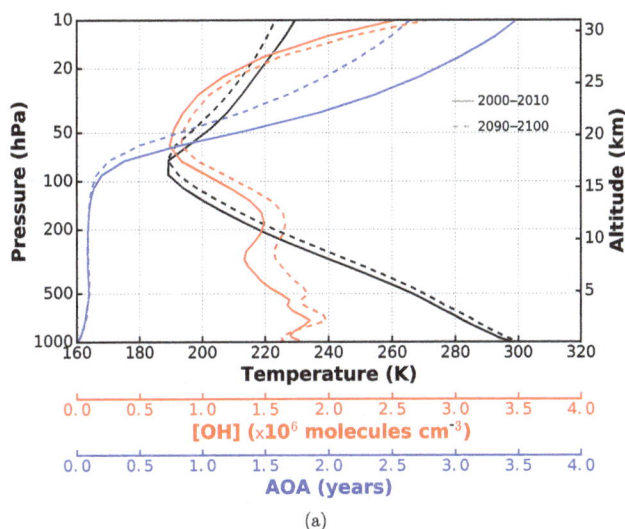

(a)

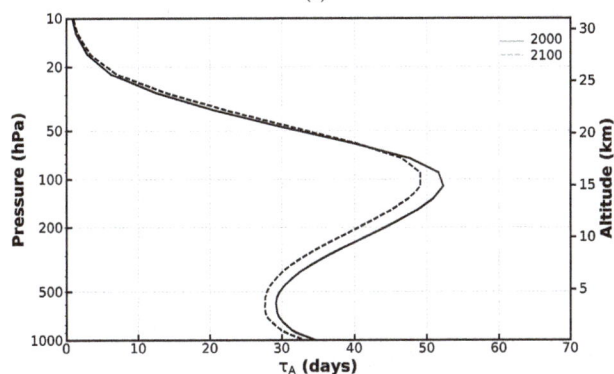

(b)

Figure 6. Tropical average profiles from ESCiMo RC2-base-05 simulation of changing variables in Eq. (3) for present day (solid lines) and future (dashed lines). **(a)** Temperature, OH concentration, and age of air; **(b)** lifetime of $CHBr_3$.

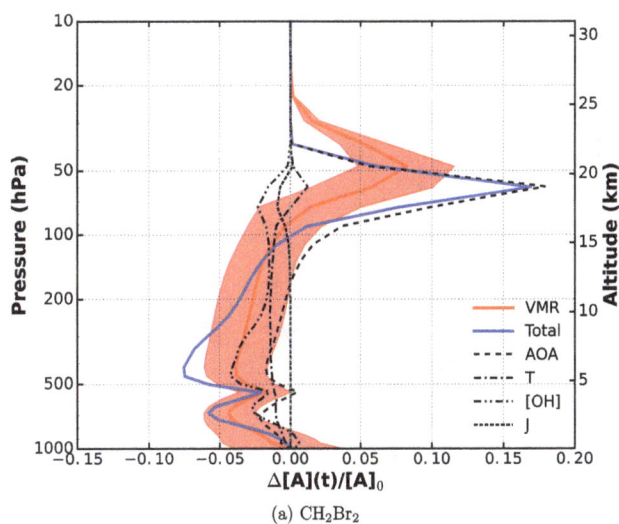

(a) CH_2Br_2

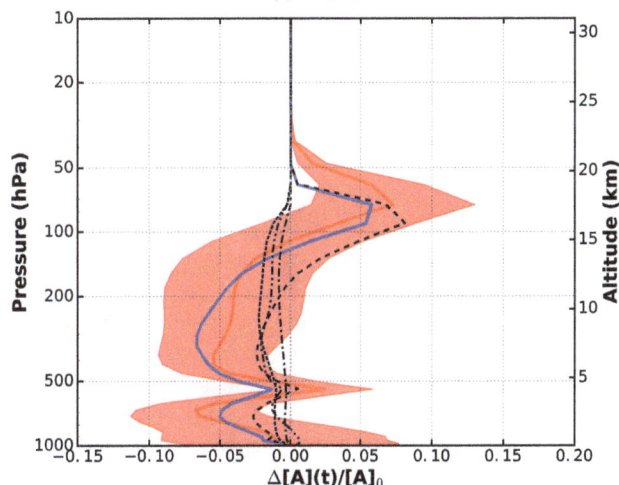

(b) $CHBr_3$

Figure 7. Relative difference of VSLS vertical profiles for 2000 and 2100. Major influences on lifetime have been separated. Shown are resulting profiles by varying the denoted variables, mean temperature T, OH concentration [OH], photolysis frequency J, and age of air (AOA), in Eq. (3) one by one.

By varying the variables in Eq. (3) one by one, the impact of each on the resulting profile difference ($\Delta[A](t)/[A]_0$) has been calculated (Fig. 7). The increase of [OH] in the RCP6.0 scenario results in a general decreasing VMR of VSLS in the troposphere and lower stratosphere. The influence is highest at 500 hPa and around the tropopause. CH_2Br_2 is affected more strongly by a change of [OH] due to chemical destruction ($\sim 5\%$) than $CHBr_3$ ($\sim 2\%$). The change in mean temperatures causes a tropospheric VMR decrease by at most 2%. In the stratosphere, decreasing temperatures increase $\Delta[A](t)/[A]_0$ by about 1%. An increased AOA in the troposphere is reflected by a decreasing $\Delta[A](t)/[A]_0$ ($\sim 2\%$), while the opposite is true for the juvenescence of air in the stratosphere (8–18%). For CH_2Br_2, the impact of AOA is apparently overestimated in the lower stratosphere by this ansatz, which might be because of the neglected AOA spectrum representing a mixing of different air masses. Since the photolytical lifetime of CH_2Br_2 in the troposphere is infinite, it has no influence on the tropospheric part of the profile.

A weak decrease on the order of 1–2% is apparent in the lower stratosphere. This has been found to be mainly driven by temperature sensitivity of photolysis. In case of $CHBr_3$, a 1–2% decrease due to changing photolysis is found in the free and upper troposphere. This change in photolysis rate is mainly due to changes in tropical ozone abundance.

If all occurring changes are included, the actual profile differences between future and present are rather well reproduced (shown in red). These VMR profiles are 10-year averages for the tropics. The corresponding standard deviation is plotted as shaded error band. The decreasing VMR of VSLS in the troposphere is on the order of 5–7% (at about 250 hPa) for CH_2Br_2 and $CHBr_3$, respectively. In the stratosphere, the

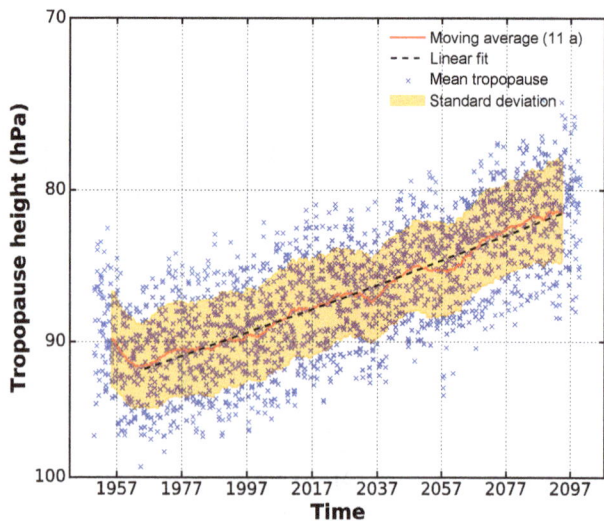

Figure 8. Model mean tropical tropopause from RC2-base-05 over a timespan of 150 years. Tropopause data have been evaluated after the apparent spinup of about 10 years. A rise of the mean tropical tropopause of (0.81 ± 0.01) hPa decade^{-1} is found by linear regression.

maximum increase, dependent on species, occurs at differing pressure levels and amounts to 7–8 %.

In summary, all occurring future changes are decreasing VMR of VSLS in the troposphere. In case of CHBr$_3$, all factors are of the same order of magnitude. The tropospheric decrease of CH$_2$Br$_2$ VMR is mainly driven by increasing [OH]. In the upper troposphere–lower stratosphere (UTLS), the impact of the juvenescence of AOA dominates, causing an increase in VMR.

4.2 Implications of a rising tropopause on VSLS mixing ratio profiles

The GHG induced warming of the troposphere and cooling of the stratosphere causes a rise of the tropopause. Model mean tropical tropopause heights from RC2-base-05 have been smoothed using a moving average with a box window size of 11 years. The corresponding standard deviation is displayed as yellow band in Fig. 8. A linear regression fit on the smoothed model mean tropical tropopause height yields a rise of (0.81 ± 0.01) hPa decade^{-1}. This is in accordance with results from ECMWF reanalysis data for the past 2 decades (Wilcox et al., 2012). As indicated by Oberländer-Hayn et al. (2016) regarding the BDC, the upward shift of the tropopause affects the interpretation of vertical profile differences between future and past. An air parcel which would have already entered the stratosphere under present-day conditions may be still considered tropospheric in the future. As pointed out earlier, profiles appear shifted by a fraction of distance between two pressure coordinate levels. We perform a spline fit to the averaged profiles and shift them ac-

cordingly with respect to the mean tropopause. The fit results have been evaluated within a valid region of ± 100 hPa around the tropopause. The results are shown in Fig. 9. Uncertainty bands have been estimated by adding or subtracting 1 standard deviation from the averaged VMR profiles and computing the corresponding splines. With respect to the mean tropopause, VMR differences show no increase of bromine from VSLS in the lower stratosphere but rather a slight decrease (Fig. 9a). A small increase of inorganic bromine from VSLS (Br$_y^{VSLS}$) is found in the tropopause region. At about 20 hPa, Br$_y^{VSLS}$ is reduced by 1–2 ppt in the future compared to 1980. Overall, a reduction of bromine in the UTLS is found at the end of the 21st century. In Fig. 9b, the amount of bromine from CH$_3$Br and halons is shown. Except for a slight increase of Br$_{tot}^{CH_3Br+halons}$ in the upper stratosphere between 1980 and 2100 of about 0.7 ppt, there is no increase of bromine from long-lived SG.

To summarize, the increase of lower-stratospheric VSLS in RC2-base-05 of about 5–10 % is due to enhanced vertical transport in the tropics. This increase is, however, counteracted by a corresponding decrease in inorganic bromine. Everything else unchanged, an increase in tropical upwelling will therefore not change the total amount of bromine in the future stratosphere. Additionally, due to enhanced future OH concentrations in RCP6.0, the tropospheric lifetime of VSLS is reduced which leads to a decrease of total bromine from VSLS. As mentioned in Sect. 3, whether the amount of inorganic PG in the UTLS is decreasing or not, strongly depends on the partitioning of Br$_y$ and conversion of soluble HBr and HOBr into insoluble BrO through heterogeneous recycling, e.g., occurring on sea-salt aerosols or ice crystals. In case insoluble species are favored, vertical transport would enhance the amount of PG in the UTLS. Otherwise, wet removal in the troposphere would decrease the amount of PG. This mechanism has not been explicitly tested in our model simulations. Taken an upward shift of the tropopause into consideration and shifting the VMR profiles accordingly with respect to the mean tropical tropopause height, a decrease of Br$_{tot}^{VSLS}$ by 0.5–2 ppt is found for a fixed $\Delta_{TP}P \approx 20$ hPa.

5 Implications on ozone depletion

In this section, the influence of brominated very short-lived source gases on ozone depletion will be discussed. Based on RT1a and RT1b, we assess the impact of VSLS on a zonally averaged ozone distribution at the end of the 21st century. For a thorough discussion of future trends, the 25-year data set is too short. However, from our long-term simulations (SC_free, SC_nudged, RC2-base-05), long-term influence of emission perturbations on ozone is not assessable. As described in Sect. 2, SC_free does not include interactive ozone chemistry, whereas RC2-base-05 incorporates prescribed fluxes based on scenario five by Warwick et al. (2006). Furthermore, results from RC2-base-05 cannot be

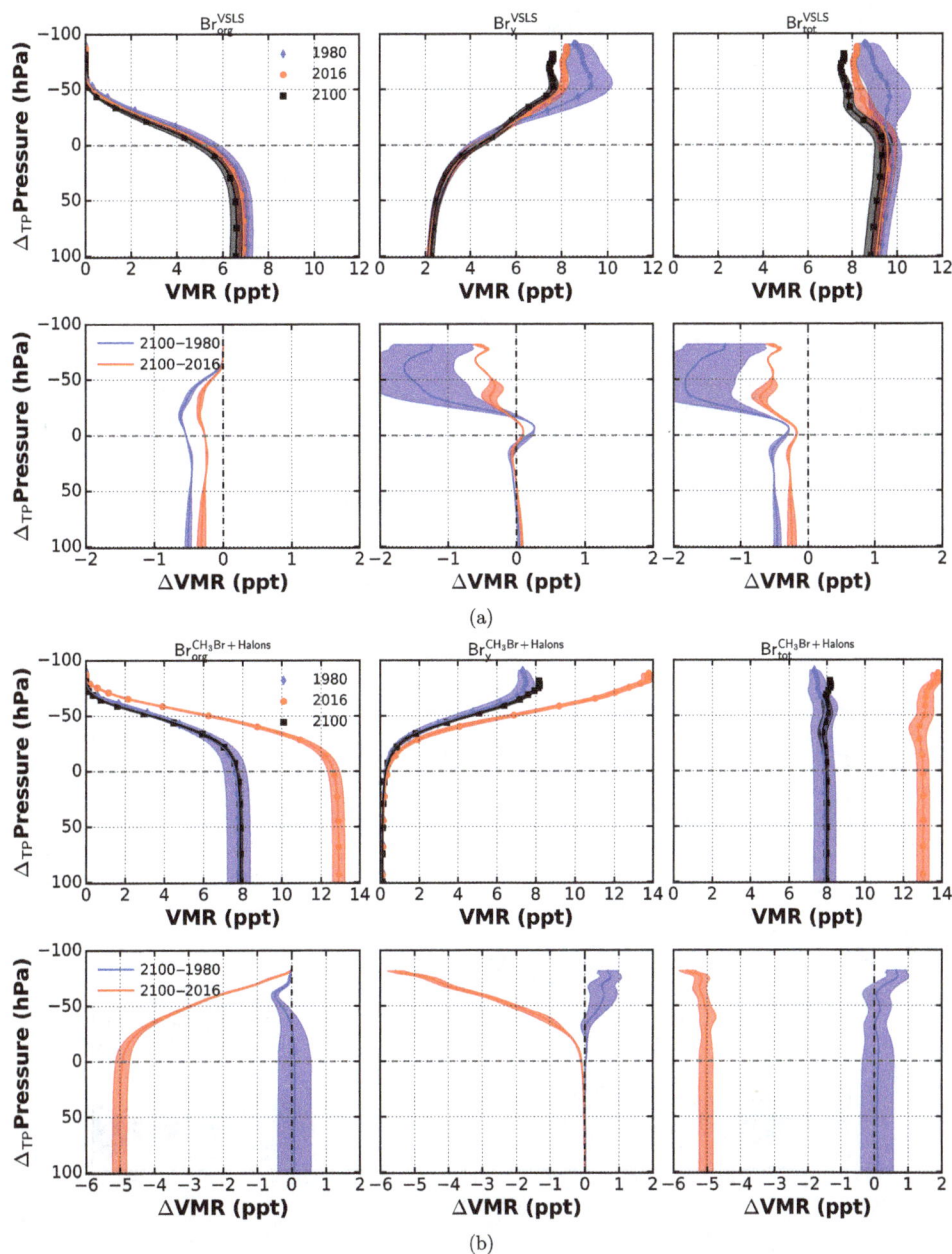

Figure 9. Spline fitted vertical profiles of brominated substances divided into SG (Br_{org}), PG (Br_y), and SG + PG (Br_{tot}) in the tropics (20° N–20° S) with respect to the mean tropical tropopause. Data from ESCiMo RC2-base-05 simulation (Jöckel et al., 2016). Absolute values of VMR are in the upper panel and the difference ΔVMR with respect to 2100 values in the lower panel. **(a)** Bromine from VSLS; **(b)** bromine from CH_3Br and halons.

compared to RT1a and RT1b directly because of significant differences in ozone distribution and amount between the differing vertical resolutions (L47MA and L90MA) of the model. This issue has been already reported by Jöckel et al. (2016).

Zonally averaged data of total column ozone have been smoothed using a moving average algorithm with box window size of 11 years (Fig. 10). In general, ozone trends at the end of the century are roughly the same for both resolutions.

The actual amount of ozone differs, with L90 showing more ozone except for the northern hemispheric polar region and midlatitudes. In the Arctic, RT1a and RT1b even indicate, in contrast to RC2-base-05, slightly decreasing total column ozone. In the case of RT1a and RT1b, this might be partially caused by interactive aerosol and accordingly added oceanic COS source. This bias in total column ozone between the vertical resolutions is larger than the difference between RT1a and RT1b.

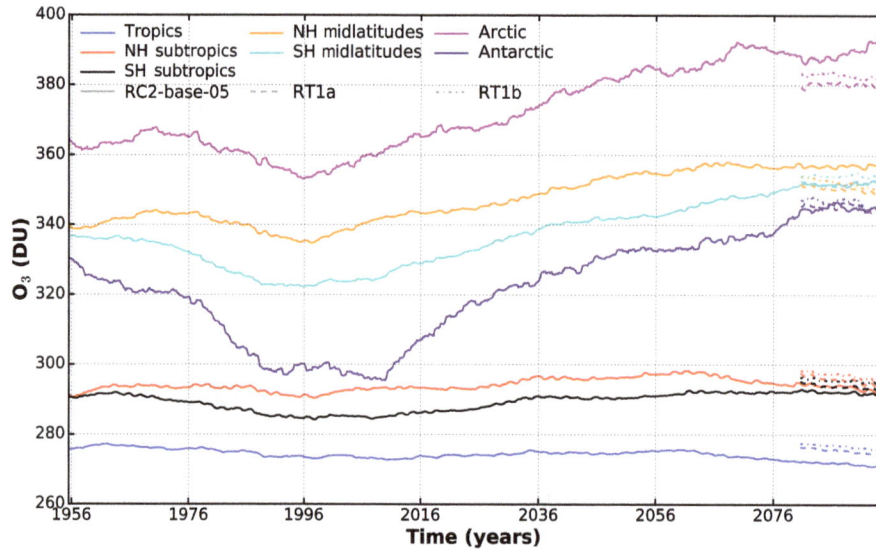

Figure 10. Zonal mean ozone trend from RC2-base-05 (1950–2100) and RT1a/RT1b (2080–2100). Smoothed with moving average box window 11 years.

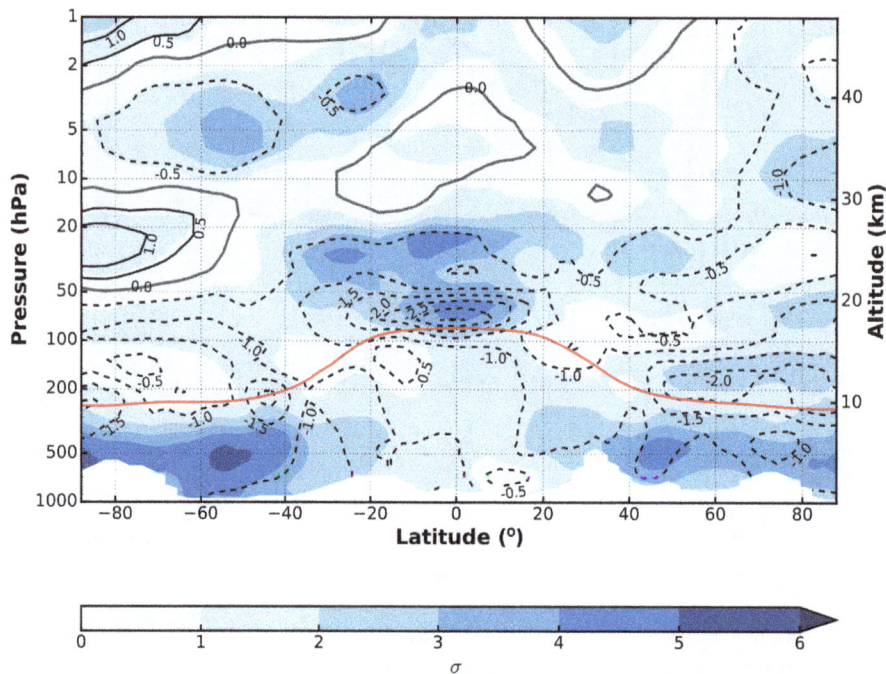

Figure 11. Ozone reduction due to VSLS. Medium-term full-chemistry future simulation with and without VSLS. Zonal mean profile of decadal mean difference in percent ($(RT1a - RT1b)/RT1a \times 100$). Dashed lines indicate a decrease of ozone in the simulation with VSLS turned on (RT1a) compared to the one with VSLS turned off (RT1b). The average tropopause is shown as red line. Significance is indicated by blue shaded areas. The significance is estimated as manifold of difference from zero in units of standard error of mean difference.

For estimating the effect of brominated VSLS on ozone depletion, the difference in zonally averaged ozone of RT1a and RT1b has been computed. A period of 20 years (2080–2100) has been used accounting for an estimated model spinup of 5 years in the beginning. In Fig. 11, the relative difference ($(RT1a - RT1b)/RT1a \times 100$) is shown as contour plot. Dashed lines indicate a decrease of ozone in the simulation with VSLS turned on (RT1a) compared to the one with VSLS turned off (RT1b), while solid lines indicate an increase. Significance has been estimated as divergence from zero in units of standard error of mean. It is indicated by shades of blue. VSLS cause a tropospheric ozone reduction on the order of 1–2 %, mainly at high latitudes. The UTLS region in the tropics is most affected; there VSLS cause a decrease of ozone of about 3 %. The decrease of ozone in the high latitude troposphere and tropical UTLS is rendered significant. Increasing amounts of ozone ($\sim$ 1 %) are found in the Antarctic middle and upper stratosphere, but these are mainly not significant. This increase in ozone abundance may be due to dynamical feedback (Braesicke et al., 2013).

While VSLS have a large impact on Antarctic ozone depletion during the ozone hole period, i.e., during times with high-stratospheric chlorine loading from about the late 1970s to the second half of the 21st century (e.g., Fernandez et al., 2017; Oman et al., 2016; Sinnhuber and Meul, 2015; Yang et al., 2014), we find that by the end of the 21st century VSLS have less impact on total Antarctic stratospheric ozone depletion under low chlorine loading, although their importance relative to the total stratospheric halogen load is increasing (about 40 % in accordance to Fernandez et al., 2017). Assuming an adherence to the Montreal Protocol, stratospheric volume mixing ratios of Cl_y will decrease exponentially in the course of the 21st century from its peak values in 2000. From the Global Ozone Research and Monitoring Project (2014, Chap. 2, Figs. 2–21), a decline of Cl_y loading at 1 hPa of about 2 ppbv between 2000 and 2080 can be deduced. In accordance, we find a reduction of zonally averaged stratospheric Cl_y at 1 hPa of 2.1 ppb by the end of the 21st century compared to year 2000 in RC2-base-05. Since RT1a and RT1b have identical chlorine load and do not include present day, we cannot assess the chlorine moderation effect from these two simulations. Sinnhuber and Meul (2015) have shown a reduction of ozone due to VSLS in the TTL region on the order of 6 % during the period 1970–1982 with significantly less stratospheric chlorine compared to the later period 1983–2005 (7 %), while Hossaini et al. (Supplement 2015, Fig. S3) have shown a change of total ozone column due to VSLS (on/off scenario) of the order of 1–3 % in preindustrial conditions. Given that the VSLS emission scenario five of Warwick et al. (2006) used in RC2-base-05 has about twice the amount of VSLS compared to RT1a and chlorine abundance in the stratosphere will drop to 1970 values by the end of the 21st century (see IPCC – Intergovernmental Panel on Climate Change, 2013, Chap. 12), our results are in good agreement with these previous studies. In more detail, Yang et al. (2014) investigated the combined influence of brominated VSLS and chlorine on ozone. For two stratospheric Cl_y loadings (3, 0.8 ppb), which correspond to 2000 and 2100, they have varied the amount of VSLS. Yang et al. (2014) have shown the more chlorine in the stratosphere the stronger ozone is affected by an increase of bromine VMR from VSLS. In concert with our results, although by doubling the initial amount of VSLS on a varying bromine background from anthropogenic sources, they have found a significant decrease of ozone in the tropical UTLS and polar region on the order of 2–4 % and slight, insignificant increases in the Antarctic mid-stratosphere (Yang et al., 2014, Fig. 1e).

6 Conclusions

We have investigated long-term changes in emission and transport of brominated VSLS under a changing climate (RCP6.0). Under the implicit assumption of constant concentrations of VSLS in the ocean waters, over a timespan of 120 years, we have found an enhancement of zonally averaged fluxes of CH_2Br_2 and $CHBr_3$ on the order of 10 % between present day and the end of the 21st century. A strong increase of flux (up to 55 % in CH_2Br_2 and 25 % in $CHBr_3$) has been found in the northern hemispheric polar region. There, the retreat of sea ice is playing a key role. Exposing almost the entire polar ocean in August–September by the end of the 21st century, sea ice does not longer act as a lid to ocean–atmosphere fluxes of VSLS. Sea ice itself has not been considered a source of VSLS in our simulations. Subsequently, an increase of organic bromine in the UTLS is found of the same order of magnitude (8–10 %).

Ocean–atmosphere fluxes are sensitive to the abundance of VSLS in the atmosphere as well as on wind speed. An increased dissociation of VSLS in the lowermost troposphere, e.g., due to increasing OH concentrations in the RCP6.0 scenario, reduces the atmospheric concentration and therefore increases the flux from the ocean to the atmosphere without necessarily increasing the actual amount which is transported to the stratosphere. The total amount of bromine from VSLS transported through the UTLS strongly depends on the washout of inorganic PG (Br_y^{VSLS}) and hence on the partitioning and heterogeneous reactions converting Br_y^{VSLS} between soluble (e.g., HBr, HOBr) and insoluble (e.g., BrO) species. But these mechanisms have not been subject to our study.

For prescribed, constant VSLS fluxes, an increase of lower-stratospheric VSLS of about 5–10 % is due to enhanced vertical transport in the tropics. This increase is counteracted by a corresponding decrease in inorganic bromine. Everything else unchanged, an increase in tropical upwelling will not change the total amount of bromine in the future stratosphere. Additionally, due to enhanced future OH concentrations in the RCP6.0 emission scenario, the tropospheric lifetime of VSLS is reduced, which leads to a decrease of total bromine from VSLS. Furthermore, we have diagnosed a decrease of Br_{tot}^{VSLS} by 0.5–2 ppt for a fixed pressure level with respect to the mean tropical tropopause $\Delta_{TP}P \approx 20$ hPa, if the upward shift of the mean tropical tropopause of 0.81 hPa decade^{-1} is taken into consideration.

The impact of enhanced fluxes of brominated VSLS on future ozone abundance has been evaluated by comparing two experiments of which one has no VSLS emission and the other interactively computed fluxes from constant ocean concentrations of VSLS. We have found a significant reduction of ozone in the tropical UTLS of about 3 %. In the troposphere the largest significant decrease of ozone amounts to 1–2 %. Thus, bromine from VSLS may not act as a major source to future stratospheric ozone depletion. While interactive emissions from constant ocean concentrations have been taken into consideration, the actual climate change inflicted change in the production of VSLS by macroalgae in the ocean remains an open question. Whether the found increase of ocean–atmosphere fluxes of VSLS and a future decrease of VSLS in the troposphere will cancel out or overcompensate would need further simulation studies.

Code availability. The Modular Earth Submodel System (MESSy) is continuously further developed and applied by a consortium of institutions. The usage of MESSy and access to the source code is licensed to all affiliates of institutions, which are members of the MESSy Consortium. Institutions can become a member of the MESSy Consortium by signing the MESSy Memorandum of Understanding. More information can be found on the MESSy Consortium website (http://www.messy-interface.org).

Author contributions. SF performed most of the analyses and wrote the paper. BMS conceived this study and provided advice through discussion of the analysis and results. GK developed and performed the simplified chemistry simulations as well as part of the corresponding data analysis in Sect. 3. STL provided advise on the ocean–atmosphere gas exchange. PJ provided advice as project leader of the ESCiMo consortial project and coordinator of overall EMAC model development; preparation of the ESCiMo model setups and realization of the ESCiMo simulations of ESCiMo Consortium, with VSLS boundary conditions and implementation of the online Br budget diagnostics for EMAC prepared by PG. All coauthors contributed to the discussion of the results.

Competing interests. The authors declare that they have no conflict of interest.

Special issue statement. This article is part of the special issue "The Modular Earth Submodel System (MESSy) (ACP/GMD interjournal SI)". It is not associated with a conference.

Acknowledgements. Parts of this work were supported by the Deutsche Forschungsgemeinschaft (DFG) through the research unit SHARP (SI1044/1-2), the German Bundesministerium für Bildung und Forschung (BMBF) through the project ROMIC-THREAT (01GL1217B), the European Union through the Horizon 2020 project GAIA-CLIM, and by the Helmholtz Association through its research program ATMO.

The CNRM data were produced in the framework of the CCMI project, with support of Météo-France. We particularly acknowledge the support of M. Michou and D. Saint-Martin and of the entire team in charge of the CNRM/CERFACS climate model.

NOAA Optimum Interpolation (OI) V2 fields were provided by the National Centers for Environmental Prediction/National Weather Service/NOAA/US Department of Commerce, and National Climatic Data Center/NESDIS/NOAA/US Department of

Commerce research Data Archive at the National Center for Atmospheric Research, Computational and Information Systems Laboratory. http://rda.ucar.edu/datasets/ds277.0/. Accessed 6 January 2016).

The ESCiMo (Earth System Chemistry integrated Modelling) model simulations have been performed at the German Climate Computing Centre (DKRZ) through support from the BMBF. DKRZ and its scientific steering committee are gratefully acknowledged for providing the HPC and data archiving resources for this consortial project.

EMAC simulations RT1a and b have been performed at Steinbuch Center for Computing at KIT. Thanks to Stefan Versick and Oliver Kirner (KIT SimLab Climate and Environment) for their technical support.

Special thanks to C. Brühl (MPI-Mainz) for his help in implementing additional sulfur reactions and usage of gmxe in context of the ROMIC-THREAT simulations.

S. Lennartz likes to thank B. Quack, C. Marandino, S. Tegtmeier (all Helmholtz-Centre for Ocean Research Kiel), and K. Krüger (University of Oslo) for their support.

Edited by: Michel Van Roozendael

References

Aschmann, J., Sinnhuber, B.-M., Atlas, E. L., and Schauffler, S. M.: Modeling the transport of very short-lived substances into the tropical upper troposphere and lower stratosphere, Atmos. Chem. Phys., 9, 9237–9247, https://doi.org/10.5194/acp-9-9237-2009, 2009.

Austin, J., Wilson, J., Li, F., and Vomel, H.: Evolution of water vapor concentrations and stratospheric age of air in coupled chemistry-climate model simulations, J. Atmos. Sci., 64, 905–921, https://doi.org/10.1175/JAS3866.1, 2007.

Austin, J., Horowitz, L. W., Schwarzkopf, M. D., Wilson, R. J., and Levy, H.: Stratospheric ozone and temperature simulated from the preindustrial era to the present day, J. Climate, 26, 3528–3543, https://doi.org/10.1175/JCLI-D-12-00162.1, 2013.

Braesicke, P., Keeble, J., Yang, X., Stiller, G., Kellmann, S., Abraham, N. L., Archibald, A., Telford, P., and Pyle, J. A.: Circulation anomalies in the Southern Hemisphere and ozone changes, Atmos. Chem. Phys., 13, 10677–10688, https://doi.org/10.5194/acp-13-10677-2013, 2013.

Brühl, C., Lelieveld, J., Crutzen, P. J., and Tost, H.: The role of carbonyl sulphide as a source of stratospheric sulphate aerosol and its impact on climate, Atmos. Chem. Phys., 12, 1239–1253, https://doi.org/10.5194/acp-12-1239-2012, 2012.

Brühl, C., Lelieveld, J., Tost, H., Höpfner, M., and Glatthor, N.: Stratospheric sulfur and its implications for radiative forcing simulated by the chemistry climate model EMAC, J. Geophys. Res.-Atmos., 120, 2103–2118, https://doi.org/10.1002/2014JD022430, 2015.

Butchart, N., Scaife, A. A., Bourqui, M., de Grandpre, J., Hare, S. H. E., Kettleborough, J., Langematz, U., Manzini, E., Sassi, F., Shibata, K., Shindell, D., and Sigmond, M.: Simulations of anthropogenic change in the strength of the Brewer-Dobson circulation, Clim. Dynam., 27, 727–741, https://doi.org/10.1007/s00382-006-0162-4, 2006.

Fernandez, R. P., Kinnison, D. E., Lamarque, J.-F., Tilmes, S., and Saiz-Lopez, A.: Impact of biogenic very short-lived bromine on the Antarctic ozone hole during the 21st century, Atmos. Chem. Phys., 17, 1673–1688, https://doi.org/10.5194/acp-17-1673-2017, 2017.

Fujino, J., Nair, R., Kainuma, M., Masui, T., and Matsuoka, Y.: Multi-gas mitigation analysis on stabilization scenarios using aim global model, Energ. J., 343–353, 2006.

Glatthor, N., Höpfner, M., Baker, I. T., Berry, J., Campbell, J. E., Kawa, S. R., Krysztofiak, G., Leyser, A., Sinnhuber, B.-M., Stiller, G. P., Stinecipher, J., and von Clarmann, T.: Tropical sources and sinks of carbonyl sulfide observed from space, Geophys. Res. Lett., 2015GL066293, https://doi.org/10.1002/2015GL066293, 2015.

Global Ozone Research and Monitoring Project: Scientific Assessment of Ozone Depletion: 2010, 2011.

Global Ozone Research and Monitoring Project: Scientific Assessment of Ozone Depletion: 2014, 2014.

Hijioka, Y., Matsuoka, Y., Nishimoto, H., Masui, M., and Kainuma, M.: Global GHG emissions scenarios under GHG concentration stabilization targets, J. Glob. Env. Eng., 13, 97–108, 2008.

Hossaini, R., Chipperfield, M. P., Monge-Sanz, B. M., Richards, N. A. D., Atlas, E., and Blake, D. R.: Bromoform and dibromomethane in the tropics: a 3-D model study of chemistry and transport, Atmos. Chem. Phys., 10, 719–735, https://doi.org/10.5194/acp-10-719-2010, 2010.

Hossaini, R., Chipperfield, M. P., Dhomse, S., Ordonez, C., Saiz-Lopez, A., Abraham, N. L., Archibald, A., Braesicke, P., Telford, P., Warwick, N., Yang, X., and Pyle, J.: Modelling future changes to the stratospheric source gas injection of biogenic bromocarbons, Geophys. Res. Lett., 39, https://doi.org/10.1029/2012GL053401, 2012.

Hossaini, R., Chipperfield, M. P., Montzka, S. A., Rap, A., Dhomse, S., and Feng, W.: Efficiency of short-lived halogens at influencing climate through depletion of stratospheric ozone, Nat. Geosci., 8, 186–190, https://doi.org/10.1038/NGEO2363, 2015.

Hossaini, R., Patra, P. K., Leeson, A. A., Krysztofiak, G., Abraham, N. L., Andrews, S. J., Archibald, A. T., Aschmann, J., Atlas, E. L., Belikov, D. A., Bönisch, H., Carpenter, L. J., Dhomse, S., Dorf, M., Engel, A., Feng, W., Fuhlbrügge, S., Griffiths, P. T., Harris, N. R. P., Hommel, R., Keber, T., Krüger, K., Lennartz, S. T., Maksyutov, S., Mantle, H., Mills, G. P., Miller, B., Montzka, S. A., Moore, F., Navarro, M. A., Oram, D. E., Pfeilsticker, K., Pyle, J. A., Quack, B., Robinson, A. D., Saikawa, E., Saiz-Lopez, A., Sala, S., Sinnhuber, B.-M., Taguchi, S., Tegtmeier, S., Lidster, R. T., Wilson, C., and Ziska, F.: A multi-model intercomparison of halogenated very short-lived substances (TransCom-VSLS): linking oceanic emissions and tropospheric transport for a reconciled estimate of the stratospheric source gas injection of bromine, Atmos. Chem. Phys., 16, 9163-9187, https://doi.org/10.5194/acp-16-9163-2016, 2016.

Hughes, C., Johnson, M., Utting, R., Turner, S., Malin, G., Clarke, A., and Liss, P. S.: Microbial control of bromocarbon concentrations in coastal waters of the western Antarctic Peninsula, Mar. Chem., 151, 35–46,

https://doi.org/10.1016/j.marchem.2013.01.007, 2013.

IPCC – Intergovernmental Panel on Climate Change: Climate Change 2013: The Physical Science Basis, 2013.

Jöckel, P., Kerkweg, A., Buchholz-Dietsch, J., Tost, H., Sander, R., and Pozzer, A.: Technical Note: Coupling of chemical processes with the Modular Earth Submodel System (MESSy) submodel TRACER, Atmos. Chem. Phys., 8, 1677–1687, https://doi.org/10.5194/acp-8-1677-2008, 2008.

Jöckel, P., Kerkweg, A., Pozzer, A., Sander, R., Tost, H., Riede, H., Baumgaertner, A., Gromov, S., and Kern, B.: Development cycle 2 of the Modular Earth Submodel System (MESSy2), Geosci. Model Dev., 3, 717–752, https://doi.org/10.5194/gmd-3-717-2010, 2010.

Jöckel, P., Tost, H., Pozzer, A., Kunze, M., Kirner, O., Brenninkmeijer, C. A. M., Brinkop, S., Cai, D. S., Dyroff, C., Eckstein, J., Frank, F., Garny, H., Gottschaldt, K.-D., Graf, P., Grewe, V., Kerkweg, A., Kern, B., Matthes, S., Mertens, M., Meul, S., Neumaier, M., Nützel, M., Oberländer-Hayn, S., Ruhnke, R., Runde, T., Sander, R., Scharffe, D., and Zahn, A.: Earth System Chemistry integrated Modelling (ESCiMo) with the Modular Earth Submodel System (MESSy) version_2.51, Geosci. Model Dev., 9, 1153–1200, https://doi.org/10.5194/gmd-9-1153-2016, 2016.

Kerkweg, A., Buchholz, J., Ganzeveld, L., Pozzer, A., Tost, H., and Jöckel, P.: Technical Note: An implementation of the dry removal processes DRY DEPosition and SEDImentation in the Modular Earth Submodel System (MESSy), Atmos. Chem. Phys., 6, 4617–4632, https://doi.org/10.5194/acp-6-4617-2006, 2006a.

Kerkweg, A., Sander, R., Tost, H., and Jöckel, P.: Technical note: Implementation of prescribed (OFFLEM), calculated (ONLEM), and pseudo-emissions (TNUDGE) of chemical species in the Modular Earth Submodel System (MESSy), Atmos. Chem. Phys., 6, 3603–3609, https://doi.org/10.5194/acp-6-3603-2006, 2006b.

Kettle, A. J., Kuhn, U., von Hobe, M., Kesselmeier, J., and Andreae, M. O.: Global budget of atmospheric carbonyl sulfide: Temporal and spatial variations of the dominant sources and sinks, J. Geophys. Res.-Atmos., 107, 1–16, https://doi.org/10.1029/2002JD002187, 2002.

Laufkötter, C., Vogt, M., Gruber, N., Aita-Noguchi, M., Aumont, O., Bopp, L., Buitenhuis, E., Doney, S. C., Dunne, J., Hashioka, T., Hauck, J., Hirata, T., John, J., Le Quéré, C., Lima, I. D., Nakano, H., Seferian, R., Totterdell, I., Vichi, M., and Völker, C.: Drivers and uncertainties of future global marine primary production in marine ecosystem models, Biogeosciences, 12, 6955–6984, https://doi.org/10.5194/bg-12-6955-2015, 2015.

Laufkötter, C., Vogt, M., Gruber, N., Aumont, O., Bopp, L., Doney, S. C., Dunne, J. P., Hauck, J., John, J. G., Lima, I. D., Seferian, R., and Völker, C.: Projected decreases in future marine export production: the role of the carbon flux through the upper ocean ecosystem, Biogeosciences, 13, 4023–4047, https://doi.org/10.5194/bg-13-4023-2016, 2016.

Lennartz, S. T., Krysztofiak, G., Marandino, C. A., Sinnhuber, B.-M., Tegtmeier, S., Ziska, F., Hossaini, R., Krüger, K., Montzka, S. A., Atlas, E., Oram, D. E., Keber, T., Bönisch, H., and Quack, B.: Modelling marine emissions and atmospheric distributions of halocarbons and dimethyl sulfide: the influence of prescribed water concentration vs. prescribed emissions, Atmos. Chem. Phys., 15, 11753–11772, https://doi.org/10.5194/acp-15-11753-2015, 2015.

Li, F., Austin, J., and Wilson, J.: The strength of the Brewer-Dobson circulation in a changing climate: coupled chemistry-climate model simulations, J. Climate, 21, 40–57, https://doi.org/10.1175/2007JCLI1663.1, 2008.

Liang, Q., Stolarski, R. S., Kawa, S. R., Nielsen, J. E., Douglass, A. R., Rodriguez, J. M., Blake, D. R., Atlas, E. L., and Ott, L. E.: Finding the missing stratospheric Br$_y$: a global modeling study of CHBr$_3$ and CH$_2$Br$_2$, Atmos. Chem. Phys., 10, 2269–2286, https://doi.org/10.5194/acp-10-2269-2010, 2010.

Liang, Q., Atlas, E., Blake, D., Dorf, M., Pfeilsticker, K., and Schauffler, S.: Convective transport of very short lived bromocarbons to the stratosphere, Atmos. Chem. Phys., 14, 5781–5792, https://doi.org/10.5194/acp-14-5781-2014, 2014.

Lin, C. Y. and Manley, S. L.: Bromoform production from seawater treated with bromoperoxidase, Limnol. Oceanogr., 57, 1857–1866, https://doi.org/10.4319/lo.2012.57.6.1857, 2012.

Löffler, M., Brinkop, S., and Jöckel, P.: Impact of major volcanic eruptions on stratospheric water vapour, Atmos. Chem. Phys., 16, 6547–6562, https://doi.org/10.5194/acp-16-6547-2016, 2016.

Michou, M., Saint-Martin, D., Teyssèdre, H., Alias, A., Karcher, F., Olivié, D., Voldoire, A., Josse, B., Peuch, V.-H., Clark, H., Lee, J. N., and Chéroux, F.: A new version of the CNRM Chemistry-Climate Model, CNRM-CCM: description and improvements from the CCMVal-2 simulations, Geosci. Model Dev., 4, 873–900, https://doi.org/10.5194/gmd-4-873-2011, 2011.

Moore, R. M., Webb, M., Tokarczyk, R., and Wever, R.: Bromoperoxidase and iodoperoxidase enzymes and production of halogenated methanes in marine diatom cultures, J. Geophys. Res.-Oceans, 101, 20899–20908, https://doi.org/10.1029/96JC01248, 1996.

Morgenstern, O., Hegglin, M. I., Rozanov, E., O'Connor, F. M., Abraham, N. L., Akiyoshi, H., Archibald, A. T., Bekki, S., Butchart, N., Chipperfield, M. P., Deushi, M., Dhomse, S. S., Garcia, R. R., Hardiman, S. C., Horowitz, L. W., Jöckel, P., Josse, B., Kinnison, D., Lin, M., Mancini, E., Manyin, M. E., Marchand, M., Marécal, V., Michou, M., Oman, L. D., Pitari, G., Plummer, D. A., Revell, L. E., Saint-Martin, D., Schofield, R., Stenke, A., Stone, K., Sudo, K., Tanaka, T. Y., Tilmes, S., Yamashita, Y., Yoshida, K., and Zeng, G.: Review of the global models used within phase 1 of the ChemistryClimate Model Initiative (CCMI), Geosci. Model Dev., 10, 639-671, https://doi.org/10.5194/gmd-10-639-2017, 2017.

Muthers, S., Kuchar, A., Stenke, A., Schmitt, J., Anet, J. G., Raible, C. C., and Stocker, T. F.: Stratospheric age of air variations between 1600 and 2100, Geophys. Res. Lett., 43, 5409–5418, https://doi.org/10.1002/2016GL068734, 2016.

Nightingale, P. D., Liss, P. S., and Schlosser, P.: Measurements of air-sea gas transfer during an open ocean algal bloom, Geophys. Res. Lett., 27, 2117–2120, https://doi.org/10.1029/2000GL011541, 2000.

Oberländer-Hayn, S., Gerber, E. P., Abalichin, J., Akiyoshi, H., Kerschbaumer, A., Kubin, A., Kunze, M., Langematz, U., Meul, S., Michou, M., Morgenstern, O., and Oman, L. D.: Is the Brewer-Dobson circulation increasing or moving upward?, Geophys. Res. Lett., 43, 1772–1779, https://doi.org/10.1002/2015GL067545, 2016.

Oman, L. D., Douglass, A. R., Salawitch, R. J., Canty, T. P., Ziemke, J. R., and Manyin, M.: The effect of represent-ing bromine from VSLS on the simulation and evolution of Antarctic ozone, Geophys. Res. Lett., 43, 9869–9876, https://doi.org/10.1002/2016GL070471, 2016.

Ordóñez, C., Lamarque, J.-F., Tilmes, S., Kinnison, D. E., Atlas, E. L., Blake, D. R., Sousa Santos, G., Brasseur, G., and Saiz-Lopez, A.: Bromine and iodine chemistry in a global chemistry-climate model: description and evaluation of very short-lived oceanic sources, Atmos. Chem. Phys., 12, 1423–1447, https://doi.org/10.5194/acp-12-1423-2012, 2012.

Pozzer, A., Jöckel, P., Sander, R., Williams, J., Ganzeveld, L., and Lelieveld, J.: Technical Note: The MESSy-submodel AIRSEA calculating the air-sea exchange of chemical species, Atmos. Chem. Phys., 6, 5435–5444, https://doi.org/10.5194/acp-6-5435-2006, 2006.

Pringle, K. J., Tost, H., Message, S., Steil, B., Giannadaki, D., Nenes, A., Fountoukis, C., Stier, P., Vignati, E., and Lelieveld, J.: Description and evaluation of GMXe: a new aerosol submodel for global simulations (v1), Geosci. Model Dev., 3, 391–412, https://doi.org/10.5194/gmd-3-391-2010, 2010.

Reynolds, R. W., Rayner, N. A., Smith, T. M., Stokes, D. C., and Wang, W. Q.: An improved in situ and satellite SST analysis for climate, J. Climate, 15, 1609–1625, https://doi.org/10.1175/1520-0442(2002)015<1609:AIISAS>2.0.CO;2, 2002.

Sander, R., Baumgaertner, A., Gromov, S., Harder, H., Jöckel, P., Kerkweg, A., Kubistin, D., Regelin, E., Riede, H., Sandu, A., Taraborrelli, D., Tost, H., and Xie, Z.-Q.: The atmospheric chemistry box model CAABA/MECCA-3.0, Geosci. Model Dev., 4, 373–380, https://doi.org/10.5194/gmd-4-373-2011, 2011a.

Sander, S. P., Burkholder, J. B., Abbatt, J. P. D., Barker, J. R., Huie, R. E., Kolb, C. E., Kurylo, M. J., Orkin, V. L., Wilmouth, D. M., and Wine, P. H.: Chemical Kinetics and Photochemical Data for Use in Atmospheric Studies, Tech. Rep. 17, National Aeronautics and Space Administration, Jet Propulsion Laboratory, 2011b.

Sander, R., Jöckel, P., Kirner, O., Kunert, A. T., Landgraf, J., and Pozzer, A.: The photolysis module JVAL-14, compatible with the MESSy standard, and the JVal PreProcessor (JVPP), Geosci. Model Dev., 7, 2653–2662, https://doi.org/10.5194/gmd-7-2653-2014, 2014.

Sinnhuber, B.-M. and Meul, S.: Simulating the impact of emissions of brominated very short lived substances on past stratospheric ozone trends, Geophys. Res. Lett., 42, 2449–2456, https://doi.org/10.1002/2014GL062975, 2015.

Sinnhuber, B.-M., Sheode, N., Sinnhuber, M., Chipperfield, M. P., and Feng, W.: The contribution of anthropogenic bromine emissions to past stratospheric ozone trends: a modelling study, Atmos. Chem. Phys., 9, 2863–2871, https://doi.org/10.5194/acp-9-2863-2009, 2009.

Stemmler, I., Hense, I., and Quack, B.: Marine sources of bromoform in the global open ocean – global patterns and emissions, Biogeosciences, 12, 1967–1981, https://doi.org/10.5194/bg-12-1967-2015, 2015.

Tost, H., Jöckel, P., Kerkweg, A., Sander, R., and Lelieveld, J.: Technical note: A new comprehensive SCAVenging submodel for global atmospheric chemistry modelling, Atmos. Chem. Phys., 6, 565–574, https://doi.org/10.5194/acp-6-565-2006, 2006.

Voldoire, A., Sanchez-Gomez, E., Salas y Melia, D., Decharme, B., Cassou, C., Senesi, S., Valcke, S., Beau, I., Alias, A., Chevallier, M., Deque, M., Deshayes, J., Douville, H., Fernandez, E., Madec, G., Maisonnave, E., Moine, M.-P., Planton, S., Saint-Martin, D., Szopa, S., Tyteca, S., Alkama, R., Belamari, S., Braun, A., Coquart, L., and Chauvin, F.: The CNRM-CM5.1 global climate model: description and basic evaluation, Clim. Dynam., 40, 2091–2121, https://doi.org/10.1007/s00382-011-1259-y, 2013.

Wanninkhof, R.: Relationship between wind speed and gas exchange over the ocean, J. Geophys. Res.-Oceans, 97, 7373–7382, https://doi.org/10.1029/92JC00188, 1992.

Wanninkhof, R. and McGillis, W. R.: A cubic relationship between air-sea CO_2 exchange and wind speed, Geophys. Res. Lett., 26, 1889–1892, https://doi.org/10.1029/1999GL900363, 1999.

Warwick, N. J., Pyle, J. A., Carver, G. D., Yang, X., Savage, N. H., O'Connor, F. M., and Cox, R. A.: Global modeling of biogenic bromocarbons, J. Geophys. Res.-Atmos., 111, https://doi.org/10.1029/2006JD007264, 2006.

Wilcox, L. J., Hoskins, B. J., and Shine, K. P.: A global blended tropopause based on ERA data. Part II: Trends and tropical broadening, Q. J. Roy. Meteor. Soc., 138, 576–584, https://doi.org/10.1002/qj.910, 2012.

Yang, X., Abraham, N. L., Archibald, A. T., Braesicke, P., Keeble, J., Telford, P. J., Warwick, N. J., and Pyle, J. A.: How sensitive is the recovery of stratospheric ozone to changes in concentrations of very short-lived bromocarbons?, Atmos. Chem. Phys., 14, 10431–10438, https://doi.org/10.5194/acp-14-10431-2014, 2014.

Ziska, F., Quack, B., Abrahamsson, K., Archer, S. D., Atlas, E., Bell, T., Butler, J. H., Carpenter, L. J., Jones, C. E., Harris, N. R. P., Hepach, H., Heumann, K. G., Hughes, C., Kuss, J., Krüger, K., Liss, P., Moore, R. M., Orlikowska, A., Raimund, S., Reeves, C. E., Reifenhäuser, W., Robinson, A. D., Schall, C., Tanhua, T., Tegtmeier, S., Turner, S., Wang, L., Wallace, D., Williams, J., Yamamoto, H., Yvon-Lewis, S., and Yokouchi, Y.: Global sea-to-air flux climatology for bromoform, dibromomethane and methyl iodide, Atmos. Chem. Phys., 13, 8915–8934, https://doi.org/10.5194/acp-13-8915-2013, 2013.

Ziska, F., Quack, B., Tegtmeier, S., Stemmler, I., and Krüger, K.: Future emissions of marine halogenated very-short lived substances under climate change, J. Atmos. Chem., 74, 245–260, https://doi.org/10.1007/s10874-016-9355-3, 2017.

5

Light-induced protein nitration and degradation with HONO emission

Hannah Meusel[1], **Yasin Elshorbany**[2,8], **Uwe Kuhn**[1], **Thorsten Bartels-Rausch**[3], **Kathrin Reinmuth-Selzle**[1], **Christopher J. Kampf**[4], **Guo Li**[1], **Xiaoxiang Wang**[1], **Jos Lelieveld**[5], **Ulrich Pöschl**[1], **Thorsten Hoffmann**[6], **Hang Su**[7,1], **Markus Ammann**[3], and **Yafang Cheng**[1,7]

[1]Max Planck Institute for Chemistry, Multiphase Chemistry Department, Mainz, Germany
[2]NASA Goddard Space Flight Center, Greenbelt, Maryland, USA
[3]Paul Scherrer Institute, Villigen, Switzerland
[4]Johannes Gutenberg University of Mainz, Institute for Organic Chemistry, Mainz, Germany
[5]Max Planck Institute for Chemistry, Atmospheric Chemistry Department, Mainz, Germany
[6]Johannes Gutenberg University of Mainz, Institute for Inorganic and Analytical Chemistry, Mainz, Germany
[7]Institute for Environmental and Climate Research, Jinan University, Guangzhou, China
[8]Earth System Science Interdisciplinary Center, University of Maryland, College Park, Maryland, USA

Correspondence to: Y. Cheng (yafang.cheng@mpic.de) and H. Su (h.su@mpic.de)

Abstract. Proteins can be nitrated by air pollutants (NO_2), enhancing their allergenic potential. This work provides insight into protein nitration and subsequent decomposition in the presence of solar radiation. We also investigated light-induced formation of nitrous acid (HONO) from protein surfaces that were nitrated either online with instantaneous gas-phase exposure to NO_2 or offline by an efficient nitration agent (tetranitromethane, TNM). Bovine serum albumin (BSA) and ovalbumin (OVA) were used as model substances for proteins. Nitration degrees of about 1 % were derived applying NO_2 concentrations of 100 ppb under VIS/UV illuminated conditions, while simultaneous decomposition of (nitrated) proteins was also found during long-term (20 h) irradiation exposure. Measurements of gas exchange on TNM-nitrated proteins revealed that HONO can be formed and released even without contribution of instantaneous heterogeneous NO_2 conversion. NO_2 exposure was found to increase HONO emissions substantially. In particular, a strong dependence of HONO emissions on light intensity, relative humidity, NO_2 concentrations and the applied coating thickness was found. The 20 h long-term studies revealed sustained HONO formation, even when concentrations of the intact (nitrated) proteins were too low to be detected after the gas exchange measurements. A reaction mechanism for the NO_2 conversion based on the Langmuir–Hinshelwood kinetics is proposed.

1 Introduction

Primary biological aerosols, or bioaerosols, including proteins, from different sources and with distinct properties are known to influence atmospheric cloud microphysics and public health (Lang-Yona et al., 2016; D'Amato et al., 2007; Pummer et al., 2015). Bioaerosols represent a diverse subset of atmospheric particulate matter that is directly emitted in form of active or dead organisms, or fragments, like bacteria, fungal spores, pollens, viruses and plant debris. Proteins are found ubiquitously in the atmosphere as part of these airborne, typically coarse-sized biological particles (diameter > 2.5 µm), as well as in fine particulate matter (diameter < 2.5 µm) associated with a host of different constituents such as polymers derived from biomaterials and proteins dissolved in hydrometeors, mixed with fine dust and other particles (Miguel et al., 1999; Riediker et al., 2000; Zhang and Anastasio, 2003). Proteins contribute up to 5 % of par-

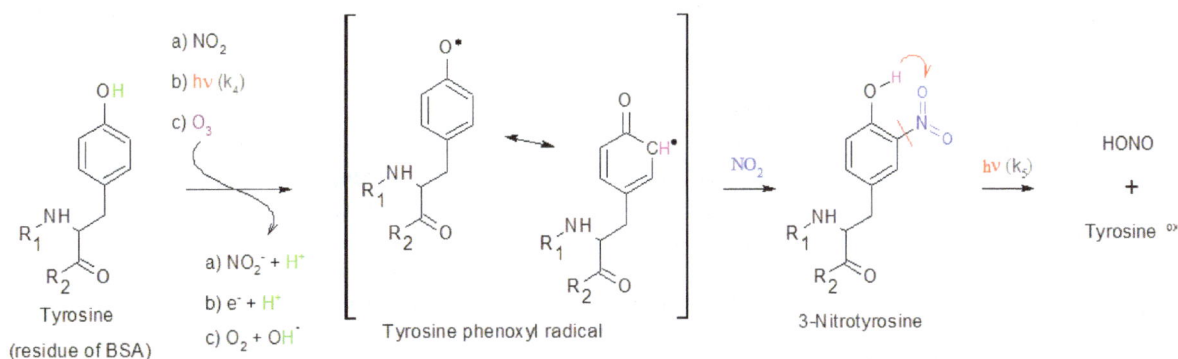

Figure 1. Overview on possible reaction mechanisms of atmospheric BSA nitration and subsequent HONO emission. The tyrosine phenoxyl radical intermediate is formed by the reaction of tyrosine with either **(a)** NO_2, **(b)** light or **(c)** ozone. A second reaction with NO_2 forms 3-nitrotyrosine (adapted from Houée-Levin et al., 2015, and Shiraiwa et al., 2012). Subsequent intramolecular H transfer initiated by irradiation decompose the protein and HONO is emitted (adapted from Bejan et al., 2006).

ticle mass in airborne particles (Franze et al., 2003a; Staton et al., 2015; Menetrez et al., 2007) and are also found at surfaces of soils and plants. Proteins can be nitrated and are then likely to enhance allergic responses (Gruijthuijsen et al., 2006). Nitrogen dioxide ($^{\bullet}NO_2$) has emerged as an important biological reactant and has been shown to be capable of electron (or H atom) abstraction from the amino acid tyrosine (Tyr) to form TyrO$^{\bullet}$ in aqueous solutions (tyrosine phenoxyl radical, also called tyrosyl radical; Prütz et al., 1984, 1985; Alfassi, 1987; Houée-Lévin et al., 2015), which subsequently can be nitrated by a second NO_2 molecule. Shiraiwa et al. (2012) observed nitration of protein aerosol, but not solely with NO_2 in the gas phase, and demonstrated that simultaneous O_3 exposure of airborne proteins in dark conditions can significantly enhance NO_2 uptake and consequent protein nitration (3-nitrotyrosine formation) by way of direct O_3 mediated formation of the TyrO$^{\bullet}$ intermediate. A connection between increased allergic diseases and elevated environmental pollution, especially traffic-related air pollution has been proposed (Ring et al., 2001). Tyrosine is one of the photosensitive amino acids and it is subject of direct and indirect photo-degradation under solar-simulated conditions (Boreen et al., 2008), especially mediated by both UV-B (λ 280–320 nm) and UV-A (λ 320–400 nm) radiation (Houee-Levin et al., 2015; Bensasson et al., 1993). Direct light absorption or absorption by adjacent endogenous or exogenous chromophores and subsequent energy transfer results in an electronically excited state of tyrosine (for details see Houée-Lévin et al., 2015, and references therein). If the triplet state of tyrosine is generated, it can undergo electron transfer reactions and deprotonation to yield TyrO$^{\bullet}$ (Fig. 1; Bensasson, 1993; Davies, 1991; Berto et al., 2016). Regardless of how the tyrosyl radical is generated, it can be nitrated by reaction with NO_2, as well as hydroxylated or dimerized (Shiraiwa et al., 2012; Reinmuth-Selzle et al., 2014; Kampf et al., 2015).

With respect to atmospheric chemistry, Bejan et al. (2006) have shown that photolysis of ortho-nitrophenols (as is the case for 3-nitrotyrosine) can generate nitrous acid (HONO). HONO is of great interest for atmospheric composition, as its photolysis forms OH radicals, which are the key oxidant for degradation of most air pollutants in the troposphere (Levy, 1971). In the lower atmosphere, up to 30 % of the primary OH radical production can be attributed to photolysis of HONO, especially during the early morning when other photochemical OH sources are still small (Reaction R1, Kleffmann et al., 2005; Alicke et al., 2002; Ren et al., 2006; Su et al., 2008; Meusel et al., 2016).

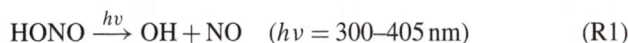

$$HONO \xrightarrow{h\nu} OH + NO \quad (h\nu = 300\text{–}405 \text{ nm}) \quad \text{(R1)}$$

HONO can be directly emitted by combustion of fossil fuels (Kurtenbach et al., 2001) or formed by gas-phase reactions of NO and OH (the backwards reaction of Reaction R1) and heterogeneous reactions of NO_2 on wet surfaces according to Reaction (R2). On carbonaceous surfaces (soot, phenolic compounds) HONO is formed via electron or H transfer reactions (Reactions R3 and R4–R6; Kalberer et al., 1999; Kleffmann et al., 1999; Gutzwiller et al., 2002; Aubin and Abbatt, 2007; Han et al., 2013; Arens et al., 2001, 2002; Ammann et al., 1998, 2005).

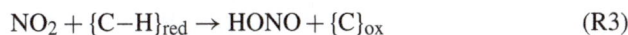

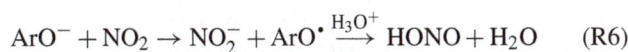

$$2NO_2 + H_2O \rightarrow HONO + HNO_3 \quad \text{(R2)}$$

$$NO_2 + \{C-H\}_{red} \rightarrow HONO + \{C\}_{ox} \quad \text{(R3)}$$

$$ArOH + NO_2 \rightarrow ArO^{\bullet} + HONO \quad \text{(R4)}$$

$$ArOH + H_2O \rightarrow ArO^- + H_3O^+ \quad \text{(R5)}$$

$$ArO^- + NO_2 \rightarrow NO_2^- + ArO^{\bullet} \xrightarrow{H_3O^+} HONO + H_2O \quad \text{(R6)}$$

Previous atmospheric measurements and modeling studies have shown unexpected high HONO concentrations during daytime, which can also contribute to aerosol formation through enhanced oxidation of precursor gases (Elshorbany et al., 2014). Measured mixing ratios are typically about 1 order of magnitude higher than simulated ones, and an additional source of 200–800 ppt h^{-1} would be required to explain observed mixing ratios (Kleffmann et al., 2005; Acker et al., 2006; Sörgel et al., 2011; Li et al., 2012; Su et al., 2008; Elshorbany et al., 2012; Meusel et al., 2016), indicating that estimates of daytime HONO sources are still under debate. It was suggested that HONO arises from the photolysis of nitric acid and nitrate or by heterogeneous photochemistry of NO$_2$ on organic substrates and soot (Zhou et al., 2001, 2002 and 2003; Villena et al., 2011; Ramazan et al., 2004; George et al., 2005; Sosedova et al., 2011; Monge et al., 2010; Han et al., 2016). Stemmler et al. (2006, 2007) found HONO formation on light-activated humic acid, and field studies showed that HONO formation correlates with aerosol surface area, NO$_2$ and solar radiation (Su et al., 2008; Reisinger, 2000; Costabile et al., 2010; Wong et al., 2012; Sörgel et al., 2015) and is increased during foggy periods (Notholt et al., 1992). Another proposed source of HONO is the soil, where it has been found to be co-emitted with NO by soil biological activities (Oswald et al., 2013; Su et al., 2011; Weber et al., 2015).

In view of light-induced nitration of proteins and HONO formation by photolysis of nitrophenols, light-enhanced production of HONO on protein surfaces can be anticipated, which, to the best of our knowledge, has not been studied before.

This work aims to provide insight into protein nitration, the atmospheric stability of the nitrated protein and respective formation of HONO from protein surfaces that were nitrated either offline in liquid phase prior to the gas exchange measurements or online with instantaneous gas-phase exposure to NO$_2$, with particular emphasis on environmental parameters like light intensity, relative humidity (RH) and NO$_2$ concentrations. Bovine serum albumin (BSA), a globular protein with a molecular mass of 66.5 kDa and 21 tyrosine residues per molecule, was chosen as a well-defined model substance for proteins. Nitrated ovalbumin (OVA) was used to study the light-induced degradation of proteins that were nitrated prior to gas exchange measurements. This well-studied protein has a molecular mass of 45 kDa and 10 tyrosine residues per molecule.

2 Materials and methods

2.1 Protein preparation and analysis

BSA (Cohn V fraction, lyophilized powder, ≥ 96 %; Sigma Aldrich, St. Louis, Missouri, USA) or nitrated OVA was solved in pure water (18.2 MΩ cm) and coated onto the glass tube.

The nitration of OVA was described previously (Yang et al., 2010; Zhang et al., 2011). Briefly, OVA (grade V, A5503-5G, Sigma Aldrich, Germany) was dissolved in phosphate-buffered saline PBS (P4417-50TAB, Sigma

Aldrich, Germany) to a concentration of 10 mg ml^{-1}. 50 µL tetranitromethane (TNM; T25003-5G, Sigma Aldrich, Germany) dissolved in methanol 4 % (v/v) were added to a 2.5 mL aliquot of the OVA solution and stirred for 180 min at room temperature. Please note that TNM is toxic if swallowed, can cause skin, eye and respiration irritation, is suspected to cause cancer and causes fires or explosions. Size exclusion chromatography columns (PD-10 Sephadex G-25 M, 17-0851-01, GE Healthcare, Germany) were used for cleanup. The eluate was dried in a freeze dryer and stored in a refrigerator at 4 °C.

After the flow-tube experiments (see below) the proteins were extracted with water from the tube and analyzed with liquid chromatography (HPLC-DAD; Agilent Technologies 1200 series) according to Selzle et al. (2013). This method provides a straightforward and efficient way to determine the nitration of proteins. Briefly, a monomerically bound C18 column (Vydac 238TP, 250 mm × 2.1 mm inner diameter, 5 µm particle size; Grace Vydac, Alltech) was used for chromatographic separation. Eluents were 0.1 % (v/v) trifluoroacetic acid in water (LiChrosolv) (eluent A) and acetonitrile (ROTISOLV HPLC gradient grade, Carl Roth GmbH + Co. KG, Germany) (eluent B). Gradient elution was performed at a flow rate of 200 µL min^{-1}. ChemStation software (Rev. B.03.01, Agilent) was used for system control and data analysis. For each chromatographic run, the solvent gradient started at 3 % B followed by a linear gradient to 90 % B within 15 min, flushing back to 3 % B within 0.2 min and maintaining 3 % B for additional 2.8 min. Column re-equilibration time was 5 min before the next run. Absorbance was monitored at wavelengths of 280 (tyrosine) and 357 nm (nitrotyrosine). The sample injection volume was 10–30 µL. Each chromatographic run was repeated three times. The protein nitration degree (ND), which is defined as the ratio of nitrated tyrosine to all tyrosine residues, was determined by the method of Selzle et al. (2013). Native and untreated BSA did not show any degree of nitration.

2.2 Coated-wall flow tube system

Figure 2 shows a flowchart of the setup of the experiment. NO$_2$ was provided in a gas bottle (1 ppm in N$_2$, Carbagas AG, Grümligen, Switzerland). NO$_2$ was further diluted (mass flow controller, MFC3) with humidified pure nitrogen to achieve NO$_2$ mixing ratios between 20 and 100 ppb. Impurities of HONO in the NO$_2$-gas cylinder were removed by means of a HONO scrubber. The Na$_2$CO$_3$ trap was prepared by soaking 4 mm firebrick in a saturated Na$_2$CO$_3$ in 50 % ethanol–water solution and drying for 24 h. The impregnated firebrick granules were put into a 0.8 cm inner diameter and 15 cm long glass tube, which was closed by quartz wool plugs on both sides. A constant total flow (1400 mL min^{-1}) was provided by means of another N$_2$ mass flow controller (MFC2) that compensated for changes in NO$_2$ addition. Different fractions of total surface areas (50, 70 and 100 %) of

Table 1. Details on the different experiments, aims and experimental conditions (coating, applied NO_2 concentration, number of lights switched on, relative humidity and time for each exposure step).

		Coating density (number of monolayers NML_f, thickness)	NO_2 (ppb)	No. of lamps	RH (%)	Time per step (h)
(a) Light-induced decomposition of nitrated protein and HONO formation						
1	Light and NO_2 dependency	n-OVA $21.5 \pm 0.8\,\mu g\,cm^{-2}$ (68 NML_f, 298.05 nm)	0–20	0–1–3–7 VIS	50	1
(b) Heterogeneous NO_2 transformation on BSA						
2	NO_2 dependency	BSA $16.1 \pm 0.4\,\mu g\,cm^{-2}$ (50 NML_f, 217.6 nm)	0–20–40–60–100	7 VIS	50	0.5–1
3	Light dependency	BSA $31.4 \pm 1.4\,\mu g\,cm^{-2}$ (99 NML_f, 435.2 nm)	20	0–1–3–7 VIS	50	0.5–1
4	Coating thickness	BSA $16.1 \pm 0.4\,\mu g\,cm^{-2}$ (50 NML_f, 217.6 nm), $22.5 \pm 0.8\,\mu g\,cm^{-2}$ (71 NML_f, 310.8 nm), $31.4 \pm 1.4\,\mu g\,cm^{-2}$ (99 NML_f, 435.2 nm)	20	7 VIS		0.5–3
5	RH dependency	BSA $17.5 \pm 0.4\,\mu g\,cm^{-2}$ (55 NML_f, 241.7 nm)	25	0–7 VIS	0–50–80	0.25–1
6	Time effect	BSA $17.5 \pm 0.4\,\mu g\,cm^{-2}$	100	7 VIS	75	20
7	Time effect	BSA $17.5 \pm 0.4\,\mu g\,cm^{-2}$	100	4 VIS + 3 UV	75	20

NML_f numbers of monolayers in flat orientation.

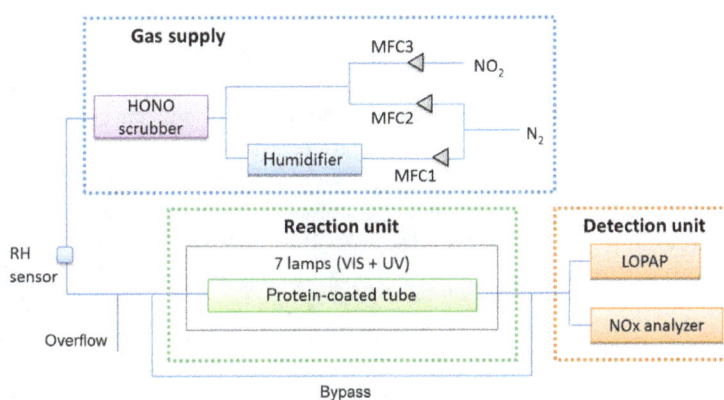

Figure 2. Flow system and setup: thin blue lines show the flow of the gas mixture, which direction is indicated by the grey triangles of the mass flow controllers (MFC). Nitrogen passes a heated water bath to humidify the gas and a HONO scrubber to eliminate any HONO impurities of the NO_2 supply. The overflow maintains a constant pressure through the reaction tube and the detection unit. The dotted boxes (blue, green, orange) indicate the three different parts: the gas supply, reaction unit and detection unit.

the reaction tube (50 cm × 0.81 cm i.d.) were coated with 2 mg BSA or nitrated OVA, respectively. Therefore 2 mg protein was dissolved in 600 μL pure water, injected into the tube and then gently dried in a low-humidity N_2 flow (RH ∼ 30–40 %) with continuous rotation of the tube. The coated reaction tube was exposed to the generated gas mixture and irradiated with either (i) one, three or seven visible (VIS) lights (400–700 nm; L 15 W/954, Lumilux de Luxe daylight, Osram, Augsburg, Germany), which is 0, 23, 69 or 161 W m^{-2}, respectively; or (ii) four VIS and three UV lights (340–400 nm; UV-A, TL-D 15 W/10, Philips, Hamburg, Germany).

An overview of the experiments performed during this study is shown in Table 1. Light-induced decomposition of nitrated proteins was studied on OVA. Instantaneous NO_2

transformation and its light and RH dependence on heterogeneous HONO formation were studied on BSA in short-term experiments. Extended studies on BSA were performed to explore the persistence of the surface reactivity and respective catalytic effects.

A commercial long-path absorption photometry instrument (LOPAP, QUMA) was used for HONO analysis. The measurement technique was introduced by Heland et al. (2001). This wet chemical analytical method has an unmatched low detection limit of 3–5 ppt with high HONO collection efficiency ($\geq 99\%$). HONO is continuously trapped in a stripping coil flushed with an acidic solution of sulfanilamide. In a second reaction with n-(1-naphthyl)ethylenediamine-dihydrochloride an azo dye is formed, whose concentration is determined by absorption photometry in a long Teflon tubing. LOPAP has two stripping coils in series to reduce known interferences. In the first stripping coil HONO is quantitatively collected. Due to the acidic stripping solution, interfering species are collected less efficiently but in both channels. The true concentration of HONO is obtained by subtracting the interferences quantified in the second channel from the total signal obtained in the first channel. The accuracy of the HONO measurements was 10 %, based on the uncertainties of liquid and gas flow, concentration of calibration standard and regression of calibration.

The reagents were all high-purity-grade chemicals, i.e., hydrochloric acid (37 %, ACS reagent, Sigma Aldrich, St. Louis, Missouri, USA), sulfanilamide (for analysis, > 99 %; Sigma Aldrich) and N-(1-naphthyl)-ethylenediamine dihydrochloride (> 98 %; ACS reagent, Fluka by Sigma Aldrich). For calibration Titrisol® 1000 mg NO_2^- ($NaNO_2$ in H_2O; Merck) was diluted to 0.001 mg L^{-1} NO_2^-. For preparation of all solutions and for cleaning of the absorption tubes 18 MΩ H_2O was used.

NO_x concentrations were analyzed by means of a commercial chemiluminescence detector from EcoPhysics (CLD 77 AM, Duernten, Switzerland).

3 Results and discussion

3.1 BSA nitration and degradation

Nitrated proteins can trigger allergic response. The nitration of proteins can be enhanced by O_3 activation (in the dark). In the atmospheric environment, about half the time sunlight is present. What happens with irradiated proteins when exposed to NO_2? Can they be nitrated efficiently? To investigate the degree of protein nitration under illuminated conditions, BSA coated on the reaction tube (17.5 μg cm^{-2}) was exposed to seven VIS lamps (40 % of a clear-sky irradiance for a solar zenith of 48°; Stemmler et al., 2006) and 100 ppb NO_2 at 70 % RH. After 20 h the BSA ND (concentration of nitrated tyrosine residues divided by the total concentration

of tyrosine residues) investigated by means of the HPLC-DAD method was $(1.0 \pm 0.1)\%$, significantly higher than the ND of untreated BSA (0 %). Introducing UV radiation (four VIS plus three UV lamps) resulted in a slightly higher ND of $(1.1 \pm 0.1)\%$. Note that no intact protein (nitrated and non-nitrated) could be detected by HPLC-DAD after another 20 h of irradiation without NO_2, indicating light-induced decomposition of proteins. However, the applied HPLC-DAD technique only detects (nitro-)tyrosine residues in proteins and does not provide information about protein fragments or single nitrated or non-nitrated tyrosine residues. Hence, proteins might have been decomposed while tyrosine remains in its nitrated form, not detectable by our analysis method. Similarly, proteins (here OVA) that were nitrated with TNM in aqueous phase prior to coating (21.5 μg cm^{-2}) to an extent of 12.5 % also decomposed when illuminated about 6 h (one to seven VIS lights; with and without 20 ppb NO_2). Thus the nitration of proteins by light and NO_2 was confirmed, but with simultaneous gradual decomposition of the proteins. Effects of UV irradiation (240–340 nm) on proteins containing aromatic amino acids were reviewed previously (Neves-Peterson et al., 2012). It was shown that triplet state tryptophan and tyrosine can transfer electron to a nearby disulfide bridge to form the tryptophan and tyrosine radical. The disulfide bridge could break leading to conformational changes in the protein but not necessarily resulting in inactivation of the protein. In strong UV light (≈ 200 nm) the peptide bond could also break (Nikogosyan and Görner, 1999).

Franze et al. (2005) analyzed a variety of natural samples (road dust, window dust and particulate matter $PM_{2.5}$) collected in the metropolitan area of Munich, containing 0.08–21 g kg^{-1} proteins, and revealed equivalent degrees of nitration (EDN, concentration of nitrated protein divided by concentration of all proteins) between 0.01 and 0.1 % only. Such low nitration degree is in line with light-induced decomposition of (nitrated) proteins. In contrast, an EDN up to 10 % (average 5 %) was found for BSA and birch pollen extract exposed to Munich ambient air for 2 weeks under dark conditions, with daily mean NO_2 (O_3) concentration of 17–50 ppb (7–43 ppb) in the same study, possibly suggesting the deficiency of decomposition without being irradiated. BSA and OVA loaded on syringe filters and exposed to 200 ppb NO_2 / O_3 for 6 days under dark conditions were nitrated to 6 and 8 %, respectively (Yang et al., 2010). Reinmuth-Selzle et al. (2014) found similar ND for major birch pollen allergen Bet v 1 loaded on syringe filters exposed to 80–470 ppb NO_2 and O_3. When exposed for 3–72 h to NO_2 / O_3 at RH < 92 % the ND was 2–4 %, while at condensing conditions (RH > 98 %) the ND increased to 6 % after less than 1 day (19 h). The ND of Bet v 1 was considerably increased to 22 % for proteins solved in the aqueous phase (0.16 mg mL^{-1}) when bubbling with a 120 ppb NO_2 / O_3 gas mixture for a similar period of time (17 h). Shiraiwa et al. (2012) performed kinetic modeling and found that maximum 30 % (conservative upper limit) of N uptake on BSA could be explained by NO_3 or N_2O_5, which are generated by the reaction of NO_2 and O_3, while overall nitration was governed by an indirect mechanism in which a radical intermediate was formed by the reaction of BSA with ozone, which then reacted with NO_2. On NaCl surface N uptake was dominated by NO_3 and N_2O_5. Furthermore, NO_3 radicals, which in this study could be formed by photolysis of NO_2 (> 410 nm, disproportionation of excited NO_2), are not stable under the light conditions applied (400–700 nm) (Johnston et al., 1996). Therefore, in the present study reactions with NO_3 were neglected. Photolysis of NO_2 forming NO (< 400 nm) can also be neglected (Gardner et al., 1987; Roehl et al., 1994). A photolysis frequency for NO_2 of up to 5×10^{-4} s^{-1} under similar experimental light conditions was determined by Stemmler et al., 2007. Other nitration methods investigated by Reinmuth-Selzle et al. (2014), e.g., nitration of Bet v 1 with peroxynitrite ($ONOO^-$, formed by reaction of NO with O_2^-) or TNM, lead to ND between 10 and 72 % depending on reaction time, reagent concentration and temperature. Similarly, high NDs of 45–50 % were obtained by aqueous-phase TNM nitration of BSA and OVA by Yang et al. (2010).

3.2 HONO formation

3.2.1 HONO formation from nitrated proteins

To study HONO emission from nitrated proteins, OVA was nitrated with TNM (see Sect. 2.1) in liquid phase. The nitrated OVA (2 mg; ND = 12.5 %) was coated onto the reaction tube and exposed to VIS lights under either pure nitrogen flow or 20 ppb NO_2 gas. Strong HONO emissions were found. A high correlation between HONO emission and light intensity was observed (50 % RH; Fig. 3). Initially, we did not apply NO_2. Thus the observed HONO formation (up to 950 ppt) originated from decomposing nitrated proteins rather than from heterogeneous conversion of NO_2. However, when exposed to 20 ppb of NO_2 in dark conditions, HONO formation increased 4-fold (50–200 ppt) and about 2-fold with seven VIS lamps turned on (950–1800 ppt). After 7 h of flow tube experiments (4.5 h irradiation with varying light intensities (0, 1, 3, 7 lights) + 2.5 h irradiation/20 ppb NO_2 (7, 3, 0 lights)), no intact protein was found according to the analysis of HPLC-DAD.

As proteins can efficiently be nitrated by O_3 and NO_2 in polluted air (Franze et al., 2005; Shiraiwa et al., 2012; Reinmuth-Selzle et al., 2014), the emission of HONO from light-induced decomposing nitrated proteins could play an important role in the HONO budget. As proteins are nitrated at their tyrosine residues (at the ortho position to the OH group on the aromatic ring) the underlying mechanism of this HONO formation should be very similar to the HONO formation by photolysis of ortho-nitrophenols described by Bejan et al. (2006). This starts with a photo-induced hydrogen transfer from the OH group to the vicinal NO_2 group (Fig. 1),

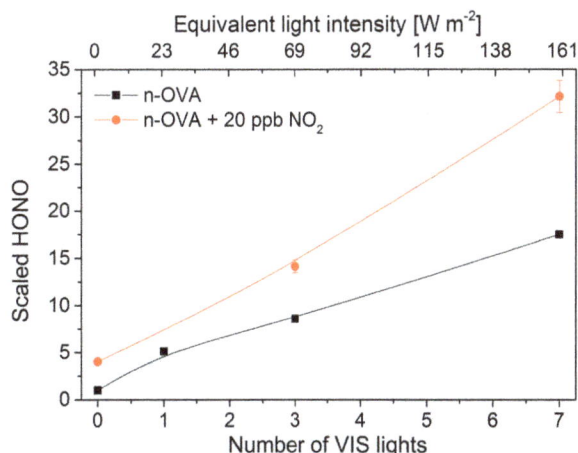

Figure 3. Light-enhanced HONO formation from TNM-nitrated proteins (n-OVA: ND 12.5 %, coating 21.5 μg cm^{-2}). Black squares indicate HONO formation via decomposition from nitrated proteins (without NO_2) while red squares indicate additional HONO formation via heterogeneous NO_2 conversion (20 ppb NO_2) at 50 % RH (HONO is scaled to the HONO concentration measured without NO_2 and no light ([HONO]$_{lights; NO_2}$/[HONO]$_{dark; NO_2=0}$)).

which leads to an excited intermediate from which HONO is eliminated subsequently.

3.2.2 Light dependency

To investigate HONO formation on unmodified BSA coating (31.4 μg cm^{-2}) dependent on light conditions, the radiation intensity (number of VIS lamps) was changed under otherwise constant conditions of exposure at 20 ppb NO_2 and 50 % RH. Decreasing light intensity revealed a linearly decreasing trend in HONO formation from about 1000 to 140 ppt (red symbols in Fig. 4). After re-illumination to the initial high light intensity the HONO formation was reduced by 32 % (blue symbol in Fig. 4). Stemmler et al. (2006) and Sosedova et al. (2011) also observed a similar saturation of HONO formation on humic, tannic and gentisic acid at higher light intensities. Stemmler et al. (2006) argued that surface sites activated for NO_2 heterogeneous conversion by light (Reaction R3) would become de-activated by competition with photo-induced oxidants (X^*, Reactions R7–R8), e.g., primary chromophores or electron donors are oxidized by surface*, which is in line with the observed decomposition of the native protein presented above.

$$\text{surface} \xrightarrow{h\nu} \text{surface}^* \xrightarrow{NO_2} \text{HONO} + \text{surface}_{ox} \quad \text{(R7)}$$

$$X \xrightarrow{h\nu} X^* \xrightarrow{\text{surface}^*} \text{surface–}X \quad \text{(R8)}$$

In other studies the NO_2 uptake coefficient on soot, mineral dust, humic acid and other solid organic compounds similarly increased at increasing light intensities (George et al., 2005; Stemmler et al., 2007; Ndour et al., 2008; Monge

(a)

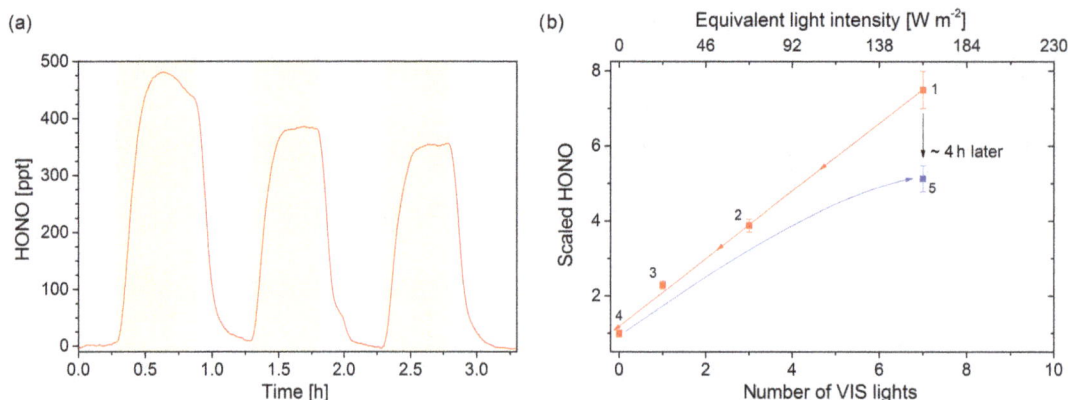

(b)

Figure 4. Light-induced HONO formation on BSA. **(a)** HONO formation under alternating dark and light conditions on BSA surface $(22.5\,\mu g\,cm^{-2})$; yellow shaded areas indicate periods in which seven VIS lamps were switched on (RH $= 50\,\%$, $NO_2 = 20\,ppb$). **(b)** Dependency of HONO formation on radiation intensity at 20 ppb NO_2 and 50 % RH (BSA $= 31.4\,\mu g\,cm^{-2}$). The experiment started with seven VIS lights switched on, sequentially decreasing the number of lights (red symbols, nominated 1–4), prior to applying the initial irradiance again (blue symbol, 5). HONO was scaled to the HONO concentration in darkness $([HONO]_{lights}/[HONO]_{dark})$. Error bars indicate SD of 20–30 min measurements; SD of point 5 covers 2.75 h measurement.

et al., 2010; Han et al., 2016; Brigante et al., 2008). Note that the HONO yield (ratio of HONO formed to NO_2 lost) was found to be constant at light intensities in the range of 60–$200\,W\,m^{-2}$ in the work of Han et al. (2016) but has shown a linear dependence on light for nitrated phenols (Bejan et al., 2006).

3.2.3 NO_2 dependency

At about 50 % relative humidity and high illumination intensities (seven VIS lamps, $\sim 161\,W\,m^{-2}$), heterogeneous formation of HONO strongly correlated with the applied NO_2 concentration (Fig. 5). On a BSA surface of about $16.1\,\mu g\,cm^{-2}$ (Table 1) the produced HONO concentration increased from 56 ppt at 20 ppb NO_2 to 160 ppt at 100 ppb NO_2. Only at a threshold NO_2 level well above those typically observed in natural environments ($\gg 150\,ppb$) did this increasing trend slow down to some extent, indicative of saturation of active surface sites. A similar pattern of NO_2 dependence was also observed for light-induced HONO formation from humic acid (Stemmler et al., 2006) and phenolic compounds like gentisic and tannic acid (Sosedova et al., 2011) or polycyclic aromatic hydrocarbons (Brigante et al., 2008) and for heterogeneous NO_2 conversion on soot under dark conditions (Stadler and Rossi, 2000; Salgado and Rossi, 2002; Arens et al., 2001).

For better comparison of the different studies the HONO concentration measured at different NO_2 concentrations was scaled to the HONO concentration at 20 ppb NO_2 $([HONO]_{NO_2}/[HONO]_{NO_2=20\,ppb})$ in Fig. 5, as variable absolute amounts of HONO were found in different studies and matrices. A cease of the NO_2 dependency on heterogeneous HONO formation can be assessed for most of the studies at NO_2 concentrations $\geq 200\,ppb$. A very similar correlation

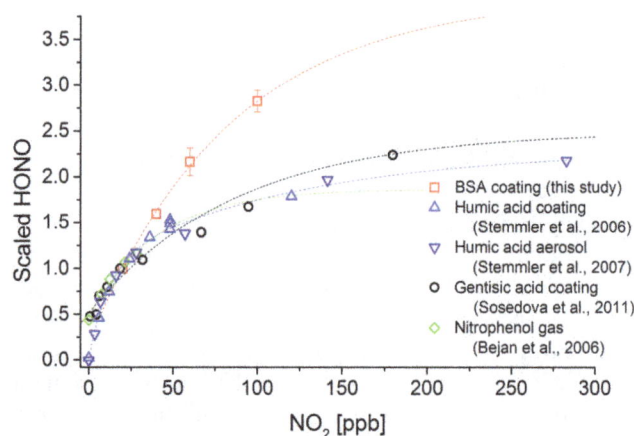

Figure 5. Comparison of HONO formation dependency on NO_2 at different organic surfaces. HONO concentrations are scaled to the HONO concentration at 20 ppb NO_2 $([HONO]_{NO_2}/[HONO]_{NO_2=20\,ppb})$. The red squares indicate BSA coating $(16\,\mu g\,cm^{-2})$ at $161\,W\,m^{-2}$ and 50 % RH (this study). Blue triangles pointing up are humic acid coating $(8\,\mu g\,cm^{-2})$ at $162\,W\,m^{-2}$ and 20 % RH (Stemmler et al., 2006), while the blue triangles pointing down are the humic acid aerosol with 100 nm diameter and a surface of $0.151\,m^2\,m^{-3}$ at 26 % RH and 1×10^{17} photons $cm^{-2}\,s^{-1}$ (Stemmler et al., 2007). The black circles are gentisic acid coating $(160–200\,\mu g\,cm^{-2})$ at 40–45 % RH and light intensity similar to that in the humic acid aerosol study (Sosedova et al., 2011). Green diamonds are ortho-nitrophenol in gas phase (ppm level) illuminated with UV/VIS light. Dotted lines are exponential fittings of the measured data points and are meant to guide the eyes.

(up to 40 ppb NO_2) was observed when NO_2 was applied additionally during the gas-phase photolysis of nitrophenols (Fig. 5; Bejan et al., 2006). Even though the matrix (nitrophenols) and conditions (illuminated) of the latter is comparable to the experiment presented here, for BSA no clear indication of saturation was found up to 160 ppb of NO_2, pointing to a highly reactive surface of BSA for NO_2 under illuminated conditions. As shown with Reactions (R7) and (R8), the concentration dependence depends on the competing channel (Reaction R8); therefore, this is strongly matrix dependent, both in terms of chemical and physical properties.

3.2.4 Impact of coating thickness

Strong differences in HONO concentrations were found for experiments with different coating thicknesses applying otherwise similar conditions (20 ppb of NO_2, seven VIS lamps and 50 % RH). While only 55 ppt of HONO concentration was observed for a shallow homogeneous coating of 16.1 µg cm^{-2} (217.6 nm thickness, see below) applied on the whole length of the tube, up to 2 ppb was found for a thick (more uneven) coating of 31.44 µg cm^{-2} (435.2 nm thickness) covering only 50 % of the tube (Fig. 6). Potential explanations are that thicker coating leads to (1) more bulk reactions producing HONO or (2) different morphologies, e.g., higher effective reaction surfaces. Exposing (20 %) different coated surface areas in the flow tube, potentially introduced bias comparing different data sets. Emitted HONO might be re-adsorbed differently by proteins and glass surface. However, as the protein is slightly acidic, a low uptake efficiency of HONO by BSA can be anticipated, which should not differ too much from the uncovered glass tube surface (Syomin and Finlayson-Pitts, 2003). Accordingly, NO_2 uptake on glass is assumed to be significantly lower than on proteins. A strong increase in NO_2 uptake coefficients with increasing coating thickness was also observed for humic acid coatings (Han et al., 2016). However, they found an upper threshold value of 2 µg cm^{-2} of cover load (20 nm absolute thickness, assuming a humic acid density of 1 g cm^{-3}), above which uptake coefficients were found to be constant. The authors also proposed that NO_2 can diffuse deeper into the coating and below 2 µg cm^{-2} the full cover depth would react with NO_2, respectively.

For proteins the number of molecules per monolayer depends on their orientation and respective layer thickness can vary accordingly. One (dry, crystalline) BSA molecule has a volume of about 154 nm^3 (Bujacz, 2012). In a flat orientation (4.4 nm layer height and a projecting area of 35 nm^2 molecule^{-1}) 3.64×10^{14} molecules (40.5 µg; 0.32 µg cm^{-2}) of BSA are needed to form one complete monolayer in the flow tube (i.d. of 0.81, 50 cm length, 100 % surface coating). Hence, the thinnest BSA coating applied in the experiment (16.1 µg cm^{-2}) would consist of 50 monolayers, revealing a total coating thickness of 217.6 nm, and the thickest BSA coating (31 µg cm^{-2}) would have 99 monolayers and an absolute thickness of 435.1 nm. At the other extreme (non-flat) orientation, more BSA molecules are needed to sustain one monolayer. With 21.7 nm^2 of pro-

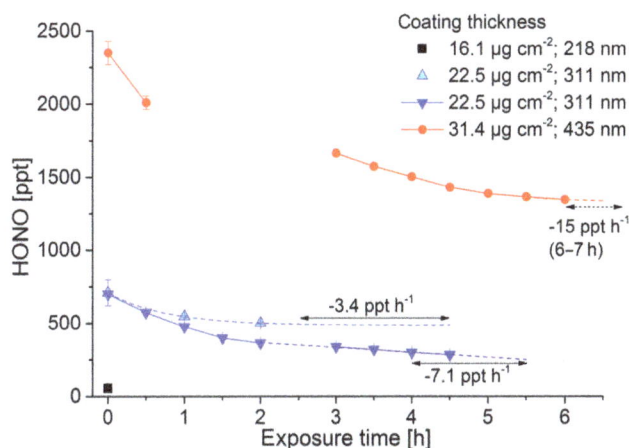

Figure 6. HONO formation on three different BSA coating thicknesses, exposed to 20 ppb of NO_2 under illuminated conditions (seven VIS lamps). The HONO concentrations were scaled to reaction tube coverage (black: 100 % of reaction tube was covered with BSA; light blue: 70 % of tube was covered; red: 50 % of tube was covered with BSA). The middle thick coating (22.46 µg cm^{-2}) was replicated and studied with different reaction times (cyan and blue triangle). Solid lines (with circles or triangles) present continuous measurements; when those are interrupted, other conditions (e.g., light intensity, NO_2 concentration) prevailed. Dotted lines show interpolations and are meant to guide the eyes. Arrows indicate the intervals in which the shown decay rates were determined. Error bars indicates SDs from 10 to 20 measuring points (5–10 min).

jected area of one molecule and 7.1 nm monolayer height, 5.86×10^{14} molecules of BSA are needed to form one complete monolayer in the flow tube. The coatings would consist of between 31 (thinnest) and 61 (thickest) monolayers of BSA. With a flat orientation 1–2 % (number or weight) of BSA molecules would build the uppermost surface monolayer, whereas in an upright molecule orientation 1.6–3.3 % would be in direct contact with surface ambient air.

In the crystalline form several molecules of water stick tightly to BSA. As BSA is highly hygroscopic, more water molecules are adsorbed at higher relative humidity. At 35 % RH BSA is deliquesced (Mikhailov et al., 2004). Therefore the above described number of monolayers and the absolute layer thickness are a lower bound estimate.

In conclusion, the thickness dependence on HONO formation is extremely complex. Activation and photolysis of nitrated Tyr occurs throughout the BSA layer. The heterogeneous reaction of NO_2 may or may be not limited to the surface depending on solubility and diffusivity of NO_2. Also the release of HONO may be limited by diffusion. The observed dependence on the coating thickness suggests the involvement of the bulk reactions, but the reactions can happen in both surface and bulk phase.

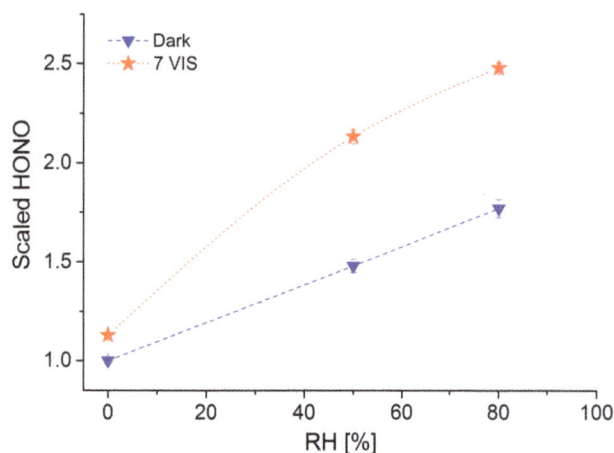

Figure 7. Dependency of relative humidity on HONO formation. 25 ppb NO_2 was applied on BSA surface (17.5 µg cm^{-2}) either in darkness (blue triangle) or at seven VIS lights (red star). HONO was scaled to HONO concentrations in darkness under dry conditions ([HONO]$_{lights\ on-off}$; RH/[HONO]$_{dark;\ RH=0}$). Dotted lines are meant to guide the eyes.

3.2.5 RH dependency

The dependence of HONO emission on relative humidity is shown in Fig. 7. Here about 25 ppb of NO_2 was applied to a (not nitrated) BSA-coated flow tube (17.5 µg cm^{-2}) both in dark and illuminated conditions (seven VIS lights). HONO formation scaled with relative humidity. Kleffmann et al. (1999) proposed that higher humidity inhibits the self-reaction of HONO ($2\ HONO_{(s,g)} \rightarrow NO_2 + NO + H_2O$), which leads to higher HONO yield from heterogeneous NO_2 conversion.

The RH dependence of HONO formation on proteins is different to other surfaces. For example, no influence of RH has been observed for dark heterogeneous HONO formation on soot particles sampled on filters (Arens et al., 2001). No impact of humidity on NO_2 uptake coefficients on pyrene was detected (Brigante et al., 2008). For HONO formation on tannic acid coatings (both at dark and irradiated conditions) a linear but relatively weak dependence has been reported between 10 and 60 % RH, while below 10 % and above 60 % RH the correlation between HONO formation and RH was much stronger (Sosedova et al., 2011). Similar results were obtained for anthrarobin coatings by Arens et al. (2002). This type of dependence of HONO formation on phenolic surfaces on RH equals the HONO formation on glass, following the BET water uptake isotherm of water on polar surfaces (Finnlayson-Pitts et al., 2003; Summer et al., 2004). For humic acid surfaces the NO_2 uptake coefficients also weakly increased below 20 % RH and were found to be constant between 20 and 60 % (Stemmler et al., 2007).

While on solid matter chemical reactions are essentially confined to the surface rather than in the bulk, proteins can adopt an amorphous solid or semisolid state, influencing the rate of heterogeneous reactions and multiphase processes. Molecular diffusion in the non-solid phase affects the gas uptake and respective chemical transformation. Shiraiwa et al. (2011) could show that the ozonolysis of amorphous protein is kinetically limited by bulk diffusion. The reactive gas uptake exhibits a pronounced increase with relative humidity, which can be explained by a decrease of viscosity and increase of diffusivity, as the uptake of water transforms the amorphous organic matrix from a glassy to a semisolid state (moisture-induced phase transition). The viscosity and diffusivity of proteins depend strongly on the ambient relative humidity because water can act as a plasticizer and increase the mobility of the protein matrix (for details see Shiraiwa et al., 2011, and references therein). Shiraiwa et al. (2011) further showed that the BSA phase changes from solid through semisolid to viscous liquid as RH increases, while trace gas diffusion coefficients increased about 10 orders of magnitude. This way, characteristic times for heterogeneous reaction rates can decrease from seconds to days as the rate of diffusion in semisolid phases can decrease by multiple orders of magnitude in response to both low temperature (not investigated in here) and/or low relative humidity. Accordingly, we propose that HONO formation rate depends on the condensed-phase diffusion coefficients of NO_2 diffusing into the protein bulk, HONO released from the bulk and mobility of excited intermediates.

3.2.6 Long-term exposure with NO_2 under irradiated conditions

To study long-term effects of irradiation on HONO formation from proteins, flow tubes were coated with 2 mg BSA (17.5 ± 0.4 µg cm^{-2}; 90 % of total length) and exposed to 100 ppb NO_2, at 80 % RH at illuminated conditions for a time period of up to 20 h (Fig. 8). Samples illuminated with VIS light only (red and orange colored lines in Fig. 8) showed persistent HONO emissions over the whole measurement period. For unknown reasons, and even though the observed HONO concentrations were within the expected range with regard to the applied NO_2 concentrations, RH and cover characteristics, one sample (orange in Fig. 8) showed a sharp short-term increase in the initial phase followed by respective decrease, not in line with all other samples (compare Fig. 6). However, after 4 h both VIS irradiated samples showed virtually constant HONO emissions (-3.8 and $+1.6$ ppt h^{-1}, respectively). The sample illuminated with UV and VIS light (three UV and four VIS lamps) showed a sustained sharp increase in the first 4 h, followed by persistent and very stable (decay rate as low as -0.5 ppt h^{-1}) HONO emissions at an about 3-fold higher level compared to samples irradiated with VIS only. HONO formation by photolysis of (adsorbed) HNO_3 is assumed to be insignificant in this study. With N_2 as carrier gas, gas-phase reactions of NO_2 do not produce HNO_3. Even when small amounts of HNO_3 would

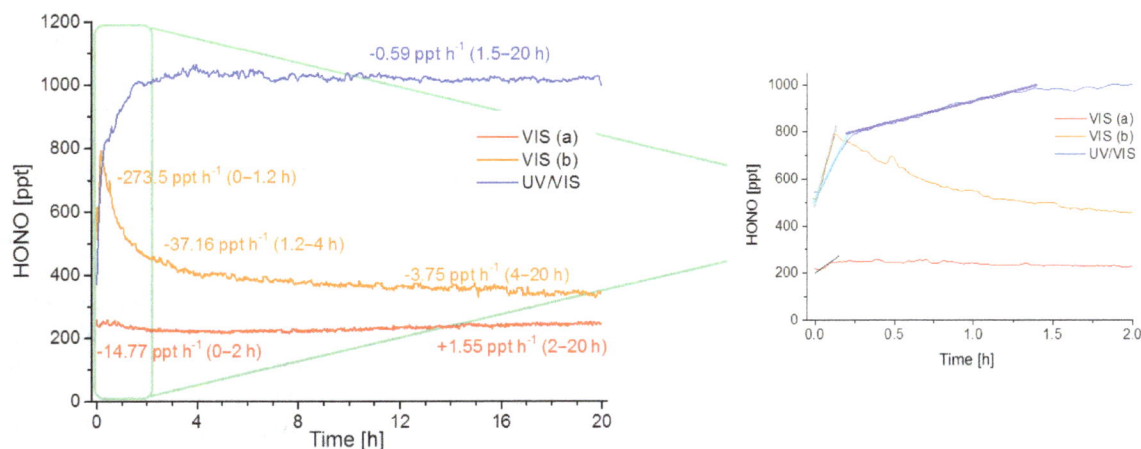

Figure 8. Extended measurements (20 h) of light-enhanced HONO formation on BSA (three coatings of $17.5 \, \mu g \, cm^{-2}$) at 80 % RH, 100 ppb NO_2. HONO formation under VIS light is shown in red and orange, under UV/VIS light in blue. HONO decay rates ($ppt \, h^{-1}$) are shown with time periods (in brackets) in which they were calculated, suggesting a stable HONO formation after 4 h. Right: magnification of the first 2 h. Straight lines (black, grey, light and dark blue) show the slopes of which $d[HONO]/dt$ were used in the kinetic studies.

be formed by unknown heterogeneous reactions, photolysis of HNO_3 is only significant at wavelengths < 350 nm, which is close to the lowest limit of the UV wavelength applied in this study. Likewise, the respective photolysis frequency recently proposed by Laufs and Kleffmann (2016) of about $2.4 \times 10^{-7} \, s^{-1}$ is very low.

Integrating the 20 h experiments, 9.23×10^{15} (4.6 ppb h, VISa), 1.53×10^{16} (7.7 ppb h, VISb) and 4.01×10^{16} (20 ppb h, UV/VIS) molecules of HONO were produced. This means between 7.7×10^{13} and 3.3×10^{14} molecules of HONO per cm^2 of BSA geometric surface were formed. With respect to the different experimental conditions concerning cover thickness, RH, and NO_2 concentrations, this is in a similar order of magnitude as found for humic acid (2×10^{15} molecules cm^{-2} in 13 h) by Stemmler et al. (2006).

If BSA acts like a catalytic surface as in a Langmuir–Hinshelwood reaction each BSA molecule can react several times with NO_2 to heterogeneously form HONO. As described in 3.1, BSA nitration is in competition with NO_2 surface reactions and only a limited number of NO_2 molecules could react with BSA forming HONO via nitration of proteins and subsequent decomposition of nitrated proteins. A BSA molecule contains 21 tyrosine residues, which could react with NO_2. However, even a strong nitration agent such as TNM is not capable of nitrating all tyrosine residues and a mean ND of 19 % was found (Peterson et al., 2001; Yang et al., 2010); i.e., four tyrosine residues of one BSA molecule can be nitrated to form HONO. As 2 mg of BSA was applied for each flow tube coating, a total of 1.8×10^{16} protein molecules can be inferred. In 20 h of irradiating with VIS light 13–22 % of the accessible Tyr residues (four Tyr per BSA molecule) would have been reacted. Irradiating with additional UV lights at least 56 % of the tyrosine residues would have been nitrated and decomposed. However, as NO_2

is a much weaker nitrating agent and nitration of only one tyrosine residue is probable (ND of BSA with O_3 / NO_2 6 %; Yang et al., 2010) up to 85 % BSA molecules would have been reacted when irradiated with VIS lights and even more HONO molecules as coated BSA molecules would have been generated under UV/VIS light conditions. Other amino acids of the protein like tryptophan or phenylalanine might also be nitrated but without formation of HONO (Goeschen et al., 2011). Hence, a contribution of heterogeneous conversion of NO_2 can be anticipated.

3.3 Kinetic studies

The experimental results (especially the stability over a long time) indicate that the formation of HONO from NO_2 on protein surfaces likely underlies the Langmuir–Hinshelwood mechanism in which the protein would act as a catalytic surface (Fig. 9). The first step is the fast, reversible physical adsorption of NO_2 (k_1) and water followed by the slow conversion into HONO.

There are two possible processes for the HONO formation. HONO is formed by heterogeneous NO_2 conversion (k_2) but also via nitration and decomposition of nitrated proteins (k_4, k_5). The final step of the mechanism is the release of the generated HONO into the air. Since proteins are in general slightly acidic, the desorption of HONO (k_3) should be fairly fast. Pseudo-first-order kinetics are assumed for the reaction of NO_2 to HONO (Stemmler et al., 2007) and the reaction can be described as follows (Eq. 1).

$$\frac{d[HONO]_g}{dt} = k_{eff} \cdot [NO_2]_g, \tag{1}$$

with k_{eff} the effective pseudo-first-order rate constant (for more detailed information check the Supplement).

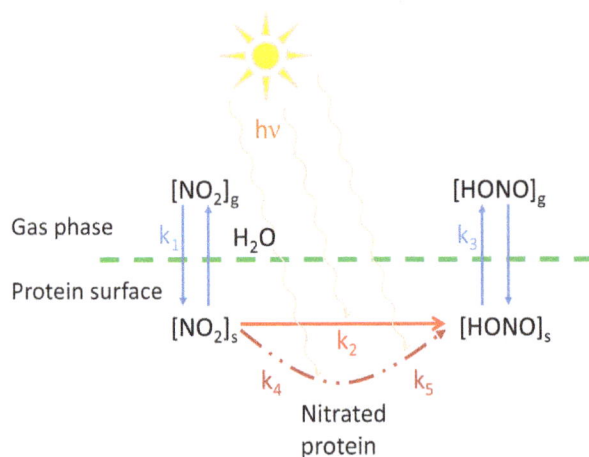

Figure 9. Schematic illustration of the underlying Langmuir–Hinshelwood mechanism of light-induced HONO formation on protein surface. Reaction constants for NO_2 uptake, direct NO_2 conversion, protein nitration, HONO formation from decomposing nitrated proteins and HONO release are indicated by k_1, k_2, k_4, k_5, and k_3.

In this study, neither HONO nor NO_2 photolysis is considered, as the overlap of the applied UV/VIS or VIS range (340–700 or 400–700 nm) and the HONO and NO_2 photolysis spectrum (< 400 nm) is low. Furthermore, the applied light intensity is lower compared to clear-sky irradiance and the respective UV light is partly absorbed by the reaction tube although quartz glass was used (transmission $\sim 90\%$) and the photolysis frequency would decrease down to $10^{-4}\,s^{-1}$. Hence, the photolysis is assumed to be not significant.

In the first 5–10 min of the long-term experiments, HONO increased (Fig. 8 – zoomed in range). This slope was taken as $d[HONO]_g/dt$ in Eq. (6). Effective rate constants between $1.48 \times 10^{-6}\,s^{-1}$ (VISa) and $7.40 \times 10^{-6}\,s^{-1}$ (VISb) were calculated. When irradiating with VIS light only, the concentration of HONO was either constant or decreased for 2 h after this first 10 min. When irradiating with additional UV light, the HONO signal showed an enhancement in two steps. In the first 10 min it was strongly increasing ($1327\,ppt\,h^{-1}$) and then in the next hour it increased less with $170\,ppt\,h^{-1}$ prior to stabilization. Therefore two rate constants of 4.10×10^{-6} and $5.2 \times 10^{-7}\,s^{-1}$ were obtained, respectively.

Reactive uptake coefficients for NO_2 were calculated according to Li et al. (2016). For both irradiation types the uptake coefficient γ was in the range of 7×10^{-6} at the very beginning of each experiment. After a few minutes they decreased to a mean of 1×10^{-7}. The calculated k_{eff} values and uptake coefficient are in the same range and match the NO_2 uptake coefficients on irradiated humic acid surfaces (coatings) and aerosols obtained by Stemmler et al. (2006/07) which were in between 2×10^{-6} and 2×10^{-5} (coatings) and 1×10^{-6} and 6×10^{-6} (aerosols), depending on NO_2 concen-

trations and light intensities. Similar NO_2 uptake coefficients on humic acid were observed by Han et al. (2016). George et al. (2005) reported about a 2-fold increased NO_2 uptake coefficients for irradiated organic substrates (benzophenone, catechol, anthracene) compared to dark conditions, in the order of $(0.6–5) \times 10^{-6}$. NO_2 uptake coefficients on gentisic acid and tannic acid were in between $(3.3–4.8) \times 10^{-7}$ (Sosedova et al., 2011), still higher than on fresh soot or dust (about 1×10^{-7}; Monge et al., 2010; Ndour et al., 2008). The NO_2 uptake coefficients on BSA in the presence of O_3 (1×10^{-5}, for 26 ppb NO_2 and 20 ppb O_3) published by Shiraiwa et al. (2012) were somewhat higher than the values calculated here without O_3 but with light.

It was not possible to extract a set of parameters for a Langmuir–Hinshelwood mechanism (like Langmuir equilibrium constant, surface accommodation coefficient or second-order rate constant) from the presented data. The saturating behavior of photochemical HONO production may be due to either the adsorbed precursor on the surface or due to a photochemical competition process, which also leads to a Lindemann–Hinshelwood type kinetic expression (Minero, 1999).

4 Summary and conclusion

Photochemical nitration of proteins accompanied by formation of HONO by (i) heterogeneous conversion of NO_2 and (ii) decomposition of nitrated proteins was studied under relevant atmospheric conditions. NO_2 concentrations ranged from 20 ppb (typical for urban regions in Europe and USA) up to 100 ppb (representative for highly polluted industrial regions). The applied relative humidity of up to 80 % and light intensities of up to $161\,W\,m^{-2}$ are common on cloudy days. Under illuminated conditions very low nitration of proteins or even no native protein was observed, indicating a light-induced decomposition of nitrated proteins to shorter peptides. These might still include nitrated residues of which potential health effects are not yet known. An average effective rate constant of the total NO_2-HONO conversion of $3.3 \times 10^{-6}\,s^{-1}$ (for about $120\,cm^2$ of protein surface, layer thickness 240 nm and a layer volume of $0.003\,cm^3$; surface/volume ratio $\sim 40\,000\,cm^{-1}$) or $8.25 \times 10^{-8}\,s^{-1}\,cm^{-2}$ BSA layer was obtained. At 20 ppb NO_2 HONO formation of $19.8\,ppb\,h^{-1}\,m^{-2}$ on a pure BSA surface could be estimated. While heterogeneous HONO formation of BSA exposed to NO_2 revealed light saturation at intensities higher than $161\,W\,m^{-2}$, the HONO formation from previously nitrated OVA was linearly increasing over the whole light intensity range investigated. The latter let assume even higher HONO formation under sunny (clear-sky) ambient atmospheric conditions. No data about representative protein surface areas on atmospheric aerosol particles are available. However, the number and mass concentration of primary biological aerosol particles such as pollen, fungal spores and bacteria, containing proteins, are in the range of 10–10^4 m^{-3} and 10^{-3}–1 µg m^{-3}, respectively (Despres et al., 2012; Shiraiwa et al., 2012). Typical aerosol surface concentrations in rural regions are about 100 µm^2 cm^{-3}. Stemmler et al. (2007) estimated a HONO formation of 1.2 ppt h^{-1} on pure humic acid aerosols in environmental conditions. As NO$_2$ uptake coefficients and HONO formation rates on proteins are similar to humic acid, but only about 5 % of the aerosol mass can be assumed to consist of proteins, it can be anticipated that HONO formation on aerosol is not a significant HONO source in ambient environmental settings. However, proteins on ground surfaces (soil, plants, etc.) might play a more important role. Accordingly, Stemmler et al. (2006 and 2007) suggested that NO$_2$ conversion on soil covered with humic acid would be sufficient to explain missing HONO sources up to 700 ppt h^{-1}. Therefore it is difficult to estimate the importance of HONO formation on protein surface and its contribution to the HONO budget. In many studies the calculated unknown source strength of daytime HONO formation is within a range of about 200–800 ppt h^{-1} (Kleffmann et al., 2005; Acker et al., 2006; Li et al., 2012).

Competing interests. The authors declare that they have no conflict of interest.

Edited by: Alexander Laskin

References

Acker, K., Moller, D., Wieprecht, W., Meixner, F. X., Bohn, B., Gilge, S., Plass-Dulmer, C., and Berresheim, H.: Strong daytime production of OH from HNO$_2$ at a rural mountain site, Geophys. Res. Lett., 33, L02809, https://doi.org/10.1029/2005GL024643, 2006.

Alfassi, Z. B.: Selective oxidation of tyrosine oxidation by NO$_2$ and ClO$_2$ at basic pH, Radiat. Phys. Chem., 29, 405–406, 1987.

Alicke, B., Platt, U., and Stutz, J.: Impact of nitrous acid photolysis on the total hydroxyl radical budget during the Limitation of Oxidant Production/Pianura Padana Produzione di Ozono study in Milan, J. Geophys. Res., 107, 8196, https://doi.org/10.1029/2000JD000075, 2002.

Ammann, M., Kalberer, M., Jost, D. T., Tobler, L., Rossler, E., Piguet, D., Gaggeler, H. W., and Baltensperger, U.: Heterogeneous production of nitrous acid on soot in polluted air masses, Nature, 395, 157–160, 1998.

Ammann, M., Rossler, E., Strekowski, R., and George, C.: Nitrogen dioxide multiphase chemistry: uptake kinetics on aqueous solutions containing phenolic compounds, Phys. Chem. Chem. Phys., 7, 2513–2518, https://doi.org/10.1039/B501808K, 2005.

Arens, F., Gutzwiller, L., Baltensperger, U., Gaggeler, H. W., and Ammann, M.: Heterogeneous reaction of NO$_2$ on

diesel soot particles, Environ. Sci. Technol., 35, 2191–2199, https://doi.org/10.1021/es000207s, 2001.

Arens, F., Gutzwiller, L., Gaggeler, H. W., and Ammann, M.: The reaction of NO$_2$ with solid anthrarobin (1,2,10-trihydroxy-anthracene), Phys. Chem. Chem. Phys., 4, 3684–3690, https://doi.org/10.1039/B201713J, 2002.

Aubin, D. G. and Abbatt, J. P. D.: Interaction of NO$_2$ with hydrocarbon soot: focus on HONO yield, surface modification, and mechanism, J. Phys. Chem.-US, 111, 6263–6273, 2007.

Bensasson, R. V., Land, E. J., and Truscott, T. G.: Excited States and Free Radicals in Biology and Medicine, Oxford University Press, Oxford, 1993.

Bejan, I., Abd El Aal, Y., Barnes, I., Benter, T., Bohn, B., Wiesen, P., and Kleffmann, J.: The photolysis of ortho-nitrophenols: a new gas phase source of HONO, Phys. Chem. Chem. Phys., 8, 2028–2035, 2006.

Berto, S., De Laurentiis, E., Tota, T., Chiavazza, E., Daniele, P. G., Minella, M., Isaia, M., Brigante, M., and Vione, D.: Properties of the humic-like material arising from the photo-transformation of l-tyrosine, Sci. Total Environ., 545, 434–444, https://doi.org/10.1016/j.scitotenv.2015.12.047, 2016.

Boreen, A. L., Edhlund, B. L., Cotner, J. B., and McNeill, K.: Indirect Photodegradation of Dissolved Free Amino Acids: The Contribution of Singlet Oxygen and the Differential Reactivity of DOM from Various Sources, Environ. Sci. Technol., 42, 5492–5498, https://doi.org/10.1021/es800185d, 2008.

Brigante, M., Cazoir, D., D'Anna, B., George, C., and Donaldson, D. J.: Photoenhanced uptake of NO$_2$ by pyrene solid films, J. Phys. Chem.-US, 112, 9503–9508, https://doi.org/10.1021/jp802324g, 2008.

Bujacz, A.: Structures of bovine, equine and leporine serum albumin, Acta Crystallogr. D, 68, 1278–1289, https://doi.org/10.1107/S0907444912027047, 2012.

Costabile, F., Amoroso, A., and Wang, F.: Sub-μm particle size distributions in a suburban Mediterranean area. Aerosol populations and their possible relationship with HONO mixing ratios, Atmos. Environ., 44, 5258–5268, 2010.

D'Amato, G., Cecchi, L., Bonini, S., Nunes, C., Annesi-Maesano, I., Behrendt, H., Liccardi, G., Popov, T., and Van Cauwenberge, P.: Allergenic pollen and pollen allergy in Europe, Allergy, 62, 976–990, https://doi.org/10.1111/j.1398-9995.2007.01393.x, 2007.

Davies, M. J.: Identification of a globin free radical in equine myoglobin treated with peroxides, Biochim. Biophys. Acta, 1077, 86-90, https://doi.org/10.1016/0167-4838(91)90529-9, 1991.

Després, V., Huffman, J. A., Burrows, S. M., Hoose, C., Safatov, A., Buryak, G., Fröhlich-Nowoisky, J., Elbert, W., Andreae, M., Pöschl, U., and Jaenicke, R.: Primary biological aerosol particles in the atmosphere: a review, Tellus B, 64, 15598, https://doi.org/10.3402/tellusb.v64i0.15598, 2012.

Elshorbany, Y. F., Steil, B., Brühl, C., and Lelieveld, J.: Impact of HONO on global atmospheric chemistry calculated with an empirical parameterization in the EMAC model, Atmos. Chem. Phys., 12, 9977–10000, https://doi.org/10.5194/acp-12-9977-2012, 2012.

Elshorbany, Y. F., Crutzen, P. J., Steil, B., Pozzer, A., Tost, H., and Lelieveld, J.: Global and regional impacts of HONO on the chemical composition of clouds and aerosols, Atmos. Chem. Phys., 14, 1167–1184, https://doi.org/10.5194/acp-14-1167-2014, 2014.

Finlayson-Pitts, B. J., Wingen, L. M., Sumner, A. L., Syomin, D., and Ramazan, K. A.: The heterogeneous hydrolysis of NO_2 in laboratory systems and in outdoor and indoor atmospheres: an integrated mechanism, Phys. Chem. Chem. Phys., 5, 223–242, https://doi.org/10.1039/b208564j, 2003.

Franze, T., Krause, K., Niessner, R., and Pöeschl, U.: Proteins and amino acids in air particulate matter, J. Aerosol Sci., 34, S777–S778, 2003.

Franze, T., Weller, M. G., Niessner, R., and Pöschl, U.: Protein nitration by polluted air, Environ. Sci. Technol., 39, 1673–1678, https://doi.org/10.1021/es0488737, 2005.

Gardner, E. P., Sperry, P. D., and Calvert, J. G.: Primary quantum yields of NO_2 photodissociation, J. Geophys. Res., 92, 6642–6652, https://doi.org/10.1029/JD092iD06p06642, 1987.

George, C., Strekowski, R. S., Kleffmann, J., Stemmler, K., and Ammann, M.: Photoenhanced uptake of gaseous NO_2 on solid-organic compounds: a photochemical source of HONO?, Faraday Discuss., 130, 195–210, 2005.

Goeschen, C., Wibowo, N., White, J. M., and Wille, U.: Damage of aromatic amino acids by the atmospheric free radical oxidant $NO_3^{Radical}$ in the presence of $NO_2^{•}$, N_2O_4, O_3 and O_2, Org. Biomol. Chem., 9, 3380–3385, https://doi.org/10.1039/C0OB01186J, 2011.

Gruijthuijsen, Y. K., Grieshuber, I., Stoecklinger, A., Tischler, U., Fehrenbach, T., Weller, M. G., Vogel, L., Vieths, S., Poeschl, U., and Duschl, A.: Nitration enhances the allergenic potential of proteins, Int. Arch. Allergy Imm., 141, 265–275, 2006.

Gutzwiller, L., Arens, F., Baltensperger, U., Gäggeler, H. W., and Ammann, M.: Significance of Semivolatile Diesel Exhaust Organics for Secondary HONO Formation, Environ. Sci. Technol., 36, 677–682, https://doi.org/10.1021/es015673b, 2002.

Han, C., Liu, Y., and He, H.: Role of Organic Carbon in Heterogeneous Reaction of $NO2$ with Soot, Environ. Sci. Technol., 47, 3174–3181, https://doi.org/10.1021/es304468n, 2013.

Han, C., Yang, W. J., Wu, Q. Q., Yang, H., and Xue, X. X.: Heterogeneous photochemical conversion of NO_2 to HONO on the humic acid surface under simulated sunlight, Environ. Sci. Technol., 50, 5017–5023, 2016.

Heland, J., Kleffmann, J., Kurtenbach, R., and Wiesen, P.: A new instrument to measure gaseous nitrous acid (HONO) in the atmosphere, Environ. Sci. Technol., 35, 3207–3212, 2001.

Houee-Levin, C., Bobrowski, K., Horakova, L., Karademir, B., Schoneich, C., Davies, M. J., and Spickett, C. M.: Exploring oxidative modifications of tyrosine: an update on mechanisms of formation, advances in analysis and biological consequences, Free Radical Res., 49, 347–373, https://doi.org/10.3109/10715762.2015.1007968, 2015.

Johnston, H. S., Davis, H. F., and Lee, Y. T.: NO_3 photolysis product channels: quantum yields from observed energy thresholds, J. Phys. Chem.-US, 100, 4713–4723, https://doi.org/10.1021/jp952692x, 1996.

Kalberer, M., Ammann, M., Arens, F., Gaggeler, H. W., and Baltensperger, U.: Heterogeneous formation of nitrous acid (HONO) on soot aerosol particles, J. Geophys. Res., 104, 13825–13832, 1999.

Kampf, C. J., Liu, F., Reinmuth-Selzle, K., Berkemeier, T.,

Meusel, H., Shiraiwa, M., and Pöschl, U.: Protein cross-linking and oligomerization through dityrosine formation upon exposure to ozone, Environ. Sci. Technol., 49, 10859–10866, https://doi.org/10.1021/acs.est.5b02902, 2015.

Kleffmann, J., H. Becker, K., Lackhoff, M., and Wiesen, P.: Heterogeneous conversion of NO_2 on carbonaceous surfaces, Phys. Chem. Chem. Phys., 1, 5443–5450, 1999.

Kleffmann, J., Gavriloaiei, T., Hofzumahaus, A., Holland, F., Koppmann, R., Rupp, L., Schlosser, E., Siese, M., and Wahner, A.: Daytime formation of nitrous acid: a major source of OH radicals in a forest, Geophys. Res. Lett., 32, L05818, https://doi.org/10.1029/2005GL022524, 2005.

Kurtenbach, R., Becker, K. H., Gomes, J. A. G., Kleffmann, J., Lorzer, J. C., Spittler, M., Wiesen, P., Ackermann, R., Geyer, A., and Platt, U.: Investigations of emissions and heterogeneous formation of HONO in a road traffic tunnel, Atmos. Environ., 35, 3385–3394, 2001.

Lang-Yona, N., Shuster-Meiseles, T., Mazar, Y., Yarden, O., and Rudich, Y.: Impact of urban air pollution on the allergenicity of Aspergillus fumigatus conidia: Outdoor exposure study supported by laboratory experiments, Sci. Total Environ., 541, 365–371, https://doi.org/10.1016/j.scitotenv.2015.09.058, 2016.

Laufs, S. and J. Kleffmann: Investigations on HONO formation from photolysis of adsorbed HNO3 on quartz glass surfaces, Phys. Chem. Chem. Phys., 18, 9616–9625, 2016.

Levy, H.: Normal atmosphere: large radical and formaldehyde concentrations predicted, Science, 173, 141–143, 1971.

Li, X., Brauers, T., Häseler, R., Bohn, B., Fuchs, H., Hofzumahaus, A., Holland, F., Lou, S., Lu, K. D., Rohrer, F., Hu, M., Zeng, L. M., Zhang, Y. H., Garland, R. M., Su, H., Nowak, A., Wiedensohler, A., Takegawa, N., Shao, M., and Wahner, A.: Exploring the atmospheric chemistry of nitrous acid (HONO) at a rural site in Southern China, Atmos. Chem. Phys., 12, 1497–1513, https://doi.org/10.5194/acp-12-1497-2012, 2012.

Li, G., Su, H., Li, X., Kuhn, U., Meusel, H., Hoffmann, T., Ammann, M., Pöschl, U., Shao, M., and Cheng, Y.: Uptake of gaseous formaldehyde by soil surfaces: a combination of adsorption/desorption equilibrium and chemical reactions, Atmos. Chem. Phys., 16, 10299–10311, https://doi.org/10.5194/acp-16-10299-2016, 2016.

Menetrez, M. Y., Foarde, K. K., Dean, T. R., Betancourt, D. A., and Moore, S. A.: An evaluation of the protein mass of particulate matter, Atmos. Environ., 41, 8264–8274, https://doi.org/10.1016/j.atmosenv.2007.06.021, 2007.

Meusel, H., Kuhn, U., Reiffs, A., Mallik, C., Harder, H., Martinez, M., Schuladen, J., Bohn, B., Parchatka, U., Crowley, J. N., Fischer, H., Tomsche, L., Novelli, A., Hoffmann, T., Janssen, R. H. H., Hartogensis, O., Pikridas, M., Vrekoussis, M., Bourtsoukidis, E., Weber, B., Lelieveld, J., Williams, J., Pöschl, U., Cheng, Y., and Su, H.: Daytime formation of nitrous acid at a coastal remote site in Cyprus indicating a common ground source of atmospheric HONO and NO, Atmos. Chem. Phys., 16, 14475–14493, https://doi.org/10.5194/acp-16-14475-2016, 2016.

Miguel, A. G., Cass, G. R., Glovsky, M. M., and Weiss, J.: Allergens in paved road dust and airborne particles, Environ. Sci. Technol., 33, 4159–4168, 1999.

Mikhailov, E., Vlasenko, S., Niessner, R., and Pöschl, U.: Interaction of aerosol particles composed of protein and saltswith water vapor: hygroscopic growth and microstructural rearrangement,

Atmos. Chem. Phys., 4, 323–350, https://doi.org/10.5194/acp-4-323-2004, 2004.

Minero, C.: Kinetic analysis of photoinduced reactions at the water semiconductor interface, Catal. Today, 54, 205–216, 1999.

Monge, M. E., D'Anna, B., Mazri, L., Giroir-Fendler, A., Ammann, M., Donaldson, D. J., and George, C.: Light changes the atmospheric reactivity of soot, P. Natl. Acad. Sci. USA, 107, 6605–6609, https://doi.org/10.1073/pnas.0908341107, 2010.

Ndour, M., D'Anna, B., George, C., Ka, O., Balkanski, Y., Kleffmann, J., Stemmler, K., and Ammann, M.: Photoenhanced uptake of NO_2 on mineral dust: laboratory experiments and model simulations, Geophys. Res. Lett., 35, L05812, https://doi.org/10.1029/2007gl032006, 2008.

Neves-Petersen, M. T., Petersen, S., and Gajula, G. P.: UV Light Effects on Proteins: From Photochemistry to Nanomedicine, in: Molecular Photochemistry – Various Aspects, edited by: Saha, S., InTech, Chapter 7, 125–158, https://doi.org/10.5772/37947, 2012.

Nikogosyan, D. N. and Gorner, H.: Laser-induced photodecomposition of amino acids and peptides: extrapolation to corneal collagen, IEEE J. Sel. Top. Quant., 5, 1107–1115, https://doi.org/10.1109/2944.796337, 1999.

Notholt, J., Hjorth, J., and Raes, F.: Formation of HNO_2 on aerosol surfaces during foggy periods in the presence of NO and NO_2, Atmos. Environ. A-Gen., 26, 211–217, 1992.

Oswald, R., Behrendt, T., Ermel, M., Wu, D., Su, H., Cheng, Y., Breuninger, C., Moravek, A., Mougin, E., Delon, C., Loubet, B., Pommerening-Roeser, A., Soergel, M., Poeschl, U., Hoffmann, T., Andreae, M. O., Meixner, F. X., and Trebs, I.: HONO emissions from soil bacteria as a major source of atmospheric reactive nitrogen, Science, 341, 1233–1235, 2013.

Petersson, A.-S., Steen, H., Kalume, D. E., Caidahl, K., and Roepstorff, P.: Investigation of tyrosine nitration in proteins by mass spectrometry, J. Mass Spectrom., 36, 616–625, https://doi.org/10.1002/jms.161, 2001.

Prutz, W. A.: Tyrosine oxidation by NO_2 in aqueous-solution, Z. Naturforsch. C, 39, 725–727, 1984.

Prutz, W. A., Monig, H., Butler, J., and Land, E. J.: Reactions of nitrogen dioxide in aqueous model systems – oxidation of tyrosine units in peptides and proteins, Arch. Biochem. Biophys., 243, 125–134, https://doi.org/10.1016/0003-9861(85)90780-5, 1985.

Pummer, B. G., Budke, C., Augustin-Bauditz, S., Niedermeier, D., Felgitsch, L., Kampf, C. J., Huber, R. G., Liedl, K. R., Loerting, T., Moschen, T., Schauperl, M., Tollinger, M., Morris, C. E., Wex, H., Grothe, H., Pöschl, U., Koop, T., and Fröhlich-Nowoisky, J.: Ice nucleation by water-soluble macromolecules, Atmos. Chem. Phys., 15, 4077–4091, https://doi.org/10.5194/acp-15-4077-2015, 2015.

Ramazan, K. A., Syomin, D., and Finlayson-Pitts, B. J.: The photochemical production of HONO during the heterogeneous hydrolysis of NO_2, Phys. Chem. Chem. Phys., 6, 3836–3843, 2004.

Reinmuth-Selzle, K., Ackaert, C., Kampf, C. J., Samonig, M., Shiraiwa, M., Kofler, S., Yang, H., Gadermaier, G., Brandstetter, H., Huber, C. G., Duschl, A., Oostingh, G. J., and Pöschl, U.: Nitration of the Birch Pollen Allergen Bet v 1.0101: Efficiency and site-selectivity of liquid and gaseous nitrating agents, J. Proteome Res., 13, 1570–1577, 2014.

Reisinger, A. R.: Observations of HNO_2 in the polluted winter atmosphere: possible heterogeneous production on aerosols, Atmos. Environ., 34, 3865–3874, 2000.

Ren, X., Brune, W. H., Oliger, A., Metcalf, A. R., Simpas, J. B., Shirley, T., Schwab, J. J., Bai, C., Roychowdhury, U., Li, Y., Cai, C., Demerjian, K. L., He, Y., Zhou, X., Gao, H., and Hou, J.: OH, HO_2, and OH reactivity during the PMTACS-NY Whiteface Mountain 2002 campaign: observations and model comparison, J. Geophys. Res., 111, D10S03, https://doi.org/10.1029/2005JD006126, 2006.

Riediker, M., Koller, T., and Monn, C.: Differences in size selective aerosol sampling for pollen allergen detection using high-volume cascade impactors, Clin. Exp. Allergy, 30, 867–873, https://doi.org/10.1046/j.1365-2222.2000.00792.x, 2000.

Ring, J., Kramer, U., Schafer, T., and Behrendt, H.: Why are allergies increasing?, Curr. Opin. Immunol., 13, 701–708, 2001.

Roehl, C. M., Orlando, J. J., Tyndall, G. S., Shetter, R. E., Vazquez, G. J., Cantrell, C. A., and Calvert, J. G.: Temperature dependence of the quantum yields for the photolysis of NO_2 near the dissociation limit, J. Phys. Chem.-US, 98, 7837–7843, https://doi.org/10.1021/j100083a015, 1994.

Salgado, M. S. and Rossi, M. J.: Flame soot generated under controlled combustion conditions: Heterogeneous reaction of NO_2 on hexane soot, Int. J. Chem. Kinet., 34, 620–631, https://doi.org/10.1002/kin.10091, 2002.

Selzle, K.; Ackaert, C.; Kampf, C. J., Kunert, A. T., Duschl, A., Oostingh, G. J., and Pöschl, U.: Determination of nitration degrees for the birch pollen allergen Bet v 1, Anal. Bioanal. Chem., 405, 8945–8949, 2013.

Shiraiwa, M., Ammann, M., Koop, T., and Pöschl, U.: Gas uptake and chemical aging of semisolid organic aerosol particles, P. Natl. Acad. Sci. USA, 108, 11003–11008, https://doi.org/10.1073/pnas.1103045108, 2011.

Shiraiwa, M., Selzle, K., Yang, H., Sosedova, Y., Ammann, M., and Poeschl, U.: Multiphase chemical kinetics of the nitration of aerosolized protein by ozone and nitrogen dioxide, Environ. Sci. Technol., 46, 6672–6680, 2012.

Sörgel, M., Regelin, E., Bozem, H., Diesch, J.-M., Drewnick, F., Fischer, H., Harder, H., Held, A., Hosaynali-Beygi, Z., Martinez, M., and Zetzsch, C.: Quantification of the unknown HONO daytime source and its relation to NO_2, Atmos. Chem. Phys., 11, 10433–10447, https://doi.org/10.5194/acp-11-10433-2011, 2011.

Sörgel, M., Trebs, I., Wu, D., and Held, A.: A comparison of measured HONO uptake and release with calculated source strengths in a heterogeneous forest environment, Atmos. Chem. Phys., 15, 9237–9251, https://doi.org/10.5194/acp-15-9237-2015, 2015.

Sosedova, Y., Rouviere, A., Bartels-Rausch, T., and Ammann, M.: UVA/Vis-induced nitrous acid formation on polyphenolic films exposed to gaseous NO_2, Photochem. Photobiol. Sci., 10, 1680–1690, 2011.

Stadler, D. and Rossi, M. J.: The reactivity of NO_2 and HONO on flame soot at ambient temperature: the influence of combustion conditions, Phys. Chem. Chem. Phys., 2, 5420–5429, https://doi.org/10.1039/b005680o, 2000.

Staton, S. J. R., Woodward, A., Castillo, J. A., Swing, K., and Hayes, M. A.: Ground level environmental protein concentrations in various ecuadorian environments: potential uses of aerosolized protein for ecological research, Ecol. Indic., 48, 389–395, https://doi.org/10.1016/j.ecolind.2014.08.036, 2015.

Stemmler, K., Ammann, M., Donders, C., Kleffmann, J., and George, C.: Photosensitized reduction of nitrogen dioxide on humic acid as a source of nitrous acid, Nature, 440, 195–198, 2006.

Stemmler, K., Ndour, M., Elshorbany, Y., Kleffmann, J., D'Anna, B., George, C., Bohn, B., and Ammann, M.: Light induced conversion of nitrogen dioxide into nitrous acid on submicron humic acid aerosol, Atmos. Chem. Phys., 7, 4237–4248, https://doi.org/10.5194/acp-7-4237-2007, 2007.

Su, H., Cheng, Y. F., Shao, M., Gao, D. F., Yu, Z. Y., Zeng, L. M., Slanina, J., Zhang, Y. H., and Wiedensohler, A.: Nitrous acid (HONO) and its daytime sources at a rural site during the 2004 PRIDE-PRD experiment in China, J. Geophys. Res., 113, D14312, https://doi.org/10.1029/2007JD009060, 2008.

Su, H., Cheng, Y., Oswald, R., Behrendt, T., Trebs, I., Meixner, F. X., Andreae, M. O., Cheng, P., Zhang, Y., and Poeschl, U.: Soil nitrite as a source of atmospheric HONO and OH radicals, Science, 333, 1616–1618, 2011.

Sumner, A. L., Menke, E. J., Dubowski, Y., Newberg, J. T., Penner, R. M., Hemminger, J. C., Wingen, L. M., Brauers, T., and Finlayson-Pitts, B. J.: The nature of water on surfaces of laboratory systems and implications for heterogeneous chemistry in the troposphere, Phys. Chem. Chem. Phys., 6, 604–613, https://doi.org/10.1039/B308125G, 2004.

Syomin, D. A. and Finlayson-Pitts, B. J.: HONO decomposition on borosilicate glass surfaces: implications for environmental chamber studies and field experiments, Phys. Chem. Chem. Phys., 5, 5236–5242, 2003.

Villena, G., Wiesen, P., Cantrell, C. A., Flocke, F., Fried, A., Hall, S. R., Hornbrook, R. S., Knapp, D., Kosciuch, E., Mauldin, R. L., McGrath, J. A., Montzka, D., Richter, D., Ullmann, K., Walega, J., Weibring, P., Weinheimer, A., Staebler, R. M., Liao, J., Huey, L. G., and Kleffmann, J.: Nitrous acid (HONO) during polar spring in Barrow, Alaska: a net source of OH radicals?, J. Geophys. Res., 116, D00R07, https://doi.org/10.1029/2011JD016643, 2011.

Weber, B., Wu, D., Tamm, A., Ruckteschler, N., Rodriguez-Caballero, E., Steinkamp, J., Meusel, H., Elbert, W., Behrendt, T., Soergel, M., Cheng, Y., Crutzen, P. J., Su, H., and Poeschi, U.: Biological soil crusts accelerate the nitrogen cycle through large NO and HONO emissions in drylands, P. Natl. Acad. Sci. USA, 112, 15384–15389, 2015.

Wong, K. W., Tsai, C., Lefer, B., Haman, C., Grossberg, N., Brune, W. H., Ren, X., Luke, W., and Stutz, J.: Daytime HONO vertical gradients during SHARP 2009 in Houston, TX, Atmos. Chem. Phys., 12, 635–652, https://doi.org/10.5194/acp-12-635-2012, 2012.

Yang, H., Zhang, Y. Y., and Pöschl, U.: Quantification of nitrotyrosine in nitrated proteins, Anal. Bioanal. Chem., 397, 879–886, 2010.

Zhang, Q. and Anastasio, C.: Free and combined amino compounds in atmospheric fine particles ($PM_{2.5}$) and fog waters from Northern California, Atmos. Environ., 37, 2247–2258, 2003.

Zhang, Y. Y., Yang, H., and Pöschl, U.: Analysis of nitrated proteins and tryptic peptides by HPLC-chip-MS/MS: site-specific quantification, nitration degree, and reactivity of tyrosine residues, Anal. Bioanal. Chem., 399, 459–471, 2011.

Zhou, X. L., Beine, H. J., Honrath, R. E., Fuentes, J. D., Simpson, W., Shepson, P. B., and Bottenheim, J. W.: Snowpack photochemical production of HONO: a major source of OH in the Arctic boundary layer in springtime, Geophys. Res. Lett., 28, 4087–4090, 2001.

Zhou, X. L., Civerolo, K., Dai, H. P., Huang, G., Schwab, J., and Demerjian, K.: Summertime nitrous acid chemistry in the atmospheric boundary layer at a rural site in New York State, J. Geophys. Res., 107, 4590, https://doi.org/10.1029/2001JD001539, 2002.

Zhou, X. L., Gao, H. L., He, Y., Huang, G., Bertman, S. B., Civerolo, K., and Schwab, J.: Nitric acid photolysis on surfaces in low-NO_x environments: significant atmospheric implications, Geophys. Res. Lett., 30, 2217, https://doi.org/10.1029/2003GL018620, 2003.

6

Modeling of the chemistry in oxidation flow reactors with high initial NO

Zhe Peng and Jose L. Jimenez

Cooperative Institute for Research in Environmental Sciences and Department of Chemistry, University of Colorado, Boulder, CO 80309, USA

Correspondence to: Zhe Peng (zhe.peng@colorado.edu) and Jose L. Jimenez (jose.jimenez@colorado.edu)

Abstract. Oxidation flow reactors (OFRs) are increasingly employed in atmospheric chemistry research because of their high efficiency of OH radical production from low-pressure Hg lamp emissions at both 185 and 254 nm (OFR185) or 254 nm only (OFR254). OFRs have been thought to be limited to studying low-NO chemistry (in which peroxy radicals (RO_2) react preferentially with HO_2) because NO is very rapidly oxidized by the high concentrations of O_3, HO_2, and OH in OFRs. However, many groups are performing experiments by aging combustion exhaust with high NO levels or adding NO in the hopes of simulating high-NO chemistry (in which $RO_2 + NO$ dominates). This work systematically explores the chemistry in OFRs with high initial NO. Using box modeling, we investigate the interconversion of N-containing species and the uncertainties due to kinetic parameters. Simple initial injection of NO in OFR185 can result in more RO_2 reacted with NO than with HO_2 and minor non-tropospheric photolysis, but only under a very narrow set of conditions (high water mixing ratio, low UV intensity, low external OH reactivity (OHR_{ext}), and initial NO concentration (NO^{in}) of tens to hundreds of ppb) that account for a very small fraction of the input parameter space. These conditions are generally far away from experimental conditions of published OFR studies with high initial NO. In particular, studies of aerosol formation from vehicle emissions in OFRs often used OHR_{ext} and NO^{in} several orders of magnitude higher. Due to extremely high OHR_{ext} and NO^{in}, some studies may have resulted in substantial non-tropospheric photolysis, strong delay to RO_2 chemistry due to peroxynitrate formation, VOC reactions with NO_3 dominating over those with OH, and faster reactions of OH–aromatic adducts with NO_2 than those with O_2, all of which are irrelevant to ambient VOC photooxidation chemistry. Some of the negative effects are the worst for alkene and aromatic precursors. To avoid undesired chemistry, vehicle emissions generally need to be diluted by a factor of > 100 before being injected into an OFR. However, sufficiently diluted vehicle emissions generally do not lead to high-NO chemistry in OFRs but are rather dominated by the low-NO $RO_2 + HO_2$ pathway. To ensure high-NO conditions without substantial atmospherically irrelevant chemistry in a more controlled fashion, new techniques are needed.

1 Introduction

The oxidation of gases that are emitted into the atmosphere, in particular volatile organic compounds (VOCs), is one of the most important atmospheric chemistry processes (Haagen-Smit, 1952; Chameides et al., 1988). VOC oxidation is closely related to radical production and consumption (Levy II, 1971), O_3 production, and the formation of secondary aerosols (Odum et al., 1996; Hoffmann et al., 1997; Volkamer et al., 2006; Hallquist et al., 2009), which have impacts on air quality and climate (Lippmann, 1991; Nel, 2005; Stocker et al., 2014).

Chemical reactors are critical tools for research of VOC oxidation. Oxidation reactions of interest often have typical timescales of hours to weeks. Studying these processes in ambient air can be confounded by dispersion and changes in ambient conditions, which often occur on similar timescales. Chemical reactors allow for the decoupling of these two types of processes. Also, they should be able to simulate the different regimes of reactions occurring in the atmosphere, e.g., VOC oxidation under the low- and high-NO conditions (peroxy radical fate dominated by reaction with HO_2 or NO) representing remote and urban areas, respectively (Orlando and Tyndall, 2012).

Large environmental chambers are a commonly used reactor type (Carter et al., 2005; Wang et al., 2011). They typically employ actinic wavelength (> 300 nm) light sources (e.g., outdoor solar radiation and UV black lights) to produce oxidants and radicals and have large volumes (on the order of several cubic meters or larger). However, the capability of generating sustained elevated levels of OH, the most important tropospheric oxidant, is usually limited in chambers, resulting in OH concentrations similar to those in the atmosphere (10^6–10^7 molecules cm^{-3}; Mao et al., 2009; Ng et al., 2010) and consequently long simulation times (typically hours) to reach OH equivalent ages of atmospheric relevance (George et al., 2007; Kang et al., 2007; Carlton et al., 2009; Seakins, 2010; Wang et al., 2011). The partitioning of gases and aerosols to chamber walls (usually made of Teflon) in timescales of tens of minutes to hours makes it difficult to conduct very long experiments that simulate high atmospherically relevant photochemical ages (Cocker et al., 2001; Matsunaga and Ziemann, 2010; Zhang et al., 2014; Krechmer et al., 2016). In addition, the long simulation times and large size of chambers and auxiliary equipment are logistically difficult for field deployment, and their cost limits the number of laboratories equipped with them.

Given the limitations of environmental chambers, a growing number of experimenters have instead employed oxidation flow reactors (OFRs). OFRs have a much smaller size (on the order of 10 L), efficiently generate OH via photolysis of H_2O and/or O_3 by more energetic 185 and 254 nm photons from low-pressure Hg lamps, and overcome the abovementioned shortcomings of chambers due to a much shorter residence time (George et al., 2007; Kang et al., 2007, 2011; Lambe et al., 2011). Moreover, OFRs are able to rapidly explore a wide range of OH equivalent ages within a short period (~ 2 h) during which significant changes in ambient conditions can usually be avoided in the case of field deployment (Ortega et al., 2016; Palm et al., 2016, 2017). Because of these advantages, OFRs have recently been widely used to study atmospheric chemistry, in particular secondary organic aerosol (SOA) formation and aging, in both the laboratory and the field (Kang et al., 2011; Li et al., 2013; Ortega et al., 2013, 2016; Tkacik et al., 2014; Palm et al., 2016).

In addition to experimental studies using OFRs, there has also been some progress in the characterization of OFR chemistry by modeling. Li et al. (2015) and Peng et al. (2015) developed a box model for OFR HO_x chemistry that predicts measurable quantities (e.g., OH exposure, OH_{exp}, in molecules cm^{-3} s and O_3 concentration, abbr. O_3 hereinafter, in ppm) in good agreement with experiments. This model has been used to characterize HO_x chemistry as a function of H_2O mixing ratio (abbr. H_2O hereinafter, unitless), UV light intensity (abbr. UV hereinafter, in photons cm^{-2} s^{-1}),

and external OH reactivity (in s^{-1}, $OHR_{ext} = \sum k_i c_i$, i.e., the sum of the products of concentrations of externally introduced OH-consuming species, c_i, and rate constants of their reactions with OH, k_i). Based on this characterization, Peng et al. (2015) found that OH suppression, i.e., the reduction of OH concentration caused by OHR_{ext}, is a common feature under many typical OFR operation conditions. Peng et al. (2016) systematically examined the relative importance of non-OH (including non-tropospheric) reactants in the fate of VOCs over a wide range of conditions and provided guidelines for OFR operation to avoid non-tropospheric VOC photolysis, i.e., VOC photolysis at 185 and 254 nm.

In previous OFR modeling studies, NO_x chemistry was not investigated in detail since in such in typical OFR experiments with large amounts of oxidants (e.g., OH, HO_2, and O_3) NO would be very rapidly oxidized and thus unable to compete with HO_2 for reaction with peroxy radicals (RO_2). Li et al. (2015) estimated an NO (NO_2) lifetime of ~ 0.5 (~ 1.5) s under a typical OFR condition. From these estimates, OFRs processing ambient air or laboratory air without a large addition of NO_x were assumed to not be suitable for studying oxidation mechanisms relevant to polluted conditions under higher NO concentrations. OFRs have recently been used to conduct laboratory experiments with very high initial NO_x levels (Liu et al., 2015) and deployed to an urban tunnel where NO_x was high enough to be a major OH reactant (Tkacik et al., 2014). The former study reported evidence for the incorporation of nitrogen into SOA. OFRs have also been increasingly employed to process emissions of vehicles, biomass burning, and other combustion sources (Table 1) in which NO can often be hundreds of ppm (Ortega et al., 2013; Martinsson et al., 2015; Karjalainen et al., 2016; Link et al., 2016; Schill et al., 2016; Alanen et al., 2017; Simonen et al., 2017). It can be expected that such a high NO input together with very high VOC concentrations would cause a substantial deviation from the good OFR operation conditions identified in Peng et al. (2016). Very recently, N_2O injection has been proposed by Lambe et al. (2017) as a way to study the oxidation of VOCs under high NO conditions in an OFR. As more OFR studies at high NO_x levels are conducted, there is a growing need to understand the chemistry of N-containing species in OFRs and whether it proceeds along atmospherically relevant channels.

In this study, we present the first comprehensive model of OFR NO_y chemistry. We extend the model of Li et al. (2015) and Peng et al. (2015) by including a scheme for NO_y species. Then this model is used to investigate (i) whether an OFR with initial NO injection results in NO significantly reacting with RO_2 under any conditions, (ii) whether previously published OFR experiments with high initial NO concentrations led to $RO_2 + NO$ being dominant in VOC oxidation without negative side effects (e.g., non-tropospheric reactions), and iii) how to avoid undesired chemistry in future studies. The results can provide insights into the design and

Table 1. Experimental conditions of several OFR studies with high NO injection.

Study	Source type	Temperature (K)	Relative humidity (%)	Dilution factor	External OH reactivity of undiluted source (s^{-1})	Source NO$_x$ concentration (ppm)
Link et al. (2016)	Diesel vehicle emission		50	45–110	~ 5000[1]	436[1]
Martinsson et al. (2015)	Biomass burning emission			1700	156 400[1]	154
Karjalainen et al. (2016)	Gasoline vehicle emission	295	60	12	$\sim 73\,000$[2,a]	~ 400[1,b]
Liu et al. (2015)	Purified gas	293	13	1	~ 1400[1,a]	10[1,b]
Tkacik et al. (2014)	Tunnel air	293	42	1	~ 60[1,a]	~ 0.8[1]
Ortega et al. (2013)	Biomass burning emission	290	30	~ 500	$\sim 250\,000$[1]	~ 0.2

[1] Maximum value in the study; [2] value at the moment of maximum NO emission; [a] NO$_y$ species excluded; [b] NO only.

interpretation of future OH-oxidation OFR experiments with large amounts of NO$_x$ injection.

2 Methods

The physical design of the OFR modeled in the present work, the chemical kinetics box model, and the method of propagating and analyzing the parametric uncertainties on the model have already been introduced previously (Kang et al., 2007; Li et al., 2015; Peng et al., 2015). We only provide brief descriptions for them below.

2.1 Potential aerosol mass flow reactor

The OFR modeled in this study is the potential aerosol mass (PAM) flow reactor first introduced by Kang et al. (2007). The PAM OFR is a cylindrical vessel with a volume of ~ 13 L, equipped with low-pressure Hg lamps (model no. 82-9304-03; BHK Inc.) to generate 185 and 254 nm UV light. This popular design is being used by many atmospheric chemistry research groups, particularly those studying SOA (Lambe and Jimenez, 2017 and references therein). When the lamps are mounted inside Teflon sleeves, photons at both wavelengths are transmitted and contribute to OH production (OFR185 mode). In OFR185, H$_2$O photolyzed at 185 nm produces OH and HO$_2$, while O$_2$ photolyzed at the same wavelengths results in O$_3$ formation. O(^{1}D) is produced via O$_3$ photolysis at 254 nm and generates additional OH through its reaction with H$_2$O. The 185 nm lamp emissions can be filtered by mounting the lamps inside quartz sleeves, leaving only 254 nm photons to produce OH (OFR254 mode). In this mode, injection of externally formed O$_3$ is necessary to ensure OH production. As the amount of O$_3$ injected is a key parameter under some conditions (Peng et al., 2015), we adopt the notation OFR254-X to denote OFR254 experiments with X ppm initial O$_3$ (O$_{3,in}$). In this study, we investigate OFR experiments with NO injected and thus utilize "OFR185-iNO" to describe the OFR185 mode of operation with initially (at the reactor entrance) injected NO.

The same terminology is used for the OFR254 mode. For instance, the initial NO injection into OFR254-7 is denoted as OFR254-7-iNO.

2.2 Model description

The basic framework of the box model used in this study, a standard chemical kinetics model, is the same as in Peng et al. (2015). Plug flow is assumed in the model, since approximately taking residence time distribution into account leads to similar results under most conditions but at much higher computational expense (Peng et al., 2015). In addition to the reactions in the model of Peng et al. (2015) including all HO$_x$ reactions available in the JPL Chemical Kinetic Data Evaluation (Sander et al., 2011), all gas-phase NO$_y$ reactions available in the JPL database except those of organic nitrates and peroxynitrates are also considered in the current reaction scheme. An updated JPL evaluation was published recently (Burkholder et al., 2015) with slightly different ($\sim 20\,\%$) rate constants for NO$_2 +$ HO$_2 +$ M $\rightarrow$ HO$_2$NO$_2 +$ M and NO$_2 +$ NO$_3 \rightarrow$ N$_2$O$_5$. The updated rate constants only result in changes of ~ 10–20 % in the concentrations of the species directly consumed or produced by these reactions. These changes are smaller than the parametric uncertainties of the model (see Sect. 3.1.3). For other species, concentration changes are negligible. HO$_2$NO$_2 +$ M $\rightarrow$ HO$_2 +$ NO$_2 +$ M and N$_2$O$_5 +$ M $\rightarrow$ NO$_2 +$ NO$_3 +$ M, are also included in the scheme with kinetic parameters from the IUPAC Task Group on Atmospheric Chemical Kinetic Data Evaluation (Ammann et al., 2016). As in Peng et al. (2015, 2016), SO$_2$ is used as a surrogate for external OH reactants (e.g., VOCs). NO$_y$ species, although also external OH reactants, are explicitly treated in the model and *not* counted in OHR$_{ext}$ in this work. Therefore, OHR$_{ext}$ stands for *non*-NO$_y$ OHR$_{ext}$ only hereinafter unless otherwise stated.

Also, particle-phase chemistry and physical and chemical interactions of gas-phase species with particles are not considered in this study. We have made this assumption because of the following.

i. The presence of aerosols has typically negligible impacts on the gas-phase chemistry of radicals, NO_y, and the OH reactants studied here. The condensational sink (CS) of ambient aerosols can rarely exceed $1 s^{-1}$ even in polluted areas and is usually 1–3 orders of magnitude lower (Donahue et al., 2016; Palm et al., 2016). Thus, even under the assumption of unity uptake coefficient, CS cannot compete with OHR_{ext} (usually on the order of $10 s^{-1}$ or higher) in OH loss. Uptake of NO onto aerosols only occurs through the reaction with RO_2 on the particle surface (Richards-Henderson et al., 2015), which is formed very slowly (see below) compared to gas-phase HO_x and NO_x chemistry. Uptake of HO_2, O_3, NO_3, etc. is even more unlikely to be of importance due to lower uptake coefficients (Moise and Rudich, 2002; Moise et al., 2002; Hearn and Smith, 2004; Lakey et al., 2015). Combustion exhausts can have high aerosol loadings with condensational sinks on the order of 10^2–$10^3 s^{-1}$ (Matti Maricq, 2007). Even if these exhausts are directly injected into the reactor without any pretreatment, uptake onto the particles still cannot play a major role in the fate of gas-phase radical and NO_x species since VOCs and NO_x in raw exhausts, which are proportionally orders of magnitude higher, still dominate the fate of oxidants. Dilution of combustion emissions simultaneously lowers condensational sinks and the sinks of oxidants due to chemical reactions with their relative importance remaining the same as in undiluted emissions.

ii. Gas-phase radical and NO_y species only has limited impacts on OA *chemistry* in this study. The heterogeneous oxidation of OA by OH is generally slow. Significant OA loss due to heterogeneous oxidation can only be seen at photochemical ages as high as weeks (Hu et al., 2016). The enhancement of heterogeneous oxidation due to NO is remarkable only at OH concentrations close to the ambient values but not at typical values in an OFR (Richards-Henderson et al., 2015).

It is well known that the aerosol concentration can have a major impact on the physical uptake of semivolatile and low-volatility gas-phase species. However, this process is not explicitly modeled in this study.

As OHR_{ext} plays a major and even dominant role in OH loss, it is an important approximation that the *real* OHR_{ext} decay (due to primary VOC oxidation and subsequent oxidation of higher-generation products) is surrogated by that of SO_2 (see Fig. S2 of Peng et al., 2015). Gas-phase measurements in the literature on laboratory studies revealed that there is large variability in the evolution of total OHR_{ext} during the oxidation of primary VOCs and subsequent oxidation of their intermediate products depending on the type of precursors (Nehr et al., 2014; Schwantes et al., 2017). This variability is clearly mainly due to the formation of different types and amounts of oxidation intermediates and products

contributing to OHR_{ext}. This variation is highly complex due to the large number of possible oxidation intermediates and the limited knowledge of detailed higher-generation mechanisms, and thus it is difficult to accurately capture even if modeling with a mechanism as explicit as the Master Chemical Mechanism (Schwantes et al., 2017). Therefore, it is justified to use a lumped surrogate to model the OHR_{ext} decay for simplicity and efficiency. This approximation is a substantial contributor to the uncertainty in our model. The uncertainties due to the types of oxidation intermediates and products are very likely larger than those due to mass transfer processes between gas and particle phases, wall losses, etc., which are not considered in this study.

A residence time of 180 s and the typical temperature (295 K) and atmospheric pressure (835 mbar) in Boulder, CO, USA are assumed for all model cases. The lower-than-sea-level pressure only leads to minor differences in the outputs (Li et al., 2015). We explore physical input cases evenly spaced in a logarithmic scale over very wide ranges: H_2O of 0.07–2.3 %, i.e., relative humidity (RH) of 2–71 % at 295 K; 185 nm UV of 1.0×10^{11}–1.0×10^{14} and 254 nm UV of 4.2×10^{13}–8.5×10^{15} photons $cm^{-2} s^{-1}$; OHR_{ext} of 1–$16000 s^{-1}$; $O_{3,in}$ of 2.2–70 ppm for OFR254; and initial NO mixing ratio (NO^{in}) from 10 ppt to 40 ppm. Conditions with $OHR_{ext} = 0$ are also explored. UV at 254 nm is estimated from that at 185 nm according to the relationship determined by Li et al. (2015). Several typical cases within this range and their corresponding four- or two-character labels (e.g., MM0V and HL) are defined in Table 2. Literature studies are modeled by adopting all reported parameters (e.g., residence time, H_2O, and $O_{3,in}$) and estimating any others that may be needed (e.g., UV) from the information provided in the papers.

In this study, OH equivalent ages are calculated under the assumption of an ambient OH concentration of 1.5×10^6 molecules cm^{-3} (Mao et al., 2009). Conditions leading to a ratio of RO_2 reacted with NO over the entire residence time $[r(RO_2 + NO)]$ to that with HO_2 $[r(RO_2 + HO_2)]$ larger than 1 are regarded as "high NO" (under the assumption of constant OHR_{ext} from VOCs; see Sect. S1 for more details), where $[r(X)]$ is the total reactive flux for reaction X over the entire residence time. $F185_{exp}/OH_{exp}$ and $F254_{exp}/OH_{exp}$ are used as measures of the relative importance of VOC photolysis at 185 and 254 nm to their reactions with OH, respectively; $F185_{exp}$ ($F254_{exp}$) are 185 (254) nm photon flux exposure, i.e., the product of 185 (254) nm photon flux and time. Readers may refer to Figs. 1 and 2 of Peng et al. (2016) for the determination of the relative importance of non-tropospheric (185 and 254 nm) photolysis of individual VOCs. Although the relative importance of non-tropospheric photolysis depends on individual VOCs, in the present work, we set criteria on $F185_{exp}/OH_{exp} < 3 \times 10^3$ cm s^{-1} and $F254_{exp}/OH_{exp} < 4 \times 10^5$ cm s^{-1} to define "good" conditions and $F185_{exp}/OH_{exp} < 1 \times 10^5$ cm s^{-1} and

Table 2. Code of the labels of typical cases. A case label can be composed of four characters denoting the water mixing ratio, the photon flux, the external OH reactivity excluding N-containing species, and the initial NO mixing ratio. A case label can also be composed of two characters denoting the water mixing ratio and the photon flux.

	Water mixing ratio	Photon flux	External OH reactivity (no ON)	Initial NO mixing ratio
Options	$L = $ low (0.07 %)	$L = $ low (10^{11} photons cm^{-2} s^{-1} at 185 nm; 4.2×10^{13} photons cm^{-2} s^{-1} at 254 nm)	0	0
	$M = $ medium (1 %)	$M = $ medium (10^{13} photons cm^{-2} s^{-1} at 185 nm; 1.4×10^{15} photons cm^{-2} s^{-1} at 254 nm)	$L = $ low (10 s^{-1})	$L = $ low (10 ppb)
	$H = $ high (2.3 %)	$H = $ high (10^{14} photons cm^{-2} s^{-1} at 185 nm; 8.5×10^{15} photons cm^{-2} s^{-1} at 254 nm)	$H = $ high (100 s^{-1})	$H = $ high (316 ppb)
			$V = $ very high (1000 s^{-1})	$V = $ very high (10 ppm)
Example	LH0V:	low water mixing ratio, high photon flux, no external OH reactivity (excluding ON), very high initial NO mixing ratio		
	ML:	medium water mixing ratio, low photon flux		

Table 3. Definition of condition types in this study (good/risky/bad high-/low-NO conditions).

Condition	Good	Risky	Bad
Criterion	F185$_{\text{exp}}$/OH$_{\text{exp}} < 3 \times 10^3$ cm s^{-1} and F254$_{\text{exp}}$/OH$_{\text{exp}} < 4 \times 10^5$ cm s^{-1}	F185$_{\text{exp}}$/OH$_{\text{exp}} < 1 \times 10^5$ cm s^{-1} and F254$_{\text{exp}}$/OH$_{\text{exp}} < 1 \times 10^7$ cm s^{-1} (excluding good conditions)	F185$_{\text{exp}}$/OH$_{\text{exp}} \geq 1 \times 10^5$ cm s^{-1} or F254$_{\text{exp}}$/OH$_{\text{exp}} \geq 1 \times 10^7$ cm s^{-1}

Condition	High NO	Low NO
Criterion*	$\frac{r(\text{RO}_2 + \text{NO})}{r(\text{RO}_2 + \text{HO}_2)} > 1$	$\frac{r(\text{RO}_2 + \text{NO})}{r(\text{RO}_2 + \text{HO}_2)} \leq 1$

* See Sect. S1 for details.

F254$_{\text{exp}}$/OH$_{\text{exp}} < 1 \times 10^7$ cm s^{-1} (excluding good conditions) to define "risky" conditions. Conditions with higher F185$_{\text{exp}}$/OH$_{\text{exp}}$ or F254$_{\text{exp}}$/OH$_{\text{exp}}$ are defined as "bad". Under good conditions, the photolysis of most VOCs has a relative contribution $< 20\,\%$ to their fate; under bad conditions, non-tropospheric photolysis is likely to be significant in all OFR experiments since it can hardly be avoided for oxidation intermediates, even if the precursor(s) does not photolyze at all. Under risky conditions, some species photolyzing slowly and/or reacting with OH rapidly (e.g., alkanes, aldehydes, and most biogenics) still have a relative photolysis contribution of $< 20\,\%$ to their fates, while species photolyzing more rapidly and/or reacting with OH more slowly (e.g., aromatics and other highly conjugated species and some saturated carbonyls) will undergo substantial non-tropospheric photolysis. Note that these definitions are slightly different than in Peng et al. (2016). All definitions of the types of conditions are summarized in Table 3.

2.3 Uncertainty analysis

We apply the same method as in Peng et al. (2014, 2015) to calculate and analyze the output uncertainties due to uncertain kinetic parameters in the model. Random samples following lognormal distributions are generated for all rate constants and photoabsorption cross sections in the model using uncertainty data available in the JPL database (Sander et al., 2011) or estimated based on IUPAC data (Ammann et al., 2016). Then, Monte Carlo uncertainty propagation (BIPM et al., 2008) is performed for these samples through the model to obtain the distributions of outputs. Finally, we compute squared correlation coefficients between corresponding input and output samples and apportion the relative contributions of individual kinetic parameters to the output uncertainties based on these coefficients (Saltelli et al., 2005).

3 Results and discussion

In this section, we study the NO_y chemistry in an OFR while considering relevant experimental issues. Based on these results, we propose some guidelines for OFR operation for high-NO OH oxidation of VOCs.

3.1 NO_y chemistry in typical OFR cases with initial NO injection

NO was thought to be unimportant (i.e., unable to significantly react with RO_2) in OFRs with initial NO injection (OFR-iNO) based on the argument that its lifetime is too short due to large amounts of O_3 OH, and HO_2 to compete with $RO_2 + HO_2$ (Li et al., 2015). We evaluate this issue below by calculating NO effective lifetime (τ_{NO}; in s), defined as NO exposure (NO_{exp}, in molecules cm^{-3} s) divided by initial NO concentration, under various conditions. This definition cannot effectively capture the true NO average lifetime if it is close to or longer than the residence time. In this case, τ_{NO} close to the residence time will be obtained, which is still long enough for our characterization purposes.

3.1.1 OFR185-iNO

In OFR185-iNO, NO is *not* oxidized extremely quickly under *all* conditions. For instance, under a typical condition in the midrange of the phase space shown in Fig. 1a, $\tau_{NO} \sim 13$ s. This lifetime is much shorter than the residence time but long enough for OH_{exp} to reach $\sim 3 \times 10^{10}$ molecules cm^{-3} s, which is equivalent to an OH equivalent age of ~ 6 h. Such an OH equivalent age is already sufficient to allow some VOC processing and even SOA formation to occur (Lambe et al., 2011; Ortega et al., 2016). Within τ_{NO}, NO suppresses HO_2 through the reaction $NO + HO_2 \rightarrow NO_2 + OH$, leading to NO_{exp}/HO_{2exp} of ~ 700 during this period, which is high enough for RO_2 to dominantly react with NO. Meanwhile, $NO + HO_2 \rightarrow NO_2 + OH$ enhances OH production, which helps OH_{exp} build up in a relatively short period. In addition, non-tropospheric photolysis of VOCs at 185 and 254 nm is minor ($F185_{exp}/OH_{exp} \sim 600$ cm s^{-1}, Fig. 1a) because of enhanced OH production and moderate UV. Therefore, such an OFR condition may be of some interest for high-NO VOC oxidation. We thus analyze the NO_y chemistry in OFR185-iNO in more detail below by taking the case shown in Fig. 1a as a representative example.

In OFR185-iNO, HO_x concentrations are orders of magnitude higher than in the atmosphere, while the amount of O_3 produced is relatively small during the first several seconds after the flow enters the reactor. As a result, NO is not oxidized almost exclusively by O_3 as in the troposphere, but also by OH and HO_2 to form HONO and NO_2, respectively (Fig. 1a). The large concentration of OH present then oxidizes HONO to NO_2 and NO_2 to HNO_3. Photolysis only plays a negligible role in the fate of HONO and NO_2 in OFRs

in contrast to the troposphere where it is the main fate of these species. This is because the reactions of HONO and NO_2 with OH are greatly accelerated in OFR compared to those in the troposphere, while photolysis not (Peng et al., 2016). The interconversion between NO_2 and HO_2NO_2 is also greatly accelerated (Fig. 1a) since a large amount of HO_2 promotes the formation of HO_2NO_2, the reaction with OH and thermal decomposition of which in turn enhance the recycling of NO_2. Though not explicitly modeled in this study, RO_2 is expected to undergo similar reactions with NO_2 to form reservoir species, i.e., peroxynitrates (Orlando and Tyndall, 2012). Peroxynitrates that decompose on timescales considerably longer than OFR residence times may serve as effectively permanent NO_y sinks in OFRs (see Sect. 3.4.1).

Interestingly but not surprisingly, the NO_y chemistry shown in Fig. 1a is far from temporally uniform during the OFR residence time (Fig. S1a in the Supplement). Within τ_{NO}, NO undergoes an e-fold decay as it is rapidly converted into NO_2 and HONO with concentrations that reach maxima around that time. After most NO is consumed, HONO and NO_2 also start to decrease, but significantly more slowly than NO, since they do not have as many or efficient loss pathways as NO. The reaction of OH with HONO, the dominant fate of HONO, is slower than that with NO (Fig. 1a). The net rate of the NO_2-to-HO_2NO_2 conversion becomes low because of the relatively fast reverse reaction (Fig. 1a). The total loss of NO_2 is also partially offset by the production from HONO. The generally stable concentrations of HONO and NO_2 (Fig. S1a) result in their respective reaction rates with OH that are comparable during and after τ_{NO} (Fig. 1a), as OH variation is also relatively small during the entire residence time (Fig. S1b). However, the NO_2-to-HO_2NO_2 conversion after τ_{NO} is much faster than during it (Fig. 1a) as a result of substantially decreased NO and HO_2 concomitantly increasing by > 1 order of magnitude after τ_{NO} (Fig. S1a and b). HNO_3 and HO_2NO_2, which are substantially produced only after NO_2 is built up, have much higher concentrations later than within τ_{NO}.

Under other OFR185-iNO conditions than in Fig. 1a, the major reactions interconverting NO_y species are generally the same, although their relative importance may vary. At lower NO^{in}, the perturbation of HO_x chemistry caused by NO_y species is smaller. The effects of NO^{in} less than 1 ppb (e.g., typical non-urban ambient concentrations) are generally negligible regarding HO_x chemistry. Regarding NO_y species, the pathways in Fig. 1a are still important under those conditions. At higher NO^{in} (e.g., > 1 ppm), one might expect NO_3 and N_2O_5 to play a role (as in OFR254-iNO; see Sect. 3.1.2 below) since high NO_y concentrations might enhance self- or cross- reactions of NO_y. However, this would not occur unless OH production is high since relatively low O_3 concentrations in OFR185-iNO cannot oxidize NO_2 to NO_3 rapidly. Also, a large amount of NO_y can lead to significant OH suppression. That would in turn slow down the NO_3 production from HNO_3 by OH. This is especially true

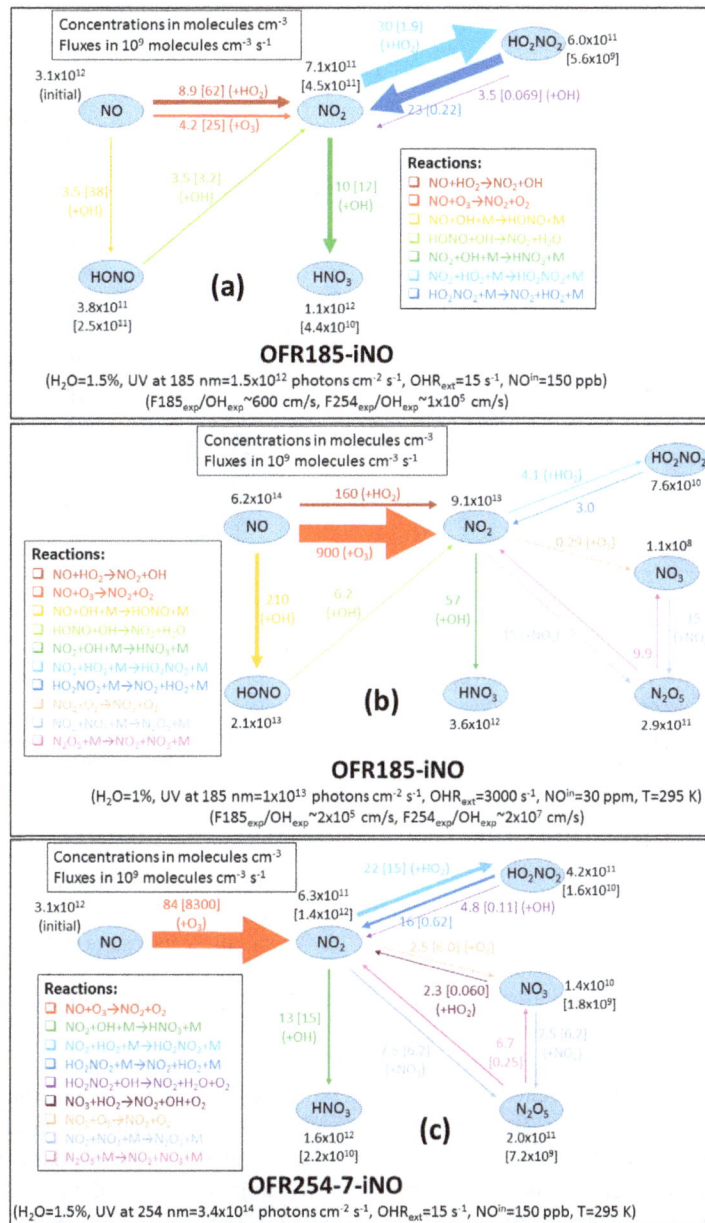

Figure 1. Schematics of main N-containing species and their major interconversion pathways under typical input conditions for **(a)** OFR185-iNO with $NO^{in} = 150$ ppb, **(b)** OFR254-7-iNO with $NO^{in} = 150$ ppb, and **(c)** OFR185-iNO with $NO^{in} = 30$ ppm. Species average concentrations (in molecules cm^{-3}) are shown in black beside species names. Arrows denote directions of the conversions. Average reaction fluxes (in units of 10^9 molecules cm^{-3} s^{-1}) are calculated according to the production rate and shown on or beside the corresponding arrows and in the same color. Within each schematic, the thickness of the arrows is a measure of their corresponding species flux. Multiple arrows in the same color and pointing to the same species should be counted only once for reaction flux on a species. Note that all values in these schematics are average ones over the residence time, except for those in square brackets in panels **(a)** and **(b)**, which are average values within approximate NO effective lifetime (τ_{NO}, or more accurately, an integer multiple of the model's output time step closest to NO effective lifetime). All concentrations and fluxes have two significant digits.

when an OFR is used to oxidize the output of highly concentrated sources (e.g., from vehicle exhausts). When sources corresponding to OHR$_{ext}$ of thousands of s^{-1} and NO^{in} of tens of ppm are injected into OFR185 (Fig. 1b), they essentially inhibit active chemistry except NO consumption, as all

subsequent products are much less abundant compared to remaining NO (Fig. S1c).

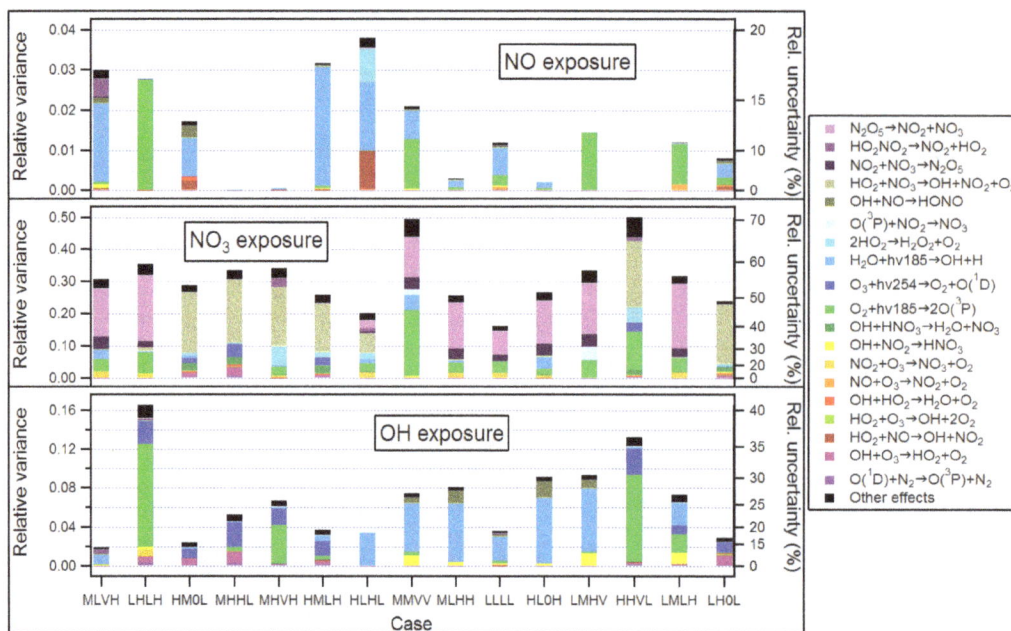

Figure 2. Relative variances (left axes) and uncertainties (right axes) in several outputs (i.e., NO, NO$_3$, and OH exposures) of Monte Carlo uncertainty propagation and relative contributions of key reactions to these relative variances in several typical cases (denoted in four-character labels; see Table 2 for the typical case label code) in OFR185-iNO. Relative variances are shown in linear scales (left axis), while corresponding relative uncertainties equal to relative variances' square roots are indicated by the nonlinear right axis. Only the reactions with a contribution of no less than 0.04 to at least one relative variance are shown.

3.1.2 OFR254-iNO

The ppm level O$_{3,\text{in}}$ used in the OFR254-iNO mode of operation has a strong impact on its NO$_y$ chemistry. An O$_{3,\text{in}}$ of 2.2 ppm (lowest in this study) is already enough to shorten τ_{NO} to ~ 1 s, preventing NO from playing a role in the chemistry under most explored conditions. The reaction fluxes under a typical O$_{3,\text{in}}$ of 7 ppm are shown in Fig. 1c. A reactive flux from NO + O$_3$ $\rightarrow$ NO$_2$ makes the reaction of NO with other oxidants (OH, HO$_2$, etc.) negligible. The HNO$_3$ production pathway from NO$_2$ is similar to that in OFR185-iNO. The interconversion between NO$_2$ and HO$_2$NO$_2$ is also fast over the residence time and even faster than in OFR185-iNO during τ_{NO} since a high concentration of O$_3$ also controls the OH–HO$_2$ interconversion and makes HO$_2$ more resilient against suppression due to high NO (Fig. S1f; Peng et al., 2015). A major difference in the NO$_y$ chemistry in OFR254-iNO (Fig. 1c) compared to OFR185-iNO (Fig. 1a) is significant NO$_3$/N$_2$O$_5$ chemistry due to high O$_3$ in OFR254-iNO, which accelerates the oxidation of NO$_2$ to NO$_3$. Interconversion between NO$_2$ + NO$_3$ and N$_2$O$_5$ also occurs to a significant extent because of high NO$_2$. Under the conditions of Fig. 1c, NO$_3$ can also be significantly consumed by HO$_2$. Unlike OFR185-iNO, OFR254-iNO can substantially form NO$_3$ from HNO$_3$ under conditions that are not on the extremes of the explored physical condition space, e.g., at higher UV and lower NO^{in} (e.g., Fig. S2). In the case of very high NO^{in}

(equal to or higher than O$_{3,\text{in}}$), all O$_3$ can be rapidly destroyed by NO. As a consequence, OH production is shut down and these cases are of little practical interest (Fig. S3h).

3.1.3 Uncertainty analysis

The results of uncertainty propagation confirm that the output uncertainties due to uncertain kinetic parameters are relatively low compared to other factors (e.g., non-plug flow in OFR; Peng et al., 2015) and the overall model accuracy compared to experimental data (a factor of 2–3; Li et al., 2015). For OFR185-iNO, NO, NO$_3$, and OH exposures have relative uncertainties of ~ 0–20, ~ 40–70, and ~ 15–40 %, respectively. The uncertainties in OH exposure are very similar to those in the cases without NO$_x$ (Peng et al., 2015). The contribution of NO$_y$ reactions to OH$_{\text{exp}}$ uncertainty is negligible, except for some contribution of OH + NO $\rightarrow$ HONO in a few cases with high NO^{in} (Fig. 2). The uncertainties in NO$_{\text{exp}}$ are dominated by the reactions producing HO$_x$ and O$_3$, i.e., the major consumers of NO. For NO$_3$ exposure, a few major production and loss pathways (e.g., NO$_2$ + NO$_3$ $\rightarrow$ N$_2$O$_5$, N$_2$O$_5$ $\rightarrow$ NO$_2$ + NO$_3$, and HO$_2$ + NO$_3$ $\rightarrow$ OH + NO$_2$ + O$_2$) dominate its uncertainties. OFR254-iNO has a simpler picture of parametric uncertainties in terms of composition. O$_3$ controls the NO oxidation under most conditions and this reaction contributes most of the output uncertainties for NO exposures. HO$_2$ + NO$_3$ $\rightarrow$ OH + NO$_2$ + O$_2$ dominates the uncertainty in NO$_3$ exposure. The levels of those uncertainties are lower than in OFR185-iNO ($< 2\%$ for NO exposure, $< 60\%$ in all cases, and $< 25\%$ in most cases for NO_3 exposure). Thus, model uncertainties in OFR254-iNO are not shown in detail.

3.2 Different conditions types

Having illustrated the main NO_y chemical pathways for typical cases, we present the results of the exploration of the entire physical parameter space (see Sect. 2.2). Note that the explored space is indeed very large and gridded logarithmically uniformly in every dimension. Therefore, the statistics of the exploration results can be useful to determine the relative importance of the condition types defined in Sect. 2.2 and Table 3.

It has been shown that during τ_{NO}, RO_2 can react dominantly with NO (Sect. 3.1.1), while the entire residence time is considered to determine if a condition is high-NO (see Table 3). This is done because for VOC oxidation systems of interest, there will be significant oxidation of the initial VOC and its products under low-NO conditions if τ_{NO} is shorter than the reactor residence time. After most NO is consumed, the longer the remaining residence time, the more RO_2 will react with HO_2 and the more likely that an input condition is classified as low NO. For a condition to be high NO, a significantly long τ_{NO} is required. Figure 3 shows the fractional occurrence distribution of good, risky, and bad conditions in the entire explored condition space over logarithm of $r(RO_2+NO)/r(RO_2 + HO_2)$, which distinguishes high- and low-NO conditions. In OFR254-iNO, τ_{NO} is so short that no good high-NO condition is found in the explored range in this study (Fig. 3a). A fraction of explored conditions are bad high-NO. These conditions result from a full consumption of O_3 by NO. Then very little HO_x is produced (right panels in Fig. S3h), but the fate of any RO_2 formed is dominated by $RO_2 + NO$ (right panels in Fig. S3i). However, also due to negligibly low OH concentration, little RO_2 is produced and non-tropospheric photolysis of VOCs is also substantial compared to their reaction with OH under these conditions, classifying all of them as "bad" (Fig. 3a).

In OFR185-iNO, in addition to the typical case shown in Fig. 1a, many other cases have a τ_{NO} of $\sim 10\,s$ or longer (Figs. S3b and S4), which allows for the possibility of high-NO conditions. Indeed, $\sim 1/3$ of explored conditions in OFR185-iNO with a residence time of 3 min are high NO (Fig. 3b). Most of these high-NO conditions are also classified as bad, similar to those in OFR254-iNO. More importantly, in contrast to OFR254-iNO, good and risky high-NO conditions also comprise an appreciable fraction of the OFR185-iNO conditions. It is easily expected that very high OHR_{ext} and NO^{in} lead to bad high-NO conditions (all panels in Fig. 4) since they strongly suppress HO_x, which yields bad conditions and in turn keeps NO destruction relatively low. The occurrence of bad high-NO conditions is reduced at high UV (bottom panels in Fig. 4), which can be explained by

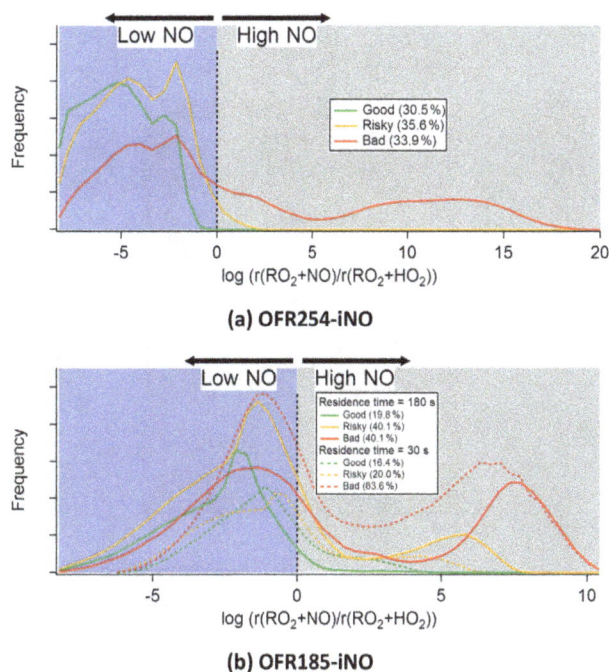

Figure 3. Frequency occurrence distributions of good, risky, and bad conditions (see Table 3) over logarithm of the ratio between RO_2 reacted with NO and with HO_2 (see Sect. S1 for more detail) for **(a)** OFR254-iNO (only the case with a residence time of 180 s) and **(b)** OFR185-iNO (including two cases with residence times of 180 and 30 s). Low- and high-NO regions (see Table 3) are colored in light blue and gray, respectively.

lowered NO due to high O_3 production and fast OH reactant loss due to high OH production. Good high-NO conditions are rare in the explored space. They are only 1.1% of total explored conditions (Fig. 3b) and present under very specific conditions, i.e., higher H_2O, lower UV, lower OHR_{ext}, and NO^{in} of tens to hundreds of ppb (Figs. 4 and S5). Since a very high NO can suppress OH, to obtain both a significant NO level and good conditions, NO^{in} can only be tens to hundreds of ppb. As NO^{in} is lower and OH is higher than under bad high-NO conditions, UV should be lower than bad high-NO conditions to keep a sufficiently long presence of NO. Thus, UV levels at 185 nm for good high-NO conditions are generally lower than 10^{12} photons cm^{-2} s^{-1} (Fig. S5). In addition, a low OHR_{ext} (generally $< 50\,s^{-1}$) and a higher H_2O (the higher the better, although there is no apparent threshold) are also required for good high-NO conditions (Fig. S5), as Peng et al. (2016) pointed out. Risky high-NO conditions often occur between good and bad high-NO conditions, e.g., at lower NO^{in} than bad conditions (e.g., Cases ML, MM, HL, and HM in Fig. 4; see Table 2 for the typical case label code), at higher OHR_{ext} and/or NO_{in} than good conditions (e.g., Cases ML and MM), and at lower H_2O than good conditions (e.g., Case LL).

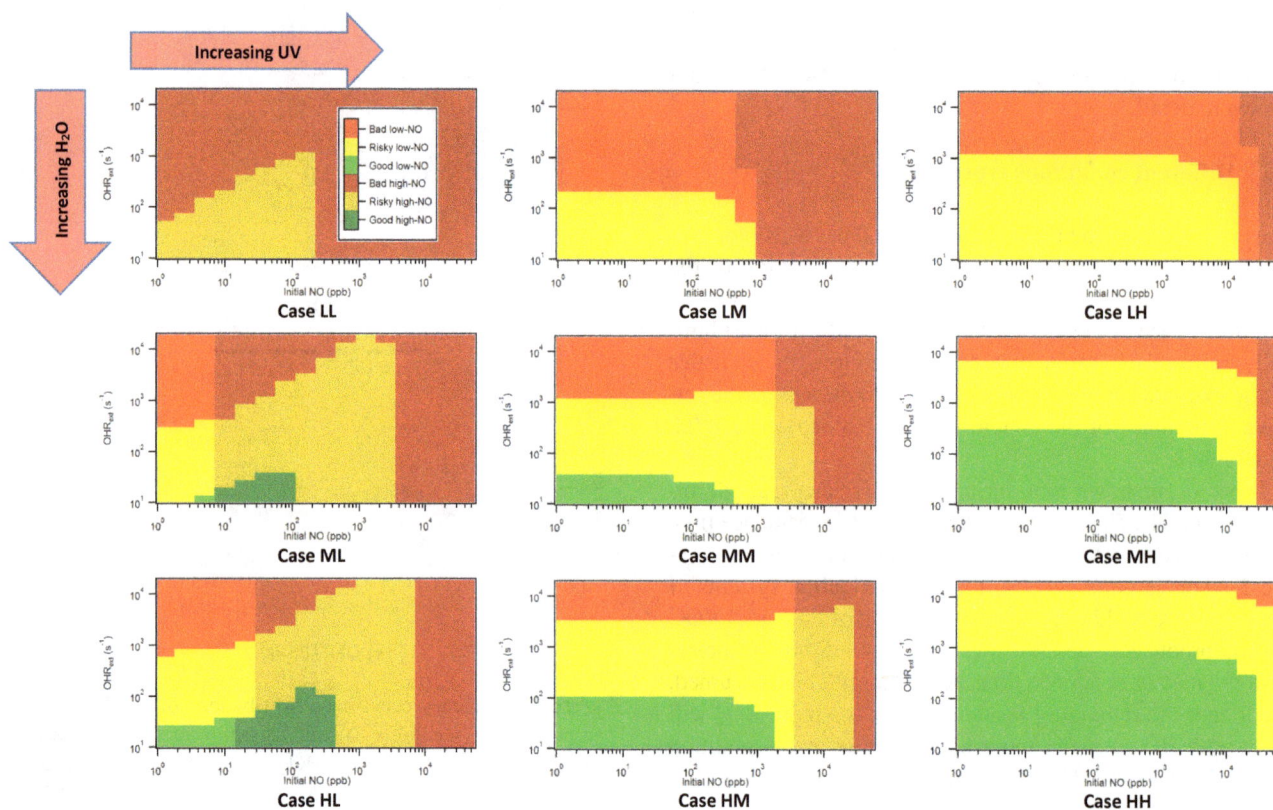

Figure 4. Image plots of the condition types defined in Table 3 vs. external OH reactivity (excluding N-containing species) and initial NO for several typical cases in OFR185-iNO (see Table 2 for the case label code).

The trend in the distributions of good, risky, and bad low-NO conditions is generally in line with the analysis in Peng et al. (2016). For low-NO conditions, NO_y species can be simply regarded as external OH reactants, as in Peng et al. (2016). As H_2O decreases and/or OHR_{ext} or NO^{in} increases, a low-NO condition becomes worse (good → risky → bad; Figs. 4 and 5). In OFR185-iNO, increasing UV generally makes a low-NO condition better because of an OH production enhancement (Fig. 4), while in OFR254-iNO, increasing UV generally makes a low-NO condition worse (Fig. 5) since at a higher UV more O_3 is destroyed and the resilience of OH to suppression is reduced.

As discussed above, the fraction of high-NO conditions also depends on OFR residence time. A shorter residence time is expected to generally lead to a larger fraction of high-NO conditions since the time spent in the reaction for $t > \tau_{NO}$ is significantly smaller. Thus, we also investigate an OFR185-iNO case with a residence time of 30 s. In Fig. 3b, compared to the case with a residence time of 3 min, the distributions of all condition types (good, risky, and bad) of the 30 s residence time case shift toward higher $r(RO_2 + NO)/r(RO_2 + HO_2)$. Nevertheless, shortening the residence time also removes the period when the condition is better (i.e., less non-tropospheric photolysis), when external OH reactants have been partially consumed, and

OH suppression due to OHR_{ext} has been reduced later in the residence time. As a result, the fractions of good and risky conditions decrease (Fig. 3b). With the two effects (higher $r(RO_2 + NO)/r(RO_2 + HO_2)$) and more significant non-tropospheric photolysis) combined, the fraction of good high-NO conditions increases by a factor of ~ 3. An even shorter residence time does not result in a larger good high-NO fraction since the effect of enhancing non-tropospheric photolysis is even more apparent.

3.3 Effect of non-plug flow

We performed model runs in which the only change with respect to our box model introduced in Sect. 2.2 is that the plug flow assumption is replaced by the residence time distribution (RTD) measured by Lambe et al. (2011; see also Fig. S8 of Peng et al., 2015). The chemistry of different air parcels with different residence times is simulated by our box model and outputs are averaged over the RTD. Lateral diffusion between different air parcels is neglected in these simulations.

OH_{exp} calculated from the mode with RTD ($OH_{exp,RTD}$) is higher than that calculated from the plug flow model ($OH_{exp,PF}$) in both OFR185-iNO and OFR254-iNO (Table 4 and Fig. S6). Under most explored conditions deviations are relatively small, which leads to an overall posi-

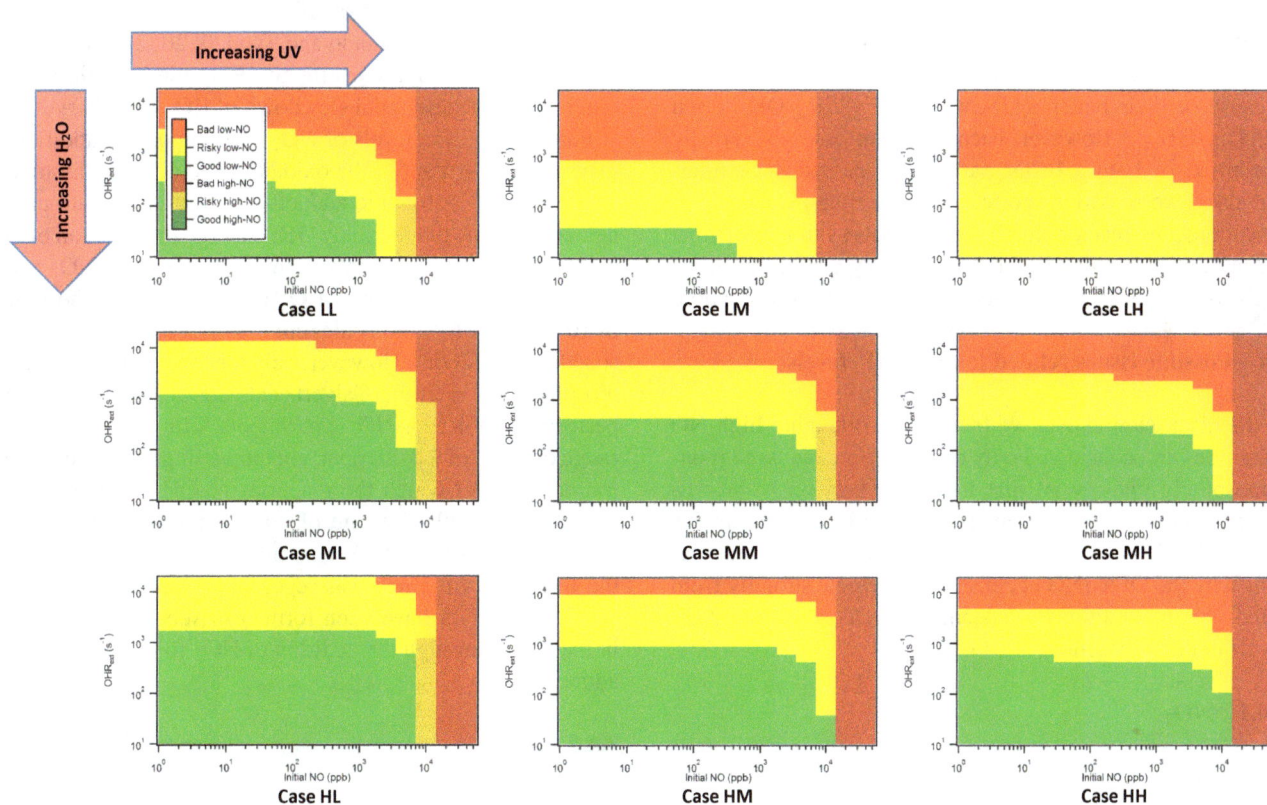

Figure 5. Same format as Fig. 4 but for OFR254-22-iNO.

tive deviation of $OH_{exp,RTD}$ from $OH_{exp,PF}$ by $\sim \times 2$ (within the uncertainties of the model and its application to real experimental systems). For OFR185-iNO, most conditions ($\sim 90\%$) in the explored space lead to $< \times 3$ differences between $OH_{exp,PF}$ and $OH_{exp,RTD}$, while for a small fraction of cases the differences can be larger (Fig. S6). The larger deviations are mainly present at high UV, OHR_{ext}, and NO^{in} in which conditions are generally bad and in which experiments are of little atmospheric relevance. Under these specific conditions, external OH reactants and NO_y can be substantially destroyed for the air parcels with residence times longer than the average, while this is not the case for the average residence time. This feature was already described by Peng et al. (2015; see Fig. S10 of that study). Although only non-NO_y external OH reactants were considered in that study, the results are the same. In the present study, a higher upper limit of the explored OHR_{ext} range (compared to Peng et al., 2015, due to trying to simulate the extremely high OHR_{ext} used in some recent literature studies) leads to large amounts of NO_y and causes somewhat larger deviations. In OFR254-iNO, OH is less suppressed at high OHR_{ext} and NO^{in} than in OFR185-iNO because of high O_3 (Peng et al., 2015); $OH_{exp,RTD}$ deviations from $OH_{exp,PF}$ are also smaller (Table 4).

Based on the outputs of the model with RTD, a similar mapping of the physical input space as in Figs. 4 and 5 can be done (Figs. S7 and S8). Overall, the mapping of

Table 4. Statistics of the ratio between OH exposures calculated in the model with the Lambe et al. (2011) residence time distribution ($OH_{exp,RTD}$) and in the plug flow model ($OH_{exp,PF}$). The geometric mean, uncertainty factor (geometric standard deviation), and percentage of outlier cases (> 3 or $< 1/3$) are shown for OFR185-iNO, OFR254-70-iNO, and OFR254-7-iNO.

	Geometric mean	Uncertainty factor	Outlier cases (%)
OFR185-iNO	1.91	1.64	11
OFR254-7-iNO	1.59	1.51	7
OFR254-70-iNO	1.48	1.29	3

the RTD model results is very similar to that of the plug flow model. The conditions appear to be only slightly better in a few places of the explored space than those from the plug flow model, which can be easily explained by the discussions above. The mapping in Figs. S7 and S8 also appear to be slightly more low NO for the same reasons discussed above. After NO is destroyed at long residence times, HO_2 suppressed by NO also recovers as OH, and $r(RO_2 + NO)/r(RO_2 + HO_2)$ is clearly expected to be smaller than in the plug flow model in general.

Note that most conditions that appear to be better in the RTD model results are already identified as bad by the plug flow model. Those conditions look slightly better only because of their better *RTD-averaged* $F185_{exp}/OH_{exp}$ and $F254_{exp}/OH_{exp}$. However, each of those cases is actually composed of both a better part at longer residence times and also a worse part at shorter residence times. Under those conditions, the reactor simultaneously works in two distinct regimes, one of which is bad due to heavy OH suppression. Such conditions are clearly not desirable for OFR operation.

3.4 Possible issues related to high NO_x levels

In the discussion above, we focused on obtaining high-NO conditions and considered only one experimental issue (non-tropospheric photolysis) that had been previously investigated in Peng et al. (2016) and is not specific for experiments with high NO injection. We discuss additional potential reasons why the OFR-iNO chemistry can deviate strongly from tropospheric conditions as specifically related to high NO_x levels in this subsection.

3.4.1 NO_2

NO_2 reacts with RO_2 to form peroxynitrates, which are generally regarded as reservoir species in the atmosphere as most of them thermally decompose very quickly compared to atmospheric timescales. However, in OFRs with residence times on the order of minutes, some peroxynitrates may no longer be considered as fast decomposing. This is especially true for acylperoxy nitrates with lifetimes that can be hours at room temperature (Orlando and Tyndall, 2012). Acylperoxy nitrates are essentially sinks instead of reservoirs in OFRs for both NO_2 and RO_2. RO_2 is estimated to be as high as several ppb in OFRs by our model (e.g., ~ 6 ppb RO_2 in OFR185 at $H_2O = 1\%$, UV at 185 nm $= 1 \times 10^{13}$ photons cm^{-2} s^{-1}, $OHR_{ext} = 1000$ s^{-1}, and $NO^{in} = 0$), while high-NO experiments can yield far higher NO_2. If all RO_2 were acylperoxy, the RO_2 chemistry could be rapidly shut down by NO_2, as rate constants of these $RO_2 + NO_2$ reactions are around 10^{-11} cm^3 molecule^{-1} s^{-1} (Orlando and Tyndall, 2012). Nevertheless, acyl peroxynitrates are not expected to typically be the dominant component of peroxynitrates since acyl radicals are not a direct oxidation product of most common VOCs and can only be formed after several steps of oxidation (Atkinson and Arey, 2003; Ziemann and Atkinson, 2012). Most alkylperoxy nitrates retain their short-lived reservoir characteristics in OFRs due to their relatively short thermal decomposition timescales (on the order of 0.1 s; Orlando and Tyndall, 2012). Even so, OFR experiments can be seriously hampered at extremely high NO_2. If NO_2 reaches ppm levels, the equilibrium between $RO_2 + NO_2$ and alkylperoxy nitrate ($RO_2 + NO_2 \leftrightarrow RO_2NO_2$) is greatly shifted toward the alkylperoxy nitrate side, as the forward and reverse rate constants are on the order of 10^{-12} cm^3 molecule^{-1} s^{-1} and 1 s^{-1}, respectively (Orlando and Tyndall, 2012). This results in a substantial decrease in effective RO_2 concentration, or in other words, a substantial slowdown of RO_2 chemistry.

Parts per million levels of NO_2 may impose an additional experimental artifact in the oxidation chemistry of aromatic precursors. OH–aromatic adducts, i.e., the immediate products of aromatic oxidation by OH, undergo the addition of O_2 and NO_2 at comparable rates under ppm levels of NO_2 (rate constants of the additions of O_2 and NO_2 are on the order of 10^{-16} and 10^{-11} molecules cm^{-3} s^{-1}, respectively; Atkinson and Arey, 2003). However, only the former addition is atmospherically relevant (Calvert et al., 2002). Liu et al. (2015) performed OFR254-iNO experiments with toluene over a range of NO^{in} of 2.5–10 ppm, encompassing the NO concentration range at which the reactions of OH–toluene adduct with O_2 and with NO_2 are of equal importance (~ 5 ppm; Atkinson and Arey, 2003). This suggests that nitroaromatics, the formation of which was reported in the study of Liu et al. (2015), might have been formed in substantial amounts in that study through the addition of NO_2 to the OH–toluene adduct.

3.4.2 NO_3

As discussed in Sect. 3.1, NO_3 can be formed in significant amounts in OFRs with high NO injection. Although NO_3 is also present in the atmosphere, especially during nighttime, significant VOC oxidation by both OH and NO_3 results in more complex chemistry that may complicate the interpretation of experimental results. NO_3 oxidation-only OFRs have been previously realized experimentally via the thermal dissociation of injected N_2O_5 (Palm et al., 2017). We discuss below how to avoid significant VOC oxidation by NO_3 and achieve OH-dominated VOC oxidation in OFRs with high NO injection.

If $NO_{3exp}/OH_{exp} > 0.1$, NO_3 can be a competitive reactant for biogenic alkenes and dihydrofurans, which have a C–C bond for NO_3 addition, and phenols, which have activated hydroxyl for fast hydrogen abstraction by NO_3 (Atkinson and Arey, 2003). For lower NO_{3exp}/OH_{exp}, OH is expected to dominate the oxidation of all VOCs, as shown in Fig. 6. Oxidation for VOCs without alkene C–C bonds and phenol hydroxyl (such as alkanes and (alkyl)benzenes) is dominated by OH unless $NO_{3exp}/OH_{exp} > 1000$. Despite its double bond, ethene reacts as slowly with NO_3 as alkanes, likely due to lack of alkyl groups enriching the electron density on the C–C bond, which slows NO_3 addition. We calculate NO_{3exp}/OH_{exp} for OFR185-iNO and OFR254-iNO and plot histograms of this ratio in Fig. 6. Many experimental conditions lead to high enough NO_{3exp}/OH_{exp} that NO_3 is a competitive sink for alkenes, while only under very extreme conditions can NO_3 be a competitive sink for species without C–C bonds. High-NO conditions in OFR185-iNO have lower NO_{3exp}/OH_{exp} ($\sim 10^{-2}–10^2$) than in OFR254-

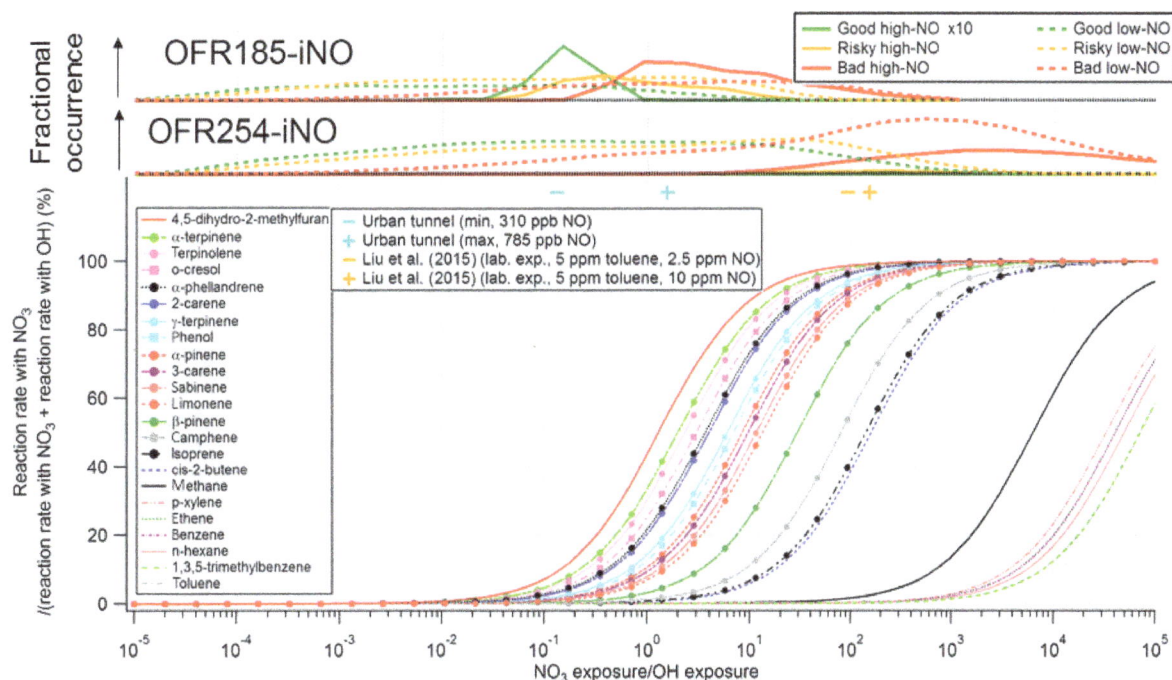

Figure 6. Fractional importance of the reaction rate of several species of interest with NO_3 vs. that with OH as a function of the ratio of exposure to NO_3 and OH. The curves of biogenics and phenols are highlighted by solid dots and squares, respectively. The turquoise and orange markers show the ranges of modeled exposure ratios between NO_3 and OH of a source study in an urban tunnel (Tkacik et al., 2014) and a laboratory study (Liu et al., 2015), respectively, using an OFR. In the upper part of the figure, the modeled frequency distributions of ratios of NO_3 exposure to OH exposure under good/risky/bad high-/low-NO conditions for OFR185-iNO and OFR254-iNO are also shown. See Table 3 for the definitions of the three types of conditions. All curves, markers, and histograms in this figure share the same abscissa.

iNO ($\sim 10^1 - 10^5$) (Figs. 6 and S3d, g, j). This difference in NO_{3exp}/OH_{exp} is due to the different levels of O_3 in the two modes, as high O_3 promotes NO_2-to-NO_3 oxidation. Note that low-NO conditions in both OFR185-iNO and OFR254-iNO can also reach high NO_{3exp}/OH_{exp} as some high-NO conditions have. This is because in OFR185-iNO a large part of NO_3 is formed by OH oxidation, resulting in NO_{3exp}/OH_{exp} being largely influenced by NO^{in} but not by other factors mainly governing OH (Fig. S3d); under low-NO conditions in OFR254-iNO, NO_3 can form rapidly from $NO_2 + O_3$, while OH can be heavily suppressed by high OHR_{ext} (Fig. S3g and j).

Most of the species shown in Fig. 6 are primary VOCs, except phenols and a dihydrofuran, which can be intermediates of the atmospheric oxidation of (alkyl)benzenes (Atkinson and Arey, 2003) and long-chain alkanes (Aimanant and Ziemann, 2013; Strollo and Ziemann, 2013; Ranney and Ziemann, 2016), respectively. Nevertheless, only the phenol production may occur in high-NO OFRs, as the particle-phase reaction in the photochemical formation of dihydrofurans from alkanes is too slow compared to typical OFR residence times (Ranney and Ziemann, 2016). Therefore, the impact of NO_3 oxidation on VOC fate needs to be considered only if the OFR input flow contains high NO mixed with biogenics and/or aromatics, e.g., (alkyl)benzenes and/or phenols.

However, (alkyl)benzenes were likely to be major SOA precursors in, to our knowledge, the only literature OFR studies with high NO levels (Ortega et al., 2013; Tkacik et al., 2014; Liu et al., 2015). In the study of the air in a traffic tunnel (OFR185-iNO mode; Tkacik et al., 2014), where toluene is usually a major anthropogenic SOA precursor as in other urban environments (Dzepina et al., 2009; Borbon et al., 2013; Hayes et al., 2015; Jathar et al., 2015), NO_x was several hundred ppb. This resulted in an estimated NO_{3exp}/OH_{exp} range of ~ 0.1–1 in which up to $\sim 30\%$ of cresols (intermediates of toluene oxidation) may have been consumed by NO_3. Dihydrofurans may also have formed in the tunnel air (but outside the OFR) in the presence of NO_x (Aimanant and Ziemann, 2013; Strollo and Ziemann, 2013), and after entering the OFR they would have been substantially (up to $\sim 50\%$) consumed by NO_3. In the laboratory experiment of Liu et al. (2015) with toluene, the injection of as much as 10 ppm of NO elevated NO_{3exp}/OH_{exp} to ~ 100, at which cresols from toluene oxidation reacted almost exclusively with NO_3 in addition to being photolyzed.

3.4.3 A case study

We use a case study of an OFR254-13-iNO laboratory experiment with a large amount of toluene (5 ppm) and NO^{in}

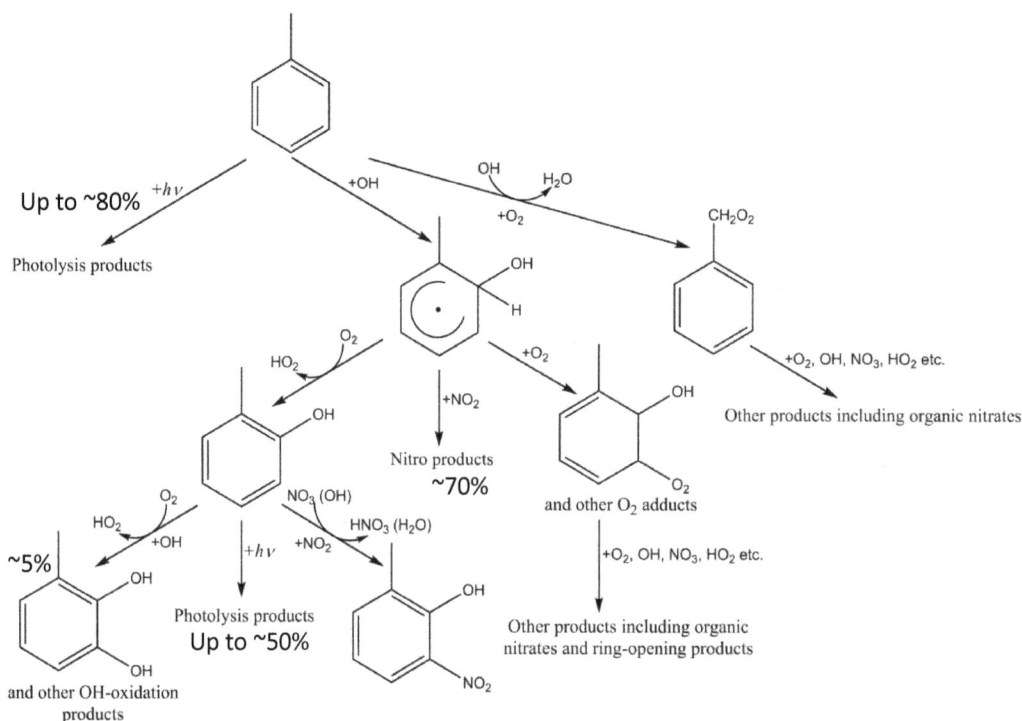

Scheme 1. Possible major reactions in an OFR254-13-iNO with 5 ppm toluene and 10 ppm initial NO. Branching ratios in red are estimated by the model and/or according to Calvert et al. (2002), Atkinson and Arey (2003), Ziemann and Atkinson (2012), and Peng et al. (2016). Note that addition or substitution on the aromatic ring may occur at other positions. Intermediates and products shown here are the isomers that are most likely to form. Branching ratios shown in red are not overall but from immediate reactant.

(10 ppm) to illustrate how very high VOC and NO concentrations cause multiple types of atmospherically irrelevant reactions in an OFR. Due to very high OHR_{ext} and NO^{in}, the photolysis of toluene at 254 nm may have been important (Peng et al., 2016). In the case of a high (close to 1) quantum yield, up to $\sim 80\%$ of the consumed toluene in their experiments could have been photolyzed (Scheme 1). Of the rest of reacted toluene, $\sim 10\%$ undergoes H abstraction by OH from the methyl group in the model, leading to an RO_2 similar to alkyl RO_2 and likely proceeding with normal RO_2 chemistry. Approximately 90 % of the toluene formed an OH adduct (Calvert et al., 2002). As discussed above, 70 % of this adduct (depending on NO^{in}) is predicted to recombine with NO_2, producing nitroaromatics because of the ppm level NO_x. The adduct could also react with O_2 via two types of pathways, one of which was addition forming a special category of RO_2 (OH–toluene–O_2 adducts) potentially undergoing ring-opening (Atkinson and Arey, 2003; Orlando and Tyndall, 2012; Ziemann and Atkinson, 2012). The other was H elimination by O_2, producing cresols. Again, like toluene, cresols may have been substantially photolyzed. As a result of $NO_{3exp}/OH_{exp} \sim 100$, only a minor portion of cresol could have undergone OH addition and then H elimination again. This pathway leads to the formation of methyldihydroxybenzenes and other OH-oxidation products (Atkinson and Arey,

2003). The rest of the cresols may have formed methylphenoxy radicals but nevertheless dominantly via H abstraction by NO_3, since H abstraction by OH was a minor pathway compared to the OH-addition one (Atkinson et al., 1992). In summary, the model results suggest that there were two possible routes leading to nitroaromatic formation. However, one of them (recombination of OH–aromatic adducts with NO_2) is likely of little atmospheric relevance due to very high NO_x needed, and the other (H abstraction from cresol) occurs in the atmosphere but is not a major fate of aromatics (Calvert et al., 2002).

3.5 Implications for OFR experiments with combustion emissions as input

Emissions from combustion sources, e.g., vehicles and biomass burning, usually contain VOCs and NO_x at very high concentrations (Table 1). An injection of this type of emissions (typically with OHR_{ext} of thousands of s^{-1} or larger and NO^{in} of tens of ppm or larger) in OFRs without any pretreatment is likely to cause all experimental issues discussed in Peng et al. (2016) and this paper, i.e., strong OH suppression, substantial non-tropospheric photolysis, strong RO_2 suppression by NO_2 whether RO_2 is acyl RO_2 or not, fast reactions of NO_2 with OH–aromatic hydrocarbon adducts, substantial NO_3 contribution to VOC fate,

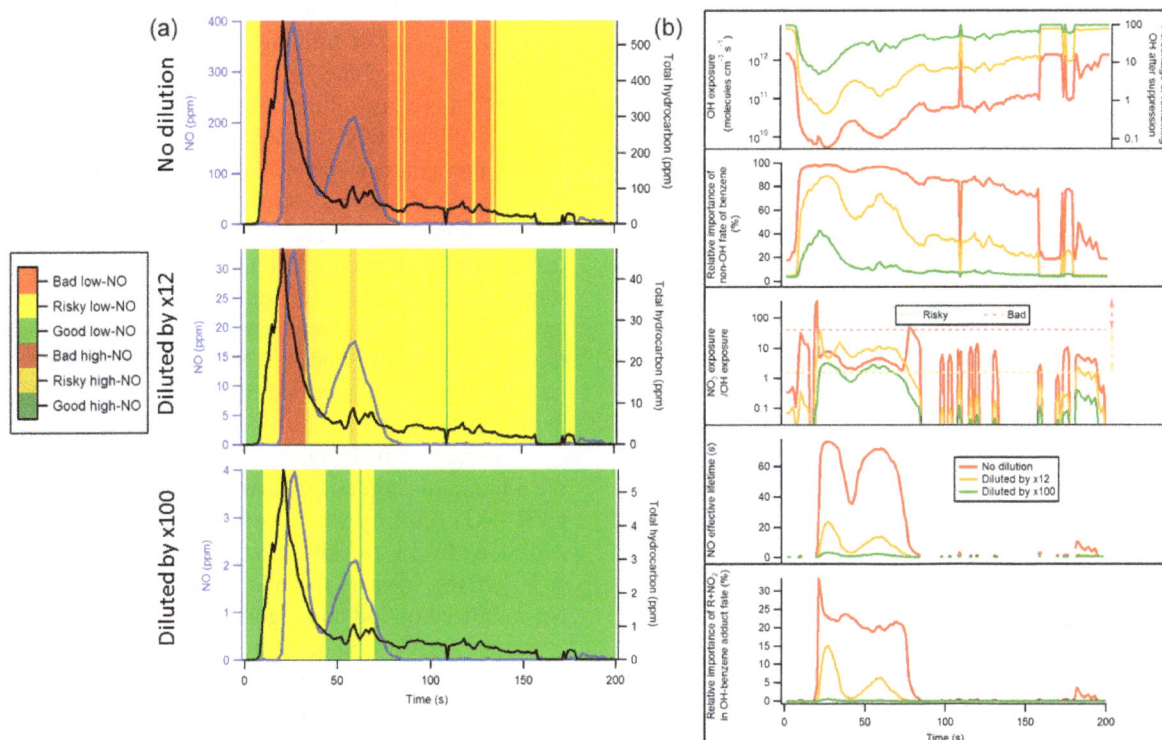

Figure 7. (a) NO and total hydrocarbon during the first 200 s of the test of Karjalainen et al. (2016) in the cases of no dilution, dilution by a factor of 12 (as actually done in that study), and dilution by a factor of 100. Different periods of time are colored according to corresponding emissions (i.e., input conditions for OFR) classified as good/risky/bad high-/low-NO emissions. **(b)** OH exposure and percentage of remaining OH after suppression, relative importance of non-OH fate of benzene, exposure ratio of NO_3 to OH, NO effective lifetime, and relative importance of reaction of OH–toluene adduct with NO_2 in the fate of this adduct in the OFR of Karjalainen et al. (2016) during the first 200 s of their test in the cases of no dilution, dilution by a factor of 12, and dilution by a factor of 100. Horizontal orange and red dashed lines in the middle right panel denote "risky" and "bad" regions for exposure ratio of NO_3 to OH, respectively. Above the orange (red) dashed line, reaction with NO_3 contributes > 20 % to the fate of phenol (isoprene).

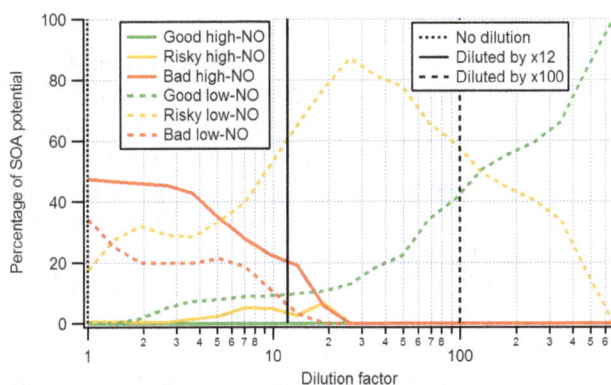

Figure 8. Secondary organic aerosol (SOA) potential (estimated from the total hydrocarbon measurement) in the OFR of Karjalainen et al. (2016) formed during periods of time in the OFR corresponding to good/risky/bad high-/low-NO conditions as a function of dilution factor. Vertical lines denoting dilution factors of 1, 12 (as actually used in that study), and 100 are also shown.

and even a nearly total inhibition of OFR chemistry due to complete titration of O_3 by NO in the case of OFR254. We take the study of Karjalainen et al. (2016), who used an OFR to oxidize diluted car exhaust in real time, as a case study to investigate the extent to which these issues may affect typical combustion source studies and to explore approaches to mitigate the problems.

During the first 200 s of their experiment (defined as the "cold start" period when the catalyst is cold and emissions are high), NO and total hydrocarbon in the emissions of the test vehicle reached $\sim$ 400 and $\sim$ 600 ppm, respectively. We first simulate the oxidation of those emissions without any dilution (even though $\times$ 12 dilution was used in their experiments) to explore the most extreme conditions. Our model simulation indicates that such an extremely concentrated source would generally lead to bad high- or low-NO conditions (depending on NO concentration) in their OFR (Fig. 7), even though it was run at relatively high H_2O and UV. OH suppression can be as high as 3 orders of magnitude, VOC fates by non-tropospheric photolysis and reactions of alkenes and phenols with NO_3 can be nearly 100 %,

(a) No dilution (background: case HH)

(b) Dilution by a factor of 100 (background: case HH)

(c) Dilution by a factor of 100 (background: case HL)

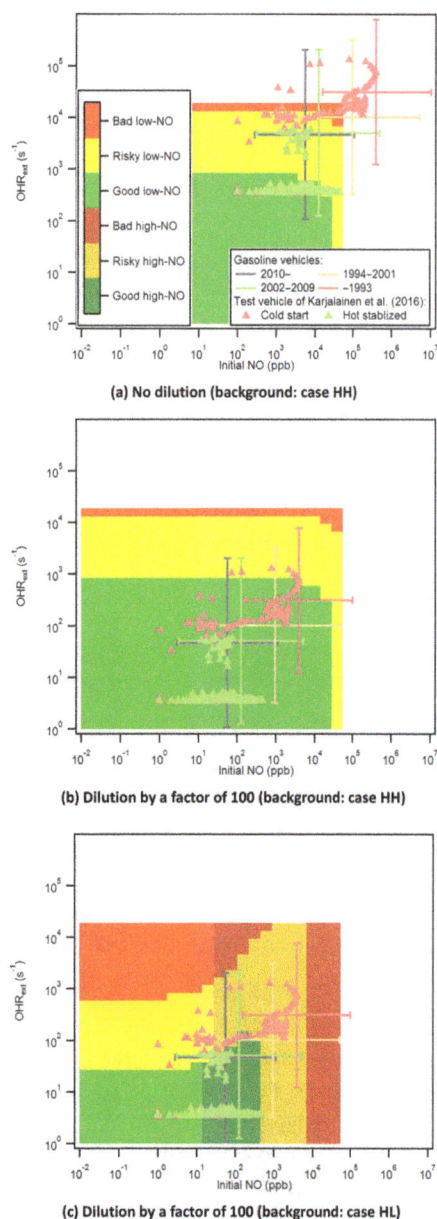

Figure 9. Location of individual 1 s data points vs. OFR185-iNO reaction conditions. Data points are shown from the test vehicle of Karjalainen et al. (2016) and average exhaust from gasoline vehicle on-road emissions measured by Bishop and Stedman (2013). On-road emissions are classified by vehicle year and the distribution of each category is shown as a cross representing 1 SD (standard deviation; with lognormal distribution assumed). The x and y axes are NO and external OH reactivity (excluding N-containing species) due to vehicle emissions in an OFR in the cases of **(a)** no dilution and **(b, c)** dilution by a factor of 100. The Karjalainen et al. (2016) points are classified as cold start (during the first 200 s) and hot stabilized (during 200–1000 s). In addition, the same image plots as the panels of cases HH (high H_2O and high UV; see Table 2 for the case label code) and HL in Fig. 4 (OFR185-iNO) are shown as background for comparison.

and up to $\sim 1/3$ of OH–toluene adduct may be recombined with NO_2 instead of forming an adduct with O_2. After the test vehicle entered the "hot stabilized" stage (200–1000 s), its VOC emissions (on the order of ppm) were still too high for an undiluted OFR to yield a good condition (Fig. S9). OH suppression can still reach 2 orders of magnitude, non-tropospheric photolysis and sometimes reactions with NO_3 can still dominate over reactions with OH in VOC fates, and reactions of OH–toluene adduct with NO_2 can still be substantial at some small NO emission spikes. Moreover, although NO emissions were roughly at ppm level even during the hot stabilized period, the NO effective lifetime may be very short during that period, leading to low-NO conditions in their OFR.

As suggested in Peng et al. (2016) for a low-NO OFR, the dilution of sources can also mitigate strong deviations in OFR-iNO chemistry vs. atmospherically relevant conditions. A dilution by a factor of 12, as actually used by Karjalainen et al. (2016), appears to be sufficient to bring most of the hot stabilized period under good conditions (Fig. S9). However, most VOC, or in other words most SOA formation potential, was emitted during the cold start period when risky and bad conditions still prevailed (Figs. 7 and 8). Even if the emissions are diluted by x100, the cold-start emission peak (Fig. 7) is still under risky conditions. Although bad conditions are eliminated and a good condition is present most of time, this emission peak under risky conditions may contribute $> 50\%$ to total SOA formation potential (Fig. 8). For SOA formed under good condition to be dominant, a dilution factor of > 400 would be needed. Note that a strong dilution lowers aerosol mass loading in vehicle emissions. As a result, the condensation of gases onto particles is slower than in raw exhausts. However, condensational sinks after dilution may still be significantly higher than typical ambient values (Matti Maricq, 2007; Donahue et al., 2016).

Note that the emissions of the test vehicle of Karjalainen et al. (2016) are rather clean compared to the typical 2013 US on-road fleet (i.e., all at the hot stabilized stage) measured by Bishop and Stedman (2013; Figs. 9 and S10). For the emissions of an average on-road fleet, a dilution by a factor of 100 or larger would be necessary to ensure that most emissions are processed in OFR185 under good conditions at the highest H_2O and UV in this study (Figs. 9b and S10b, e, h). In the case of lower H_2O and/or UV, an even larger dilution factor would be required.

Conducting OFR185-iNO experiments at high UV lowers the dilution factor needed for good conditions. However, it also renders good high-NO conditions impossible (see Sect. 3.2 and Fig. S4). If one wants to oxidize vehicle exhausts in a high-NO environment in an OFR, as in an urban atmosphere, OFR185 at low UV is necessary. Consequently, a much stronger dilution is in turn necessary to keep the operation conditions good. Nevertheless, not all vehicle emissions can be moved into the good high-NO region through a simple dilution (Figs. 9c and S10c, f, i). Furthermore, a low UV would seriously limit the highest OH_{exp} that OFRs can achieve ($\sim 3 \times 10^{11}$ molecules cm^{-3} s for modeled good high-NO conditions in this study), while a much higher OH_{exp} would be desirable to fully convert SOA formation potential into measurable SOA mass. If both good high-NO conditions and high OH_{exp} are required, new techniques (e.g., injection of N_2O at the percent level as proposed by Lambe et al., 2017) may be necessary.

4 Conclusions

In this study, OFR chemistry involving NO_y species was systematically investigated over a wide range of conditions. NO initially injected into the OFR was found to be rapidly oxidized under most conditions. In particular, due to high O_3 concentrations, the NO lifetime in OFR254-iNO was too short to result in significant RO_2 consumption by NO compared to that by HO_2 under all conditions with active chemistry. Nevertheless, it is not completely impossible for OFR185-iNO to have a significant RO_2 fate by NO and minor non-tropospheric photolysis at the same time (good high-NO conditions). According to our simulations, these conditions are most likely present at high H_2O, low UV, low OHR_{ext}, and NO^{in} of tens to hundreds of ppb.

However, many past OFR studies with high NO injection were conducted under conditions remarkably different from the abovementioned very narrow range. NO^{in} and/or OHR_{ext} in those studies were often much higher than good high-NO conditions require (particularly, > 3 orders of magnitude in some OFR studies using combustion emissions as input). In addition to non-tropospheric organic photolysis, OFR oxidation of highly concentrated sources can cause multiple large deviations from tropospheric OH oxidation, i.e., RO_2 suppression by high NO_2, substantial nitroaromatic formation from the recombination of NO_2 and OH–aromatic adducts, and fast reactions of VOCs with NO_3 compared to those with OH.

Working at lower NO_x (sub ppm level) and VOC concentrations or dilution can mitigate these experimental problems. In general, a strong dilution (by a factor of > 100) is needed for an OFR that processes typical on-road vehicle emissions. Humidification can also make good conditions more likely. By using these measures, good conditions can be guaranteed as long as NO and/or precursor concentrations are sufficiently low, while high-NO conditions cannot be ensured. To aid the design and interpretation of OFR experiments with high NO injection, we provide our detailed modeling results in a visualized form (Fig. S3). For OFR users in need of both high OH_{exp} and high NO, simple NO injection is not a good option. New techniques (e.g., the injection of N_2O as proposed by Lambe et al. (2017) or other innovations) may be necessary to meet this need.

Competing interests. The authors declare that they have no conflict of interest.

Acknowledgements. This work was partially supported by DOE (BER/ASR) DE-SC0011105 and DE-SC0016559, EPA STAR 83587701-0, and NSF AGS-1360834. We thank Pengfei Liu, Andrew Lambe, and Daniel Tkacik for providing some OFR experimental data, the authors of Karjalainen et al. (2016) and their project IEA-AMF Annex 44 for providing the data and information for the vehicle tests, Gary Bishop for providing on-road vehicle emission data, and Andrew Lambe and William Brune for useful discussions.

Edited by: Dwayne Heard

References

Aimanant, S. and Ziemann, P. J.: Chemical Mechanisms of Aging of Aerosol Formed from the Reaction of n-Pentadecane with OH Radicals in the Presence of NO_x, Aerosol Sci. Tech., 47, 979–990, https://doi.org/10.1080/02786826.2013.804621, 2013.

Alanen, J., Simonen, P., Saarikoski, S., Timonen, H., Kangasniemi, O., Saukko, E., Hillamo, R., Lehtoranta, K., Murtonen, T., Vesala, H., Keskinen, J., and Rönkkö, T.: Comparison of primary and secondary particle formation from natural gas engine exhaust and of their volatility characteristics, Atmos. Chem. Phys. Discuss., https://doi.org/10.5194/acp-2017-44, in review, 2017.

Ammann, M., Cox, R. A., Crowley, J. N., Jenkin, M. E., Mellouki, A., Rossi, M. J., Troe, J., Wallington, T. J., Cox, B., Atkinson, R., Baulch, D. L., and Kerr, J. A.: IUPAC Task Group on Atmospheric Chemical Kinetic Data Evaluation, available at: http://iupac.pole-ether.fr/# (last access: February 2017), 2016.

Atkinson, R. and Arey, J.: Atmospheric degradation of volatile organic compounds, Chem. Rev., 103, 4605–4638, https://doi.org/10.1021/cr0206420, 2003.

Atkinson, R., Aschmann, S. M., and Arey, J.: Reactions of hydroxyl and nitrogen trioxide radicals with phenol, cresols, and 2-nitrophenol at 296 ± 2 K, Environ. Sci. Technol., 26, 1397–1403, https://doi.org/10.1021/es00031a018, 1992.

BIPM, IEC, IFCC, ILAC, ISO, IUPAC and IUPAPOIML: JCGM 101: 2008 Evaluation of measurement data – Supplement 1 to the "Guide to the expression of uncertainty in measurement" – Propagation of distributions using a Monte Carlo method, http://www.bipm.org/utils/common/documents/jcgm/JCGM_101_2008_E.pdf (last access: December 2016), 2008.

Bishop, G. A. and Stedman, D. H.: Fuel Efficiency Automobile Test: Light-Duty Vehicles, available at: http://www.feat.biochem.du.edu/light_duty_vehicles.html (last access: 1 February 2017), 2013.

Borbon, A., Gilman, J. B., Kuster, W. C., Grand, N., Chevaillier, S., Colomb, A., Dolgorouky, C., Gros, V., Lopez, M., Sarda-Esteve, R., Holloway, J., Stutz, J., Petetin, H., McKeen, S., Beekmann, M., Warneke, C., Parrish, D. D., and De Gouw, J. A.: Emission ratios of anthropogenic volatile organic compounds in northern mid-latitude megacities: Observations versus emission inventories in Los Angeles and Paris, J. Geophys. Res.-Atmos., 118, 2041–2057, https://doi.org/10.1002/jgrd.50059, 2013.

Burkholder, J. B., Sander, S. P., Abbatt, J., Barker, J. R., Huie, R. E., Kolb, C. E., Kurylo, M. J., Orkin, V. L., Wilmouth, D. M., and Wine, P. H.: Chemical Kinetics and Photochemical Data for Use in Atmospheric Studies, Evaluation Number 18, Pasadena, CA, USA, available at: http://jpldataeval.jpl.nasa.gov/ (last access: February 2017), 2015.

Calvert, J. G., Atkinson, R., Becker, K. H., Kamens, R. M., Seinfeld, J. H., Wallington, T. H., and Yarwood, G.: The Mechanisms of Atmospheric Oxidation of the Aromatic Hydrocarbons, Oxford University Press, USA, available at: https://books.google.com/books?id=P0basaLrxDMC (last access: February 2017), 2002.

Carlton, A. G., Wiedinmyer, C., and Kroll, J. H.: A review of Secondary Organic Aerosol (SOA) formation from isoprene, Atmos. Chem. Phys., 9, 4987–5005, https://doi.org/10.5194/acp-9-4987-2009, 2009.

Carter, W. P. L., Cocker, D. R., Fitz, D. R., Malkina, I. L., Bumiller, K., Sauer, C. G., Pisano, J. T., Bufalino, C., and Song, C.: A new environmental chamber for evaluation of gas-phase chemical mechanisms and secondary aerosol formation, Atmos. Environ., 39, 7768–7788, https://doi.org/10.1016/j.atmosenv.2005.08.040, 2005.

Chameides, W., Lindsay, R., Richardson, J., and Kiang, C.: The role of biogenic hydrocarbons in urban photochemical smog: Atlanta as a case study, Science, 241, 1473–1475, https://doi.org/10.1126/science.3420404, 1988.

Cocker, D. R., Flagan, R. C., and Seinfeld, J. H.: State-of-the-Art Chamber Facility for Studying Atmospheric Aerosol Chemistry, Environ. Sci. Technol., 35, 2594–2601, https://doi.org/10.1021/es0019169, 2001.

Donahue, N. M., Posner, L. N., Westervelt, D. M., Li, Z., Shrivastava, M., Presto, A. A., Sullivan, R. C., Adams, P. J., Pandis, S. N., and Robinson, A. L.: Where Did This Particle Come From? Sources of Particle Number and Mass for Human Exposure Estimates, in: Airborne Particulate Matter: Sources, Atmospheric Processes and Health, edited by: Harrison, R. M., Hester, R. E., and Querol, X., Royal Society of Chemistry, Cambridge, UK, 35–71, 2016.

Dzepina, K., Volkamer, R. M., Madronich, S., Tulet, P., Ulbrich, I. M., Zhang, Q., Cappa, C. D., Ziemann, P. J., and Jimenez, J. L.: Evaluation of recently-proposed secondary organic aerosol models for a case study in Mexico City, Atmos. Chem. Phys., 9, 5681–5709, https://doi.org/10.5194/acp-9-5681-2009, 2009.

George, I. J., Vlasenko, A., Slowik, J. G., Broekhuizen, K., and Abbatt, J. P. D.: Heterogeneous oxidation of saturated organic aerosols by hydroxyl radicals: uptake kinetics, condensed-phase products, and particle size change, Atmos. Chem. Phys., 7, 4187–4201, https://doi.org/10.5194/acp-7-4187-2007, 2007.

Haagen-Smit, A. J.: Chemistry and Physiology of Los Angeles Smog, Ind. Eng. Chem., 44, 1342–1346, https://doi.org/10.1021/ie50510a045, 1952.

Hallquist, M., Wenger, J. C., Baltensperger, U., Rudich, Y., Simpson, D., Claeys, M., Dommen, J., Donahue, N. M., George, C., Goldstein, A. H., Hamilton, J. F., Herrmann, H., Hoffmann, T., Iinuma, Y., Jang, M., Jenkin, M. E., Jimenez, J. L., Kiendler-Scharr, A., Maenhaut, W., McFiggans, G., Mentel, Th. F., Monod, A., Prévôt, A. S. H., Seinfeld, J. H., Surratt, J. D., Szmigielski, R., and Wildt, J.: The formation, properties and im-

pact of secondary organic aerosol: current and emerging issues, Atmos. Chem. Phys., 9, 5155–5236, https://doi.org/10.5194/acp-9-5155-2009, 2009.

Hayes, P. L., Carlton, A. G., Baker, K. R., Ahmadov, R., Washenfelder, R. A., Alvarez, S., Rappenglück, B., Gilman, J. B., Kuster, W. C., de Gouw, J. A., Zotter, P., Prévôt, A. S. H., Szidat, S., Kleindienst, T. E., Offenberg, J. H., Ma, P. K., and Jimenez, J. L.: Modeling the formation and aging of secondary organic aerosols in Los Angeles during CalNex 2010, Atmos. Chem. Phys., 15, 5773–5801, https://doi.org/10.5194/acp-15-5773-2015, 2015.

Hearn, J. D. and Smith, G. D.: Kinetics and Product Studies for Ozonolysis Reactions of Organic Particles Using Aerosol CIMS, J. Phys. Chem. A, 108, 10019–10029, https://doi.org/10.1021/jp0404145, 2004.

Hoffmann, T., Odum, J. R., Bowman, F., Collins, D., Klockow, D., Flagan, R. C., and Seinfeld., J. H.: Formation of Organic Aerosols from the Oxidation of Biogenic Hydrocarbons, J. Atmos. Chem., 26, 189–222, https://doi.org/10.1023/A:1005734301837, 1997.

Hu, W., Palm, B. B., Day, D. A., Campuzano-Jost, P., Krechmer, J. E., Peng, Z., de Sá, S. S., Martin, S. T., Alexander, M. L., Baumann, K., Hacker, L., Kiendler-Scharr, A., Koss, A. R., de Gouw, J. A., Goldstein, A. H., Seco, R., Sjostedt, S. J., Park, J.-H., Guenther, A. B., Kim, S., Canonaco, F., Prévôt, A. S. H., Brune, W. H., and Jimenez, J. L.: Volatility and lifetime against OH heterogeneous reaction of ambient isoprene-epoxydiols-derived secondary organic aerosol (IEPOX-SOA), Atmos. Chem. Phys., 16, 11563–11580, https://doi.org/10.5194/acp-16-11563-2016, 2016.

Jathar, S. H., Cappa, C. D., Wexler, A. S., Seinfeld, J. H., and Kleeman, M. J.: Multi-generational oxidation model to simulate secondary organic aerosol in a 3-D air quality model, Geosci. Model Dev., 8, 2553–2567, https://doi.org/10.5194/gmd-8-2553-2015, 2015.

Kang, E., Root, M. J., Toohey, D. W., and Brune, W. H.: Introducing the concept of Potential Aerosol Mass (PAM), Atmos. Chem. Phys., 7, 5727–5744, https://doi.org/10.5194/acp-7-5727-2007, 2007.

Kang, E., Toohey, D. W., and Brune, W. H.: Dependence of SOA oxidation on organic aerosol mass concentration and OH exposure: experimental PAM chamber studies, Atmos. Chem. Phys., 11, 1837–1852, https://doi.org/10.5194/acp-11-1837-2011, 2011.

Karjalainen, P., Timonen, H., Saukko, E., Kuuluvainen, H., Saarikoski, S., Aakko-Saksa, P., Murtonen, T., Bloss, M., Dal Maso, M., Simonen, P., Ahlberg, E., Svenningsson, B., Brune, W. H., Hillamo, R., Keskinen, J., and Rönkkö, T.: Time-resolved characterization of primary particle emissions and secondary particle formation from a modern gasoline passenger car, Atmos. Chem. Phys., 16, 8559–8570, https://doi.org/10.5194/acp-16-8559-2016, 2016.

Krechmer, J. E., Pagonis, D., Ziemann, P. J., and Jimenez, J. L.: Quantification of Gas-Wall Partitioning in Teflon Environmental Chambers Using Rapid Bursts of Low-Volatility Oxidized Species Generated in Situ, Environ. Sci. Technol., 50, 5757–5765, https://doi.org/10.1021/acs.est.6b00606, 2016.

Lakey, P. S. J., George, I. J., Whalley, L. K., Baeza-Romero, M. T. and Heard, D. E.: Measurements of the HO2 Uptake Coefficients onto Single Component Organic Aerosols, Environ. Sci. Technol., 49, 4878–4885, https://doi.org/10.1021/acs.est.5b00948, 2015.

Lambe, A., Massoli, P., Zhang, X., Canagaratna, M., Nowak, J., Daube, C., Yan, C., Nie, W., Onasch, T., Jayne, J., Kolb, C., Davidovits, P., Worsnop, D., and Brune, W.: Controlled nitric oxide production via $O(^1D) + N_2O$ reactions for use in oxidation flow reactor studies, Atmos. Meas. Tech., 10, 2283–2298, https://doi.org/10.5194/amt-10-2283-2017, 2017.

Lambe, A. T. and Jimenez, J. L.: PAM Wiki: Publications Using the PAM Oxidation Flow Reactor, available at: https://sites.google.com/site/pamwiki/publications, last access: 27 September 2017.

Lambe, A. T., Ahern, A. T., Williams, L. R., Slowik, J. G., Wong, J. P. S., Abbatt, J. P. D., Brune, W. H., Ng, N. L., Wright, J. P., Croasdale, D. R., Worsnop, D. R., Davidovits, P., and Onasch, T. B.: Characterization of aerosol photooxidation flow reactors: heterogeneous oxidation, secondary organic aerosol formation and cloud condensation nuclei activity measurements, Atmos. Meas. Tech., 4, 445–461, https://doi.org/10.5194/amt-4-445-2011, 2011.

Levy II, H.: Normal atmosphere: large radical and formaldehyde concentrations predicted, Science, 173, 141–143, https://doi.org/10.1126/science.173.3992.141, 1971.

Li, R., Palm, B. B., Borbon, A., Graus, M., Warneke, C., Ortega, A. M., Day, D. A., Brune, W. H., Jimenez, J. L., and de Gouw, J. A.: Laboratory Studies on Secondary Organic Aerosol Formation from Crude Oil Vapors, Environ. Sci. Technol., 47, 12566–12574, https://doi.org/10.1021/es402265y, 2013.

Li, R., Palm, B. B., Ortega, A. M., Hu, W., Peng, Z., Day, D. A., Knote, C., Brune, W. H., de Gouw, J., and Jimenez, J. L.: Modeling the radical chemistry in an Oxidation Flow Reactor (OFR): radical formation and recycling, sensitivities, and OH exposure estimation equation, J. Phys. Chem. A, 119, 4418–4432, https://doi.org/10.1021/jp509534k, 2015.

Link, M. F., Friedman, B., Fulgham, R., Brophy, P., Galang, A., Jathar, S. H., Veres, P., Roberts, J. M. and Farmer, D. K.: Photochemical processing of diesel fuel emissions as a large secondary source of isocyanic acid (HNCO), Geophys. Res. Lett., 43, 4033–4041, https://doi.org/10.1002/2016GL068207, 2016.

Lippmann, M.: Health effects of tropospheric ozone, Environ. Sci. Technol., 25, 1954–1962, https://doi.org/10.1021/es00024a001, 1991.

Liu, P. F., Abdelmalki, N., Hung, H.-M., Wang, Y., Brune, W. H., and Martin, S. T.: Ultraviolet and visible complex refractive indices of secondary organic material produced by photooxidation of the aromatic compounds toluene and m-Xylene, Atmos. Chem. Phys., 15, 1435–1446, https://doi.org/10.5194/acp-15-1435-2015, 2015.

Mao, J., Ren, X., Brune, W. H., Olson, J. R., Crawford, J. H., Fried, A., Huey, L. G., Cohen, R. C., Heikes, B., Singh, H. B., Blake, D. R., Sachse, G. W., Diskin, G. S., Hall, S. R., and Shetter, R. E.: Airborne measurement of OH reactivity during INTEX-B, Atmos. Chem. Phys., 9, 163–173, https://doi.org/10.5194/acp-9-163-2009, 2009.

Martinsson, J., Eriksson, A. C., Nielsen, I. E., Malmborg, V. B., Ahlberg, E., Andersen, C., Lindgren, R., Nyström, R., Nordin, E. Z., Brune, W. H., Svenningsson, B., Swietlicki, E., Boman, C., and Pagels, J. H.: Impacts of Combustion Conditions and Photochemical Processing on the Light Absorption of Biomass

Combustion Aerosol, Environ. Sci. Technol., 49, 14663–14671, https://doi.org/10.1021/acs.est.5b03205, 2015.

Matsunaga, A. and Ziemann, P. J.: Gas-Wall Partitioning of Organic Compounds in a Teflon Film Chamber and Potential Effects on Reaction Product and Aerosol Yield Measurements, Aerosol Sci. Tech., 44, 881–892, https://doi.org/10.1080/02786826.2010.501044, 2010.

Matti Maricq, M.: Chemical characterization of particulate emissions from diesel engines: A review, J. Aerosol Sci., 38, 1079–1118, https://doi.org/10.1016/j.jaerosci.2007.08.001, 2007.

Moise, T. and Rudich, Y.: Reactive Uptake of Ozone by Aerosol-Associated Unsaturated Fatty Acids: Kinetics, Mechanism, and Products, J. Phys. Chem. A, 106, 6469–6476, https://doi.org/10.1021/jp025597e, 2002.

Moise, T., Talukdar, R. K., Frost, G. J., Fox, R. W., and Rudich, Y.: Reactive uptake of NO_3 by liquid and frozen organics, J. Geophys. Res., 107, 4014, https://doi.org/10.1029/2001JD000334, 2002.

Nehr, S., Bohn, B., Fuchs, H., Häseler, R., Hofzumahaus, A., Li, X., Rohrer, F., Tillmann, R., and Wahner, A.: Atmospheric photochemistry of aromatic hydrocarbons: OH budgets during SAPHIR chamber experiments, Atmos. Chem. Phys., 14, 6941–6952, https://doi.org/10.5194/acp-14-6941-2014, 2014.

Nel, A.: Air Pollution-Related Illness: Effects of Particles, Science, 308, 804–806, https://doi.org/10.1126/science.1108752, 2005.

Ng, N. L., Canagaratna, M. R., Zhang, Q., Jimenez, J. L., Tian, J., Ulbrich, I. M., Kroll, J. H., Docherty, K. S., Chhabra, P. S., Bahreini, R., Murphy, S. M., Seinfeld, J. H., Hildebrandt, L., Donahue, N. M., DeCarlo, P. F., Lanz, V. A., Prévôt, A. S. H., Dinar, E., Rudich, Y., Worsnop, D. R.: Organic aerosol components observed in Northern Hemispheric datasets from Aerosol Mass Spectrometry, Atmos. Chem. Phys., 10, 4625–4641, https://doi.org/10.5194/acp-10-4625-2010, 2010.

Odum, J. R., Hoffmann, T., Bowman, F., Collins, D., Flagan Richard, C., and Seinfeld, J. H.: Gas particle partitioning and secondary organic aerosol yields, Environ. Sci. Technol., 30, 2580–2585, https://doi.org/10.1021/es950943+, 1996.

Orlando, J. J. and Tyndall, G. S.: Laboratory studies of organic peroxy radical chemistry: an overview with emphasis on recent issues of atmospheric significance, Chem. Soc. Rev., 41, 6294, https://doi.org/10.1039/c2cs35166h, 2012.

Ortega, A. M., Day, D. A., Cubison, M. J., Brune, W. H., Bon, D., de Gouw, J. A., and Jimenez, J. L.: Secondary organic aerosol formation and primary organic aerosol oxidation from biomass-burning smoke in a flow reactor during FLAME-3, Atmos. Chem. Phys., 13, 11551–11571, https://doi.org/10.5194/acp-13-11551-2013, 2013.

Ortega, A. M., Hayes, P. L., Peng, Z., Palm, B. B., Hu, W., Day, D. A., Li, R., Cubison, M. J., Brune, W. H., Graus, M., Warneke, C., Gilman, J. B., Kuster, W. C., de Gouw, J., Gutiérrez-Montes, C., and Jimenez, J. L.: Real-time measurements of secondary organic aerosol formation and aging from ambient air in an oxidation flow reactor in the Los Angeles area, Atmos. Chem. Phys., 16, 7411–7433, https://doi.org/10.5194/acp-16-7411-2016, 2016.

Palm, B. B., Campuzano-Jost, P., Ortega, A. M., Day, D. A., Kaser, L., Jud, W., Karl, T., Hansel, A., Hunter, J. F., Cross, E. S., Kroll, J. H., Peng, Z., Brune, W. H., and Jimenez, J. L.: In situ secondary organic aerosol formation from ambient pine forest air using an oxidation flow reactor, Atmos. Chem. Phys., 16, 2943–2970, https://doi.org/10.5194/acp-16-2943-2016, 2016.

Palm, B. B., Campuzano-Jost, P., Day, D. A., Ortega, A. M., Fry, J. L., Brown, S. S., Zarzana, K. J., Dube, W., Wagner, N. L., Draper, D. C., Kaser, L., Jud, W., Karl, T., Hansel, A., Gutiérrez-Montes, C. and Jimenez, J. L.: Secondary organic aerosol formation from in situ OH, O_3, and NO_3 oxidation of ambient forest air in an oxidation flow reactor, Atmos. Chem. Phys., 17, 5331–5354, https://doi.org/10.5194/acp-17-5331-2017, 2017.

Peng, Z., Carrasco, N., and Pernot, P.: Modeling of synchrotron-based laboratory simulations of Titan's ionospheric photochemistry, Geo. Res. J., 1–2, 33–53, https://doi.org/10.1016/j.grj.2014.03.002, 2014.

Peng, Z., Day, D. A., Ortega, A. M., Palm, B. B., Hu, W., Stark, H., Li, R., Tsigaridis, K., Brune, W. H., and Jimenez, J. L.: Non-OH chemistry in oxidation flow reactors for the study of atmospheric chemistry systematically examined by modeling, Atmos. Chem. Phys., 16, 4283–4305, https://doi.org/10.5194/acp-16-4283-2016, 2016.

Peng, Z., Day, D. A., Stark, H., Li, R., Lee-Taylor, J., Palm, B. B., Brune, W. H., and Jimenez, J. L.: HO_x radical chemistry in oxidation flow reactors with low-pressure mercury lamps systematically examined by modeling, Atmos. Meas. Tech., 8, 4863–4890, https://doi.org/10.5194/amt-8-4863-2015, 2015.

Ranney, A. P. and Ziemann, P. J.: Kinetics of Acid-Catalyzed Dehydration of Cyclic Hemiacetals in Organic Aerosol Particles in Equilibrium with Nitric Acid Vapor, J. Phys. Chem. A, 120, 2561–2568, https://doi.org/10.1021/acs.jpca.6b01402, 2016.

Richards-Henderson, N. K., Goldstein, A. H., and Wilson, K. R.: Large Enhancement in the Heterogeneous Oxidation Rate of Organic Aerosols by Hydroxyl Radicals in the Presence of Nitric Oxide, J. Phys. Chem. Lett., 6, 4451–4455, https://doi.org/10.1021/acs.jpclett.5b02121, 2015.

Saltelli, A., Ratto, M., Tarantola, S., and Campolongo, F.: Sensitivity Analysis for Chemical Models, Chem. Rev., 105, 2811–2828, https://doi.org/10.1021/cr040659d, 2005.

Sander, S. P., Friedl, R. R., Barker, J. R., Golden, D. M., Kurylo, M. J., Wine, P. H., Abbatt, J. P. D., Burkholder, J. B., Kolb, C. E., Moortgat, G. K., Huie, R. E., and Orkin, V. L.: Chemical Kinetics and Photochemical Data for Use in Atmospheric Studies, Evaluation Number 7, Pasadena, CA, USA, available at: http://jpldataeval.jpl.nasa.gov/pdf/JPL10-6Final (last access: June 2016), 2011.

Schill, G. P., Jathar, S. H., Kodros, J. K., Levin, E. J. T., Galang, A. M., Friedman, B., Link, M. F., Farmer, D. K., Pierce, J. R., Kreidenweis, S. M., and DeMott, P. J.: Ice-nucleating particle emissions from photochemically aged diesel and biodiesel exhaust, Geophys. Res. Lett., 43, 5524–5531, https://doi.org/10.1002/2016GL069529, 2016.

Schwantes, R. H., Schilling, K. A., McVay, R. C., Lignell, H., Coggon, M. M., Zhang, X., Wennberg, P. O., and Seinfeld, J. H.: Formation of highly oxygenated low-volatility products from cresol oxidation, Atmos. Chem. Phys., 17, 3453–3474, https://doi.org/10.5194/acp-17-3453-2017, 2017.

Seakins, P. W.: A brief review of the use of environmental chambers for gas phase studies of kinetics, chemical mechanisms and characterisation of field instruments, EPJ Web Conf., 9, 143–163, https://doi.org/10.1051/epjconf/201009012, 2010.

Simonen, P., Saukko, E., Karjalainen, P., Timonen, H., Bloss, M., Aakko-Saksa, P., Rönkkö, T., Keskinen, J., and Dal Maso, M.: A new oxidation flow reactor for measuring secondary aerosol formation of rapidly changing emission sources, Atmos. Meas. Tech., 10, 1519–1537, https://doi.org/10.5194/amt-10-1519-2017, 2017.

Stocker, T. F., Qin, D., Plattner, G.-K., Tignor, M., Allen, S. K., Boschung, J., Nauels, A., Xia, Y., Bex, V., and Midgley, P. M.: Climate Change 2013 – The Physical Science Basis, edited by: Intergovernmental Panel on Climate Change, Cambridge University Press, Cambridge, 2014.

Strollo, C. M. and Ziemann, P. J.: Products and mechanism of secondary organic aerosol formation from the reaction of 3-methylfuran with OH radicals in the presence of NO_x, Atmos. Environ., 77, 534–543, https://doi.org/10.1016/j.atmosenv.2013.05.033, 2013.

Tkacik, D. S., Lambe, A. T., Jathar, S., Li, X., Presto, A. A., Zhao, Y., Blake, D., Meinardi, S., Jayne, J. T., Croteau, P. L., and Robinson, A. L.: Secondary Organic Aerosol Formation from in-Use Motor Vehicle Emissions Using a Potential Aerosol Mass Reactor, Environ. Sci. Technol., 48, 11235–11242, https://doi.org/10.1021/es502239v, 2014.

Volkamer, R., Jimenez, J. L., San Martini, F., Dzepina, K., Zhang, Q., Salcedo, D., Molina, L. T., Worsnop, D. R., and Molina, M. J.: Secondary organic aerosol formation from anthropogenic air pollution: Rapid and higher than expected, Geophys. Res. Lett., 33, L17811, https://doi.org/10.1029/2006GL026899, 2006.

Wang, J., Doussin, J. F., Perrier, S., Perraudin, E., Katrib, Y., Pangui, E., and Picquet-Varrault, B.: Design of a new multi-phase experimental simulation chamber for atmospheric photosmog, aerosol and cloud chemistry research, Atmos. Meas. Tech., 4, 2465–2494, https://doi.org/10.5194/amt-4-2465-2011, 2011.

Zhang, X., Cappa, C. D., Jathar, S. H., McVay, R. C., Ensberg, J. J., Kleeman, M. J., and Seinfeld, J. H.: Influence of vapor wall loss in laboratory chambers on yields of secondary organic aerosol, P. Natl. Acad. Sci. USA, 111, 5802–5807, https://doi.org/10.1073/pnas.1404727111, 2014.

Ziemann, P. J. and Atkinson, R.: Kinetics, products, and mechanisms of secondary organic aerosol formation, Chem. Soc. Rev., 41, 6582, https://doi.org/10.1039/c2cs35122f, 2012.

7

Particulate pollutants in the Brazilian city of São Paulo: 1-year investigation for the chemical composition and source apportionment

Guilherme Martins Pereira[1,4], Kimmo Teinilä[2], Danilo Custódio[1,3], Aldenor Gomes Santos[4,5,6], Huang Xian[7], Risto Hillamo[2], Célia A. Alves[3], Jailson Bittencourt de Andrade[4,5,6], Gisele Olímpio da Rocha[4,5,6], Prashant Kumar[8,9], Rajasekhar Balasubramanian[7], Maria de Fátima Andrade[10], and Pérola de Castro Vasconcellos[1,4]

[1]Institute of Chemistry, University of São Paulo, São Paulo – SP, 05508-000, Brazil
[2]Finnish Meteorological Institute, P.O. Box 503, 00101 Helsinki, Finland
[3]CESAM & Department of Environment, University of Aveiro, Aveiro, 3810-193, Portugal
[4]INCT for Energy and Environment, Federal University of Bahia, Salvador – BA, 40170-115, Brazil
[5]CIEnAm, Federal University of Bahia, Salvador – BA, 40170-115, Brazil
[6]Institute of Chemistry, Federal University of Bahia, Salvador – BA, 40170-115, Brazil
[7]Department of Civil and Environmental Engineering, National University of Singapore, E1A 07-03, 117576, Singapore
[8]Global Centre for Clean Air Research (GCARE), Department of Civil and Environmental Engineering, Faculty of Engineering and Physical Sciences, University of Surrey, Guildford GU2 7XH, UK
[9]Environmental Flow Research Centre, Faculty of Engineering and Physical Sciences, University of Surrey, Guildford GU2 7XH, UK
[10]Institute of Astronomy, Geophysics and Atmospheric Sciences, University of São Paulo, São Paulo – SP, 05508-090, Brazil

Correspondence to: Guilherme Martins Pereira (martinspereira2@hotmail.com)

Abstract. São Paulo in Brazil has relatively relaxed regulations for ambient air pollution standards and often experiences high air pollution levels due to emissions of particulate pollutants from local sources and long-range transport of air masses impacted by biomass burning. In order to evaluate the sources of particulate air pollution and related health risks, a year-round sampling was done at the University of São Paulo campus (20 m a.g.l.), a green area near an important expressway. The sampling was performed for $PM_{2.5}$ ($\leq 2.5\,\mu m$) and PM_{10} ($\leq 10\,\mu m$) in 2014 through intensive (everyday sampling in wintertime) and extensive campaigns (once a week for the whole year) with 24 h of sampling. This year was characterized by having lower average precipitation compared to meteorological data, and high-pollution episodes were observed all year round, with a significant increase in pollution level in the intensive campaign, which was performed during wintertime. Different chemical constituents, such as carbonaceous species, polycyclic aromatic hydrocarbons (PAHs) and derivatives, water-soluble ions, and biomass burning tracers were identified in order to evaluate health risks and to apportion sources. The species such as PAHs, inorganic and organic ions, and monosaccharides were determined using chromatographic techniques and carbonaceous species using thermal-optical analysis. Trace elements were determined using inductively coupled plasma mass spectrometry. The risks associated with particulate matter exposure based on PAH concentrations were also assessed, along with indexes such as the benzo[*a*]pyrene equivalent (BaPE) and lung cancer risk (LCR). High BaPE and LCR were observed in most of the samples, rising to critical values in the wintertime. Also, biomass burning tracers and PAHs were higher in this season, while secondarily formed ions presented low variation throughout the year. Meanwhile, vehicular tracer species were also higher in the intensive campaign, suggesting the influence of lower dispersion conditions in that period. Source apportionment was performed using positive matrix factorization (PMF), which indicated five different factors: road dust, industrial emissions, vehicular exhaust, biomass burning and secondary processes. The results highlighted the contribution of vehicular emissions and the significant input from biomass combustion in wintertime, suggesting that most of the particulate matter is due to local sources, in addition to the influence of pre-harvest sugarcane burning.

---

1 Introduction

Air pollution caused by atmospheric particulate matter (PM) is one of the major environmental problems encountered in Latin American cities such as São Paulo (Brazil), Mexico City (Mexico), Bogotá (Colombia) and Santiago (Chile) (Romero-Lankao et al., 2013; Vasconcellos et al., 2010, 2011a; Villalobos et al., 2015). The air pollution thresholds in most of the Latin American cities are not very stringent compared to international standards or guidelines (Alvarez et al., 2013; Kumar et al., 2016). Several studies have highlighted a statistical relation between PM and health problems, including respiratory and cardiovascular diseases and genotoxic risks (Newby et al., 2015; de Oliveira Alves et al., 2014; Pope, 2000). In this context, $PM_{2.5}$ (PM with an aerodynamic diameter smaller than 2.5 μm) and PM_{10} (PM with an aerodynamic diameter smaller than 10 μm) are particles that are able to penetrate the respiratory system, with $PM_{2.5}$ reaching alveoli in the lungs, and induce adverse impacts on human health (Cai et al., 2015; Kumar et al., 2014). The elderly and the children are more susceptible to the health effects resulting from $PM_{2.5}$ (Cançado et al., 2006; Segalin et al., 2017). Considering that the elderly population has grown in São Paulo over the last decades (SEADE, 2016; Segalin et al., 2017), the PM health-related issues can become more relevant. PM also plays an important role in ecosystem biogeochemistry, the hydrological cycle, cloud formation and atmospheric circulation (Pöschl, 2005).

Carbonaceous species such as organic and elemental carbon (OC and EC) represent a large fraction of PM and play an important role in the formation of haze, interaction with climate and adverse human health effects (Bisht et al., 2015; Liu et al., 2016; Seinfeld and Pandis, 2006). Water-soluble ions (WSIs) account for another major fraction of aerosols in urban areas and are able to affect visibility, particle hygroscopicity and cloud formation; they also influence acidity in rainwater and impact climate (Cheng et al., 2011; Jung et al., 2009; Khoder and Hassan, 2008; Tan et al., 2009; Tang et al., 2016; Yang et al., 2015).

Particulate organic carbon includes key species including polycyclic aromatic hydrocarbons (PAHs) and monosaccharides. The last are considered biomass burning tracers (such as levoglucosan, mannosan and galactosan) (Simoneit et al., 1999). PAHs have natural sources (synthesis by plants and bacteria, degradation of plants, forest fires, and volcanic emissions) but are mostly emitted by anthropogenic sources

at urban sites (such as domestic, mobile, industrial and agricultural sources) (Abdel-shafy and Mansour, 2016; Ravindra et al., 2008). They have been studied because of their carcinogenic properties (de Oliveira Alves et al., 2014; Seinfeld and Pandis, 2006). The nitrated and oxygenated PAHs (nitro- and oxy-PAHs) are emitted as primary species or are formed in situ as secondary compounds (Kojima et al., 2010; Souza et al., 2014b; Zhou and Wenger, 2013; Zimmermann et al., 2013). They are potentially more mutagenic and/or carcinogenic than their PAH precursors (Franco et al., 2010).

Chemical speciation and PAH risk assessment have been performed at several Latin American sites, specifically in urban São Paulo, Bogotá, Buenos Aires (Vasconcellos et al., 2011a, b) and forested areas such as the Amazon region (de Oliveira Alves et al., 2015). Biomass burning tracers have been detected in high concentrations in São Paulo during the dry season and are attributed to the long-range transport of aerosols from areas affected by sugarcane burning. Source apportionment studies have been carried out in São Paulo (Table 1) in the last 3 decades, but not in as much detail as in other megacities. Detailed characterization of the organic fraction of aerosols is still scarce.

A previous study performed in São Paulo in 1989 highlights the relative importance of the emissions from residual oil and diesel in $PM_{2.5}$ and soil dust in the coarse grain size (Andrade et al., 1994). Da Rocha et al. (2012) studied the emission sources of fuel and biomass burning, the gas-to-particle conversion, and sea spray emissions in PM in São Paulo, in a 1-year period (between 2003 and 2004). Another study conducted in the winter of 2003 pointed out a strong impact of local sources at three sites in the state of São Paulo, in addition to the influence of remote sources (Vasconcellos et al., 2007). A source apportionment for PAHs in the winter of 2002 reported a predominance of diesel emissions for the polyaromatics in $PM_{2.5}$ (Bourotte et al., 2005). In turn, Castanho and Artaxo (2001), in their study of 1997 and 1998 in São Paulo city, reported no significant differences in the main air pollution sources (i.e., automobile traffic and soil dust) between wintertime and summertime. The main sources for $PM_{2.5}$ were automobile traffic and soil dust. However, biomass burning was not considered as a potential source by the authors.

The current study presents a more comprehensive study that should lead to a better understanding of the main PM sources and atmospheric processes occurring in the São Paulo megacity than previous studies reported in the literature. A year of extensive sampling of aerosol ($PM_{2.5}$ and PM_{10}) and an intensive wintertime campaign were performed. Different classes of chemical components in PM were determined such as carbonaceous species, WSIs, monosaccharides, PAHs and their derivatives. Meteorological data were also collected during the sampling days. Moreover, the benzo[a]pyrene equivalent (BaPE) and lung cancer risk (LCR) indexes were calculated in order to assess the potential toxicity of PAHs. Positive matrix factorization (PMF)

Table 1. Results of previous source apportionment studies in São Paulo city.

Site	Year	Instruments	Species	Range	Identified sources	Source
University of São Paulo – Atmospheric Sciences Department (campus)	1989 (winter)	PIXE	Elements	$PM_{2.5}$	Factor analysis – five sources: industrial emissions (13 %), emissions from residual oil and diesel (41 %), resuspended soil dust (28 %) and emissions of Cu and Mg (18 %).	Andrade et al. (1994)
				$PM_{2.5-15}$	Four sources: soil dust (59 %), industrial emissions (19 %), oil burning (8 %) and sea salt aerosol (14 %).	
University of São Paulo – Medicine School (downtown)	1994 (winter)	PIXE, reflectance	Elements, black carbon	PM_2	Factor analysis – five sources: vehicles, garbage incineration, vegetation, suspended soil dust and burning of fuel oil.	Sánchez-Ccoyllo and Andrade (2002)
University of São Paulo – Medicine School building and Atmospheric Sciences Department	1997 (winter) and 1998 (summer)	TEOM; PIXE; ACPM; Aethalometer	OC, EC, elements and gaseous species	$PM_{2.5}$	Factor analysis – five sources: motor vehicle (28 and 24 %, for winter and summer), resuspended soil dust (25 and 30 %), oil combustion source (18 and 21 %), sulfates (23 and 17 %) and industrial emissions (5 and 6 %).	Castanho and Artaxo (2001)
				$PM_{2.5-10}$	Resuspended soil dust represented a large fraction (75–78 %).	
University of São Paulo – Atmospheric Sciences Department	2002 (winter)	GC-MS	PAHs	$PM_{2.5}$	Factor analysis – four factors: diesel emissions, stationary combustion source, vehicular emissions, natural gas combustion and biomass burning.	Bourotte et al. (2005)
				$PM_{2.5-10}$	Two factors: vehicular emissions and mixture of combustion sources (natural gas combustion, incineration emissions and oil combustion).	
University of São Paulo – Atmospheric Sciences Department	2003 (winter)	IC; ICPMS	WSI and elements	PM_{10}	Principal component analysis – two factors (48.5 % of variance): local and remote sources.	Vasconcellos et al. (2007)
University of São Paulo – Atmospheric Sciences Department	2003–2004 (year round)	IC; CCD ICP	WSI and elements	PM_{10}	Principal component analysis – three principal components: biomass burning and/or automobile fuel burning (40.3 %), gas-to-particle conversion (12.7 %) and sea spray contribution (11.7 %).	da Rocha et al. (2012)
University of São Paulo – Medicine School building	2007–2008 (year round)	X-ray spectrometry, reflectance	Elements and black carbon	$PM_{2.5}$	APCA – four factors: crustal emission (soil and construction) (13 %); oil-burning boilers, industrial emissions and secondary aerosol formation (13 %); light-duty vehicle emissions (12 %) and heavy-duty diesel fleet (28 %).	Andrade et al. (2012)

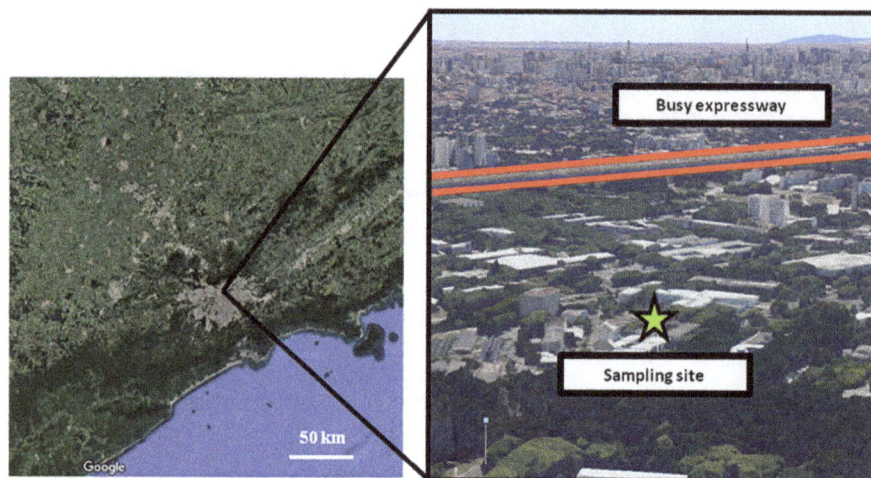

Figure 1. Location of the sampling site. Maps are a courtesy of Google Maps.

analysis was also used for the source apportionment of PM_{10} during the extensive campaign.

2 Methodology

2.1 Sampling campaigns

Aerosol samples were collected at a São Paulo site (SPA, 23° 33′34″ S and 46°44′01″ W) located on the rooftop of the Atmospheric Sciences Department (about 20 m a.g.l.), at the Institute of Astronomy, Geophysics and Atmospheric Sciences (IAG-USP) building, within the campus of the University of São Paulo. The location is inside a green area and approximately 2 km away from an important expressway (Marginal Pinheiros) (Fig. 1). Aerosols were collected in intensive (every day) and extensive campaigns (once a week) throughout 2014. Firstly, the extensive campaign was performed weekly. Accordingly, samples were collected every Tuesday for $PM_{2.5}$ (termed $Ext_{2.5}$ in this study) and PM_{10} (termed Ext_{10}). However, due to equipment breaking down, the $PM_{2.5}$ sampling was stopped in September while the PM_{10} sampling continued until December ($n = 32$ and 38, respectively). Secondly, the intensive campaign (termed $Int_{2.5}$) took place between 1 and 18 July 2014 ($n = 12$), only for $PM_{2.5}$ due to problems with PM_{10} equipment. However, there were 4 days (between 8 and 11 July) for which data were not collected due to heavy rain.

PM samples were collected for a period of 24 h, with high-volume air samplers (HiVol), with a flow rate of $1.13 \, m^3 \, min^{-1}$, with 2.5 and 10 μm size selective inlets (Thermo Andersen, USA). Prior to sampling, quartz fiber filters (20 cm × 25 cm, Millipore, USA) were baked for 8 h at 800 °C to remove the organics. In addition, filters were equilibrated at room temperature and weighed in a microbalance before and after the sampling in order to estimate the PM concentration. After sampling and weighing, the filters were wrapped in aluminum foil and stored in a refrigerator at 5 °C until chemical analyses were performed.

2.2 Meteorological data

The meteorological data (ambient temperature, relative humidity, precipitation and wind speed) were collected from the climatological bulletin of the IAG-USP meteorological station (IAG, 2014). The climate of São Paulo is often classified as humid subtropical (Andrade et al., 2012a). The wintertime in the city is characterized by a slight decrease in temperatures, together with considerably lower relative humidity and precipitations, with more thermodynamic stability, often resulting in accumulation of air pollutants in the lower troposphere, and is also subjected to thermal inversion episodes (Miranda et al., 2012). The local air circulation is mainly associated with the Atlantic Ocean breeze and cold fronts in wintertime often intensify that, with winds generally coming from the southeast (Vasconcellos et al., 2003). In Fig. S1 in the Supplement a comparison between the average climatological temperature and the data for 2014 is presented (IAG, 2014). During the 2014 campaign, the summer was atypically warmer and drier.

In order to analyze the long-range transport of air pollutants, backward air mass trajectories (96 h) were run using the HYSPLIT model (Draxler and Rolph, 2003), through the READY (Real-time Environmental Applications and Display sYstem) platform from NOAA. The heights considered were 500, 1500 and 3000 m, corresponding to trajectories near the ground, upper boundary layer and low free troposphere, respectively (Cabello et al., 2016; Toledano et al., 2009).

Table 2. Details of the analyzed species, analytical methods and detection limits.

Analytical method	Detection limits ($\mathrm{ng\,m^{-3}}$)	Determined species
Thermal-optical analysis	14 (EC) / 262 (OC)	Carbonaceous species: OC and EC.
GC-MS	0.01–0.06	PAHs: naphthalene (Nap), acenaphthene (Ace), acenaphthylene (Acy), Fluorene (Flu), phenanthrene (Phe), anthracene (Ant), fluoranthene (Flt), pyrene (Pyr), benzo[*a*]anthracene (BaA), chrysene (Chr), benzo[*b*]fluoranthene (BbF), benzo[*k*]fluoranthene (BkF), benzo[*e*]pyrene (BeP), benzo[*a*]pyrene (BaP), perylene (Per), indeno[1,2,3-*cd*]pyrene (InP), dibenzo[*ah*]anthracene (DBA), benzo[*ghi*]perylene (BPe) and coronene (Cor).
	0.01–0.50	Nitro-PAHs: 1-nitronaphthalene (1-NNap), 2-nitronaphthalene (2-NNap), 1-methyl-4-nitronaphthalene (1-Methyl-4-NNap), 1-methyl-5-nitronaphthalene (1-Methyl-5-NNap), 1-methyl-6-nitronaphthalene (1-Methyl-6-NNap), 2-methyl-4-nitronaphthalene (2- methyl-4-NNap), 2-nitrobiphenyl (2-NBP), 3-nitrobiphenyl (3-NBP), 4-nitrobiphenyl (4-NBP), 5-nitroacenaphthene (5-NAce), 2-nitrofluorene (2-Nflu), 2-nitrophenanthrene (2-NPhe), 3-nitrophenanthrene (3-NPhe), 9-nitrophenanthrene (9-NPhe), 2-nitroanthracene (2-NAnt), 9-nitroanthracene (9-NAnt), 2-nitrofluoranthene (2-NFlt), 3-nitrofluoranthene (3-NFlt), 1-nitropyrene (1-NPyr), 2-nitropyrene (2-NPyr), 4-nitropyrene (4-NPyr), 6-nitrochrysene (6-NChr), 7-nitrobenz[*a*]anthracene (7-NBaA), 3-nitrobenzanthrone (3- NBA), 6-nitrobenzo[*a*]pyrene (6-NBaPyr), 1-nitrobenzo[*e*]pyrene (1-NBePyr), and 3-nitrobenzo[*e*]pyrene (3-NBePyr).
	0.3–10.3	Oxy-PAHs: 1,4-benzoquinone (1,4-BQ), 9,10-phenanthraquinone (9,10-PQ), 9,10-anthraquinone (9,10-AQ), 1,2-naphthoquinone (1,2-NQ) and 1,4-naphthoquinone (1,4-NQ).
IC	1.3–1.3	WSI: Cl^-, NO_3^-, SO_4^{2-}, $C_2O_4^{2-}$, methylsulfonate (MSA^-), Na^+, K^+, and NH_4^+.
HPAEC-MS	1.3–2.5	Monosaccharides: levoglucosan (Lev), mannosan (Man) and galactosan (Gal).
ICP-MS	0.0002–2.3	Elements: Li, Mg, Al, K, Ca, Cr, Mn, Fe, Co, Ni, Cu, Zn, As, Se, Rb, Sr, Cd, Sn, Cs, Tl, Pb, and Bi.

2.3 Analytical procedures, reagents and standards

After sampling, the filters were punched for the chemical analysis, as shown in Table 2, which lists all substances determined and their respective analytical techniques as well as their detection limits (DLs).

Carbonaceous species were determined at the University of Aveiro, with two punches of 9 mm diameter. Firstly, the carbonates were removed with hydrochloric acid fumes and then OC and EC were determined using a thermal-optical transmission equipment developed at the university. The system comprises a quartz tube with two heating zones, a pulsed laser and a nondispersive infrared CO_2 analyzer (NDIR). The filters were placed into the first heating zone of the quartz tube then heated to 600 °C in a nitrogen atmosphere for the organic fraction to vaporize, which was quantified as OC. EC was determined with a sequential heating at 850 °C in an atmosphere containing 4 % O_2. The other heating zone was filled with cupric oxide and was maintained at 650 °C in a 4 % O_2 atmosphere to assure that all carbon was oxidized to

CO_2, which was quantified using a NDIR analyzer (Alves et al., 2015).

The determination of polycyclic aromatic hydrocarbons and their derivatives was performed at the Federal University of Bahia, Brazil, and is summarized in Table 2. Briefly, samples were extracted for 23 min in an ultrasonic bath (4.2 cm² punches) with a 500 μL solution of 18 % of acetonitrile in dichloromethane, employing miniaturized extraction devices (Whatman Mini™ UniPrep Filters, Whatman, USA). Their quantification was carried out using gas chromatography with high-resolution mass spectrometer detection (GC-MS). The procedure is described in more detail in Santos et al. (2016). BeP was quantified with the same calibration curve as BaP since they have a similar fragmentation pattern in the MS detector (Robbat and Wilton, 2014).

The US Environmental Protection Agency (EPA) 610 PAH mix in methanol : dichloromethane (1 : 1), containing 2000 μg mL⁻¹ each, was purchased from Supelco (St. Louis, USA). Individual standards of 50 μg mL⁻¹ coronene (Cor) and 1000 μg mL⁻¹ perylene (Per) and two deuterated compounds, pyrene D10 (Pyr d10) and fluorene D10 (Flu d10), were purchased from Sigma-Aldrich (St. Louis, USA). Quinones investigated in this study were purchased from Fluka (St. Louis, USA). Nitro-PAH certified standard solutions SRM 2264 (aromatic hydrocarbons nitrated in methylene chloride I) and SRM 2265 (polycyclic aromatic hydrocarbons nitrated in methylene chloride II) were purchased from the National Institute of Standards and Technology (NIST, USA).

Monosaccharides and WSIs were determined at the Finnish Meteorological Institute. From quartz fiber filter samples, $1 \, cm^2$ filter pieces were punched for both analyses. Concentrations of monosaccharides were determined using a Dionex ICS-3000 system coupled to a quadrupole mass spectrometer (Dionex MSQ^{TM}) using high-performance anion exchange chromatography (HPAEC-MS). Levoglucosan (1,6-anhydro-β-D-glucopyranose, purity 99%; Acros Organics, NJ, USA), mannosan (1,6-anhydro-β-D-mannopyranose, purity 99%; Sigma-Aldrich Co., MO, USA) and galactosan (1,6-anhydro-β-D-galactopyranose; Sigma-Aldrich Co.) were used for the calibration. The $1 \, cm^2$ punches were extracted with 5 mL of deionized water (Milli-Q water; resistivity 18.2 MΩ cm at 25 °C; Merck Millipore, MA, USA), with methyl-β-D-arabinopyranoside as the internal standard (purity 99%; Aldrich Chemical Co., WI, USA) and 10 min of gentle rotation. The extract was filtered through an IC Acrodisc$^®$ syringe filter (13 mm, 0.45 μm Supor$^®$ (PES) membrane, Pall Sciences) (Saarnio et al., 2010).

In order to determine the WSIs (Cl^-, NO_3^-, SO_4^{2-}, $C_2O_4^{2-}$, methylsulfonate, Na^+, K^+, NH_4^+), 10 mL of deionized water was used to extract the sample aliquots, with 10 min of gentle rotation. The ions were determined using two ion chromatography systems (ICS 2000 system, Dionex) simultaneously; cations were analyzed using a CG12A/CS12A column with an electrochemical suppressor (CSRS ULTRA II, 4 mm) and anions using an AG11/AS11 column with an electrochemical suppressor (ASRS ULTRA II, 4 mm).

Finally, trace elements in the samples were extracted using a microwave digestion system (MLS-1200 mega, Milestone Inc., Italy) at the National University of Singapore. Punches of the filters were cut into small pieces and added into PTFE vessels with 4 mL HNO_3 (Merck), 2 mL H_2O_2 (Merck) and 0.2 mL HF (Merck). The vessels were then subjected to a three-stage digestion inside the microwave digester (250 W for 5 min, 400 W for 5 min, and 600 W for 2 min). Following the digestion procedure, extracts were filtered with 0.45 μm PTFE syringe filters, diluted eight times and stored in the 4 °C cold room. The concentrations of trace elements were quantified using ICP-MS (Agilent 7700, USA) in triplicates. The instrumental parameters maintained during sample runs using the ICP-MS analysis were plasma gas (15.0 L min^{-1}), auxiliary gas (1.0 L min^{-1}) and nebulizer gas (1.0 L min^{-1}). Clean ceramic scissors and forceps were used to handle all PM samples. ICP-MS standards (purchased from High-Purity Standards, USA) were used for calibration.

2.4 Statistical analysis and receptor model

Pearson coefficients were calculated to verify the correlation between all the species (software STATISTICA). It determines the extent to which values of the variables are linearly correlated. The coefficients (r) were considered significant when $p < 0.05$. Two-tailed t tests were also employed in order to evaluate equal and unequal variances ($p < 0.05$). Polar plots considered the mass concentrations as a function of wind speed and direction (software R x64 3.3.2).

The widely used source apportionment model, PMF, was applied to all sample datasets (Paatero and Tapper, 1994). In this study, specifically, the EPA PMF5.0 software was used. Variables were classified as "strong", "weak" and "bad" according to the signal-to-noise ratio (S / N), number of samples below the detection limit (Amato et al., 2016; Contini et al., 2016; Paatero and Hopke, 2003) and thermal stability of the species. The species were categorized as bad when the S / N ratios were less than 0.2 and weak when the S / N ratios were greater than 0.2 but less than 2 (Lang et al., 2015). Accordingly, species with S / N ratios higher than 2 were considered strong. Bad variables were excluded from the model and the weak ones had their uncertainty increased by a factor of 3, as described in the EPA PMF Fundamentals and User Guide (Norris et al., 2014).

When concentrations were below the DLs, they were substituted by half the DL. Missing data were replaced by the median (M) of the whole dataset for that species (Brown et al., 2015). Uncertainties were calculated by Eq. (1) according to Norris et al. (2014), when the concentrations were below the detection limits:

$$\mathrm{Unc} = 5/6 \times \mathrm{DL}. \tag{1}$$

Uncertainty for missing data (Brown et al., 2015) is given by Eq. (2):

$$\mathrm{Unc} = 4 \times M. \tag{2}$$

When the concentrations were above the detection limit, uncertainty was determined from Eq. (3):

$$\mathrm{Unc} = \left([\mathrm{EF} \times C]^2 + [0.5 \times \mathrm{DL}]^2 \right)^{1/2}, \tag{3}$$

where EF is the error fraction and C is the element concentration. Q robust value (Q_R) is the goodness-of-fit parameter computed with the exclusion of points not fitted by the model. To evaluate the number of factors, Q_R was compared to Q_T (Q theoretical value). At the point when changes in the ratio Q_R/Q_T become smaller with the increase of the number of factors, it can be demonstrative that there might be an excessive number of factors being fitted (Brown et al., 2015). Q_T was estimated as in Lang et al. (2015), given by Eq. (4):

$$Q_T = (n_s \times n_e) - ([n_s \times n_f] + [n_e \times n_f]), \tag{4}$$

where n_s is the number of samples, n_e is the number of strong elements and n_f is the number of factors.

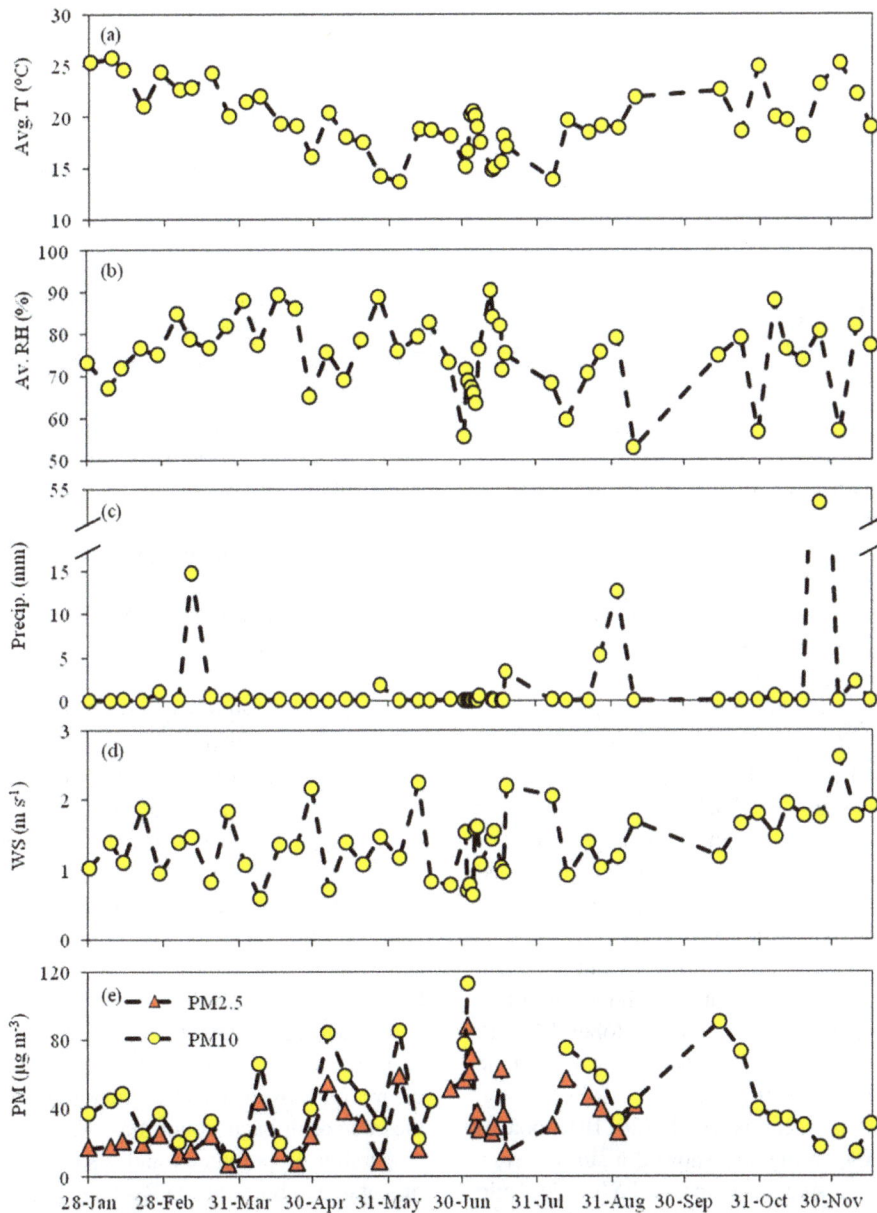

Figure 2. (a) Daily average temperatures, **(b)** relative humidity (RH), **(c)** precipitation, **(d)** wind speed, and **(e)** PM_{10} and $PM_{2.5}$ concentrations for the three campaigns.

3 Results and discussions

3.1 Concentrations of $PM_{2.5}$ and PM_{10} during extensive campaigns

The extensive campaigns ($Ext_{2.5}$ and Ext_{10}) were carried out over a whole year, during which the meteorological conditions varied largely. The average temperature during the sampling days in all campaigns ranged from 14 to 26 °C and the wind speed varied between 0.6 and 2.6 m s^{-1}; most of the sampling was carried out on days without rainfall. In Fig. 2

the meteorological variables PM_{10} and $PM_{2.5}$ concentrations for all analyzed days are presented.

There were moderate negative correlations between PM_{10} and average wind speed, average and minimum relative humidity and between $PM_{2.5}$ and average wind speed and minimum relative humidity (Table S1 in the Supplement). This observation is in agreement with the fact that days with lower relative humidity and lower wind speed present higher $PM_{2.5}$ and PM_{10} levels than in more humid and windier conditions.

In the extensive campaign, the PM mass concentrations exhibited a wide range of concentrations. For example, $Ext_{2.5}$ ranged from 8 to 78 µg m^{-3} (average 30 µg m^{-3}),

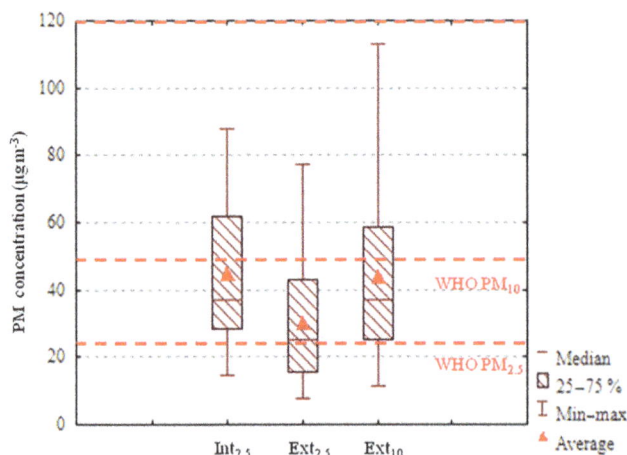

Figure 3. Box plot for particulate matter concentrations in the intensive and extensive campaigns.

whereas Ext_{10} values varied between 12 and 113 $\mu g\,m^{-3}$ (average 44 $\mu g\,m^{-3}$) (Fig. 3). The World Health Organization (WHO) recommends a daily limit of 50 $\mu g\,m^{-3}$ for PM_{10} and 25 $\mu g\,m^{-3}$ for $PM_{2.5}$, (WHO, 2006), while the Brazilian environmental agency (CONAMA) recommends a threshold of 150 $\mu g\,m^{-3}$ for PM_{10} (CONAMA, 1990; Pacheco et al., 2017). When considering the CONAMA standards, only 1 day in the extensive campaign was near the target limit. For PM, a coverage of 90 % for 1 calendar year is recommended for a proper risk assessment by the European Union ambient air quality directive, although, often a coverage of 75 % is often adopted (EEA, 2016); the data in this study did not reach these coverages. The Ext_{10} campaign was divided into two periods: dry (April–September) and rainy (October–March). It was observed that the average PM_{10} was 52 $\mu g\,m^{-3}$ in the dry period and 35 $\mu g\,m^{-3}$ in the rainy period.

A study done by Vasconcellos et al. (2011b) about a decade ago (2003–2004) in the city showed a similar average of PM_{10} (46 $\mu g\,m^{-3}$). According to CETESB (São Paulo state environmental agency), the annual average PM_{10} concentrations (considering all monitoring stations in the São Paulo metropolitan area) ranged from 33 to 41 $\mu g\,m^{-3}$, between the years of 2005 and 2014, showing no significant differences (CETESB, 2015).

The average values for $PM_{2.5}$ were higher than those obtained in a 1-year study done at traffic sites in two European metropolises: London and Madrid in 2005 (warm period: 19.40 and 20.63 $\mu g\,m^{-3}$ for $PM_{2.5}$, respectively) (Kassomenos et al., 2014). The European Union has a more restrictive control of pollutant emissions compared to Latin American countries since an annual mean of 40 $\mu g\,m^{-3}$ is established for PM_{10} and a limit value of 25 $\mu g\,m^{-3}$ is imposed for $PM_{2.5}$ (Kassomenos et al., 2014). However, these averages in São Paulo are lower than those observed in year-round studies performed in Chinese megacities, such as Shanghai (83 $\mu g\,m^{-3}$ for $PM_{2.5}$ and 123 $\mu g\,m^{-3}$ for PM_{10})

and Nanjing (222 $\mu g\,m^{-3}$ for $PM_{2.5}$ and 316 $\mu g\,m^{-3}$ for PM_{10}) (Shi et al., 2015; Wang et al., 2003; J. Wang et al., 2013). Indeed, in 2014, Zheng et al. (2016) assessed the $PM_{2.5}$ concentrations in 161 Chinese cities, reporting an annual average concentration of 62 $\mu g\,m^{-3}$.

In this study it was found that, on average, more than 60 % of the total mass PM is within the category $PM_{2.5}$; it is consistent with a previous study performed at this site (dry season, 2008) when this value was 69 % (Souza et al., 2014a). Conversely, in a 2-year study conducted at 10 urban sites in Rio de Janeiro, the coarse fraction represented from 60 to 70 % of the PM_{10} mass concentration (Godoy et al., 2009). The $PM_{2.5}/PM_{10}$ ratio found at other urban Brazilian sites with different characteristics (biomass burning, coastal environment) were close to 40 %, considerably lower than in the São Paulo metropolitan area, according to the local environmental agency (CETESB, 2015), highlighting the importance of fine PM in São Paulo city aerosol.

3.2 Concentrations of $PM_{2.5}$ during the intensive campaign

The winter campaign began with high $PM_{2.5}$ concentrations (a maximum of 88 $\mu g\,m^{-3}$ on 2 July) and low relative humidity (minimum of 21 %). The average temperatures ranged from 15 to 21 °C and the wind speed ranged between 0.6 and 2.2 $km\,h^{-1}$. The concentrations of $PM_{2.5}$ in the intensive campaign ranged from 15 to 88 $\mu g\,m^{-3}$ (average 45 $\mu g\,m^{-3}$), with a similar average to that obtained in another intensive study in 2008, 47 $\mu g\,m^{-3}$ (Souza et al., 2014a). The levels of $PM_{2.5}$ in this campaign were above the levels recommended by WHO on 90 % of the sampling days.

The average concentration of $PM_{2.5}$ was higher in $Int_{2.5}$ than in $Ext_{2.5}$, which can be explained by the fact that the campaign took place in the dry season (winter). In winter, the meteorological conditions are more unfavorable to the dispersion of pollutants and also due to the predominance of sugarcane burning (da Rocha et al., 2005, 2012; Sánchez-Ccoyllo and Andrade, 2002; Vasconcellos et al., 2010).

3.3 WSI and trace elements

The WSIs represent a large fraction in the aerosol mass and have already been suggested to present the ability to form CCN (cloud condensation nuclei) and fog (Rastogi et al., 2014). The secondary inorganic components, sulfate, nitrate and ammonium (SNA), were the most abundant ions in all campaigns (Table 3), which has already been observed in previous studies for this site (Vasconcellos et al., 2011a). SNA accounted for 74, 82 and 79 % of the total mass of inorganic species in the $Int_{2.5}$, $Ext_{2.5}$ and Ext_{10} campaigns, respectively. SNA were also found to be the major portion of the WSIs in other studies around the world. For instance, Zheng et al. (2016) assessed $PM_{2.5}$ concentrations at 17 diversified sites in China. An average contribution of SNA of

Table 3. Concentrations of WSI in all campaigns.

(ng m^{-3})	Int$_{2.5}$ Average (min–max)	Ext$_{2.5}$ Average (min–max)	Ext$_{10}$ Average (min–max)
Cl$^-$	964 (107–4549)	330 (16–1427)	641 (76–5904)
NO$_3^-$	2678 (667–6873)	1430 (183–3419)	2872(437–8880)
SO$_4^{2-}$	3266 (1252–5959)	3197 (922–6300)	3680 (569–9361)
MSA$^-$	84 (15–214)	63 (13–226)	107 (28–444)*
C$_2$O$_4^{2-}$	478 (176–753)	282 (57–726)	367 (50–1180)
Na$^+$	350 (46–869)	238 (64–512)	571 (76–1908)
NH$_4^+$	1712 (613–4075)	1370 (281–2845)	1336 (57–4436)
Nss-K$^+$	809 (237–2007)	366 (49–1137)	413 (63–1181)
SNA	7655	5997	7888
Total	10 334	7276	9986
SO$_4^{2-}$ / NO$_3^-$	1.2	2.2	1.3
Cl$^-$ / Na$^+$	2.7	1.4	1.1
SNA / total (%)	74	82	79

* Data were not determined after 21 August.

more than 90 % of total ions was obtained, which represented 50 % of PM$_{2.5}$. The levels of SNA in aerosols from urban sites are highly influenced by the anthropogenic emissions of precursors (SO$_2$, NO$_x$, and NH$_3$) (Y. Wang et al., 2013), although they may also be directly emitted for different sources, such as automobile or industrial sources.

Sulfate average concentrations appear to vary less than nitrate, comparing Int$_{2.5}$ to Ext$_{2.5}$, and were not statistically different ($p \sim 0.8$). The same trend in sulfate in this study was also observed by Villalobos et al. (2015) for Santiago, Chile, in 2013. In that study, the annual average concentration of sulfate (2000 ng m^{-3}) was considerably lower than that observed in extensive São Paulo campaigns in 2014. The sulfate concentrations in Santiago aerosols have been reduced since air quality regulations limited the sulfur content in diesel and gasoline to 15 ppm (MMA, 2014). In Brazil, since 2013 the S-10 diesel (10 ppm of sulfur) was substituted for the S-50 diesel (50 ppm), whereas in 2014 the S-50 gasoline replaced the S-800 gasoline (800 ppm), although older vehicles are still allowed to use S-500 (500 ppm) diesel (CETESB, 2015). During the studies performed at several urban sites in China, sulfate concentrations varied between 4200 and 23 000 ng m^{-3}. These values are higher than those of this study and also 5 to 10 times higher than the measured concentrations in Europe and United States (Hidy, 2009; Putaud et al., 2004; Zheng et al., 2016).

The SO$_4^{2-}$ / NO$_3^-$ ratio was nearly twice as high in the Ext$_{2.5}$ campaign as in Int$_{2.5}$. This pattern has already been observed; in warmer ambient conditions the fine NO$_3^-$ aerosols can be volatilized, increasing the ratio between these species (Rastogi and Sarin, 2009; Souza et al., 2014a). NH$_4$NO$_3$ exists in a reversible equilibrium between HNO$_3$ and NH$_3$ (Tang et al., 2016). Ammonium concentrations were not significantly different ($p \sim 0.3$) and were slightly higher in Int$_{2.5}$ (1712 ng m^{-3}) than in Ext$_{2.5}$ (1370 ng m^{-3}).

Non-sea salt potassium was calculated (nss-K$^+$) based on seawater ion ratios [ss-K$^+$] = 0.036 [Na$^+$] (Nayebare et al., 2016; Seinfeld and Pandis, 2006). Concentrations of nss-K$^+$ were significantly higher in Int$_{2.5}$ than in Ext$_{2.5}$, with average concentrations of 809 and 366 ng m^{-3}, respectively ($p < 0.01$). The higher concentrations found in the intensive campaigns have already been attributed to biomass burning in previous studies (Pereira et al., 2017; Vasconcellos et al., 2011a). However, potassium ions can also come from soil resuspension (Ram et al., 2010; Tiwari et al., 2016), which becomes important in PM$_{2.5-10}$. Higher concentrations of chloride in fine particles (964 and 330 ng m^{-3} for Int$_{2.5}$ and Ext$_{2.5}$, respectively) were observed in the Int$_{2.5}$ campaign (although the value of p was slightly above 0.05), probably due to a higher influence of biomass burning (Allen et al., 2004). Conversely, chloride in coarse particles is mostly attributed to marine aerosols. Cl$^-$ / Na$^+$ ratios were below 1.8 in Ext$_{2.5}$ and Ext$_{10}$ and higher in Int$_{2.5}$; although Cl$^-$ / Na$^+$ ratios are attributed to increased sea salt contribution (Souza et al., 2014a), the higher contribution in the intensive campaign may be explained by a higher contribution of other sources of chloride, such as biomass burning, in that period.

Pearson correlations were obtained for all determined species in Ext$_{2.5}$ and Ext$_{2.5-10}$ (coarse mode), including meteorological data such as temperature, relative humidity and wind speed. Some gaseous species such as NO$_x$ and CO were obtained from the CETESB database and were also included. NO$_x$ was monitored at a station inside the university campus (IPEN, 800 m away from the sampling site, at ground level) and CO was monitored at another station (Pinheiros,

Table 4. Average, minimum and maximum concentrations of tracer elements for all campaigns.

(ng m^{-3})	$Int_{2.5}$ Average (min–max)	$Ext_{2.5}$ Average (min–max)	Ext_{10} Average (min–max)
Li	0.48 (< DL–1.12)	0.27 (< DL–0.70)	0.40 (< DL–1.25)
Mg	210 (5–469)	93 (5–356)	154 (< DL–377)
Al	1851 (< DL–2782)	691 (< DL–2712)	981 (< DL–3014)
K	1431 (191–3833)	500 (< DL–1967)	600 (< DL–1682)
Ca	1164 (< DL–3204)	397 (< DL–1671)	666 (< DL–2160)
Cr	23 (1–60)	13 (1–60)	20 (< DL–54)
Mn	30 (< DL–64)	17 (< DL–49)	33 (4–175)
Fe	962 (173–2056)	581 (140–1408)	1269 (240–3578)
Co	0.45 (0.03–1.06)	0.23 (0.01–0.78)	0.59 (0.07–1.74)
Ni	7.3 (2.3–14.8)	4.6 (< DL–16.1)	6.6 (< DL–25.9)
Cu	181 (7–390)	109 (7–308)	188 (32 –976)
Zn	284 (< DL–673)	110 (< DL–279)	193 (< DL–716)
As	2.8 (0.06–5.7)	1.9 (< DL–7.1)	2.2 (< DL–7.9)
Se	5.6 (< DL–13.2)	2.6 (< DL–7.5)	2.6 (< DL–7.9)
Rb	5.7 (0.4–12.3)	2.2 (0.1–8.9)	2.6 (0.2–8.9)
Sr	6.6 (0.4–13.4)	3.0 (0.2–12.2)	4.8 (0.4–14.3)
Cd	2.5 (0.2–15.1)	0.8 (0.1–3.0)	1.2 (0.2–10.6)
Sn	19.5 (3.2–40.2)	8.8 (0.3–35.9)	12.3 (1.6–41.8)
Cs	0.28 (0.07–1.01)	0.14 (< DL–0.51)	0.19 (0.02–0.77)
Tl	0.21 (< DL–0.75)	0.13 (< DL–0.38)	0.15 (0.03 –0.65)
Pb	54 (3–172)	31 (3–71)	42 (4–176)
Bi	0.76 (0.06–3.03)	0.47 (< DL–3.03)	0.83 (0.12–3.24)

< DL: below detection limit.

3.2 km away, at 2 m height). Previous studies have identified high correlations of pollutant concentrations between the IPEN station and the sampling site (Oyama et al., 2016; Vara-Vela et al., 2016). NH_4^+ was moderately to strongly correlated with $C_2O_4^{2-}$ (oxalate), Cl^-, NO_3^- and SO_4^{2-} in $Ext_{2.5}$ ($R = 0.66, 0.62, 0.85$ and 0.79, respectively), suggesting the neutralization of oxalic, hydrochloric, nitric and sulfuric acids by NH_3 (Table S2). The formation of $(NH_4)_2SO_4$, a nonvolatile species, could represent a gas-to-particle conversion process and can account for the formation of new particles through nucleation (Mkoma et al., 2014; da Rocha et al., 2005) and can lead to CCN formation. NH_4NO_3 and NH_4Cl also have an important influence on Earth's acid deposition (Tang et al., 2016).

Na^+ was strongly correlated with Cl^- in $Ext_{2.5}$ ($R = 0.78$) and presented a relatively moderate correlation ($R = 0.35$) in $Ext_{2.5-10}$. These species are often associated with marine aerosol, which is mainly in the coarse mode (Godoy et al., 2009; da Rocha et al., 2012). Although it was observed that ocean influence is not the only source of Na^+ at the site, this species may have vehicular sources (Vieira-Filho et al., 2016). $C_2O_4^{2-}$ was also moderately correlated with NO_3^-, SO_4^{2-} and K^+ ($R = 0.67, 0.61$ and 0.68, respectively). These species-reported sources can be biomass burning and secondary conversion of natural and anthropogenic gases (Custódio et al., 2016). The secondarily formed species were

negatively correlated with wind speed (from $R = -0.40$ to $R = -0.70$); lower wind speed can increase in the formation of secondary ionic species due to an increase in the precursor species concentrations (Yu et al., 2017).

Average, maximum and minimum trace element concentrations are presented in Table 4. Mg, Al, K, Ca, Fe, Cu and Zn were the most abundant elements in all campaigns, similar to those observed by Vasconcellos et al. (2011a) for the intensive campaign in 2008. All of them had higher concentrations in $Int_{2.5}$ than in $Ext_{2.5}$. However, crustal elements were significantly higher; Al and Ca had concentrations nearly 3 times higher in the intensive campaign ($p < 0.05$). A similar trend was observed between wintertime and summertime campaigns by Castanho and Artaxo (2001). They reported higher concentrations of soil resuspension elements during wintertime. An increase in soil resuspension is expected in drier conditions.

As observed for nss-K^+, elemental K average concentration was more than twice as high in $Int_{2.5}$ than in $Ext_{2.5}$ ($p < 0.05$). This may be explained by a higher biomass burning contribution during the intensive campaign since sugarcane burning significantly increases in this time of the year. Cu has been attributed to vehicular emissions in São Paulo (Castanho and Artaxo, 2001) because it may be present in the ethanol, which is mixed with gasoline and used in light-duty vehicles (LDVs) in Brazil. Cu has also been related to wear emissions of road traffic (Pio et al., 2013). This element was approximately 70 % higher in $Int_{2.5}$ than $Ext_{2.5}$. Although there is no significant difference in vehicular emissions all year round, the meteorological conditions are more unfavorable to pollutant dispersion in the winter season.

The enrichment factor (EF) is an approximation often used in order to identify the degree to which an element in an aerosol is enriched or depleted regarding a specific source. EFs are calculated based on a reference metal (Al as a soil tracer in this study), considering crustal element composition (Lee, 1999). A convention often adopted is to consider that when elements have EFs below 10 they have significant crustal sources and are often called non-enriched elements (NEEs), and when the elements have EFs above 10 they have a higher non-crustal character and are referred to as anomalously enriched elements (AEEs) (Pereira et al., 2007). Values were higher than 10 for Cr (except for $Int_{2.5}$), Cu, Zn, As, Se, Cd, Sn, Tl, Pb and Bi, meaning that they can be attributed to anthropogenic sources as vehicular and industry emissions (Table S3). Elements like K, Mn, Ni, Rb, Sr, Cs, Li, Mg, Ca, Fe, Co and Sr had EFs lower than 10 and could be attributed to soil resuspension (da Rocha et al., 2012).

Strong correlations were observed between Al and Li, Mg, K, Ca, Mn, Fe, Rb and Sr ($R > 0.85$) in $Ext_{2.5}$. Al also had strong correlations with Li, Mg, K, Ca and Fe in $Ext_{2.5-10}$ ($R > 0.70$). Strong correlations were observed between species like Cl^- and NO_3^- and Mg, Al, Ca and Fe ($R > 0.7$); atmospheric reactions can occur between acids (HCl and HNO_3) and soil particles that have alkaline character (Rao et al., 2016). Mg, Al, K, Sr and Fe were negatively correlated with relative humidity ($R \leq -0.60$), suggesting a strong influence of drier conditions over these species.

3.4 OC, EC and mass balance

Higher concentrations of OC and EC were observed in $Int_{2.5}$ than in $Ext_{2.5}$, with average values of $10.2\,\mu g\,m^{-3}$ for OC and $7.0\,\mu g\,m^{-3}$ for EC (Fig. 4 and Table S4). However, the difference in carbonaceous species concentrations was not considered statistically significant between the campaigns ($p \sim 0.1$). The OC / EC ratios were 1.5, 1.7 and 1.8 for $Int_{2.5}$, $Ext_{2.5}$ and Ext_{10}, respectively. Since the ratio values were similar, and the absolute OC and EC concentrations were higher in intensive than extensive campaigns, this may indicate that similar sources of OC and EC contribute all year long but with higher concentrations during $Int_{2.5}$. Ratios lower than 1 are constantly observed in roadway tunnels and are assumed to describe the composition of fresh traffic emissions (Pio et al., 2011). Amato et al. (2016) found values ranging from 1.8 to 3.7 at the urban background sites using equivalent measurement protocols. It was attributed to the distance from main roads, which can increase the influence of secondary OC (Pio et al., 2011). In this way, the values for OC / EC found in the present study may be due to

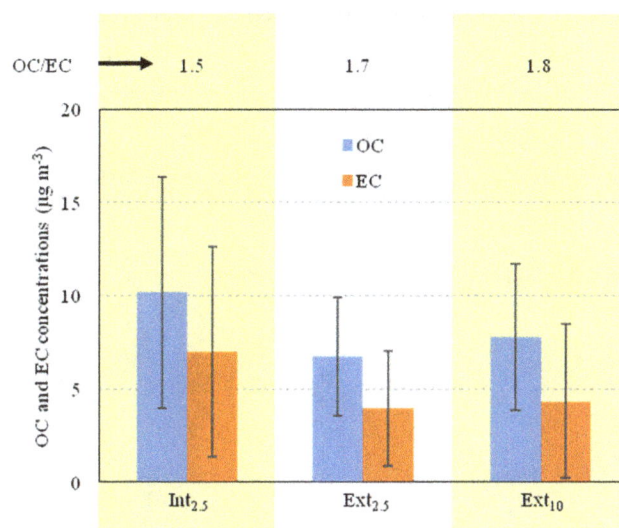

Figure 4. Carbonaceous species concentrations for all campaigns.

vehicle emissions with contributions from secondary organic aerosols.

TOM (total organic matter) was calculated by multiplying the OC content by 1.6 (Timonen et al., 2013) and represented 36, 36 and 28 % of the total PM (Fig. S2). Mass balance was determined for the aerosol by considering trace elements as if they all existed as oxides (Alves et al., 2015). The unaccounted part was 6, 15 and 26 % for $Int_{2.5}$, $Ext_{2.5}$ and Ext_{10}, respectively. This unaccounted part can be attributed to adsorbed water or the fact that abundant species as carbonates and Si were not determined, as similarly observed in Pio et al. (2013).

OC and EC were well correlated in $Ext_{2.5}$, with values above 0.8. This suggests that a large amount of OC is emitted by a dominant primary source at this site (Aurela et al., 2011; Kumar and Attri, 2016). The studied site is strongly affected by vehicle emissions and during the winter months biomass burning also contributes to these species (Pereira et al., 2017). Correlations were strong between the carbonaceous species with vehicular-emitted gases such as NO_x and CO ($R > 0.85$). OC also had good correlations with soil elements (Mg and Al) and also nss-K^+ ($R > 0.8$), suggesting an association with the resuspension of road dust and also a significant biomass burning contribution.

3.5 Polycyclic aromatic hydrocarbons and derivatives

The PAH and derivative concentrations are presented in Table 5. The total PAHs were higher in $Int_{2.5}$ than in $Ext_{2.5}$; 23.3 and $18.4\,ng\,m^{-3}$, respectively (although not significantly different, with $p > 0.05$). The total PAH concentration for the Ext_{10} was $24.3\,ng\,m^{-3}$. The lowest total PAH concentration of $2.6\,ng\,m^{-3}$ was observed in $Ext_{2.5}$, while the maximum of $115.3\,ng\,m^{-3}$ was observed in Ext_{10}. These levels were similar to those obtained in past studies at the same

Table 5. Concentrations of PAHs and derivatives for all campaigns.

($ng\,m^{-3}$)	Int$_{2.5}$ Average (min–max)	Ext$_{2.5}$ Average (min–max)	Ext$_{10}$ Average (min–max)
Nap	0.30 (0.17–0.77)	0.36 (0.02–0.77)	0.41 (0.09–0.81)
Acy	0.09 (0.06–0.12)	0.10 (0.03–0.19)	0.12 (0.05–0.34)
Ace	0.03 (0.02–0.08)	0.05 (0.02–0.16)	0.07 (0.02–0.23)
Flu	0.27 (0.15–1.03)	0.31 (0.06–1.44)	0.51 (0.10–1.75)
Phe	0.65 (0.30–2.48)	0.74 (0.12–3.55)	1.28 (0.28–4.08)
Ant	0.17 (0.10–0.44)	0.16 (0.06–0.60)	0.25 (0.08–0.67)
Flt	0.48 (0.21–0.86)	0.53 (0.06–1.40)	0.73 (0.19–2.21)
Pyr	0.52 (0.20–0.99)	0.54 (0.07–1.54)	0.71 (0.19–2.45)
BaA	1.0 (0.3–2.4)	0.9 (0.1–4.8)	1.2 (0.3–5.9)
Chr	1.8 (0.5–4.4)	1.6 (0.3–5.7)	2.1 (0.5–10.5)
BbF	3.0 (0.9–6.1)	2.3 (0.5–6.4)	3.0 (0.7–13.3)
BkF	2.5 (0.6–5.2)	1.9 (0.2–7.4)	2.5 (0.4–11.8)
BeP	2.8 (0.6–6.1)	2.2 (0.3–7.3)	2.8 (0.5–14.4)
BaP	2.3 (0.4–5.5)	1.6 (0.2–7.6)	2.0 (0.3–12.5)
Per	0.35 (0.04–0.79)	0.27 (< DL–1.27)	0.38 (0.05–1.90)
InP	2.9 (0.6–6.0)	1.8 (0.3–6.3)	2.4 (0.4–13.2)
DBA	0.8 (0.1–2.3)	0.6 (0.0–2.0)	0.9 (0.0–5.1)
BPe	2.4 (0.5–4.8)	1.6 (0.2–5.5)	2.1 (0.4–10.5)
Cor	1.0 (0.1–2.4)	0.7 (0.0–2.4)	0.9 (0.1–5.2)
Total	23.3 (6.0–48.8)	18.4 (2.6–61.6)	24.3 (5.4–115.3)
BaPE	3.4 (0.6–8.0)	2.4 (0.3–10.5)	3.2 (0.5–18.3)
1-NNap	< DL	< DL	< DL
1-Methyl-4-NNap	< DL	< DL	< DL
2-NNap	< DL	< DL	< DL
2-NBP	0.56 (< DL–1.36)	0.56 (< DL–1.36)	1.23 (0.47–2.47)
1-Methyl-5-NNap	0.18 (< DL–0.28)	< DL	< DL
1-Methyl-6-NNap	0.36 (< DL–0.40)	0.27 (< DL–0.41)	0.29 (< DL–0.86)
2-Methyl-4-NNap	0.45 (< DL–0.45)	0.36 (< DL–0.44)	0.42 (< DL–1.26)
3-NBP	0.60 (0.48–0.88)	0.52 (< DL–0.87)	0.55 (< DL–1.58)
4-NBP	< DL	< DL	0.18 (< DL–0.41)
5-NAce	< DL	< DL	0.20 (< DL–0.52)
2-NFlu	0.98 (0.78–1.39)	0.99 (0.38–1.56)	1.09 (0.54–1.79)
2-NPhe	0.43 (0.30–0.67)	0.51 (0.19–1.40)	0.61 (< DL–1.80)
3-NPhe	0.43 (< DL–0.46)	0.44 (< DL–0.68)	0.47 (< DL–1.11)
9-NPhe	0.62 (< DL–0.64)	0.52 (< DL–0.64)	0.55 (< DL–0.82)
2-Nant	0.66 (< DL–0.80)	0.56 (< DL–0.80)	0.61 (< DL–0.88)
9-Nant	0.44 (< DL–0.57)	0.42 (< DL–0.69)	0.46 (< DL–1.15)
2-NFlt	1.19 (< DL–1.35)	0.98 (< DL–1.25)	1.02 (< DL–1.43)
3-NFlt	1.45 (< DL–1.48)	1.05 (< DL–1.48)	1.02 (< DL–1.11)
1-NPyr	0.98 (< DL–1.12)	0.73 (< DL–0.88)	0.79 (< DL–1.28)
2-NPyr	0.94 (< DL–0.99)	0.76 (< DL–0.99)	0.78 (< DL–1.27)
4-NPyr	1.61 (< DL–1.67)	1.27 (< DL–1.34)	1.28 (< DL–1.72)
7-NBaA	1.19 (< DL–1.34)	0.91 (< DL–1.06)	1.01 (< DL–1.67)
6-NChr	< DL	0.60 (< DL–0.67)	0.69 (0.58–1.10)
3-NBA	< DL	< DL	< DL
6-NBaPyr	< DL	< DL	1.01 (< DL–1.19)
1-NBaPyr	< DL	< DL	< DL
3NBePyr	< DL	< DL	< DL
1,4-BQ	< DL	< DL	< DL
1,4-NQ	0.54 (0.43–0.72)	0.44 (0.28–0.67)	0.46 (0.31–1.08)
1,2-NQ	< DL	< DL	< DL
9,10-AQ	1.6 (0.8–3.7)	2.5 (0.3–8.0)	2.6 (0.4–10.9)
9,10-PQ	< DL	< DL	< DL
Total PAHs / OC (%)	0.23	0.27	0.31
ΣLMW / ΣHMW	0.32	0.41	0.43
Flt / (Flt + Pyr)	0.5	0.5	0.5
BaA / Chr	0.5	0.6	0.5
InP / (InP + BPe)	0.5	0.5	0.5
BaP / (BaP + BeP)	0.4	0.4	0.4
BPe / BaP	1	1	1
2-NFlt / 1-NPyr	1.3	1.3	1.3

site, as $25.9\,\text{ng}\,\text{m}^{-3}$ in PM_{10} samples during the intensive campaign in the winter of 2008 (Vasconcellos et al., 2011a) and $27.4\,\text{ng}\,\text{m}^{-3}$ for PM_{10} in the winter of 2003 (Vasconcellos et al., 2011b). In addition, the total PAH levels from the present study are higher than in the 2013 and 2012 intensive campaigns (8.7 and $8.2\,\text{ng}\,\text{m}^{-3}$ in PM_{10}) (Pereira et al., 2017). Total PAHs represented 0.23, 0.27 and 0.31 % of OC for $Int_{2.5}$, $Ext_{2.5}$ and Ext_{10}, respectively. In spite of accounting for a small fraction of OC, it is important to observe that PAHs are among the pollutants of major concern due to their carcinogenic and mutagenic effects.

BbF was the most abundant PAH (the BbF percentages in relation to total PAHs were 13, 12 and 12 % for $Int_{2.5}$, $Ext_{2.5}$ and Ext_{10}, respectively) in all the campaigns. This compound has carcinogenic properties already reported in other studies (Ravindra et al., 2008). Its concentrations reached the values of 6.1, 6.4 and $13.3\,\text{ng}\,\text{m}^{-3}$ in $Int_{2.5}$, $Ext_{2.5}$ and Ext_{10}, respectively. BbF was also the most abundant PAH in the 2013 intensive campaign (Pereira et al., 2017). This species was also among the most abundant PAHs in the study performed at the Jânio Quadros (JQ) tunnel, with a predominance of LDVs (Brito et al., 2013). Correlations were strong between all PAHs heavier than Flt ($R > 0.8$), suggesting different sources of the PAHs with lower molecular weight at this site. Most of the heavier PAHs appeared to have negative correlations with temperature; the condensation of organic compounds in the aerosol is influenced by lower temperatures (Bandowe et al., 2014). Coronene, a PAH often used as a vehicular fuel marker (Ravindra et al., 2006) was correlated to vehicular-related species as Cu and Pb ($R > 0.7$).

BaP, the PAH most studied due to its proven carcinogenic potential, was considerably higher in $Int_{2.5}$ than in $Ext_{2.5}$. It reached the mean values of 5.5, 7.6 and $12.5\,\text{ng}\,\text{m}^{-3}$ in $Int_{2.5}$, $Ext_{2.5}$ and Ext_{10}, respectively. In the tunnels, its presence was associated with the higher contribution of LDV emissions (Brito et al., 2013). Among the nitro-PAHs with highest concentrations were 2-NFlu and 2-NBP. The nitro-PAH 2-NFlu is a major component of diesel exhaust particles, such as the nitropyrenes, and is known as a carcinogenic nitro-PAH (Draper, 1986; Fujimoto et al., 2003). The PAH 2-NFlt was moderately correlated with Flt ($R = 0.4$); this species is produced from reactions between Flt and NO_2 (Albinet et al., 2008). The 2-NFlt / 1-NPyr ratios were close to 1; ratios lower than 5 indicate a predominance of primary emissions of nitro-PAHs (Ringuet et al., 2012). The compound 9,10-AQ was the most abundant oxy-PAH found in this study. It can be either primarily emitted or secondarily formed. A recent study showed that it can be formed from the heterogeneous reaction between NO_2 and Ant adsorbed on NaCl particles (sea salt) (Chen and Zhu, 2014). A moderate correlation was found between 9,10-AQ and Ant ($R = 0.54$).

3.5.1 PAH diagnostic ratios

The PAH diagnostic ratios (Table 5) were obtained for all the campaigns since they can point to some emission sources, such as oil products, fossil fuels, coal or biomass combustion. However, these ratios should be used with caution due to the peculiarity of fuel compositions in Brazil's car fleet. The values of PAH ratios can also be affected by changes in phase, transport and degradation (Tobiszewski and Namieœnik, 2012). The ratio of $BaP / (BaP + BeP)$ is related to the aerosol photolysis. Most of the local PAH emissions contain equal concentrations of BeP and BaP. However, BaP is more likely to undergo photolysis or oxidation (Oliveira et al., 2011). The average $BaP / (BaP + BeP)$ ratio was close to 0.4 for the three campaigns. This ratio was slightly lower than the ratio obtained in the 2013 intensive campaign, although still very close to 0.5 (Pereira et al., 2017); it is suggested that the PAHs found at the site are mostly emitted locally.

The $Flt / (Flt + Pyr)$ and $InP / (InP + BPe)$ ratios were reported to be the most conservative by Tobiszewski and Namieœnik (2012). The $Flt / (Flt + Pyr)$ ratios for all the campaigns were close to 0.5, falling within the range for fossil fuel combustion (0.4–0.5) (de la Torre-Roche et al., 2009). The $InP / (InP + BPe)$ ratios represented values close to 0.5, similar to the ratio obtained for the JQ tunnel (0.55) impacted by LDVs (the ratio found for the Maria Maluf (MM) tunnel was 0.36). The average BaA / Chr ratio ranged between 0.5 and 0.6 in the 2014 campaigns, also approaching that of the JQ tunnel (0.48) (Brito et al., 2013), whilst a value of 0.79 was obtained for the MM tunnel. The $BaA / (BaA + Chr)$ ratio was reported to be sensitive to photodegradation (Tobiszewski and Namieœnik, 2012). However, it is possible to consider that this degradation was not significant due to proximity to the emission sources (the expressway). All ratios suggested a greater contribution of LDVs to PAHs at the sampling site.

The $\Sigma LMW / \Sigma HMW$ ratios (PAHs with a low molecular weight (LMW) had three and four aromatic rings and PAHs with a high molecular weight (HMW) had more than four rings) were considerably low in all campaigns (predominance of HMW PAHs). It is known that LMW PAHs have higher concentrations in the gas phase while HMW PAHs are preferentially present in PM (Agudelo-Castañeda and Teixeira, 2014; Duan et al., 2007). The HMW PAH contribution was higher in winter, just as the ratios were lower, corroborating the results of some previous studies (Chen et al., 2016; Teixeira et al., 2013). In turn, HMW PAHs are more likely to be retained in particles due to their lower vapor pressure than LMW PAHs. LMW PAHs are also mostly associated with diesel engines, while HMW PAHs are predominantly emitted by gasoline exhaust (Chen et al., 2013; Cui et al., 2016; Miguel et al., 1998). The ratio between BPe and BaP for all campaigns (all close to 1) was very similar to that found in a study with Brazilian diesel LDV exhaust (1.13) (de Abrantes et al., 2004) and was also found in a campaign performed in São Paulo (1.11) (de Martinis et al., 2002). This may be a characteristic fingerprint for local vehicular emissions.

3.5.2 PAH risk assessment

BaPE is a parameter introduced to quantify the aerosol carcinogenicity related to all carcinogenic PAHs instead of BaP solely. BaPE values above $1.0 \, \mathrm{ng \, m^{-3}}$ represent an increased cancer risk. The carcinogenic nitro-PAHs (1-NPyr, 4-NPyr and 6-NChr) were below the detection limit in most parts of the extensive campaign samples; thus, they were not considered in the risk assessment. BaPE is calculated according to Eq. (5), given by Yassaa et al. (2001) and Vasconcellos et al. (2011a):

$$
\begin{aligned}
\mathrm{BaPE} = &([\mathrm{BaA}] \times 0.06) + ([\mathrm{BbF}] \times 0.07) + ([\mathrm{BkF}] \times 0.07) \\
&+ ([\mathrm{BaP}] \times 1) + ([\mathrm{DBA}] \times 0.6) \\
&+ ([\mathrm{InP}] \times 0.08).
\end{aligned} \tag{5}
$$

The BaPE values for the $\mathrm{Int_{2.5}}$ ranged between 0.6 and $8.0 \, \mathrm{ng \, m^{-3}}$ and for $\mathrm{Ext_{2.5}}$ between 0.3 and $10.5 \, \mathrm{ng \, m^{-3}}$, while the average of BaPE for the $\mathrm{Int_{2.5}}$ was considerably higher than in $\mathrm{Ext_{2.5}}$ (3.4 and $2.4 \, \mathrm{ng \, m^{-3}}$, respectively). In the $\mathrm{Ext_{10}}$ this index ranged between 0.5 and $18.3 \, \mathrm{ng \, m^{-3}}$. The maximum value was even higher than the value of $12.1 \, \mathrm{ng \, m^{-3}}$ in $\mathrm{PM_{10}}$, obtained in São Paulo in an intensive campaign conducted in 2008 (Vasconcellos et al., 2011a). More than 70 % of the samples in the $\mathrm{Ext_{10}}$ had BaPE indexes higher than $1 \, \mathrm{ng \, m^{-3}}$. The year 2014 was a relatively dry year, with an annual rainfall 13 % below the average (IAG, 2014). The average values for BaPE in $\mathrm{PM_{10}}$ at the site were 1.9 and $3.7 \, \mathrm{ng \, m^{-3}}$ in the intensive campaigns of 2007 and 2008, respectively. Conversely, at forested areas in São Paulo state the value can be as low as $0.1 \, \mathrm{ng \, m^{-3}}$ (Vasconcellos et al., 2010).

The lifetime LCR was assessed from the carcinogenic potential (BaP-TEQ) and mutagenic potential (BaP-MEQ) through Eqs. (6) and (7) (Jung et al., 2010):

$$
\begin{aligned}
(\mathrm{BaP\text{-}TEQ}) = &([\mathrm{BaA}] \times 0.1) + ([\mathrm{Chr}] \times 0.01) \\
&+ ([\mathrm{BbF}] \times 0.1) + ([\mathrm{BkF}] \times 0.1) \\
&+ ([\mathrm{BaP}] \times 1) + ([\mathrm{InP}] \times 0.1) \\
&+ ([\mathrm{DBA}] \times 5) + ([\mathrm{BPe}] \times 0.01)
\end{aligned} \tag{6}
$$

$$
\begin{aligned}
(\mathrm{BaP\text{-}MEQ}) = &([\mathrm{BaA}] \times 0.082) + ([\mathrm{Chr}] \times 0.017) \\
&+ ([\mathrm{BbF}] \times 0.25) + ([\mathrm{BkF}] \times 0.11) \\
&+ ([\mathrm{BaP}] \times 1) + ([\mathrm{InP}] \times 0.31) \\
&+ ([\mathrm{DBA}] \times 0.29) + ([\mathrm{BPe}] \times 0.19).
\end{aligned} \tag{7}
$$

DBA had the largest contribution to carcinogenic potential and BaP for mutagenic potential; in studies performed in urban Italian areas BaP was the compound that most contributed to total carcinogenicity in PM, although the

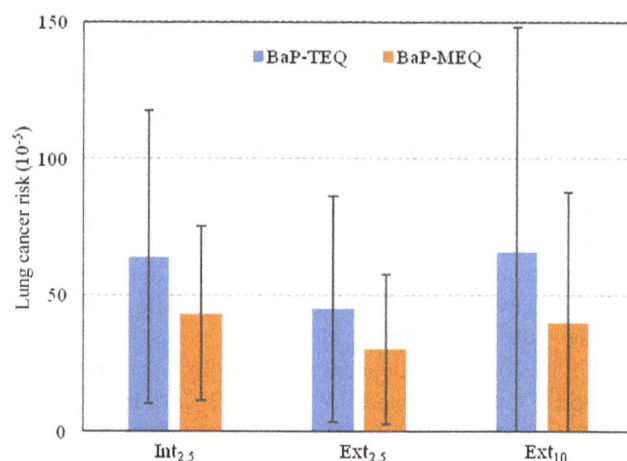

Figure 5. Lung cancer risk from the BaP-TEQ and BaP-MEQ for all three campaigns.

toxic equivalence factor (TEF) used for DBA was lower in those cases (Cincinelli et al., 2007; Gregoris et al., 2014). LCR from exposure to atmospheric PAH was estimated by multiplying BaP-TEQ and BaP-MEQ by the unit risk $(8.7 \times 10^{-5} \, (\mathrm{ng \, m^{-3}})^{-1})$ for exposure to BaP established by WHO (de Oliveira Alves et al., 2015; WHO, 2000) (Fig. 5), and it was possible to observe an increase during the intensive campaign. In all campaigns, the values observed were higher than those observed in studies performed in the Amazon during the dry season with events of biomass burning (de Oliveira Alves et al., 2015); studies performed in different seasons in other urban areas such as New York and Madrid showed carcinogenic risks within the parameters recommended by environmental and health agencies (Jung et al., 2010; Mirante et al., 2013).

3.6 Biomass burning tracers

The highest concentrations of biomass burning tracers (mean values of 509, 45 and $33 \, \mathrm{ng \, m^{-3}}$ for levoglucosan, mannosan and galactosan, respectively) were observed in the $\mathrm{Int_{2.5}}$ campaign ($p \sim 0.05$) during the biomass burning period (Fig. 6, Table S5). In the intensive campaign period, 1364 fire spots were registered in São Paulo state, with an average of 72 fires per day (INPE, 2014). In the same way, on 65 % of the sampling days the backward air masses passed through regions with biomass burning. The average concentration of levoglucosan obtained in the $\mathrm{Int_{2.5}}$ campaign ($509 \, \mathrm{ng \, m^{-3}}$) was higher than that of the intensive $\mathrm{PM_{10}}$ sampling campaigns in 2013 and 2012 (474 and $331 \, \mathrm{ng \, m^{-3}}$) (Caumo et al., 2016; Pereira et al., 2017), as well as more than twice the values obtained in the 2008 intensive campaign (Vasconcellos et al., 2010).

The Lev / Man ratios are characteristic of each type of biomass. The ratios were similar to those obtained in a chamber study with sugarcane burning (Lev / Man = 10) (Hall et

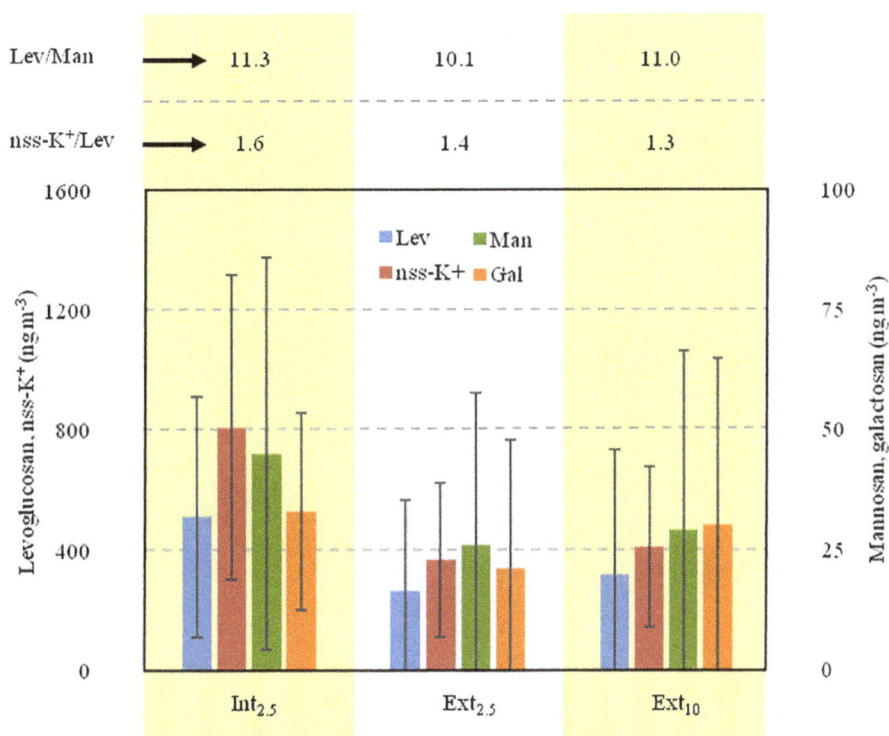

Figure 6. Concentrations of biomass burning tracers for all campaigns.

al., 2012) and also to those reported for the 2013 intensive campaign (Lev / Man = 12) (Pereira et al., 2017). Nss-K^+ / Lev ratios were 1.6, 1.4 and 1.3 for $Int_{2.5}$, $Ext_{2.5}$ and Ext_{10}, respectively. These ratios are similar to those obtained in the previous PM_{10} intensive campaign (1.4) in 2013, which was attributed to a combination of smouldering (flameless combustion) and flaming processes during the combustion of biomass (Kundu et al., 2010; Pereira et al., 2017). The flaming combustion is predominant for sugarcane leaves (Hall et al., 2012; Urban et al., 2016).

Correlations between potassium and monosaccharides in $Ext_{2.5}$ were high ($R > 0.8$), indicating that, most of the year, potassium in $PM_{2.5}$ can be linked to biomass burning. Coarse fraction potassium, more related to soil sources, did not present strong correlations with levoglucosan. Local burning can also affect the site since some restaurants use wood for cooking (pizzerias and steakhouses) (Kumar et al., 2016). There is a stronger correlation between chloride and other biomass burning tracers in $Ext_{2.5}$ than in $Ext_{2.5-10}$. Chloride is also a major emission from biomass burning in the form of KCl (Allen et al., 2004) and is also emitted as HCl in garbage burning (Calvo et al., 2013). Carbonaceous species presented high correlations ($R > 0.75$) with levoglucosan in $Ext_{2.5}$. This suggests that some of these species may also be linked to biomass burning emissions.

The highest concentrations of biomass burning tracers were found on 1 July, when the levoglucosan level reached $1263 \, ng \, m^{-3}$. On that day, about 100 fire spots (INPE, 2014)

were observed in the state of São Paulo and the back trajectories revealed air masses crossing the west and northwest of the state (Fig. 7a), where the fire spots were observed. On this same sampling day, local fire spots were observed, possibly due to landfill burning.

On 12 July, the air masses traveled over the Atlantic Ocean before reaching the site. In the same period, the $PM_{2.5}$ and biomass burning tracer concentrations dropped. Figure 7b shows the trajectories for 13 July. Some of the lowest concentrations of levoglucosan (80 and $74 \, ng \, m^{-3}$) and $PM_{2.5}$ (28 and $26 \, \mu g \, m^{-3}$) were observed on 12 and 13 July, respectively.

3.7 Distribution of species in fine and coarse particles during extensive campaigns

Figure 8 shows the mass percentage of tracers in fine ($PM_{2.5}$) and coarse particles ($PM_{2.5-10}$) in the extensive campaign; their values are presented in the Supplement (Table S6). The biomass burning tracers, levoglucosan and mannosan were present mostly in $PM_{2.5}$ mass fractions (over 75 %). In this study, 73 % of nss-K^+ mass was in $PM_{2.5}$. This species may also be attributed to biomass burning, although coarse potassium may be from soil dust resuspension (Souza et al., 2014a; Vasconcellos et al., 2011a).

Species related to vehicular emissions such as coronene and Cu (Brito et al., 2013; Ravindra et al., 2006) were also predominantly found in $PM_{2.5}$ (73 and 61 %, respectively).

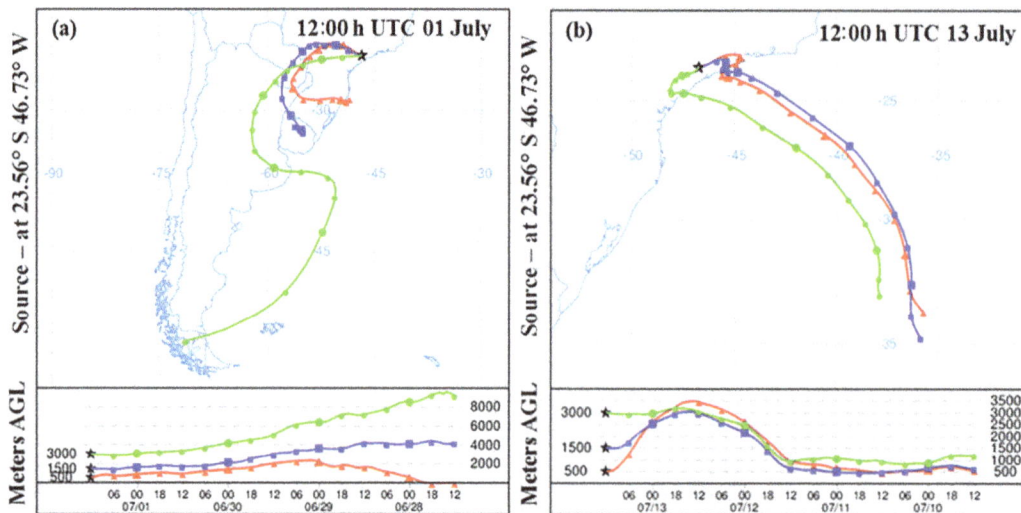

Figure 7. Backward air mass trajectories for the days **(a)** 1 July and **(b)** 13 July.

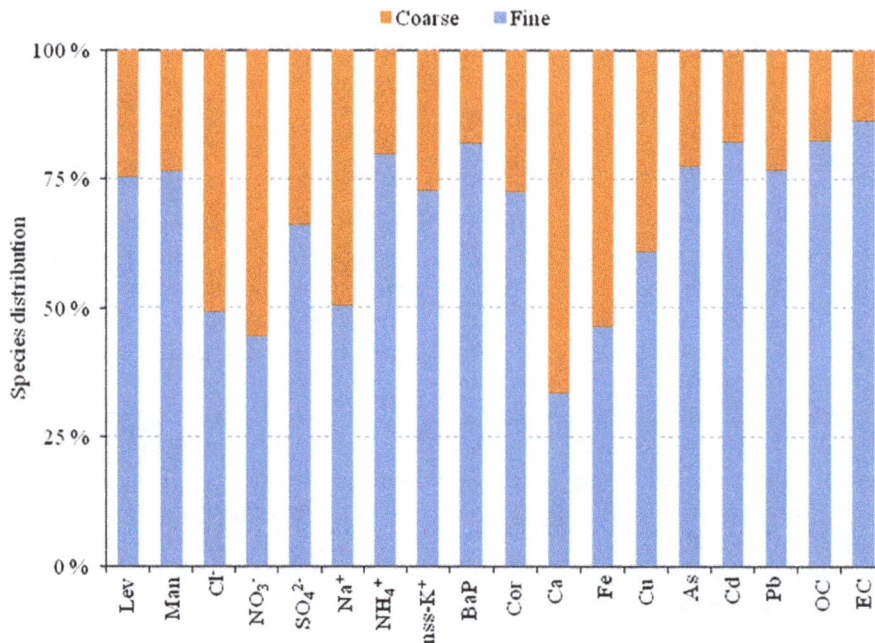

Figure 8. Mass percentage distribution of species in the fine and coarse particles.

More than 50 % of Fe and Ca (crustal elements) was found in $PM_{2.5-10}$. A previous source apportionment study in southern European cities (AIRUSE-LIFE+ project) also pointed out soil dust as a significant source, accounting for 2–7 % of $PM_{2.5}$ at suburban and urban background sites and 15 % at a traffic-impacted station. In the case of PM_{10}, these percentages increased to 7–12 and 19 %, respectively (Amato et al., 2016).

In a previous winter campaign in 2008, in São Paulo (Souza et al., 2014a) levoglucosan and mannosan were also mostly present in $PM_{2.5}$. Urban et al. (2014) found that be-

tween 58 and 83 % of levoglucosan was present in particles smaller than 1.5 μm for an agro-industrial region in São Paulo state. This is similar to the values observed in other studies done in the state of São Paulo and in the Amazon region (Decesari et al., 2006; Schkolnik et al., 2005; Urban et al., 2012).

Sulfate and ammonium were predominant in $PM_{2.5}$ (over 65 and 80 %). Sulfate was also predominant in $PM_{2.5}$ in a previous study done in São Paulo between 1997 and 1998; it was attributed to the gas-to-particle conversion of vehicular SO_2 (Castanho and Artaxo, 2001). Both ions may be present as $(NH_4)_2SO_4$ in the fine mode. Nitrate is well distributed in both phases, likely resulting from reactions of HNO_3 with soil species (Tang et al., 2016). In turn, fine-mode nitrate is often present in the form of ammonium nitrate, which is a thermally unstable species (Maenhaut et al., 2008).

Other ions, such as sodium and chloride were halved in each mode. These species are related to sea salt aerosols and are more often present in the coarse mode, as observed in the study done at urban sites in Rio de Janeiro (Godoy et al., 2009). Chloride in $PM_{2.5}$ can also originate from biomass burning emissions (Allen et al., 2004). OC and EC, which are mainly related to vehicular emissions in São Paulo (Castanho and Artaxo, 2001), were mostly in the fine particles. OC and EC were also associated with biomass burning in a recent study (Pereira et al., 2017).

BaP was found mainly in the $PM_{2.5}$ (over 80 %), which can be deposited in the tracheobronchial region of the human respiratory tract, representing an increased health risk (Sarigiannis et al., 2015). In turn, As, Cd and Pb, identified as elements that can cause carcinogenic health effects (Behera et al., 2015), were also found predominantly in the $PM_{2.5}$ (over 75 %). In this way, they may be indicative of higher carcinogenicity of fine over coarse particles.

3.8 Source apportionment using PMF and polar plots

Source apportionment was performed with PMF including all data. Then, the factor contributions were separated for each campaign ($n = 78$). Eleven strong species were considered (SO_4^{2-}, nss-K^+, Mg, Cr, Mn, Fe, Ni, Cd, Pb, OC and EC), six were considered weak (levoglucosan, mannosan, NO_3^-, NH_4^+, Ca and Cu) and the PM concentrations were set as a total variable. An extra modeling uncertainty of 25 % was added to all variables in the model. The uncertainties were increased in order to avoid discarding measurements that had poor data quality due to measurements below detection limits; this procedure was performed according to the method of Paatero and Hopke (2003). The solutions proved to be stable since the same sources could be identified in most of the solutions generated, with different additional uncertainties.

Considering the limited number of samples, a restricted number of species had to be chosen. Some variables were not considered due to high colinearity and redundancies since they would not provide more information regarding the sources. Elements already studied and attributed to sources in São Paulo were preferred. In some of the base model runs it was possible to observe a sea salt profile with Na^+ and Cl^-, but after they were removed, other profiles were clearly improved. PAHs were first included in the model but it created a factor associated with temperature conditions, increasing in the dry season since the lower dispersion conditions in the period favor the accumulation of HMW-PAHs in suspended particles (Agudelo-Castañeda and Teixeira, 2014; Ravindra et al., 2006). Levoglucosan was set as weak since it can de-

compose in the atmosphere (Pio et al., 2008), mannosan was set to weak due to concentrations below detection limit, and NO_3^- and NH_4^+ were set to weak due to their thermal instability. Ca and Cu also had to be set as weak in order to have a convergent base model run.

Solutions with three to eight factors were tested. The ratio of robust to theoretical parameters (Q_R/Q_T) reduced between simulations when increasing the number of factors. A solution with five factors was found to have more meaningful results; Q_R and Q_T values were 367 (Table S7). The source profiles obtained in the PMF analysis and the contribution of each factor to PM_{10} concentrations are found in Fig. 9. Constraints were applied, Cu was pulled up maximally in the vehicular factor and levoglucosan and mannosan were pulled up maximally in the biomass burning factor in order to have a better separation between both factors; relative change in Q was 0.4 %. The PMF result charts are presented in Fig. S3.

Factor 1 presented higher loadings for Mg, Ca and Fe, elements associated with soil resuspension in previous studies (da Rocha et al., 2012). The factor was also mixed with vehicular-related species, such as Cu and OC, which can be attributed to the resuspension of road dust by traffic. Accounting for 24.3, 12.5 and 25.7 % of $Int_{2.5}$, $Ext_{2.5}$ and Ext_{10}, respectively, it was more relevant for the PM_{10} campaign. In some runs, it was possible to observe Li and Tl in this factor, but these species were not considered in the final model. This soil contribution was similar to that obtained for PM_{10} in a year-round inventory in the city (CETESB, 2015). High loadings for ions, such as nss-K^+ and NO_3^-, were also present in the factor. Gaseous HNO_3 can interact with soil particles and form coarse nitrates (Tang et al., 2016). The factor contribution appeared to increase with wind speed from the NW and decrease with SE winds (Fig. 10). Soil dust and vegetation sources also tended to reduce with SE winds, as observed previously by Sánchez-Ccoyllo and Andrade (2002).

Factor 2 shows high loads for Ni, Pb and Cr, which are often attributed to industrial emissions (Bourotte et al., 2011; Castanho and Artaxo, 2001). This factor had some of the lowest contributions, 10.5, 9.7, and 9.5 % for $Int_{2.5}$, $Ext_{2.5}$, and Ext_{10}, and appeared to increase with SE winds, passing through nearby industrial regions (southeast of the city). The growth of industries has been limited in the last years and the vehicle fleet is expected to be a main source of atmospheric pollutants in the area (Kumar et al., 2016).

Factor 3 showed high loadings for vehicular-related tracers, such as Cu, Fe, OC and EC (with a higher load on EC and Cu). Cu and Fe were found in the LDV-impacted tunnel study in São Paulo and Cu is emitted from brake pads in stop-and-go driving in the expressways (Andrade et al., 2012b; Brito et al., 2013). Cu and Fe are also present in ethanol after the processing of copper tanks. Loadings for levoglucosan and mannosan were observed in this factor, which precluded the total separation from the biomass burning factor. On days with NW winds, both source contributions tended to increase as observed in the polar plots. This factor represented 30.9,

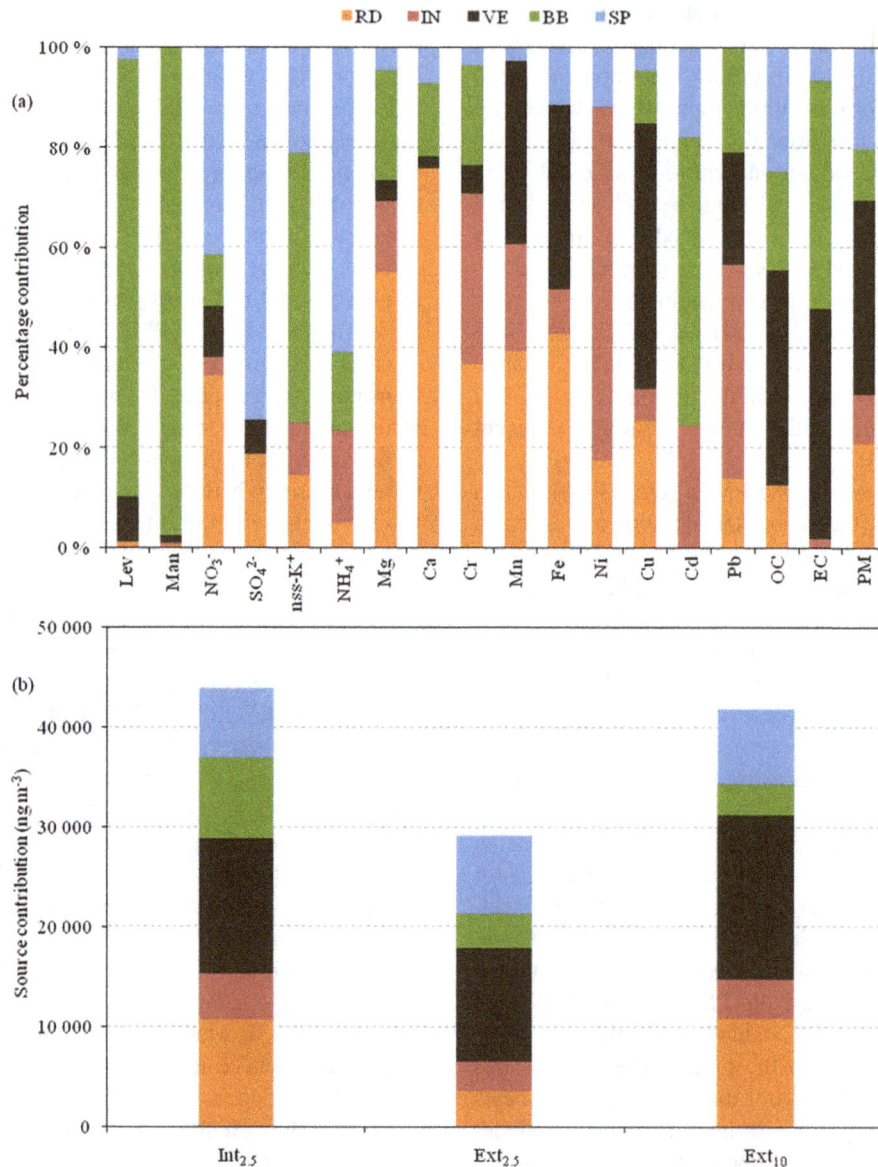

Figure 9. (a) Profile of species for each source (RD – road dust, IN – industrial, VE – vehicular, BB – biomass burning and SP – secondary processes). **(b)** Contribution of sources for each campaign.

39.1, and 39.2 % contribution for Int$_{2.5}$, Ext$_{2.5}$, and Ext$_{10}$, and had a constant contribution comparing the dry and wet periods in the Ext$_{10}$ campaign. Vehicular sources seemed to increase with winds coming from the north and northwest passing by the expressway, but decreased with SE winds, as observed previously (Sánchez-Ccoyllo and Andrade, 2002). The polar plot profiles of vehicular and road dust factors presented a different pattern since the aerosol from road dust suspension has a larger aerodynamic diameter (Karanasiou et al., 2009) and tends to increase with wind speed.

Factor 4 was associated with biomass burning due to the loadings for levoglucosan, mannosan and non-sea salt potassium, OC and EC. The loading of Cd in this factor is also

noteworthy; wood burning was pointed out as a possible source of this metal in a previous study in Belgium (Maenhaut et al., 2016), but more studies are necessary in order to explain the biomass burning contribution to this species in São Paulo. This factor represented 18.3, 11.6 and 7.6 % for Int$_{2.5}$, Ext$_{2.5}$ and Ext$_{10}$, respectively. The contributions of this factor were higher in the intensive campaign (sugarcane burning period), but were also present in the other periods, suggesting other biomass burning sources in the city, such as waste burning and wood stoves (Kumar et al., 2016). Several fire spots were registered in São Paulo state in the intensive campaign, some of them in the neighboring towns (INPE, 2014). The polar plot showed that this factor tended to in-

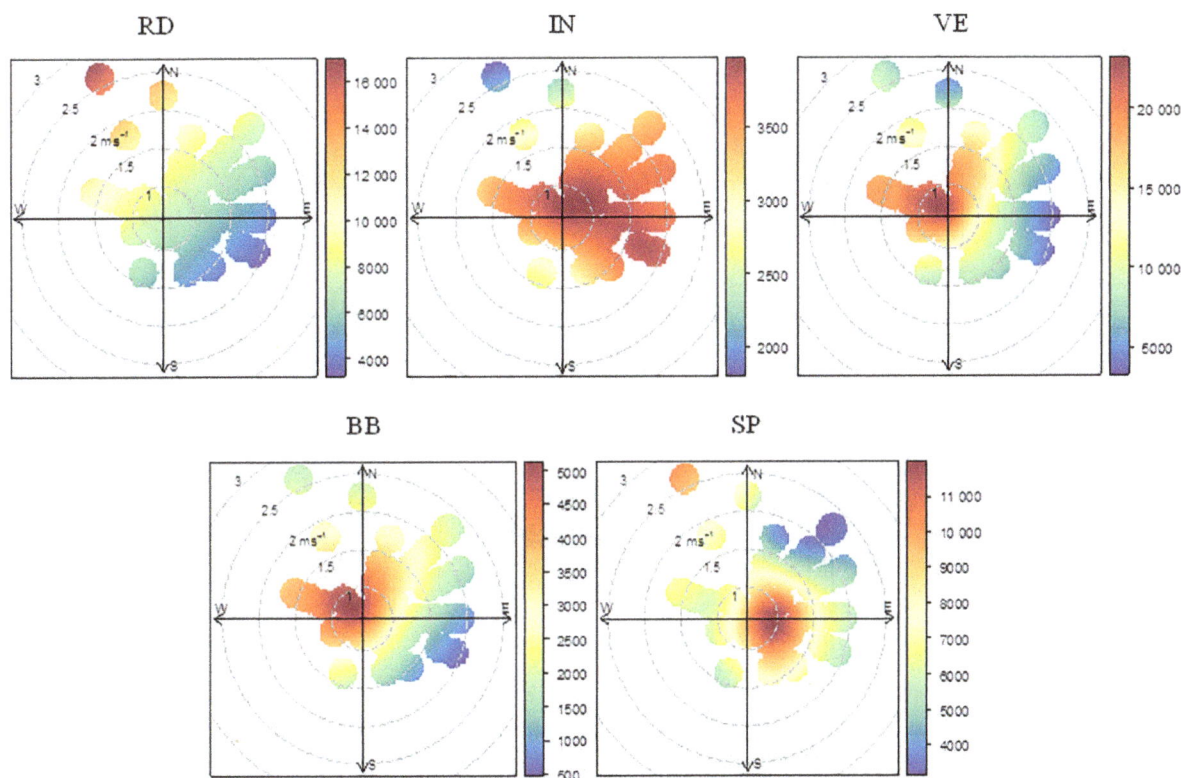

Figure 10. Polar plots of source contributions in São Paulo ($ng\,m^{-3}$ and $m\,s^{-1}$).

crease with NW winds, passing through the inland of São Paulo state, and decrease with SE (more humid) winds from the sea. High correlations ($R > 0.8$) were observed between the gases CO and NO_x and the primary source factors vehicular and biomass burning (Table S8). These gases are related to vehicular emissions (Alonso et al., 2010) and the correlations with the biomass burning factors may be due to the fact that the biomass burning factor increases with the same wind direction as the vehicular factor. No correlations were found between these gases and secondary process factor.

Factor 5 was attributed to the secondary inorganic aerosol formation processes (as seen by high mass loadings for NO_3^-, SO_4^{2-} and NH_4^+) and also OC (secondary organic carbon). The contributions were 15.9, 27.1 and 17.9 % for $Int_{2.5}$, $Ext_{2.5}$ and Ext_{10}, respectively. The contributions of this profile did not follow any seasonal trend (Fig. S3a). In 2014, 78 % of NO_x and 43 % of SO_x emissions in greater São Paulo were attributed to the vehicle fleet (CETESB, 2015). Taking into account that in São Paulo SO_x and NO_x concentrations are similar all year round, this could explain the lack of seasonality of this factor. The polar plot showed a centralized profile, increasing with lower wind speed, which suggests a local secondary process.

Other polar plots were obtained for individual species and are presented in Fig. 11. It is possible to see that Na^+ tended to increase with stronger winds coming from the sea, while Cl^- had a different pattern. Chloride in the marine aerosol

can be depleted after atmospheric reactions with acids (Calvo et al., 2013; White, 2008). It is noteworthy that MSA was associated with NW winds. This species is often associated with the decomposition of DMS emitted by the sea (Bardouki et al., 2003). More studies are needed in order to identify MSA sources at this site. Similarly, as for the biomass burning factor, levoglucosan tended to increase with NW winds. However, it is also possible to observe local sources for this species due to its high levels, even with lower wind speed. Secondarily formed species such as NO_3^- and SO_4^{2-} had a centralized profile and tended to increase with lower wind speed. EC, Chr and Cor seemed to be emitted by local sources, likely vehicular emissions (Alves et al., 2016; Ravindra et al., 2008). Conversely, Flt (a light-molecular-mass PAH), seemed to be influenced by different air masses, suggesting different sources.

4   Summary and conclusions

Particulate matter ($PM_{2.5}$ and PM_{10}) was collected throughout the year 2014 to determine different chemical constituents, including carbonaceous species, WSIs, monosaccharides, PAHs and their derivatives. The risks of PAHs for human health were assessed with levels exceeding the suggested guidelines. High concentrations of biomass burning species were found in the fine particles during the campaigns.

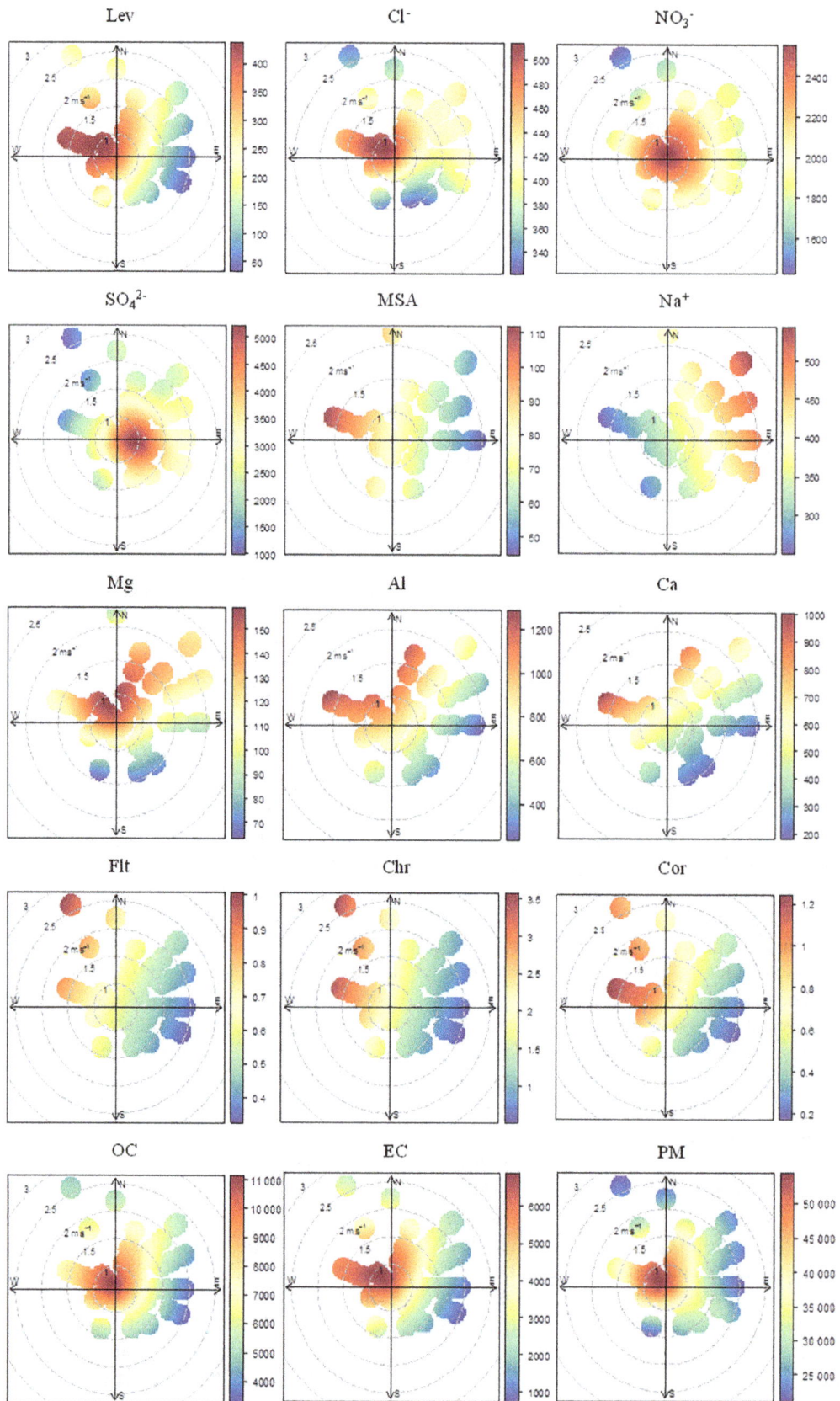

Figure 11. Polar plots for different species ($ng\,m^{-3}$ and $m\,s^{-1}$).

Good correlations were found between the monosaccharides and OC and EC, highlighting their contributions to carbonaceous species. Non-sea salt potassium was also well correlated with the biomass burning species, corroborating the input from this source.

PMF analysis was performed and source profiles were obtained for Int$_{2.5}$, Ext$_{2.5}$ and Ext$_{10}$. Five factors were identified: road dust, industrial, vehicular, biomass burning and secondary processes. Almost 20 % of biomass burning contribution was observed for the PM$_{2.5}$ intensive sampling campaign. The source apportionment led to the identification of traffic-related sources, as expected for the site, since the samples were collected during weekdays. The considerable biomass burning contribution suggests not only the importance of long-range transport of emissions from sugarcane burning but also the input from local biomass burning sources, such as waste burning and wood stoves in restaurants. More studies are needed on the impact of local sources of biomass burning in order to identify the different inputs.

Author contributions. GMP performed the PAH analysis, WSI and monosaccharide determination, and the PMF analysis, and PV is his advisor and laboratory supervisor in Brazil. KT was responsible for WSI and monosaccharide determination with GMP, and RH was the laboratory supervisor. DC was responsible for carbonaceous species determination and polar plots; CA is the department supervisor in Portugal. AGS developed the method for PAH determination; GOdR and JBdA are the supervisors at UFBA. PK and MdFA contributed to source apportionment. HX and RB were responsible for element determination.

Competing interests. The authors declare that they have no conflict of interest.

Acknowledgements. This work was supported by grants from FAPESP, São Paulo Research Foundation and CNPq (project 152601/2013-9); the National Council for Scientific and Technological Development for the postgraduate scholarship and Santander Bank for an international scholarship in Helsinki, Finland. The authors also thank INCT Energy and Environment. Prashant Kumar also acknowledges the collaborative funding received by the Universities of Surrey and São Paulo through the UGPN-funded projects BIOBURN (Towards the Treatment of Aerosol Emissions from Biomass Burning in Chemical Transport Models through a case study in the Metropolitan Area of São Paulo) and NEST-SEAS (Next-Generation Environmental Sensing for Local to Global Scale Health Impact Assessment) that allowed Guilherme Martins Pereira to work at the University of Surrey, United Kingdom. Pérola de Castro Vasconcellos, Gisele Olímpio da Rocha and Jailson Bittencourt de Andrade thank CNPq for their fellowships. Aldenor Gomes Santos, Gisele Olímpio da Rocha and Jailson Bittencourt de Andrade also thank CAPES, CNPq, FAPESB, FINEP and PETROBRAS for research funding at UFBA. Finally, Gisele Olímpio da Rocha is thankful for partial fellowship funding from Fundação Lehmann. Guilherme Martins Pereira also thanks Ioar Rivas and students Bruna Segalin and Fatima Khanun for the help with the PMF analysis.

Edited by: Veli-Matti Kerminen

References

Abdel-shafy, H. I. and Mansour, M. S. M.: A review on polycyclic aromatic hydrocarbons: Source, environmental impact, effect on human health and remediation, Egypt. J. Pet., 25, 107–123, https://doi.org/10.1016/j.ejpe.2015.03.011, 2016.

Agudelo-Castañeda, D. M. and Teixeira, E. C.: Seasonal changes, identification and source apportionment of PAH in PM$_{1.0}$, Atmos. Environ., 96, 186–200, https://doi.org/10.1016/j.atmosenv.2014.07.030, 2014.

Albinet, A., Leoz-garziandia, E., Budzinski, H., Villenave, E., and Jaffrezo, J.: Nitrated and oxygenated derivatives of polycyclic aromatic hydrocarbons in the ambient air of two French alpine valleys – Part 1: Concentrations , sources and gas/particle partitioning, Atmos. Environ., 42, 43–54, https://doi.org/10.1016/j.atmosenv.2007.10.009, 2008.

Allen, A. G., Cardoso, A. A., and da Rocha, G. O.: Influence of sugar cane burning on aerosol soluble ion composition in Southeastern Brazil, Atmos. Environ., 38, 5025–5038, https://doi.org/10.1016/j.atmosenv.2004.06.019, 2004.

Alonso, M. F., Longo, K. M., Freitas, S. R., Mello da Fonseca, R., Marécal, V., Pirre, M., and Klenner, L. G.: An urban emissions inventory for South America and its application in numerical modeling of atmospheric chemical composition at local and regional scales, Atmos. Environ., 44, 5072–5083, https://doi.org/10.1016/j.atmosenv.2010.09.013, 2010.

Alvarez, H. B., Echeverria, R. S., Alvarez, P. S., and Krupa, S.: Air Quality Standards for Particulate Matter (PM) at high altitude cities, Environ. Pollut., 173, 255–256, https://doi.org/10.1016/j.envpol.2012.09.025, 2013.

Alves, C. A., Gomes, J., Nunes, T., Duarte, M., Calvo, A., Custódio, D., Pio, C., Karanasiou, A., and Querol, X.: Size-segregated particulate matter and gaseous emissions from motor vehicles in a road tunnel, Atmos. Res., 153, 134–144, https://doi.org/10.1016/j.atmosres.2014.08.002, 2015.

Alves, C. A., Oliveira, C., Martins, N., Mirante, F., Caseiro, A., Pio, C., Matos, M., Silva, H. F., Oliveira, C., and Camões, F.: Road tunnel, roadside, and urban background measurements of aliphatic compounds in size-segregated particulate matter, Atmos. Res., 168, 139–148, https://doi.org/10.1016/j.atmosres.2015.09.007, 2016.

Amato, F., Alastuey, A., Karanasiou, A., Lucarelli, F., Nava, S., Calzolai, G., Severi, M., Becagli, S., Gianelle, V. L., Colombi, C., Alves, C., Custódio, D., Nunes, T., Cerqueira, M., Pio, C., Eleftheriadis, K., Diapouli, E., Reche, C., Minguillón, M. C., Manousakas, M.-I., Maggos, T., Vratolis, S., Harrison, R. M., and Querol, X.: AIRUSE-LIFE+: a harmonized PM speciation and source apportionment in five southern European cities, Atmos. Chem. Phys., 16, 3289–3309, https://doi.org/10.5194/acp-16-3289-2016, 2016.

Andrade, M. de F., Orsini, C., and Maenhaut, W.: Relation between aerosol sources and meteorological parameters for inhale atmospheric particles in São Paulo city, Brazil, Atmos. Environ., 28, 2307–2309, https://doi.org/10.1016/1352-2310(94)90484-7, 1994.

Andrade, M. de F., Fornaro, A., Freitas, E. D. de, Mazzoli, C. R., Martins, L. D., Boian, C., Oliveira, M. G. L., Peres, J., Carbone, S., Alvalá, P., and Leme, N. P.: Ozone sounding in the Metropolitan Area of São Paulo, Brazil: Wet and dry season campaigns of 2006, Atmos. Environ., 61, 627–640, https://doi.org/10.1016/j.atmosenv.2012.07.083, 2012a.

Andrade, M. de F., Miranda, R. M., Fornaro, A., Kerr, A., Oyama, B., de André, P. A., and Saldiva, P.: Vehicle emissions and $PM_{2.5}$ mass concentrations in six Brazilian cities, Air Qual. Atmos. Heal., 5, 79–88, https://doi.org/10.1007/s11869-010-0104-5, 2012b.

Aurela, M., Saarikoski, S., Timonen, H., Aalto, P., Keronen, P., Saarnio, K., Teinilä, K., Kulmala, M., and Hillamo, R.: Carbonaceous aerosol at a forested and an urban background sites in Southern Finland, Atmos. Environ., 45, 1394–1401, https://doi.org/10.1016/j.atmosenv.2010.12.039, 2011.

Bandowe, B. A. M., Meusel, H., Huang, R., Ho, K., Cao, J., Hoffmann, T., and Wilcke, W.: $PM_{2.5}$-bound oxygenated PAHs, nitro-PAHs and parent-PAHs from the atmosphere of a Chinese megacity: Seasonal variation, sources and cancer risk assessment, Sci. Total Environ., 473–474, 77–87, https://doi.org/10.1016/j.scitotenv.2013.11.108, 2014.

Bardouki, H., Berresheim, H., Vrekoussis, M., Sciare, J., Kouvarakis, G., Oikonomou, K., Schneider, J., and Mihalopoulos, N.: Gaseous (DMS, MSA, SO_2, H_2SO_4 and DMSO) and particulate (sulfate and methanesulfonate) sulfur species over the northeastern coast of Crete, Atmos. Chem. Phys., 3, 1871–1886, https://doi.org/10.5194/acp-3-1871-2003, 2003.

Behera, S. N., Cheng, J., Huang, X., Zhu, Q., Liu, P., and Balasubramanian, R.: Chemical composition and acidity of size-fractionated inorganic aerosols of 2013–14 winter haze in Shanghai and associated health risk of toxic elements, Atmos. Environ., 122, 259–271, https://doi.org/10.1016/j.atmosenv.2015.09.053, 2015.

Bisht, D. S., Dumka, U. C., Kaskaoutis, D. G., Pipal, A. S., Srivastava, A. K., Soni, V. K., Attri, S. D., Sateesh, M., and Tiwari, S.: Carbonaceous aerosols and pollutants over Delhi urban environment: Temporal evolution, source apportionment and radiative forcing, Sci. Total Environ., 521–522, 431–445, https://doi.org/10.1016/j.scitotenv.2015.03.083, 2015.

Bourotte, C., Forti, M.-C., Taniguchi, S., Bícego, M. C., and Lotufo, P. A.: A wintertime study of PAHs in fine and coarse aerosols in São Paulo city, Brazil, Atmos. Environ., 39, 3799–3811, https://doi.org/10.1016/j.atmosenv.2005.02.054, 2005.

Bourotte, C. L. M., Sanchez-Ccoyllo, O. R., Forti, M. C., and Melfi, A. J.: Chemical composition of atmospheric particulate matter soluble fraction and meteorological variables in São Paulo state, Brazil, Rev. Bras. Meteorol., 26, 419–432, https://doi.org/10.1590/S0102-77862011000300008, 2011.

Brito, J., Rizzo, L. V., Herckes, P., Vasconcellos, P. C., Caumo, S. E. S., Fornaro, A., Ynoue, R. Y., Artaxo, P., and Andrade, M. F.: Physical–chemical characterisation of the particulate matter inside two road tunnels in the São Paulo Metropolitan Area, Atmos. Chem. Phys., 13, 12199–12213, https://doi.org/10.5194/acp-13-12199-2013, 2013.

Brown, S. G., Eberly, S., Paatero, P., and Norris, G. A.: Methods for estimating uncertainty in PMF solutions: examples with ambient air and water quality data and guidance on reporting PMF results, Sci. Total Environ., 518–519, 626–635, https://doi.org/10.1016/j.scitotenv.2015.01.022, 2015.

Cabello, M., Orza, J. A. G., Dueñas, C., Liger, E., Gordo, E., and Cañete, S.: Back-trajectory analysis of African dust outbreaks at a coastal city in southern Spain?: Selection of starting heights and assessment of African and concurrent Mediterranean contributions, Atmos. Environ., 140, 10–21, https://doi.org/10.1016/j.atmosenv.2016.05.047, 2016.

Cai, Y., Shao, Y., and Wang, C.: The association of air pollution with the patients' visits to the department of respiratory diseases, J. Clin. Med. Res., 7, 551–555, https://doi.org/10.14740/jocmr2174e, 2015.

Calvo, A. I., Alves, C., Castro, A., Pont, V., Vicente, A. M., and Fraile, R.: Research on aerosol sources and chemical composition: Past, current and emerging issues, Atmos. Res., 120–121, 1–28, https://doi.org/10.1016/j.atmosres.2012.09.021, 2013.

Cançado, J. E. D., Saldiva, P. H. N., Pereira, L. A. A., Lara, L. B. L. S., Artaxo, P., Martinelli, L. A., Arbex, M. A., Zanobetti, A., and Braga, A. L. F.: The impact of sugar cane-burning emissions on the respiratory system of children and the elderly, Environ. Health Perspect., 114, 725–729, https://doi.org/10.1289/ehp.8485, 2006.

Castanho, A. D. A. and Artaxo, P.: Wintertime and summertime São Paulo aerosol source apportionment study, Atmos. Environ., 35, 4889–4902, https://doi.org/10.1016/S1352-2310(01)00357-0, 2001.

Caumo, S., Claeys, M., and Maenhaut, W.: Physicochemical characterization of winter PM_{10} aerosol impacted by sugarcane burning from São Paulo city , Brazi, Atmos. Environ., 145, 272–279, https://doi.org/10.1016/j.atmosenv.2016.09.046, 2016.

CETESB: Companhia de Tecnologia do Saneamento Ambiental: Relatório de qualidade do ar no Estado de São Paulo 2014, Report of air quality in the São Paulo State 2014, São Paulo, Brazil, available at: http://ar.cetesb.sp.gov.br/publicacoes-relatorios/ (last access: 1 August 2016), 2015.

Chen, F., Hu, W., and Zhong, Q.: Emissions of particle-phase polycyclic aromatic hydrocarbons (PAHs) in the Fu Guishan Tunnel of Nanjing, China, Atmos. Res., 124, 53–60, https://doi.org/10.1016/j.atmosres.2012.12.008, 2013.

Chen, W. and Zhu, T.: Formation of nitroanthracene and anthraquinone from the heterogeneous reaction between NO_2 and anthracene adsorbed on NaCl particles, Environ. Sci. Technol., 48, 8671–8678, https://doi.org/10.1021/es501543g, 2014.

Chen, Y.-C., Chiang, H.-C., Hsu, C.-Y., Yang, T.-T., Lin, T.-Y., Chen, M.-J., Chen, N.-T., and Wu, Y.-S.: Ambient $PM_{2.5}$-bound polycyclic aromatic hydrocarbons (PAHs) in Changhua County, central Taiwan: Seasonal variation, source apportionment and cancer risk assessment, Environ. Pollut., 218, 118–128, https://doi.org/10.1016/j.envpol.2016.07.016, 2016.

Cheng, S., Yang, L., Zhou, X., Xue, L., Gao, X., Zhou, Y., and Wang, W.: Size-fractionated water-soluble ions, situ pH and water content in aerosol on hazy days and the influences on visibility impairment in Jinan, China, Atmos. Environ., 45, 4631–4640, https://doi.org/10.1016/j.atmosenv.2011.05.057, 2011.

Cincinelli, A., Del, M., Martellini, T., Gambaro, A., and

Lepri, L.: Gas-particle concentration and distribution of n-alkanes and polycyclic aromatic hydrocarbons in the atmosphere of Prato (Italy), Chemosphere, 68, 472–478, https://doi.org/10.1016/j.chemosphere.2006.12.089, 2007.

CONAMA: Padrões de qualidade do Ar, Resolução CONAMA No. 3/1990, available at: http://www.mma.gov.br/cidades-sustentaveis/qualidade-do-ar/padroes-de-qualidade-do-ar (last access: 1 April 2016), 1990.

Contini, D., Cesari, D., Conte, M., and Donateo, A.: Application of PMF and CMB receptor models for the evaluation of the contribution of a large coal-fired power plant to PM_{10} concentrations, Sci. Total Environ., 560–561, 131–140, https://doi.org/10.1016/j.scitotenv.2016.04.031, 2016.

Cui, M., Chen, Y., Tian, C., Zhang, F., Yan, C., and Zheng, M.: Chemical composition of $PM_{2.5}$ from two tunnels with different vehicular fleet characteristics, Sci. Total Environ., 550, 123–132, https://doi.org/10.1016/j.scitotenv.2016.01.077, 2016.

Custódio, D., Cerqueira, M., Alves, C., Nunes, T., Pio, C., Esteves, V., Frosini, D., Lucarelli, F., and Querol, X.: A one-year record of carbonaceous components and major ions in aerosols from an urban kerbside location in Oporto, Portugal, Sci. Total Environ., 562, 822–833, https://doi.org/10.1016/j.scitotenv.2016.04.012, 2016.

da Rocha, G. O., Allen, A. G., and Cardoso, A.: Influence of Agricultural Biomass Burning on Aerosol Size Distribution and Dry Deposition in Southeastern Brazil, Environ. Sci. Technol., 39, 5293–5301, https://doi.org/10.1021/es048007u, 2005.

da Rocha, G. O., Vasconcellos, P. de C., Ávila, S. G., Souza, D. Z., Reis, E. A. O., Oliveira, P. V., and Sanchez-Ccoyllo, O.: Seasonal distribution of airborne trace elements and water-soluble ions in São Paulo Megacity, Brazil, J. Braz. Chem. Soc., 23, 1915–1924, https://doi.org/10.1590/S0103-50532012005000062, 2012.

de Abrantes, R., de Assunção, J. V., and Pesquero, C. R.: Emission of polycyclic aromatic hydrocarbons from light-duty diesel vehicles exhaust, Atmos. Environ., 38, 1631–1640, https://doi.org/10.1016/j.atmosenv.2003.11.012, 2004.

Decesari, S., Fuzzi, S., Facchini, M. C., Mircea, M., Emblico, L., Cavalli, F., Maenhaut, W., Chi, X., Schkolnik, G., Falkovich, A., Rudich, Y., Claeys, M., Pashynska, V., Vas, G., Kourtchev, I., Vermeylen, R., Hoffer, A., Andreae, M. O., Tagliavini, E., Moretti, F., and Artaxo, P.: Characterization of the organic composition of aerosols from Rondônia, Brazil, during the LBA-SMOCC 2002 experiment and its representation through model compounds, Atmos. Chem. Phys., 6, 375–402, https://doi.org/10.5194/acp-6-375-2006, 2006.

de Oliveira Alves, N., Hacon, S. de S., Galvão, M. F. de O., Peixoto, M. S., Artaxo, P., Vasconcellos, P. de C., and de Medeiros, S. R. B.: Genetic damage of organic matter in the Brazilian Amazon: A comparative study between intense and moderate biomass burning, Environ. Res., 130, 51–58, https://doi.org/10.1016/j.envres.2013.12.011, 2014.

de Oliveira Alves, N., Brito, J., Caumo, S., Arana, A., Hacon, S. de S., Artaxo, P., Hillamo, R., Teinilä, K., de Medeiros, S. R. B., and Vasconcellos, P. de C.: Biomass burning in the Amazon region: Aerosol source apportionment and associated health risk assessment, Atmos. Environ., 120, 277–285, https://doi.org/10.1016/j.atmosenv.2015.08.059, 2015.

Draper, W. M.: Quantitation of nitro- and dinitropol ycyclic aromatic hydrocarbons in diesel exhaust particulate matter, Chemosphere, 15, 437–447, https://doi.org/10.1016/0045-6535(86)90537-0, 1986.

Draxler, R. and Rolph, G.: HYSPLIT (Hybrid Single-Particle Lagrangian Integrated Trajectory) model, NOAA Air Resour. Lab., Silver Spring, MD, available at: http://www.arl.noaa.gov/ready/hysplit4.html (last access: 1 June 2016), 2003.

Duan, J., Bi, X., Tan, J., Sheng, G., and Fu, J.: Seasonal variation on size distribution and concentration of PAHs in Guangzhou city, China, Chemosphere, 67, 614–622, https://doi.org/10.1016/j.chemosphere.2006.08.030, 2007.

EEA (European Environmental Agency): Air Quality in Europe – 2016 Report, EEA Report No 28/2016, Published by Publications Office of the European Union, 2016, ISBN 978-92-9213-824-0, 2016.

Franco, A., Kummrow, F., Umbuzeiro, G. A., Vasconcellos, P. de C., and de Carvalho, L. R.: Occurrence of polycyclic aromatic hydrocarbons derivatives and mutagenicitys study in extracts of PM_{10} collected in São Paulo, Brazil, Rev. Bras. Toxicol., 23, 1–10, 2010.

Fujimoto, T., Kitamura, S., Sanoh, S., Sugihara, K., Yoshihara, S., Fujimoto, N., and Ohta, S.: Estrogenic activity of an environmental pollutant, 2-nitrofluorene, after metabolic activation by rat liver microsomes, Biochem. Biophys. Res. Commun., 303, 419–426, https://doi.org/10.1016/S0006-291X(03)00311-5, 2003.

Godoy, M. L. D. P., Godoy, J. M., Roldão, L. A., Soluri, D. S., and Donagemma, R. A.: Coarse and fine aerosol source apportionment in Rio de Janeiro, Brazil, Atmos. Environ., 43, 2366–2374, https://doi.org/10.1016/j.atmosenv.2008.12.046, 2009.

Gregoris, E., Argiriadis, E., Vecchiato, M., Zambon, S., De Pieri, S., Donateo, A., Contini, D., Piazza, R., Barbante, C., and Gambaro, A.: Science of the Total Environment Gas-particle distributions, sources and health effects of polycyclic aromatic hydrocarbons (PAHs), polychlorinated biphenyls (PCBs) and polychlorinated naphthalenes (PCNs) in Venice aerosols, Sci. Total Environ., 476–477, 393–405, https://doi.org/10.1016/j.scitotenv.2014.01.036, 2014.

Hall, D., Wu, C.-Y., Hsu, Y.-M., Stormer, J., Engling, G., Capeto, K., Wang, J., Brown, S., Li, H.-W., and Yu, K.-M.: PAHs, carbonyls, VOCs and $PM_{2.5}$ emission factors for pre-harvest burning of Florida sugarcane, Atmos. Environ., 55, 164–172, https://doi.org/10.1016/j.atmosenv.2012.03.034, 2012.

Hidy, G. M.: Surface-Level Fine Particle Mass Concentrations: From Hemispheric Distributions to Megacity Sources Surface-Level Fine Particle Mass Concentrations: From Hemispheric Distributions to Megacity Sources, J. Air Waste Manage. Assoc., 59, 770–789, https://doi.org/10.3155/1047-3289.59.7.770, 2009.

IAG: IAG/USP Annual Meteorological Bulletin – 2014, available at: http://www.estacao.iag.usp.br/Boletins/2014.pdf (last access: 1 April 2016), 2014.

INPE: INPE (Instituto Nacional de Pesquisas Espaciais) – Portal do Monitoramento de Queimadas, available at: https://queimadas.dgi.inpe.br/queimadas/ (last access: 1 April 2016), 2014.

Jung, J., Lee, H., Kim, Y. J., Liu, X., Zhang, Y., Gu, J., and Fan, S.: Aerosol chemistry and the effect of aerosol water content on visibility impairment and radiative forcing in Guangzhou during the 2006 Pearl River Delta campaign, J. Environ. Manage.,

90, 3231–3244, https://doi.org/10.1016/j.jenvman.2009.04.021, 2009.

Jung, K. H., Yan, B., Chillrud, S. N., Perera, F. P., Whyatt, R., Camann, D., Kinney, P. L., and Miller, R. L.: Assessment of Benzo(a)pyrene-equivalent Carcinogenicity and mutagenicity of residential indoor versus outdoor polycyclic aromatic hydrocarbons exposing young children in New York city, Int. J. Environ. Res. Public Health, 7, 1889–1900, https://doi.org/10.3390/ijerph7051889, 2010.

Karanasiou, A. A., Siskos, P. A., and Eleftheriadis, K.: Assessment of source apportionment by Positive Matrix Factorization analysis on fine and coarse urban aerosol size fractions, Atmos. Environ., 43, 3385–3395, https://doi.org/10.1016/j.atmosenv.2009.03.051, 2009.

Kassomenos, P. A., Vardoulakis, S., Chaloulakou, A., Paschalidou, A. K., Grivas, G., Borge, R., and Lumbreras, J.: Study of PM_{10} and $PM_{2.5}$ levels in three European cities: Analysis of intra and inter urban variations, Atmos. Environ., 87, 153–163, https://doi.org/10.1016/j.atmosenv.2014.01.004, 2014.

Khoder, M. I. and Hassan, S. K.: Weekday/weekend differences in ambient aerosol level and chemical characteristics of water-soluble components in the city centre, Atmos. Environ., 42, 7483–7493, https://doi.org/10.1016/j.atmosenv.2008.05.068, 2008.

Kojima, Y., Inazu, K., Hisamatsu, Y., Okochi, H., Baba, T., and Nagoya, T.: Comparison of Pahs, Nitro-Pahs and Oxy-Pahs Associated With Airborne Particulate Matter At Roadside and Urban Background Sites in Downtown Tokyo, Japan, Polycycl. Aromat. Compd., 30, 321–333, https://doi.org/10.1080/10406638.2010.525164, 2010.

Kumar, A. and Attri, A. K.: Biomass Combustion a Dominant Source of Carbonaceous Aerosols in the Ambient Environment of Western Himalayas, Aerosol Air Qual. Res., 16, 519–529, https://doi.org/10.4209/aaqr.2015.05.0284, 2016.

Kumar, P., Morawska, L., Birmili, W., Paasonen, P., Hu, M., Kulmala, M., Harrison, R. M., Norford, L., and Britter, R.: Ultra fine particles in cities, Environ. Int., 66, 1–10, https://doi.org/10.1016/j.envint.2014.01.013, 2014.

Kumar, P., Andrade, M. de F., Ynoue, R. Y., Fornaro, A., de Freitas, E. D., Martins, J., Martins, L. D., Albuquerque, T., Zhang, Y., and Morawska, L.: New directions: From biofuels to wood stoves: The modern and ancient air quality challenges in the megacity of São Paulo, Atmos. Environ., 140, 364–369, https://doi.org/10.1016/j.atmosenv.2016.05.059, 2016.

Kundu, S., Kawamura, K., Andreae, T. W., Hoffer, A., and Andreae, M. O.: Diurnal variation in the water-soluble inorganic ions, organic carbon and isotopic compositions of total carbon and nitrogen in biomass burning aerosols from the LBA-SMOCC campaign in Rondônia, Brazil, J. Aerosol Sci., 41, 118–133, https://doi.org/10.1016/j.jaerosci.2009.08.006, 2010.

de La Torre-Roche, R. J., Lee, W. Y., and Campos-Díaz, S. I.: Soil-borne polycyclic aromatic hydrocarbons in El Paso, Texas: analysis of a potential problem in the United States/Mexico border region, J. Hazard. Mater., 163, 946–958, https://doi.org/10.1016/j.jhazmat.2008.07.089, 2009.

Lang, Y.-H., Li, G., Wang, X.-M., and Peng, P.: Combination of Unmix and PMF receptor model to apportion the potential sources and contributions of PAHs in wetland soils from Jiaozhou Bay, China, Mar. Pollut. Bull., 90, 129–134,

https://doi.org/10.1016/j.marpolbul.2014.11.009, 2015.

Lee, J. D.: Concise Inorganic Chemistry, 5th Edn., Willey, 1070 pp., 1999.

Liu, B., Bi, X., Feng, Y., Dai, Q., Xiao, Z., Li, L., Wu, J., Yuan, J., and Zhang, Y.: Fine carbonaceous aerosol characteristics at a megacity during the Chinese Spring Festival as given by OC / EC online measurements, Atmos. Res., 181, 20–28, https://doi.org/10.1016/j.atmosres.2016.06.007, 2016.

Maenhaut, W., Raes, N., Chi, X., Cafmeyer, J., and Wang, W.: Chemical composition and mass closure for $PM_{2.5}$ and PM_{10} aerosols at K-puszta, Hungary, in summer 2006, X-Ray Spectrom., 37, 193–197, https://doi.org/10.1002/xrs.1062, 2008.

Maenhaut, W., Vermeylen, R., Claeys, M., Vercauteren, J., and Roekens, E.: Sources of the PM_{10} aerosol in Flanders, Belgium, and re-assessment of the contribution from wood burning, Sci. Total Environ., 562, 550–560, https://doi.org/10.1016/j.scitotenv.2016.04.074, 2016.

de Martinis, B. S., Okamoto, R. A., Kado, N. Y., Gundel, L. A., and Carvalho, L. R. F.: Polycyclic aromatic hydrocarbons in a bioassay-fractionated extract of PM_{10} collected in São Paulo, Brazil, Atmos. Environ., 36, 307–314, https://doi.org/10.1016/S1352-2310(01)00334-X, 2002.

Miguel, A. H., Kirchstetter, T. W., Harley, R. A., and Hering, S. V.: On-road emissions of particulate polycyclic aromatic hydrocarbons and black carbon from gasoline and diesel vehicles, Environ. Sci. Technol., 32, 450–455, https://doi.org/10.1021/es970566w, 1998.

Miranda, R. M. de, Andrade, M. de F., Fornaro, A., Astolfo, R., de André, P. A., and Saldiva, P.: Urban air pollution: A representative survey of $PM_{2.5}$ mass concentrations in six Brazilian cities, Air Qual. Atmos. Heal., 5, 63–77, https://doi.org/10.1007/s11869-010-0124-1, 2012.

Mirante, F., Alves, C., Pio, C., Pindado, O., Perez, R., Revuelta, M. A., and Artiñano, B.: Organic composition of size segregated atmospheric particulate matter, during summer and winter sampling campaigns at representative sites in Madrid, Spain, Atmos. Res., 132–133, 345–361, https://doi.org/10.1016/j.atmosres.2013.07.005, 2013.

Mkoma, S. L., da Rocha, G. O., Regis, A. C. D., Domingos, J. S. S., Santos, J. V. S., de Andrade, S. J., Carvalho, L. S., and De Andrade, J. B.: Major ions in $PM_{2.5}$ and PM_{10} released from buses: The use of diesel/biodiesel fuels under real conditions, Fuel, 115, 109–117, https://doi.org/10.1016/j.fuel.2013.06.044, 2014.

MMA: Ministerio del Medio Ambiente (MMA) Progress Report on Santiago's Pollution Prevention Plan, available at: http://www.sinia.cl/1292/articles-55841_InformeFINALSeguimientoPPDA2012_RM.pdf (last access: 1 April 2016), 2014 (in Spanish).

Nayebare, S. R., Aburizaiza, O. S., Khwaja, H. A., Siddique, A., Hussain, M. M., Zeb, J., Khatib, F., Carpenter, D. O., and Blake, D. R.: Chemical Characterization and Source Apportionment of $PM_{2.5}$ in Rabigh, Saudi Arabia, Aerosol Air Qual. Res., 16, 3114–3129, https://doi.org/10.4209/aaqr.2015.11.0658, 2016.

Newby, D. E., Mannucci, P. M., Tell, G. S., Baccarelli, A. A., Brook, R. D., Donaldson, K., Forastiere, F., Franchini, M., Franco, O. H., Graham, I., Hoek, G., Hoffmann, B., Hoylaerts, M. F., Künzli, N., Mills, N., Pekkanen, J., Peters, A., Piepoli, M. F., Rajagopalan, S., and Storey, R. F.: Expert position paper on air pollution and cardiovascular disease, Eur. Heart J., 36, 83–93, https://doi.org/10.1093/eurheartj/ehu458, 2015.

Norris, G., Duvall, R., Brown, S., and Bai, S.: EPA Positive Matrix Factorization (PMF) 5.0 Fundamentals and User Guide, available at: https://www.epa.gov/air-research/epa-positive-matrix-factorization-50-fundamentals-and-user-guide (last access: 1 August 2016), 2014.

Oliveira, C., Martins, N., Tavares, J., Pio, C., Cerqueira, M., Matos, M., Silva, H., Oliveira, C., and Camões, F.: Size distribution of polycyclic aromatic hydrocarbons in a roadway tunnel in Lisbon, Portugal, Chemosphere, 83, 1588–1596, https://doi.org/10.1016/j.chemosphere.2011.01.011, 2011.

Oyama, B. S., Andrade, M. D. F., Herckes, P., Dusek, U., Röckmann, T., and Holzinger, R.: Chemical characterization of organic particulate matter from on-road traffic in São Paulo, Brazil, Atmos. Chem. Phys., 16, 14397–14408, https://doi.org/10.5194/acp-16-14397-2016, 2016.

Paatero, P. and Hopke, P. K.: Discarding or downweighting high-noise variables in factor analytic models, Anal. Chim. Acta, 490, 277–289, https://doi.org/10.1016/S0003-2670(02)01643-4, 2003.

Paatero, P. and Tapper, U.: Positive matrix factorization: A non-negative factor model with optimal utilization of error estimates of data values, Environmetrics, 5, 111–126, https://doi.org/10.1002/env.3170050203, 1994.

Pacheco, M. T., Parmigiani, M. M. M., Andrade, M. de F., Morawska, L., and Kumar, P.: A review of emissions and concentrations of particulate matter in the three major metropolitan areas of Brazil, J. Transp. Heal., 4, 53–72, https://doi.org/10.1016/j.jth.2017.01.008, 2017.

Pereira, P. A. de P., Lopes, W. A., Carvalho, L. S., da Rocha, G. O., Carvalho, N. De, Loyola, J., Quiterio, S. L., Escaleira, V., Arbilla, G., and de Andrade, J. B.: Atmospheric concentrations and dry deposition fluxes of particulate trace metals in Salvador, Bahia, Brazil, Atmos. Environ., 41, 7837–7850, https://doi.org/10.1016/j.atmosenv.2007.06.013, 2007.

Pereira, G. M., Alves, N. O., Caumo, S. E. S., Soares, S., Teinilä, K., Custódio, D., Hillamo, R., Alves, C., and Vasconcellos, P. C.: Chemical composition of aerosol in São Paulo, Brazil: Influence of the transport of pollutants, Air Qual. Atmos. Heal., 10, 457–468, https://doi.org/10.1007/s11869-016-0437-9, 2017.

Pio, C. A., Legrand, M., Alves, C. A., Oliveira, T., Afonso, J., Caseiro, A., Puxbaum, H., Sanchez-Ochoa, A., and Gelencsér, A.: Chemical composition of atmospheric aerosols during the 2003 summer intense forest fire period, Atmos. Environ., 42, 7530–7543, https://doi.org/10.1016/j.atmosenv.2008.05.032, 2008.

Pio, C. A., Cerqueira, M., Harrison, R. M., Nunes, T., Mirante, F., Alves, C., Oliveira, C., de la Campa, A. S., Artíñano, B., and Matos, M.: OC / EC ratio observations in Europe: Rethinking the approach for apportionment between primary and secondary organic carbon, Atmos. Environ., 45, 6121–6132, https://doi.org/10.1016/j.atmosenv.2011.08.045, 2011.

Pio, C. A., Mirante, F., Oliveira, C., Matos, M., Caseiro, A., Oliveira, C., Querol, X., Alves, C., Martins, N., Cerqueira, M., Camões, F., Silva, H., and Plana, F.: Size-segregated chemical composition of aerosol emissions in an urban road tunnel in Portugal, Atmos. Environ., 71, 15–25, https://doi.org/10.1016/j.atmosenv.2013.01.037, 2013.

Pope, C. A.: Epidemiology of fine particulate air pollution and human health: Biologic mechanisms and who's at risk?, Environ. Health Perspect., 108, 713–723, https://doi.org/10.1289/ehp.00108s4713, 2000.

Pöschl, U.: Atmospheric aerosols: Composition, transformation, climate and health effects, Angew. Chem. Int. Edit., 44, 7520–7540, https://doi.org/10.1002/anie.200501122, 2005.

Putaud, J. P., Raes, F., Van Dingenen, R., Brüggemann, E., Facchini, M.-C., Decesari, S., Fuzzi, S., Gehrig, R., Hüglin, C., Laj, P., Lorbeer, G., Maenhaut, W., Mihalopoulos, N., Müller, K., Querol, X., Rodriguez, S., Schneider, J., Spindler, G., Brink, H. ten, Tørseth, K., and Wiedensohler, A.: A European aerosol phenomenology – 2:chemical characteristics of particulate matter at kerbside, urban, rural and background sites in Europe, Atmos. Environ., 38, 2579–2595, https://doi.org/10.1016/j.atmosenv.2004.01.041, 2004.

Ram, K., Sarin, M. M., and Tripathi, S. N.: A 1 year record of carbonaceous aerosols from an urban site in the Indo-Gangetic Plain: Characterization, sources, and temporal variability, J. Geophys. Res.-Atmos., 115, 1–14, https://doi.org/10.1029/2010JD014188, 2010.

Rao, P. S. P., Tiwari, S., Matwale, J. L., Pervez, S., Tunved, P., Safai, P. D., Srivastava, A. K., Bisht, D. S., Singh, S., and Hopke, P. K.: Sources of chemical species in rainwater during monsoon and non- monsoonal periods over two mega cities in India and dominant source region of secondary aerosols, Atmos. Environ., 146, 90–99, https://doi.org/10.1016/j.atmosenv.2016.06.069, 2016.

Rastogi, N. and Sarin, M. M.: Quantitative chemical composition and characteristics of aerosols over western India: One-year record of temporal variability, Atmos. Environ., 43, 3481–3488, https://doi.org/10.1016/j.atmosenv.2009.04.030, 2009.

Rastogi, N., Singh, A., Singh, D., and Sarin, M. M.: Chemical characteristics of $PM_{2.5}$ at a source region of biomass burning emissions: Evidence for secondary aerosol formation, Environ. Pollut., 184, 563–569, https://doi.org/10.1016/j.envpol.2013.09.037, 2014.

Ravindra, K., Bencs, L., Wauters, E., De Hoog, J., Deutsch, F., Roekens, E., Bleux, N., Berghmans, P., and Van Grieken, R.: Seasonal and site-specific variation in vapour and aerosol phase PAHs over Flanders (Belgium) and their relation with anthropogenic activities, Atmos. Environ., 40, 771–785, https://doi.org/10.1016/j.atmosenv.2005.10.011, 2006.

Ravindra, K., Sokhi, R., and Van Grieken, R.: Atmospheric polycyclic aromatic hydrocarbons: Source attribution, emission factors and regulation, Atmos. Environ., 42, 2895–2921, https://doi.org/10.1016/j.atmosenv.2007.12.010, 2008.

Ringuet, J., Albinet, A., Leoz-Garziandia, E., Budzinski, H., and Villenave, E.: Diurnal/nocturnal concentrations and sources of particulate-bound PAHs, OPAHs and NPAHs at traffic and suburban sites in the region of Paris (France), Sci. Total Environ., 437, 297–305, https://doi.org/10.1016/j.scitotenv.2012.07.072, 2012.

Robbat, A. and Wilton, N. M.: A new spectral deconvolution – Selected ion monitoring method for the analysis of alkylated polycyclic aromatic hydrocarbons in complex mixtures, Talanta, 125, 114–124, https://doi.org/10.1016/j.talanta.2014.02.068, 2014.

Romero-Lankao, P., Qin, H., and Borbor-Cordova, M.: Exploration of health risks related to air pollution and temperature in three Latin American cities, Soc. Sci. Med., 83, 110–118, https://doi.org/10.1016/j.socscimed.2013.01.009, 2013.

Saarnio, K., Teinilä, K., Aurela, M., Timonen, H., and Hillamo, R.: High-performance anion-exchange chromatography-mass spectrometry method for determination of levoglucosan, mannosan, and galactosan in atmospheric fine particulate matter, Anal. Bioanal. Chem., 398, 2253–2264, https://doi.org/10.1007/s00216-010-4151-4, 2010.

Sánchez-Ccoyllo, O. R. and Andrade, M. d. F.: The influence of meteorological conditions on the behavior of pollutants concentrations in São Paulo, Brazil, Environ. Pollut., 116, 257–263, https://doi.org/10.1016/S0269-7491(01)00129-4, 2002.

Santos, A. G., Regis, A. C. D., da Rocha, G. O., Bezerra, M. de A., de Jesus, R. M., and de Andrade, J. B.: A simple, comprehensive, and miniaturized solvent extraction method for determination of particulate-phase polycyclic aromatic compounds in air., J. Chromatogr. A, 1435, 6–17, https://doi.org/10.1016/j.chroma.2016.01.018, 2016.

Sarigiannis, D. A., Karakitsios, S. P., Zikopoulos, D., Nikolaki, S., and Kermenidou, M.: Lung cancer risk from PAHs emitted from biomass combustion, Environ. Res., 137, 147–156, https://doi.org/10.1016/j.envres.2014.12.009, 2015.

Schkolnik, G., Falkovich, A. H., Rudich, Y., Maenhaut, W., and Artaxo, P.: New analytical method for the determination of levoglucosan, polyhydroxy compounds, and 2-methylerythritol and its application to smoke and rainwater samples, Environ. Sci. Technol., 39, 2744–2752, https://doi.org/10.1021/es048363c, 2005.

SEADE: SP Demografico – Resenha de Estatísticas Vitais do Estado de São Paulo: Diferenciais regionais de fecundidade no município de São Paulo, available at: http://www.seade.gov.br/produtos/midia/2016/06/N.2_jun2016-final.pdf, last access: 1 August 2016.

Segalin, B., Kumar, P., Micadei, K., Fornaro, A., and Gonçalves, F. L. T.: Size-segregated particulate matter inside residences of elderly in the Metropolitan Area of São Paulo, Brazil, Atmos. Environ., 148, 139–151, https://doi.org/10.1016/j.atmosenv.2016.10.004, 2017.

Seinfeld, J. H. and Pandis, S. N.: Atmospheric Chemistry and Physics: From Air Pollution to Climate Change, 2nd Edn., John Wiley & Sons, New York, 2006.

Shi, G., Tian, Y., Ye, S., Peng, X., Xu, J., Wang, W., Han, B., and Feng, Y.: Source apportionment of synchronously size segregated fine and coarse particulate matter, using an improved three-way factor analysis model, Sci. Total Environ., 505, 1182–1190, https://doi.org/10.1016/j.scitotenv.2014.10.106, 2015.

Simoneit, B. R. T., Schauer, J. J., Nolte, C. G., Oros, D. R., Elias, V. O., Fraser, M. P., Rogge, W. F., and Cass, G. R.: Levoglucosan, a tracer for cellulose in biomass burning and atmospheric particles, Atmos. Environ., 33, 173–182, https://doi.org/10.1016/S1352-2310(98)00145-9, 1999.

Souza, D. Z., Vasconcellos, P. C., Lee, H., Aurela, M., Saarnio, K., Teinilä, K., and Hillamo, R.: Composition of $PM_{2.5}$ and PM_{10} collected at Urban Sites in Brazil, Aerosol Air Qual. Res., 14, 168–176, https://doi.org/10.4209/aaqr.2013.03.0071, 2014a.

Souza, K. F., Carvalho, L. R. F., Allen, A. G., and Cardoso, A. A.: Diurnal and nocturnal measurements of PAH, nitro-PAH, and oxy-PAH compounds in atmospheric particulate matter of a sugar cane burning region, Atmos. Environ., 83, 193–201, https://doi.org/10.1016/j.atmosenv.2013.11.007, 2014b.

Tan, J.-H., Duan, J.-C., Chen, D.-H., Wang, X.-H., Guo, S.-J., Bi, X.-H., Sheng, G.-Y., He, K.-B., and Fu, J.-M.: Chemical characteristics of haze during summer and winter in Guangzhou, Atmos. Res., 94, 238–245, https://doi.org/10.1016/j.atmosres.2009.05.016, 2009.

Tang, X., Zhang, X., Ci, Z., Guo, J., and Wang, J.: Speciation of the major inorganic salts in atmospheric aerosols of Beijing, China: Measurements and comparison with model, Atmos. Environ., 133, 123–134, https://doi.org/10.1016/j.atmosenv.2016.03.013, 2016.

Teixeira, E. C., Mattiuzi, C. D. P., Agudelo-Castañeda, D. M., Garcia, K. de O., and Wiegand, F.: Polycyclic aromatic hydrocarbons study in atmospheric fine and coarse particles using diagnostic ratios and receptor model in urban/industrial region, Environ. Monit. Assess., 185, 9587–9602, https://doi.org/10.1007/s10661-013-3276-2, 2013.

Timonen, H., Carbone, S., Aurela, M., Saarnio, K., Saarikoski, S., Ng, N. L., Canagaratna, M. R., Kulmala, M., Kerminen, V. M., Worsnop, D. R., and Hillamo, R.: Characteristics, sources and water-solubility of ambient submicron organic aerosol in springtime in Helsinki, Finland, J. Aerosol Sci., 56, 61–77, https://doi.org/10.1016/j.jaerosci.2012.06.005, 2013.

Tiwari, S., Dumka, U. C., Kaskaoutis, D. G., Ram, K., Panicker, A. S., Srivastava, M. K., Tiwari, S., Attri, S. D., Soni, V. K., and Pandey, A. K.: Aerosol chemical characterization and role of carbonaceous aerosol on radiative effect over Varanasi in central Indo-Gangetic Plain, Atmos. Environ., 125, 437–449, https://doi.org/10.1016/j.atmosenv.2015.07.031, 2016.

Tobiszewski, M. and Namieœnik, J.: PAH diagnostic ratios for the identification of pollution emission sources, Environ. Pollut., 162, 110–119, https://doi.org/10.1016/j.envpol.2011.10.025, 2012.

Toledano, C., Cachorro, V. E., Frutos, A. M. de, Torres, B., Berjon, A., Sorribas, M., and Stone, R. S.: Air-mass Classification and Analysis of Aerosol Types at El Arenosillo (Spain), J. Appl. Meteorol. Climatol., 48, 962–981, https://doi.org/10.1175/2008JAMC2006.1, 2009.

Urban, R. C., Lima-Souza, M., Caetano-Silva, L., Queiroz, M. E. C., Nogueira, R. F. P., Allen, A. G., Cardoso, A. A., Held, G., and Campos, M. L. A. M.: Use of levoglucosan, potassium, and water-soluble organic carbon to characterize the origins of biomass-burning aerosols, Atmos. Environ., 61, 562–569, https://doi.org/10.1016/j.atmosenv.2012.07.082, 2012.

Urban, R. C., Alves, C. A., Allen, A. G., Cardoso, A. A., Queiroz, M. E. C., and Campos, M. L. A. M.: Sugar markers in aerosol particles from an agro-industrial region in Brazil, Atmos. Environ., 90, 106–112, https://doi.org/10.1016/j.atmosenv.2014.03.034, 2014.

Urban, R. C., Alves, C. A., Allen, A. G., Cardoso, A. A., and Campos, M. L. A. M.: Organic aerosols in a Brazilian agro-industrial area: Speciation and impact of biomass burning, Atmos. Res., 169, 271–279, https://doi.org/10.1016/j.atmosres.2015.10.008, 2016.

Vara-Vela, A., Andrade, M. F., Kumar, P., Ynoue, R. Y., and Muñoz, A. G.: Impact of vehicular emissions on the formation of fine particles in the Sao Paulo Metropolitan Area: a numerical study with the WRF-Chem model, Atmos. Chem. Phys., 16, 777–797, https://doi.org/10.5194/acp-16-777-2016, 2016.

Vasconcellos, P. C., Zacarias, D., Pires, M. A. F., Pool, C. S., and Carvalho, L. R. F.: Measurements of polycyclic aromatic hydrocarbons in airborne particles from the metropolitan area of São Paulo City, Brazil, Atmos. Environ., 37, 3009–3018, https://doi.org/10.1016/S1352-2310(03)00181-X, 2003.

Vasconcellos, P. C., Balasubramanian, R., Bruns, R. E., Sanchez-Ccoyllo, O., Andrade, M. F., and Flues, M.: Water-soluble ions and trace metals in airborne particles over urban areas of the state of São Paulo, Brazil: Influences of local sources and long range transport, Water. Air. Soil Pollut., 186, 63–73, https://doi.org/10.1007/s11270-007-9465-2, 2007.

Vasconcellos, P. C., Souza, D. Z., Sanchez-Ccoyllo, O., Bustillos, J. O. V, Lee, H., Santos, F. C., Nascimento, K. H., Araújo, M. P., Saarnio, K., Teinilä, K., and Hillamo, R.: Determination of anthropogenic and biogenic compounds on atmospheric aerosol collected in urban, biomass burning and forest areas in São Paulo, Brazil, Sci. Total Environ., 408, 5836–5844, https://doi.org/10.1016/j.scitotenv.2010.08.012, 2010.

Vasconcellos, P. C., Souza, D. Z., Avila, S. G., Araujo, M. P., Naoto, E., Nascimento, K. H., Cavalcante, F. S., Dos Santos, M., Smichowski, P., and Behrentz, E.: Comparative study of the atmospheric chemical composition of three South American cities, Atmos. Environ., 45, 5770–5777, https://doi.org/10.1016/j.atmosenv.2011.07.018, 2011a.

Vasconcellos, P. C., Souza, D. Z., Magalhães, D., and da Rocha, G. O.: Seasonal variation of n-alkanes and polycyclic aromatic hydrocarbon concentrations in PM_{10} samples collected at urban sites of São Paulo State, Brazil, Water. Air. Soil Pollut., 222, 325–336, https://doi.org/10.1007/s11270-011-0827-4, 2011b.

Vieira-Filho, M., Pedrotti, J. J., and Fornaro, A.: Water-soluble ions species of size-resolved aerosols: Implications for the atmospheric acidity in São Paulo megacity, Brazil, Atmos. Res., 181, 281–287, https://doi.org/10.1016/j.atmosres.2016.07.006, 2016.

Villalobos, A. M., Barraza, F., Jorquera, H., and Schauer, J. J.: Chemical speciation and source apportionment of fine particulate matter in Santiago, Chile, 2013, Sci. Total Environ., 512–513, 133–142, https://doi.org/10.1016/j.scitotenv.2015.01.006, 2015.

Wang, G., Wang, H., Yu, Y., Gao, S., Feng, J., Gao, S., and Wang, L.: Chemical characterization of water-soluble components of PM_{10} and $PM_{2.5}$ atmospheric aerosols in five locations of Nanjing, China, Atmos. Environ., 37, 2893–2902, https://doi.org/10.1016/S1352-2310(03)00271-1, 2003.

Wang, J., Hu, Z., Chen, Y., Chen, Z., and Xu, S.: Contamination characteristics and possible sources of PM_{10} and $PM_{2.5}$ in different functional areas of Shanghai, China, Atmos. Environ., 68,

221–229, https://doi.org/10.1016/j.atmosenv.2012.10.070, 2013.

Wang, Y., Zhang, Q. Q., He, K., Zhang, Q., and Chai, L.: Sulfate-nitrate-ammonium aerosols over China: response to 2000–2015 emission changes of sulfur dioxide, nitrogen oxides, and ammonia, Atmos. Chem. Phys., 13, 2635–2652, https://doi.org/10.5194/acp-13-2635-2013, 2013.

White, W. H.: Chemical markers for sea salt in IMPROVE aerosol data, Atmos. Environ., 42, 261–274, https://doi.org/10.1016/j.atmosenv.2007.09.040, 2008.

WHO: Air quality guidelines for Europe, WHO Reg. Publ. Eur. Ser. No. 91, 2nd Edn., https://doi.org/10.1007/BF02986808, 2000.

WHO: WHO Air quality guidelines for particulate matter, ozone, nitrogen dioxide and sulfur dioxide: global update 2005: summary of risk assessment, Geneva World Heal. Organ., 1–22, https://doi.org/10.1016/0004-6981(88)90109-6, 2006.

Yang, Y., Zhou, R., Wu, J., Yu, Y., Ma, Z., Zhang, L., and Di, Y.: Seasonal variations and size distributions of water-soluble ions in atmospheric aerosols in Beijing, 2012, J. Environ. Sci., 34, 197–205, https://doi.org/10.1016/j.jes.2015.01.025, 2015.

Yassaa, N., Meklati, B. Y., Cecinato, A., and Marino, F.: Particulate n-alkanes, n-alkanoic acids and polycyclic aromatic hydrocarbons in the atmosphere of Algiers City Area, Atmos. Environ., 35, 1843–1851, https://doi.org/10.1016/S1352-2310(00)00514-8, 2001.

Yu, G., Zhang, Y., Cho, S., and Park, S.: Influence of haze pollution on water-soluble chemical species in $PM_{2.5}$ and size-resolved particles at an urban site during fall, J. Environ. Sci., 57, 370–382, https://doi.org/10.1016/j.jes.2016.10.018, 2017.

Zheng, J., Hu, M., Peng, J., Wu, Z., Kumar, P., Li, M., Wang, Y., and Guo, S.: Spatial distributions and chemical properties of $PM_{2.5}$ based on 21 field campaigns at 17 sites in China, Chemosphere, 159, 480–487, https://doi.org/10.1016/j.chemosphere.2016.06.032, 2016.

Zhou, S. and Wenger, J. C.: Kinetics and products of the gas-phase reactions of acenaphthylene with hydroxyl radicals, nitrate radicals and ozone, Atmos. Environ., 75, 103–112, https://doi.org/10.1016/j.atmosenv.2013.04.049, 2013.

Zimmermann, K., Jariyasopit, N., Simonich, S. L. M., Tao, S., Atkinson, R., and Arey, J.: Formation of nitro-PAHs from the heterogeneous reaction of ambient particle-bound PAHs with $N_2O_5/NO_3/NO_2$, Environ. Sci. Technol., 47, 8434–8442, https://doi.org/10.1021/es401789x, 2013.

8

Uncertainty from the choice of microphysics scheme in convection-permitting models significantly exceeds aerosol effects

Bethan White[1], Edward Gryspeerdt[2], Philip Stier[1], Hugh Morrison[3], Gregory Thompson[3], and Zak Kipling[4]

[1]Atmospheric, Oceanic and Planetary Physics, University of Oxford, Oxford, UK
[2]Institute for Meteorology, Universität Leipzig, Leipzig, Germany
[3]National Center for Atmospheric Research, Boulder, Colorado, USA
[4]European Centre for Medium-Range Weather Forecasts, Shinfield Park, Reading, UK

Correspondence to: Bethan White (bethan.white@physics.ox.ac.uk)

Abstract. This study investigates the hydrometeor development and response to cloud droplet number concentration (CDNC) perturbations in convection-permitting model configurations. We present results from a real-data simulation of deep convection in the Congo basin, an idealised supercell case, and a warm-rain large-eddy simulation (LES). In each case we compare two frequently used double-moment bulk microphysics schemes and investigate the response to CDNC perturbations. We find that the variability among the two schemes, including the response to aerosol, differs widely between these cases. In all cases, differences in the simulated cloud morphology and precipitation are found to be significantly greater between the microphysics schemes than due to CDNC perturbations within each scheme. Further, we show that the response of the hydrometeors to CDNC perturbations differs strongly not only between microphysics schemes, but the inter-scheme variability also differs between cases of convection. Sensitivity tests show that the representation of autoconversion is the dominant factor that drives differences in rain production between the microphysics schemes in the idealised precipitating shallow cumulus case and in a sub-region of the Congo basin simulations dominated by liquid-phase processes. In this region, rain mass is also shown to be relatively insensitive to the radiative effects of an overlying layer of ice-phase cloud. The conversion of cloud ice to snow is the process responsible for differences in cold cloud bias between the schemes in the Congo. In the idealised supercell case, thermodynamic impacts on the storm system using different microphysics parameterisations can equal those due to aerosol effects. These results highlight the large uncertainty in cloud and precipitation responses to aerosol in convection-permitting simulations and have important implications not only for process studies of aerosol–convection interaction, but also for global modelling studies of aerosol indirect effects. These results indicate the continuing need for tighter observational constraints of cloud processes and response to aerosol in a range of meteorological regimes.

1 Introduction

Deep convection has a significant influence on the state of the atmosphere and climate through shortwave and longwave radiative interactions, heat transfer through the release of latent heat and global heat redistribution. It also plays an important part in the hydrological cycle through the conversion of water vapour to precipitation. One major way that aerosols can influence the properties of deep convection is through their effect on cloud microphysics. By acting as cloud condensation nuclei (CCN), increased aerosol loading can lead to an increase in cloud droplet number concentration (CDNC) and a subsequent reduction in cloud droplet size, which in turn has been hypothesised to suppress warm-phase precipitation (Albrecht, 1989). Some theoretical (e.g. Rosenfeld et al., 2008; Stevens and Feingold, 2009) and cloud-resolving (or cloud-system-resolving) modelling studies (e.g. Fan et al., 2007; Tao et al., 2007; Lebo and Seinfeld, 2011, amongst many others) have suggested that under certain conditions, precipitation suppression in the liquid phase may lead to an invigoration of deep convection and a subsequent enhancement of convective precipitation. The detection of positive correlations between satellite-observed aerosol optical depth (AOD) and precipitation or convective cloud properties (e.g. Koren et al., 2005; Gryspeerdt et al., 2014) might suggest observational evidence of convective invigoration by aerosols. However, factors such as meteorological covariation and retrieval errors may contribute to or even dominate such correlations (Zhang et al., 2005; Mauger and Norris, 2007; Chand et al., 2012; Gryspeerdt et al., 2014). Complex process interactions in ice- and mixed-phase microphysics, along with coupling to surface and radiative feedbacks and dynamics over a range of spatiotemporal scales, means that understanding and quantifying aerosol impacts on deep convection remains a significant challenge (e.g. Noppel et al., 2010; Seifert et al., 2012; Tao et al., 2012).

Representing cloud microphysical processes, which occur on length scales of microns to millimetres, has always been a significant challenge for atmospheric models. Even in cloud-resolving models, horizontal grid lengths tend to be on the order of kilometres to a few hundred metres at best, so it is impossible for such models to explicitly simulate microphysical processes. There is a long history of microphysical parameterisation (see Khain et al., 2015, for a comprehensive review), and microphysics schemes today tend to fall into one of two categories: bin models, in which the size distribution of each hydrometeor class is explicitly calculated (e.g. Feingold et al., 1994; Stevens et al., 1996; Khain et al., 2004), and bulk models, in which a size distribution function is typically used to represent each hydrometeor class and one (or several) moments of the size distribution function are calculated explicitly (e.g. Kessler, 1969; Lin et al., 1983; Rutledge and Hobbs, 1983; Thompson et al., 2004; Morrison et al., 2005; Thompson et al., 2008, amongst many others). Bulk models are therefore very computationally efficient compared to bin models (often by at least 2 orders of magnitude; Jiang et al., 2000) and are used as standard in many atmospheric modelling systems today. Although certain aspects of cloud processes and aerosol indirect effects cannot be reproduced well in bulk schemes (see Khain et al., 2015, for a detailed analysis), there nevertheless remains a trade-off between how completely the hydrometeor size spectra are represented and the physical domain size that can then be used in a simulation. For most applications, full bin microphysics (which can even resolve the autoconversion process of cloud water to rain) are only feasible using small domains and idealised simulations, which then cannot represent the dynamical feedbacks that can occur on larger domains (a notable exception, proving the cost of such simulations, are the multiple month-long case study simulations using bin microphysics presented by Fan et al., 2013). Thus, studies using bulk and bin microphysics representations provide differently imperfect and thus complementary information. Indeed, bulk schemes remain as standard in global models, and successful studies of aerosol indirect effects in global models have been performed using bulk microphysics (e.g. Zhang et al., 2016; Ghan et al., 2016).

Whilst early bulk microphysics schemes were single moment only (predicting only the $k = 1$ moment of the particle size distribution equation, mass), a significant development has been predicting two moments of the size distribution equation ($k = 0$, number concentration, and $k = 1$, mass) (e.g. Meyers et al., 1997; Thompson et al., 2004, 2008; Morrison et al., 2005), which has been shown to have improved results compared to single-moment schemes (e.g. Lynn and Khain, 2007; Morrison and Grabowski, 2007; Morrison et al., 2009; Kumjian and Ryzhkov, 2012; Saleeby and van den Heever, 2013). Indeed, although not widely used at present, three-moment schemes have been shown to further improve representations of large hail (Milbrandt and Yau, 2006; Loftus and Cotton, 2014) and precipitation reflectivities (Kumjian and Ryzhkov, 2012).

However, bulk schemes make a priori assumptions about the shape of the particle size distributions (usually approximated by exponential or gamma distributions and more rarely by lognormal functions), whereas bin schemes calculate particle size distributions by solving explicit microphysical equations and make no a priori assumption about the particle size distribution shapes. This can lead to significant differences in the cloud and precipitation simulated by bin vs. bulk schemes. For example, bulk schemes have been shown to underestimate areas of weak and stratiform rain in an MCS compared to a bin scheme which performed better against observations (Lynn et al., 2005a, b). Li et al. (2009a, b) showed that a one-moment bulk scheme was shown to be worse at partitioning rain into stratiform and convective components in a continental squall line compared to a bin scheme (although many studies have shown that two-moment schemes are a significant improvement on single-moment schemes; e.g. Lynn and Khain, 2007; Morrison and Grabowski, 2007; Morrison et al., 2009; Kumjian and Ryzhkov, 2012; Saleeby and van den Heever, 2013). Lynn and Khain (2007) found that, while all schemes overestimated maximum rain rates in a simulated MCS, all bulk schemes tested overpredicted average and maximum rain rates by a factor of 2 to 3, while bin schemes overestimated maximum rain rates by about 20 %. In idealised supercell simulations, Khain and Lynn (2009) found that the Thompson et al. (2004) double-moment bulk scheme produced 2 times more accumulated surface rain than a bin scheme, while Lebo et al. (2012) found that the Morrison et al. (2005) bulk scheme also produced twice as much surface rain as the same bin scheme used by Khain and Lynn (2009) in simulations of the same supercell. Investigations of the shape of the cloud droplet size distribution in large-eddy simulations of non-precipitating shallow cumulus clouds with a bin (Igel and van den Heever, 2017a) and bulk (Igel and van den Heever, 2017b) scheme showed the importance of the cloud droplet size distribution shape parameter. In the bulk scheme, evaporation rates were much more sensitive to the value of the shape parameter than to the condensation rates, and thus the shape parameter strongly impacted cloud properties such as droplet number concentration, mean droplet diameter and cloud fraction (Igel and van den Heever, 2017b). Bin scheme simulations suggested that the shape parameter should be based on the relationship between local values of the cloud droplet concentration and the relative width of the cloud droplet size distribution rather than cloud mean values, as are traditionally used (Igel and van den Heever, 2017a). Further, Igel and van den Heever (2017c) showed that despite other fundamental differences between the bin and bulk condensation parameterisations, differences in condensation rates could be predominantly explained by accounting for the width of the cloud droplet size distributions simulated by the bin scheme.

Seifert and Beheng (2006a) found that the most important factor in achieving agreement in concentrations and mass contents between bulk and bin schemes in simulations of continental and tropical maritime clouds was accurate representation of warm-phase autoconversion. Sensitivity tests of four different autoconversion parameterisations conducted by Fan et al. (2012a, b) and Wang et al. (2013) showed that errors in predicting cloud water content in bulk schemes could be attributed to the saturation adjustment used in the calculation of evaporation and condensation. Likewise, Saleeby and van den Heever (2013) also showed, using four different types of autoconversion scheme, that saturation adjustment was the leading order factor in discrepancies of prediction of cloud water content by bulk schemes. Khain et al. (2016) found that tropical cyclones showed weak sensitivity to aerosol due to the use of saturation adjustment. In the ice phase, Li et al. (2009a, b) found artificial spikes in heating rates from deposition and sublimation due to the saturation adjustment scheme. Bryan and Morrison (2012) found that even at very high resolution, convective cores in an idealised squall line simulation remained undiluted due to the saturation adjustment used in the bulk microphysics scheme. However, saturation adjustment alone is insufficient to explain all differences between bin and bulk schemes: in idealised supercell simulations using bulk microphysics both with saturation adjustment and without (in which the scheme was modified to include an explicit representation of supersaturation predicted over each time step), Lebo et al. (2012) found that the use of saturation adjustment was able to explain differences between a bulk and bin scheme in the response of cold pool evolution and convective dynamics under polluted conditions, but was not sufficient to explain the large differences in the response of surface precipitation to aerosol loading.

Differences between bin and bulk schemes can often be traced to their different process representations. For example, some studies have found rain evaporation in bulk schemes to be too fast compared to bin schemes (Fan et al., 2012a, b; Wang et al., 2013; Li et al., 2009b; Shipway and Hill, 2012). Bulk schemes have been found to have higher condensation and evaporation rates but similar rates of freezing and melting compared to bin schemes (Li et al., 2009a, b). Shipway and Hill (2012) compared rates of diffusional growth, collisions, sedimentation and surface precipitation in several bulk schemes against results from the Tel Aviv University bin scheme (Tzivion et al., 1987) and found that precipitation peaks in the bulk schemes were too sharp and too narrow compared to the bin scheme, whereas the bin scheme produced weaker precipitation covering an overall larger area than that in the bulk schemes. Morrison and Grabowski (2007) tested three different parameterisations of the coalescence process in the Morrison et al. (2005) bulk scheme against a bin scheme, under different aerosol loadings and in both warm stratocumulus and warm cumulus clouds, and found that for both the bulk and bin scheme, each representation of the coalescence process led to different averaged rain contents and mean raindrop diameters. Fan et al. (2013) showed that, because bulk schemes do not represent size-resolved ice particle fall speeds, they were unable compared to bin schemes to simulate the reduced fall velocities of ice and snow at upper levels from clean to polluted conditions in tropical, mid-latitude coastal and mid-latitude summertime inland continental deep convective clouds. Fan et al. (2013) also suggested that bulk schemes tended to artificially freeze large raindrops due to the use of a fixed gamma distribution.

In some cases, tuning particular processes in bulk schemes has led to better agreement with bin schemes, e.g. tuning evaporation rates and fall velocities of graupel in single-moment bulk scheme simulations of a continental squall line (Li et al., 2009a, b). Similarly, although no active tuning was performed, Seifert and Beheng (2006a) found that precipitation rates and accumulated precipitation values were in close agreement between simulations of continental and tropical maritime clouds in high and low CCN conditions using a bin and bulk scheme, with agreement between the bulk and bin scheme even greater in the high CCN case compared to the low CCN case.

Not only do bin and bulk schemes often produce different results in terms of cloud and precipitation, but Fan et al. (2012a) found that the use of fixed CCN in a bulk scheme led to opposite CCN effects on convection and heavy rain compared to CCN effects when using a bin scheme. Similarly, Lebo and Seinfeld (2011) found an opposite response of accumulated surface rain to CCN in idealised supercell simulations using a bulk and bin scheme. Khain and Lynn (2009) found a difference in the response of an idealised supercell to aerosol perturbations when a bin and bulk scheme was used, with the bulk scheme producing stronger updraughts and greater average precipitation than the bin scheme and with the left-moving storm prevailing in the bulk simulation, while the right-moving storm prevailed in the bin simulation. The differences were attributed to differences in the vertical velocities in the bin vs. bulk schemes, which led to hydrometeors ascending to different altitudes with different directions of background flow.

Nevertheless, bulk schemes have shown sensitivity to aerosol. In simulations of tropical deep convection, Morrison and Grabowski (2011) found an ice-phase response to aerosol in which cloud top heights and anvil ice mixing ratios increase under polluted conditions due to increased freezing of larger numbers of cloud droplets and subsequent higher ice particle concentrations with smaller sizes and reduced fall speeds. Indeed, a similar mechanism was later confirmed in bin scheme simulations by Fan et al. (2013), who performed month-long simulations of deep convection over the tropical western Pacific, southeastern China and the US southern Great Plains. Further, Kalina et al. (2014) found that auto-conversion of cloud water to rain decreased under polluted conditions, and subsequently near-surface rain and hail particles increased in size due to enhanced collection of cloud droplets. In simulations of deep convection over Florida using a bin-emulating bulk scheme, van den Heever et al. (2006) found that updraught strengths increased and anvil areas became smaller but better organised and with increased condensate mixing ratios. Similarly, in simulations of summertime convection over Germany using a two-moment bulk scheme, Seifert et al. (2012) found a strong aerosol effect on cloud properties such as condensate amounts and glaciation.

Unlike liquid cloud and rain drops (well described by spheres of constant density), ice particles have a wide range of densities and shapes, making the representation of ice-phase microphysics in parameterisations much more difficult than the liquid phase. Traditionally, the approach in both bin (e.g. Khain et al., 2004) and bulk schemes (e.g. Meyers et al., 1997; Thompson et al., 2004; Morrison et al., 2005, etc.) was to partition ice particles into one of a fixed number of categories (e.g. cloud ice, snow, hail and graupel) each with its own specified density, shape distribution and physical parameters such as fall speeds. However, such partitioning oversimplifies the complex nature of ice-phase processes, requiring thresholds and parameters – often chosen on a relatively ad hoc basis – to determine the partitioning of ice particles into each category and for converting between categories. As such, it is unsurprising that simulations have been found to be highly sensitive to particle fall speeds and densities (e.g. McFarquhar et al., 2006), the description of dense precipitating ice as hail or graupel categories (e.g. Morrison and Milbrandt, 2011; Bryan and Morrison, 2012) and changes in thresholds or rates for converting between ice categories (e.g. Morrison and Grabowski, 2008). Differences in ice-phase microphysics in bulk schemes have been shown to affect cloud biases, especially at upper levels (Cintineo et al., 2014), and to affect ice–cloud–radiation feedbacks with impacts on tropospheric stability, triggering of deep convection and surface precipitation (Hong et al., 2009). Such limitations have led to the development in more recent years of new representations of ice microphysics in bulk schemes, such as approaches which separately prognose ice mass mixing ratios grown by riming and vapour deposition (Morrison and Grabowski, 2008), approaches where particle habit

evolution is predicted by prognosing the mixing ratios of ice crystal axes (Harrington et al., 2013) and approaches where ice-phase particles are represented by several physical properties that evolve freely in time and space (Morrison and Milbrandt, 2015). Although these developments are relatively new, they have already been shown to improve simulations of observed squall lines and orographic precipitation when compared to traditional two-moment bulk schemes (Morrison et al., 2015a).

Evaluations of microphysics schemes frequently involve comparison against observations of a real precipitation event (e.g. Morrison and Pinto, 2005). Often, multiple microphysics schemes are compared against each other and against observations (e.g. Morrison and Pinto, 2006; Gallus Jr. and Pfeifer, 2008; Rajeevan et al., 2010; Jankov et al., 2011). Another common approach is to evaluate a single microphysics scheme against observations and then use different aerosol concentrations in the model to test the sensitivity of the observed storm to aerosol processes (e.g. van den Heever et al., 2006; Seifert et al., 2012). However, studies of different convective events in different regions using different models with different microphysics schemes often produce conflicting results on the nature of the storm response to aerosol. Mesoscale studies of Florida convection found that cloud water mass, updraught strength and surface precipitation tend to increase with increased aerosol concentration, while anvil areas decreased but contained greater condensate mass (van den Heever et al., 2006). Studies of summertime convective precipitation in Germany found that increased aerosol concentrations had a strong effect on cloud microphysical (and therefore radiative) properties but that the combined effects of microphysical and dynamical processes resulted in relatively little effect on surface precipitation (Seifert et al., 2012). This is similar to the findings of Thompson and Eidhammer (2014) in idealised and continental-scale simulations.

Detailed process modelling studies of aerosol–convection interactions often focus on the sensitivity of a single idealised model configuration (without large-scale meteorology or surface and radiative interactions) to perturbations using either CCN spectra (e.g. Seifert and Beheng, 2006b; Morrison and Grabowski, 2011) or CDNC values (e.g. Thompson et al., 2004; Morrison, 2012) as a proxy variable to test the sensitivity of the microphysics to aerosol. Many types of idealised models are used, ranging from flow over a 2-D mountain (e.g. Thompson et al., 2004), to 2-D cloud-system-resolving studies of interacting convective clouds (e.g. Morrison and Grabowski, 2011) to 3-D simulations of idealised supercell storms (e.g. Khain and Lynn, 2009; Lebo and Seinfeld, 2011; Morrison, 2012; Lebo et al., 2012). With such a wide range of model configurations, convective and large-scale environments, microphysics parameterisations (bin and bulk models are both frequently used in idealised studies of aerosol–convection interactions) and proxy variables used to represent aerosol processes, it is perhaps not surprising that a consistent response of idealised convection to aerosol has not been seen; indeed, due to environment and regime dependence, it may not exist. Idealised flow over a 2-D mountain using CDNC values to represent aerosol amounts showed that cloud water content increased with CDNC and drizzle content decreased (Thompson et al., 2004), while a similar study using an idealised supercell configuration found that differences in the accumulated surface precipitation and convective mass flux between polluted and pristine values of CDNC were very small (Morrison, 2012). In studies using modified CCN spectra to represent different levels of aerosol in a two-moment scheme, 2-D ensemble simulations of interacting convective clouds have found that although cloud top heights and anvil ice increase under polluted conditions, convection actually weakens slightly compared to pristine conditions (Morrison and Grabowski, 2011). However, similar 3-D simulations also using a two-moment microphysics scheme have shown that for isolated convective cells, increased aerosol leads to reduced total precipitation and updraught velocity; for multicell systems it leads to increased secondary convection, total precipitation and updraught velocities, whilst supercell systems are relatively insensitive to aerosol (Seifert and Beheng, 2006b). Additionally, environmental wind shear has been shown to have a role in determining the response of convective systems to aerosol, with increased aerosol loading invigorating convection under weak shear conditions and suppressing convection under strong shear in simulations performed with both bin (Fan et al., 2009) and bulk (Lebo and Morrison, 2014) microphysics schemes.

The focus of this work is to show within a single modelling framework that uncertainty in cloud impacts through the choice of microphysics scheme can far exceed any aerosol effect seen within a single scheme and that this is a consistent finding across different types of convection in different environments and types of simulation (all of which are known to impact the effect of aerosol loading on cloud development, e.g. Altaratz et al., 2014). Although we use two bulk microphysics schemes to show this, there is a body of literature which identifies signals of aerosol impact on cloud in bulk schemes (e.g. Morrison and Grabowski, 2011; Morrison, 2012; Lebo et al., 2012; Kalina et al., 2014), albeit not always convective invigoration (see especially Lebo et al., 2012), and in bin-emulating bulk schemes (e.g. van den Heever et al., 2006; Lee and Feingold, 2010, 2013). Nevertheless, using a two-moment bulk scheme to simulate a single cumulonimbus in an environment characterised by high CAPE and low wind shear, Seifert and Beheng (2006a) found higher overshooting tops and larger sizes with increased aerosol loading, indicating that in some environments bulk schemes are able to produce invigoration effects. In some cases, aerosol effects may be relatively small (less than 15 %; e.g. Morrison and Grabowski, 2011). However, while some argue (fairly) that this is at least in part due to the limitations of bulk schemes to fully represent aerosol–cloud inter-

actions (such as saturation adjustment; Lebo et al., 2012), others argue that this is consistent with the concept of clouds as a buffered system hypothesised by Stevens and Feingold (2009). Month-long simulations approaching the climatological scale using bin microphysics performed by Fan et al. (2013) also showed aerosol impacts on precipitation on the order of a few percent. However, those authors showed a significant aerosol impact on rain rates rather than total rain amount, observing a shift towards heavier rain rates and fewer light rain rates under polluted conditions in two regions (a tropical environment and mid-latitude coastal environment), although the response in a mid-latitude inland summertime continental environment varied temporally over the simulation. Similarly to the environmental dependence found by Fan et al. (2013), Kalina et al. (2014) showed that even in an idealised simulation of a supercell using open boundaries and bulk microphysics, the relative humidity and shear used in the initial profile had an impact on the aerosol effects observed in the simulation.

We perform high-resolution convection-permitting simulations with the Weather Research and Forecast (WRF) model in three configurations: a real-data simulation of deep convection in the Congo basin, an idealised supercell case and a shallow convection large-eddy simulation (LES). In each case we compare hydrometeor development in two commonly used double-moment bulk schemes and investigate the response of each model configuration to CDNC perturbations. Our focus is not to provide a detailed process study of aerosol effects on convection per se (to do so in the context of multiple model configurations is beyond the scope of this paper), but rather to explore and identify uncertainty in the cloud and precipitation response to CDNC perturbations across a range of model configurations. We acknowledge that, due to a lack of fully coupled aerosol–cloud processes (e.g. supersaturation representation, droplet activation, wet deposition and buffering processes; Lebo et al., 2012; Stevens and Feingold, 2009; Lee and Feingold, 2010; Seifert et al., 2012), the magnitude of the response of bulk microphysics schemes to CDNC perturbations may differ from that in schemes that explicitly treat the cloud processing of aerosol. Our goal is therefore to highlight the large uncertainty in cloud and precipitation responses to perturbations of CDNC in convection-permitting models, even between multiple configurations of the same widely used model.

2 Experimental design

We use the Advanced Research WRF version 3 (Skamarock et al., 2008) in three different configurations: a real-data simulation of deep convection over the Congo basin, an idealised supercell simulation and a warm-rain shallow cumulus LES simulation. WRF is a nonhydrostatic, compressible, 3-D atmospheric model. We use version 3.5 of WRF in the Congo basin and the idealised supercell simulations, but ver-

Table 1. List of model configurations.

Model settings	Congo	Supercell	RICO LES
Horizontal grid length (km)	4	4	0.1
Number of grid points (W–E and S–N)	525	400	129
Number of vertical levels	30	30	100
Model top	5000 Pa	20 km	4 km
Time step (s)	12	12	1
Simulation length	10 days	2 h	24 h
LW radiation scheme	RRTM	–	–
SW radiation scheme	Goddard	–	–
PBL scheme	YSU	–	–

Table 2. List of microphysics configurations tested and the abbreviations used for each run.

Prescribed CDNC	Congo MORR	Congo THOM	Supercell MORR	Supercell THOM	RICO MORR	RICO THOM
$100\,cm^{-3}$	CONGO-M100	CONGO-T100	SUPER-M100	SUPER-T100	RICO-M100	RICO-T100
$250\,cm^{-3}$	CONGO-M250	CONGO-T250	SUPER-M250	SUPER-T250	RICO-M250	RICO-T250
$2500\,cm^{-3}$	CONGO-M2500	CONGO-T2500	SUPER-M2500	SUPER-T2500	RICO-M2500	RICO-T2500

sion 3.3.1 was utilised for the warm-rain LES simulation because the LES packages were only available for this version of the model at this time (Yamaguchi and Feingold, 2012). In order to keep the simulations as consistent with each other as possible, we therefore implement the versions of the microphysics schemes from WRF version 3.5 into version 3.3.1 of the model for the LES simulations. Each set of simulations is performed using two microphysics parameterisations at three different prescribed CDNC values, resulting in a total of six simulations for each model configuration. The model configurations used in this study are summarised in Table 1.

2.1 Microphysics parameterisations

This study is presented as an indication of the uncertainty that can arise from the choice of microphysics scheme alone, and thus we restrict our comparison to two double-moment bulk microphysics schemes rather than diversifying into a comparison of bin schemes against bulk schemes. The literature surveyed in Sect. 1 indicates the wide range of differences that may be expected when comparing bulk against bin schemes. A significant body of work has shown that two-moment bulk microphysics schemes generally represent cloud and precipitation characteristics more realistically than single-moment schemes (most recently Morrison et al., 2009; Wu and Petty, 2010; Weverberg et al., 2013, 2014; Igel et al., 2015), and thus our study is restricted to the comparison of two five-class, double-moment schemes commonly used in WRF and shown by Cintineo et al. (2014) to perform well against satellite observations of cloud in North America: that described by Morrison et al. (2005, 2009) and Morrison and Milbrandt (2011) (hereafter Morrison, or abbreviated to MORR), and that described by Thompson et al. (2004, 2008)

(hereafter Thompson, or THOM). Both schemes are two-moment in rain and ice (prognostic mass and number), while the Morrison scheme is also two-moment in snow and graupel. Both are single-moment in cloud water: mass is the only prognostic liquid cloud variable, and CDNC is prescribed at a given value. Following the method used in many previous studies including that of Morrison (2012), we prescribe CDNC values (in this study, at 100, 250 and $2500\,cm^{-3}$) as a proxy for CCN varying under conditions ranging from clean to highly polluted. The list of microphysics configurations tested and the abbreviations used to describe them are summarised in Table 2.

2.2 Model configurations

The real-data Congo simulations use a model domain covering a $2100 \times 2100\,km$ region over the Congo basin (Fig. 1) chosen due to the high frequency of isolated deep convective systems occurring in the region and also due to the presence of strong sources of biomass burning aerosol. The model initial and boundary conditions were generated from ERA-Interim reanalysis (Dee et al., 2011) starting at 00:00 UTC on 1 August 2007. The simulation start date was chosen to coincide with the onset of the seasonal peak in precipitation (Washington et al., 2013) and the simulation was integrated for 10 days (with a time step of 12 s) in order to identify the nature of the convection and its response to CDNC perturbations over timescales greater than that of the life cycle of any individual convective system. We use a horizontal grid length of 4 km and 30 vertical levels with the standard WRF stretched vertical grid. This gives a vertical grid spacing of about 100 m in the lower levels with grid spacing increasing towards the upper levels. Although 30 verti-

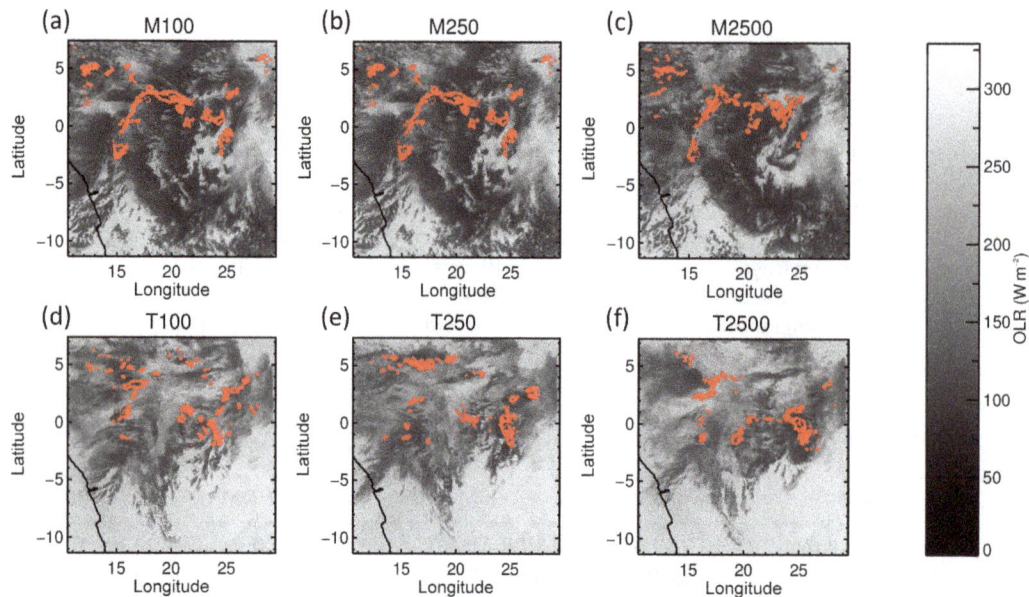

Figure 1. Congo case: instantaneous outgoing longwave radiation (W m^2, greyscale) and 5 mm h^{-1} surface precipitation rate (red contour) at 07:00 UTC on 7 August 2007 in the Congo basin configuration. **(a–c)** CONGO-MORR simulations and **(d–f)** CONGO-THOM simulations. Prescribed CDNC values of 100, 250 and 2500 cm^{-3} are shown in panels **(a, d)**, **(b, e)** and **(c, f)**, respectively.

cal levels may seem relatively coarse, it has been shown in a previous study to be sufficient to reproduce observed cloud morphology and resolve the vertical structure of aerosol and precipitation and their interactions in this region (Gryspeerdt et al., 2015). Longwave and shortwave radiation in the simulations are parameterised by the RRTM (Mlawer et al., 1997) and Goddard (Chou and Suarez, 1994) schemes, respectively. Other physics parameterisations (other than the microphysics schemes previously discussed) are the MM5 Monin–Obukhov similarity surface layer scheme available in WRF (which uses stability functions and surface fluxes from Dyer and Hicks, 1970; Paulson, 1970; Webb, 1970; Beljaars, 1994), the NOAH land surface model (Ek and Mahrt, 1991) and the YSU boundary layer scheme (Hong et al., 2006), also shown by Cintineo et al. (2014) to perform well.

The idealised supercell set-up follows the standard 3-D idealised supercell case available as part of the WRF modelling system. Boundary conditions are open on all lateral boundaries, and the model top and surface are free-slip. For consistency with the Congo basin simulations, we use a horizontal grid length of 4 km. The model domain is 1600 × 1600 km in the horizontal and, for consistency with the Congo simulations, also uses 30 vertical levels with a model lid at 20 km. A Rayleigh damper with a damping coefficient of 0.003 s^{-1} is applied in the top 5 km of the model to prevent spurious wave reflection off the model top. Following the set-up commonly used in idealised supercell studies (e.g. Morrison, 2012), surface energy fluxes, surface drag, Coriolis acceleration and radiative transfer are neglected for simplicity, and the subgrid-scale horizontal and vertical mix-

ing is calculated with a prognostic turbulent kinetic energy scheme (Skamarock et al., 2008). The model is initialised as in the idealised quarter-circle supercell test case available in WRF using the analytic sounding of Weisman and Klemp (1982, 1984) and the quarter-circle supercell hodograph of Weisman and Rotunno (2000) with the shear extended to a height of 7 km. Convection is triggered using a thermal perturbation in the centre of the domain with a maximum perturbation potential temperature of 3 K centred at a height of 1.5 km and with horizontal and vertical radii of 10 and 1.5 km, respectively. All simulations are integrated for 2 h with a time step of 12 s (the same time step used in the Congo simulations).

The warm-rain shallow cumulus set-up deviates from the other simulations in that it follows the LES intercomparison guidelines for the Precipitating Shallow Cumulus Case 1 (van Zanten et al., 2011) of the Rain in Shallow Cumulus Over the Ocean (RICO; Rauber et al., 2007) project and uses the RICO WRF LES package provided by Yamaguchi and Feingold (2012). The model domain is 12.8 × 12.8 × 4 km with a horizontal grid spacing of 100 m and uses 100 vertical levels, implying a vertical grid spacing of about 40 m. The lateral boundary conditions are doubly periodic. As in the idealised supercell simulations, surface energy fluxes, surface drag, Coriolis acceleration and radiative transfer are neglected for simplicity, and the subgrid-scale horizontal and vertical mixing is calculated with a prognostic TKE scheme. The surface conditions, wind and thermodynamic profiles, large-scale forcings and large-scale radiation, geostrophic wind, initial perturbations and translation velocity are prescribed following the RICO case guidelines (van Zanten et al., 2011). For consistency, we prescribe cloud droplet number concentrations at 100, 250 and 2500 cm^{-3}, following the other simulations in our study, instead of the 70 cm^{-3} suggested for the standard RICO case. However, we also perform an extra simulation at 50 cm^{-3}. The simulations are integrated for 24 h with a time step of 1 s.

3　Results

3.1　WRF Congo basin

Maps of simulated outgoing longwave radiation (OLR) and surface precipitation at 07:00 UTC on 7 August 2007 (7 days into the simulation) indicate that the cloud morphological and precipitation differences for different microphysics schemes are much greater than the cloud and precipitation response within each scheme to different CDNC values (Fig. 1). In the CONGO-MORR simulations, low OLR values (indicating cold, high cloud) are distributed across the domain. Precipitation at this time occurs only in cloud north of 3° S, but there is a large band of non-precipitating cold cloud across the south of the domain. There is little discernable response of the morphology of the OLR and precipitation in the CONGO-MORR simulations to different CDNC values (Fig. 1a–c). In comparison, cold cloud in the CONGO-THOM simulations occurs mostly north of 3° S (Fig. 1d–f). Less cloud forms in CONGO-THOM compared to CONGO-MORR, and the cloud generally has greater OLR values than that in CONGO-MORR. Some non-precipitating cloud occurs south of 3° S in the CONGO-THOM simulations, but the band is significantly weaker and warmer than in CONGO-MORR. The differences at this snapshot are representative of differences that persist throughout the simulation. Frequency distributions of OLR over the entire 10-day simulation period show that CONGO-MORR has a much higher frequency of occurrence of colder, higher cloud (values of about 120 W m^2) than CONGO-THOM (which increases in frequency slightly with increased CDNC), while CONGO-THOM has a much higher frequency of occurrence of warmer cloud (values of about 270 W m^2) than CONGO-MORR (Fig. 2a). When compared to observations of OLR from the Geostationary Earth Radiation Budget (GERB; Harries et al., 2005) over the same region and period, CONGO-THOM represents warm cloud more consistently with GERB than CONGO-MORR, despite overpredicting colder cloud somewhat, while CONGO-MORR overpredicts higher cloud and underpredicts warm cloud compared to the observations (Fig. 2a). However, despite a poorer prediction of cloud radiative properties, CONGO-MORR predicts surface precipitation better than CONGO-THOM when compared to observations from the Tropical Rainfall Measuring Mission (TRMM; Huffman et al., 2007) merged

product. Both schemes significantly overpredict surface precipitation compared to observations from the TRMM 3B42 product (although the spatial patterns of precipitation are reasonably similar); however, total accumulated surface precipitation over the 10-day simulation period is much greater in CONGO-THOM than CONGO-MORR (Fig. 3). Further differences are seen when the distributions of precipitation rates are compared, with CONGO-THOM overpredicting and CONGO-MORR underpredicting the occurrence of low precipitation rates compared to TRMM, CONGO-MORR overpredicting and CONGO-THOM underpredicting moderate rates, and CONGO-THOM overpredicting the frequency of occurrence of very high precipitation rates (Fig. 2b). That CONGO-MORR overpredicts the frequency of moderate rain rates and CONGO-THOM overpredicts the frequency of very high rain rates likely explains why both schemes overpredict total accumulated surface rain compared to the observations. Additionally, the overprediction of the frequency of very high precipitation rates by CONGO-THOM is likely the reason that the total accumulated surface precipitation is much greater in this scheme than in CONGO-MORR (Fig. 3a and b).

Further to the significant difference between the two schemes in their reproduction of cold cloud and precipitation rates, the updraught dynamics respond very differently to aerosol loading. Joint histograms of cloud top height in the convective updraughts and the radius of the updraughts show that the most significant dynamical difference between the simulations comes from the choice of microphysics scheme: the Morrison scheme has a tendency towards higher frequencies of wider updraught radii with higher cloud tops than the Thompson scheme (Fig. S1 in the Supplement). Under increased values of CDNC, convection in the CONGO-MORR simulation shifts towards wider cores and higher core tops for midsized cores (radius 11 to 22 km), whilst there is a reduction in the frequency of smaller cores of all core top heights (Fig. 2c). Conversely, convection under polluted conditions in the CONGO-THOM simulation shows a reduced frequency of occurrence of the highest updraught cloud tops for all updraught radii under polluted conditions with an increased frequency of occurrence of small updraught radii with lower cloud tops (Fig. 2d). Therefore, a consistent aerosol response is observed in CONGO-THOM, resulting in smaller and lower convective updraughts (i.e. weakened convection under polluted conditions). Interestingly, both of these effects contradict the findings of Morrison and Grabowski (2011), who found an ice-phase response to aerosol in which cloud top heights and anvil ice mixing ratios increase under polluted conditions due to increased freezing of larger numbers of cloud droplets and subsequent higher ice particle concentrations with smaller sizes and reduced fall speeds. However, we note that we consider different values of CDNC/CCN to Morrison and Grabowski (2011) and that responses may be nonmonotonic (Kalina et al., 2014). We also consider a different case of convec-

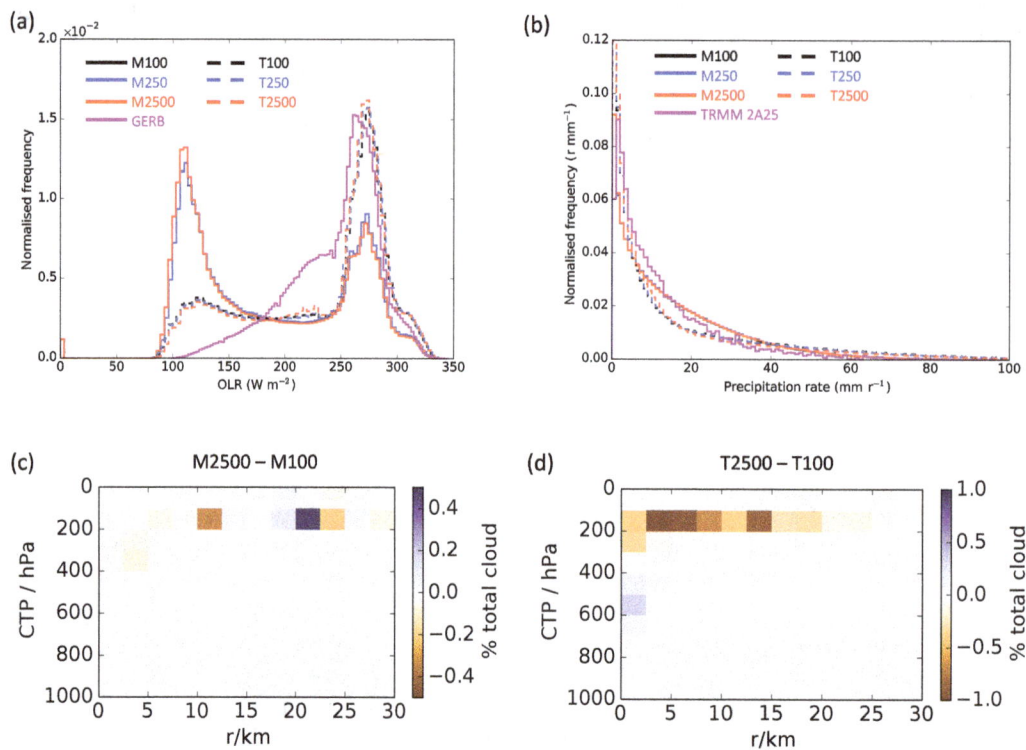

Figure 2. Congo case: **(a)** frequency distributions of OLR from the WRF simulations and observations from GERB over the period 1 to 10 August 2007; **(b)** self-weighted precipitation rate distributions from the WRF simulations and observations from the ungridded TRMM 2A25 product, which has a similar spatial resolution to the 4 km model grid length; **(c)** difference in the joint distribution of cloud top pressure in updraughts (identified by masking points where the maximum vertical velocity exceeds 1 ms^{-1} and then applying a connected-components labelling algorithm to identify unique updraught areas) and horizontal radius of updraughts when CDNC is increased from 100 to 1000 cm^{-3} using the Morrison microphysics scheme; **(d)** difference in the joint distribution of cloud top pressure in updraughts and horizontal radius of updraughts when CDNC is increased from 100 to 1000 cm^{-3} using the Thompson microphysics scheme.

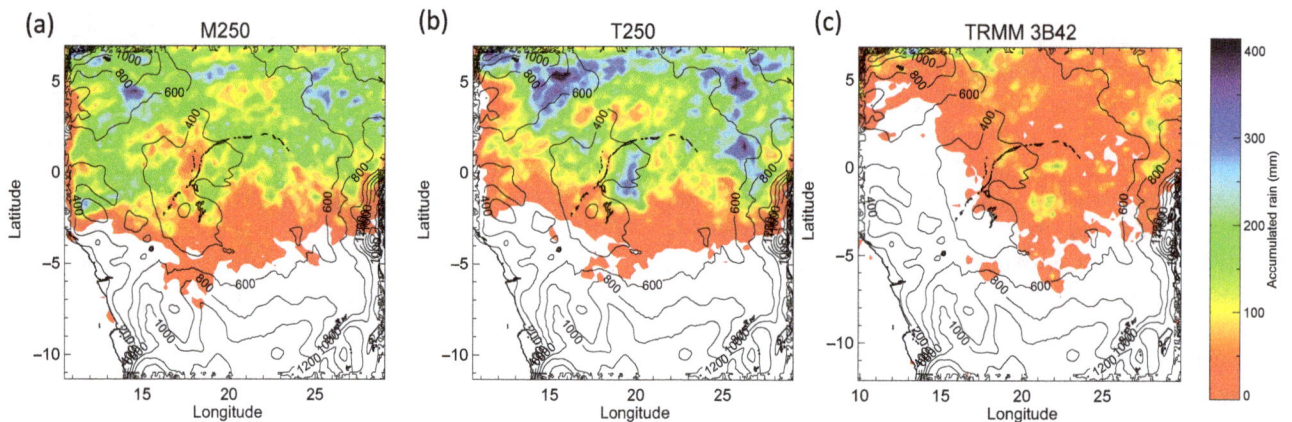

Figure 3. Congo case: accumulated surface precipitation (mm) from 1 to 10 August 2007 in the Congo basin, showing data from **(a)** CONGO-M250, **(b)** CONGO-T250 and **(c)** observations from the TRMM 3B42 gridded 3-hourly mean merged precipitation product. The simulation data shown in this figure have been coarsened to the 0.25° spatial resolution of the TRMM product.

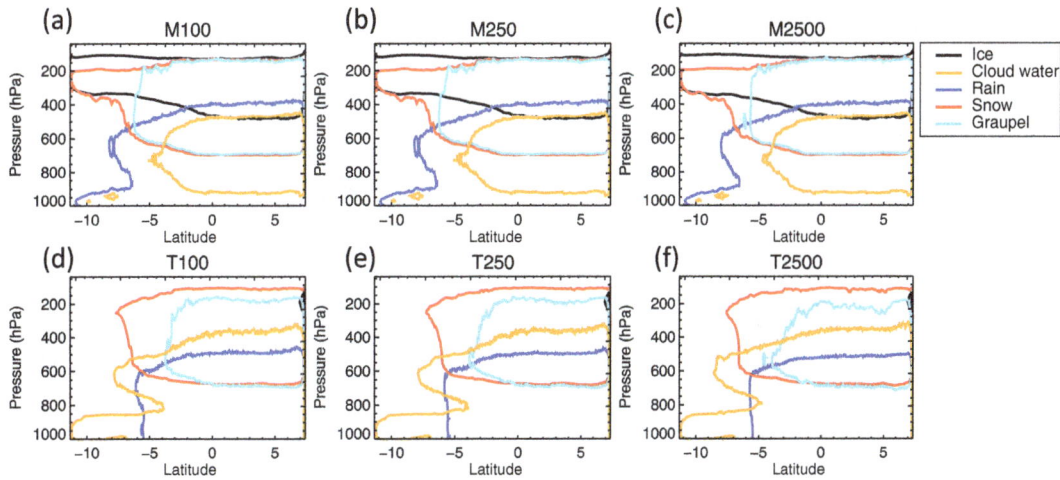

Figure 4. Congo case: zonal mean vertical sections of hydrometeor classes (colour contours) from 1 to 10 August 2007. Hydrometeor mass mixing ratios are contoured at 10^{-6} kg kg^{-1}.

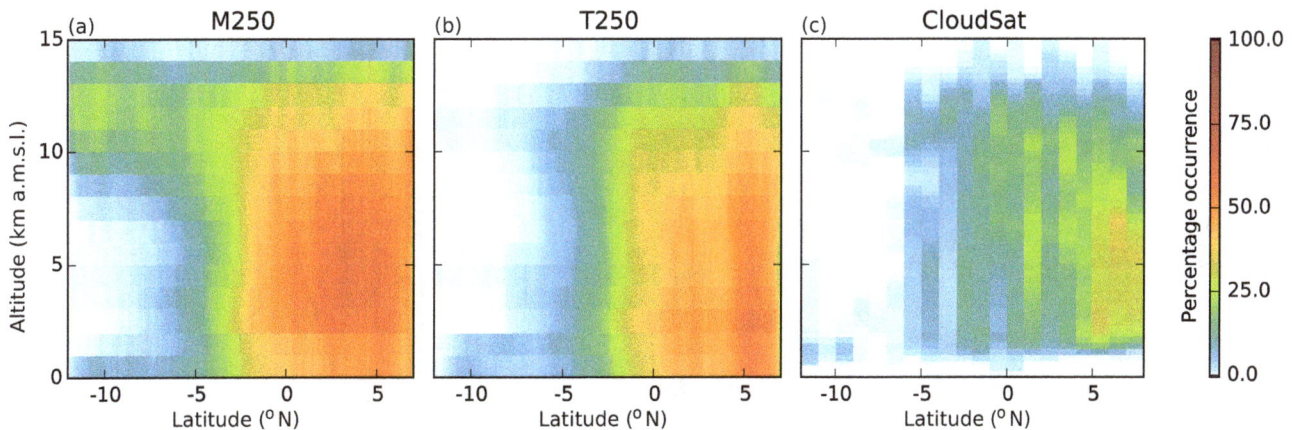

Figure 5. Congo case: 10-day histogram for the period 1–10 August 2007 of model reflectivities derived from hydrometeor fields passed through the QuickBeam radar simulator (Haynes et al., 2007); thresholded at values greater than −20 dBZ for **(a)** CONGO-M250, **(b)** CONGO-T250 and **(c)** the CloudSat 2B-GEOPROF product. In panels **(a)** and **(b)** the models have been sampled at the times of the nearest CloudSat overpasses.

tion (indeed, our 10-day Congo simulation covers many convective lifecycles). We note that the response of the convective updraughts to aerosol loading in these two bulk schemes cannot be attributed to saturation adjustment alone (the suggested effects of which on updraught invigoration are detailed in Khain et al., 2015) because both schemes use this method.

Not only does the simulated cloud and precipitation morphology differ significantly between microphysics schemes irrespective of the CDNC values used in the comparison, zonal-mean vertical sections of the mass mixing ratios of the different hydrometeor classes show significant differences in the hydrometeor classes (due to microphysics) between CONGO-MORR and CONGO-THOM (Fig. 4). The most significant difference between the two microphysics schemes is that south of 3° S, CONGO-MORR produces

a large amount of high ice cloud between 300 and 150 hPa (Fig. 4a–c). Analysis of these vertical sections at hourly intervals throughout the simulation in conjunction with hourly maps of OLR, as in Fig. 1, show that this upper-level ice is transported from the convective anvils in the north of the domain to the non-convective region in the south of the domain (not shown). In comparison, CONGO-THOM produces significantly less ice with almost no ice visible at this contour value (Fig. 4d–f). However, all three CONGO-THOM simulations form a large amount of non-precipitating low-level (950 to 850 hPa) liquid cloud south of 3° S. The bands of cloud seen south of 3° S in Fig. 1 are therefore high ice cloud in the CONGO-MORR simulations and low liquid cloud in the CONGO-THOM simulations, illustrating not only a cloud morphological difference between the microphysics schemes but also a significant difference in the

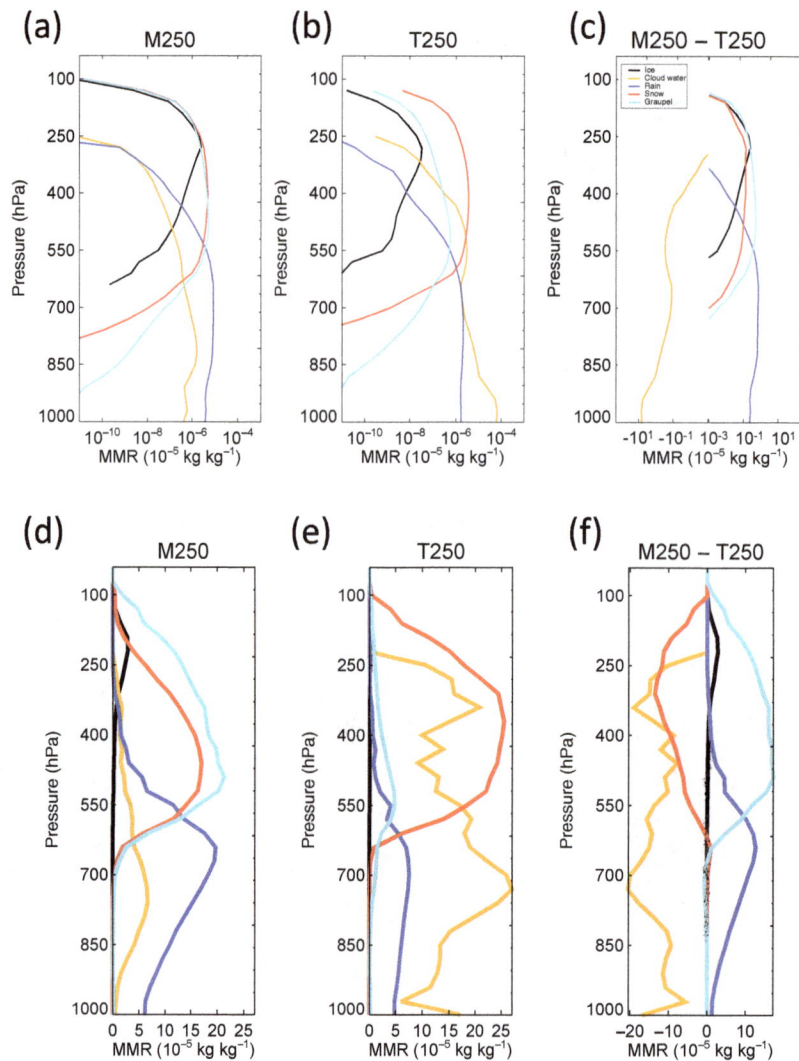

Figure 6. Congo case: mean vertical profiles of hydrometeor mass mixing ratios (MMRs) averaged over the period 1–10 August 2007. (a) CONGO-M250 cloudy column domain mean, (b) CONGO-T250 domain mean, (c) the difference in the domain-mean hydrometeor mixing ratio profiles (CONGO-M250 minus CONGO-T250), (d) CONGO-M250 mean over condensed points only, (e) CONGO-T250 mean over condensed points only for each hydrometeor class and (f) the difference in the condensate-mean hydrometeor mixing ratio profiles (CONGO-M250 minus CONGO-T250). Note the logarithmic horizontal axis used in panels (a–c) due to the total difference between the hydrometeor classes simulated by the two schemes spanning several orders of magnitude.

simulated hydrometeor classes and in the vertical distribution of hydrometeors. Even in the convective precipitating region in the north of the domain, the simulated hydrometeor classes differ significantly between the microphysics configurations with the CONGO-MORR simulations generating more ice and less liquid cloud (Fig. 4a–c) and the CONGO-THOM simulations producing less ice and more liquid cloud (Fig. 4d–f). Rain is confined to the convective region in the north in CONGO-THOM, while in CONGO-MORR it is also present at low levels in the non-convective southern region of the domain which is dominated by liquid cloud in CONGO-THOM. We explain the mechanisms behind these differences later, but here we highlight that it is clear from Fig. 4 that the

differences in the simulated hydrometeors between microphysics schemes are much greater than the differences due to different levels of CDNC.

Because the partitioning of water into liquid and ice phases in the full-physics model configuration appears to depend strongly on the microphysics scheme, vertical sections of reflectivity occurrences derived from model hydrometeor fields passed through the QuickBeam radar simulator (Haynes et al., 2007) are compared against equivalent reflectivity occurrences from the CloudSat 2B-GEOPROF product (Marchand et al., 2008) (Fig. 5). The histograms are derived from the reflectivity fields thresholded to include all values greater than -20 dBZ. The largest reflectivity values produced by the model occur in the convective region in the north of the domain where the largest reflectivity values are detected by the satellite radar (Fig. 5), which is also in agreement with the TRMM precipitation observations (Fig. 3). However, both CONGO-MORR and CONGO-THOM have a large positive bias in reflectivity compared to the observations (Fig. 5), which is indicative of limitations in the ability of both bulk microphysics schemes to represent the observed vertical cloud structure in this geographic region over this time period. In general, CONGO-MORR has a much larger positive bias in reflectivity than CONGO-THOM (Fig. 5). The CloudSat observations show a small frequency of occurrence of reflectivities detected at altitudes of 10 to 15 km in the south of the domain, which is well represented by CONGO-THOM and indicates the overproduction of ice in CONGO-MORR (Fig. 5).

Differences in the simulated hydrometeor classes between the schemes persist throughout the simulation and are illustrated by mean profiles of hydrometeor mass mixing ratios (Fig. 6). There is significantly more ice-phase condensate in the CONGO-M250 configuration (Fig. 6a), whereas the CONGO-T250 profile is dominated by a large amount of liquid cloud mass between the near surface and 750 hPa (Fig. 6b). The differences in the total cloud water mass between the schemes are very large: at 950 hPa (the altitude with the greatest liquid cloud mass in CONGO-T250; Fig. 6), cloud water mass contents are about 140 times greater in CONGO-T250. The liquid cloud mass is always greater in CONGO-T250 than CONGO-M250 (Fig. 6c) by several orders of magnitude at some levels, but despite this the liquid phase does not appear to drive differences in precipitation between the microphysics schemes: CONGO-M250 has about 4 times more rain mass in the mid-levels and 2 times more rain mass near the surface than CONGO-T250 (Fig. 6a and b). In the ice phase, CONGO-M250 has only slightly more snow mass than CONGO-T250 but up to 10 times more graupel mass (Fig. 6a and b), and while ice is a significant hydrometeor at upper levels in CONGO-M250, CONGO-T250 has almost no cloud ice at all (Fig. 6a and b). We note that the magnitude of the difference due to the choice of scheme is the same when a bin scheme is used (Figs. S2 and S3 in the Supplement).

Mean profiles over all condensed points (i.e. representing the mean values of each hydrometeor type but not accounting for changes in absolute quantities across the model domain) show that CONGO-T250 has consistently more cloud water through the depth of the mean cloud compared to CONGO-M250 (Fig. 6a and b), while CONGO-M250 produces more rain (Fig. 6a). That rain production in CONGO-M250 occurs mostly through the depths of the atmosphere where cloud water persists suggests that a significant proportion of the rain may be produced though autoconversion in CONGO-M250, although note that these mean cloud profiles are calculated over the entire domain and therefore incorporate both the deep convective region in the north and the warm-

cloud region in the south, as seen in Fig. 4. Further, the two schemes show differences in the frozen hydrometeors with the mean cloud in CONGO-M250 containing more graupel and less snow than CONGO-T250 (Fig. 6c). This may be a result of the use of distinct and different definitions of ice-phase hydrometeor categories in the two schemes, which have been shown to cause deficiencies in simulations of observed squall lines (Morrison and Milbrandt, 2015).

Not only does the partitioning of ice amongst the hydrometeor classes differ between schemes, the response of the hydrometeors to CDNC perturbations also differs between schemes (Fig. 7). First note that the scale of the hydrometeor response to CDNC perturbations in the CONGO-MORR simulations is an order of magnitude smaller than the scale of the response in the CONGO-THOM simulations. Over the entire domain, liquid cloud mass appears insensitive to CDNC perturbations in the CONGO-MORR configuration (Fig. 7a), although a reduction in mean-cloud liquid cloud mass under polluted conditions (Fig. 7c) indicates that there must be very few liquid cloud points in the CONGO-MORR simulation compared to other hydrometeor types, notably ice (Fig. 7a). Very weak decreases in domain-mean near-surface rain mass may be evident under polluted conditions in CONGO-MORR, but this difference is on the order of 10^{-8} kg kg^{-1} (Fig. 7a) A reduction in rain mass under polluted conditions is more evident in the mean rain profile (Fig. 7c), again indicating how few rainy points exist compared to other hydrometeor types in CONGO-MORR when considering the entire domain (Fig. 7a). Nearly all of the hydrometeor response in CONGO-MORR occurs in the ice-phase processes: graupel mass decreases significantly under polluted conditions (Fig. 7a and c), while ice mass increases at upper levels in both a domain mean and ice mean sense (Fig. 7a and c). In contrast, the hydrometeor response to CDNC perturbations in the CONGO-THOM configuration is an order of magnitude greater than in CONGO-MORR and the dominant hydrometeor response to CDNC perturbations in CONGO-THOM occurs in the liquid phase. Not only does the CONGO-THOM configuration generate significantly more liquid cloud than the CONGO-MORR configuration (Fig. 6c), but the liquid cloud mass also increases under polluted conditions by an order of magnitude more than any other hydrometeor response (Fig. 7b and d). Rain mass is relatively insensitive to increased CDNC in CONGO-THOM (Fig. 7b and d). The significant difference between the response of the two schemes to perturbations in CDNC, with CONGO-MORR producing less liquid cloud and rain under polluted conditions while CONGO-THOM produces more cloud water, indicates significant differences in the cloud processes represented by the two schemes in this meteorological regime.

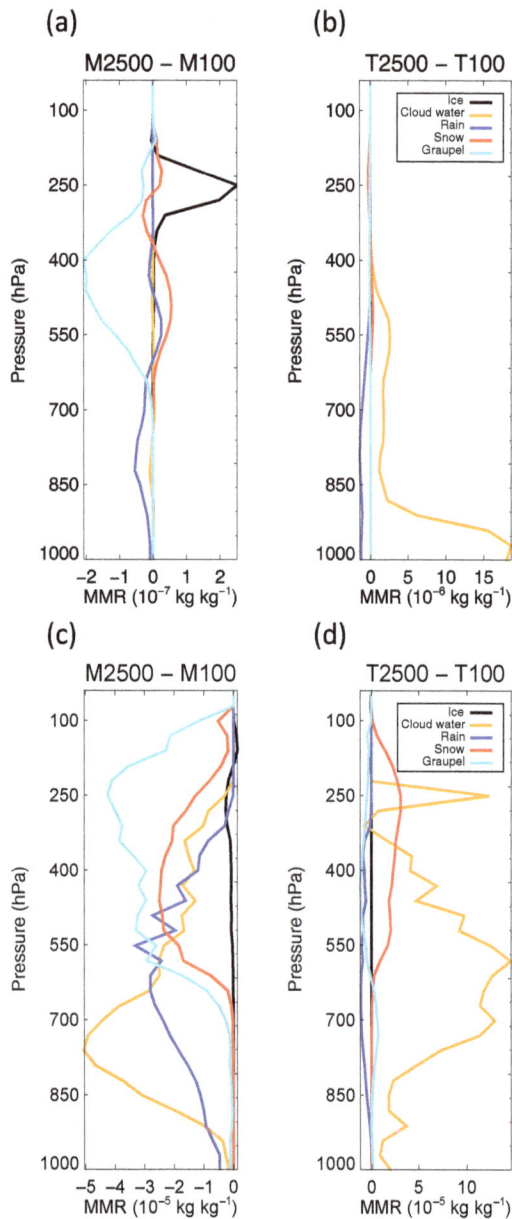

Figure 7. Congo case: difference in the mean hydrometeor mixing ratio profiles under polluted and pristine conditions averaged over the period 1–10 August 2007. **(a)** CONGO-M2500 cloudy column domain mean minus CONGO-M100 domain-mean, **(b)** CONGO-T2500 domain-mean minus CONGO-T100 domain mean, **(c)** CONGO-M2500 mean over all condensed points of each hydrometeor class minus CONGO-M100 mean over all condensed points and **(d)** ONGO-T2500 mean over all condensed points minus CONGO-T100 mean over all condensed points.

3.2   WRF idealised supercell

The results from the real-data Congo basin simulations indicate that the development of the simulated hydrometeor classes and the response of the hydrometeors to CDNC per-

turbations depend strongly on the choice of microphysics scheme. Although some previous studies have focused on the response of real-data case studies to both microphysics scheme and CDNC response (e.g. Fan et al., 2012a, 2013; Li et al., 2015), there is a much larger body of literature that investigates the response of idealised supercell simulations to CDNC (or CCN) perturbations (e.g. Seifert and Beheng, 2006b; Khain and Lynn, 2009; Lebo and Seinfeld, 2011; Morrison, 2012). We therefore place our study in the wider context of the existing literature by investigating the response of a single isolated idealised supercell under both the MORR and THOM microphysics configurations to the same CDNC perturbations used in our Congo simulations, simultaneously allowing us to explore the case dependence of the deep convective response to aerosol effects.

Figure 8 shows mean hydrometeor profiles from the idealised supercell model configurations under "moderately polluted" prescribed CDNC values of $250\,cm^{-3}$. As in the Congo basin case, it is clear that the simulated hydrometeor classes differ significantly between schemes. In contrast to the Congo basin configuration, both the SUPER-MORR and SUPER-THOM configurations show similar behaviour in the liquid phase, producing similar profiles of liquid cloud mass and rain mass in both a domain mean and hydrometeor-class mean sense (Fig. 8a, d and b, e), and instead the most significant differences occur in the ice phase. Graupel dominates as the frozen precipitating hydrometeor in the SUPER-M250 configuration, amounting to about 4 times the snow and ice masses at their peak amounts (Fig. 8a and d). In contrast, snow is the dominant frozen precipitating hydrometeor in the SUPER-T250 configuration, amounting to about 1.5 times the graupel mass at peak amounts and virtually no ice present (Fig. 8b and e). Although there is very little difference between the SUPER-MORR and SUPER-THOM configurations in the liquid phase (except for the SUPER-MORR configuration producing about $2 \times 10^{-7}\,kg\,kg^{-1}$ less domain-mean rain mass at the surface than SUPER-THOM; Fig. 8c), the SUPER-MORR configuration forms significantly more ice, more graupel and less snow than SUPER-THOM (highlighting that the partitioning of ice-phase hydrometeors into categories is very different, by design, in different microphysics schemes). Greater total quantities of frozen hydrometeors are present between 600 and about 150 hPa in SUPER-MORR compared to SUPER-THOM (Fig. 8c and f). This is a significant difference from the Congo real-data configuration in which the dominant contribution to the difference between the CONGO-MORR and CONGO-THOM configurations came from the liquid cloud (Fig. 6c).

There is a more significant aerosol impact on hydrometeor mass in the supercell case than in the Congo case for both microphysics schemes, with mean responses over each hydrometeor type an order of magnitude greater in the supercell case (Fig. 7 compared to Fig. 9). Although many past studies have shown that aerosol impacts depend on cloud dynamics and thermodynamics (e.g. Khain and Lynn, 2009;

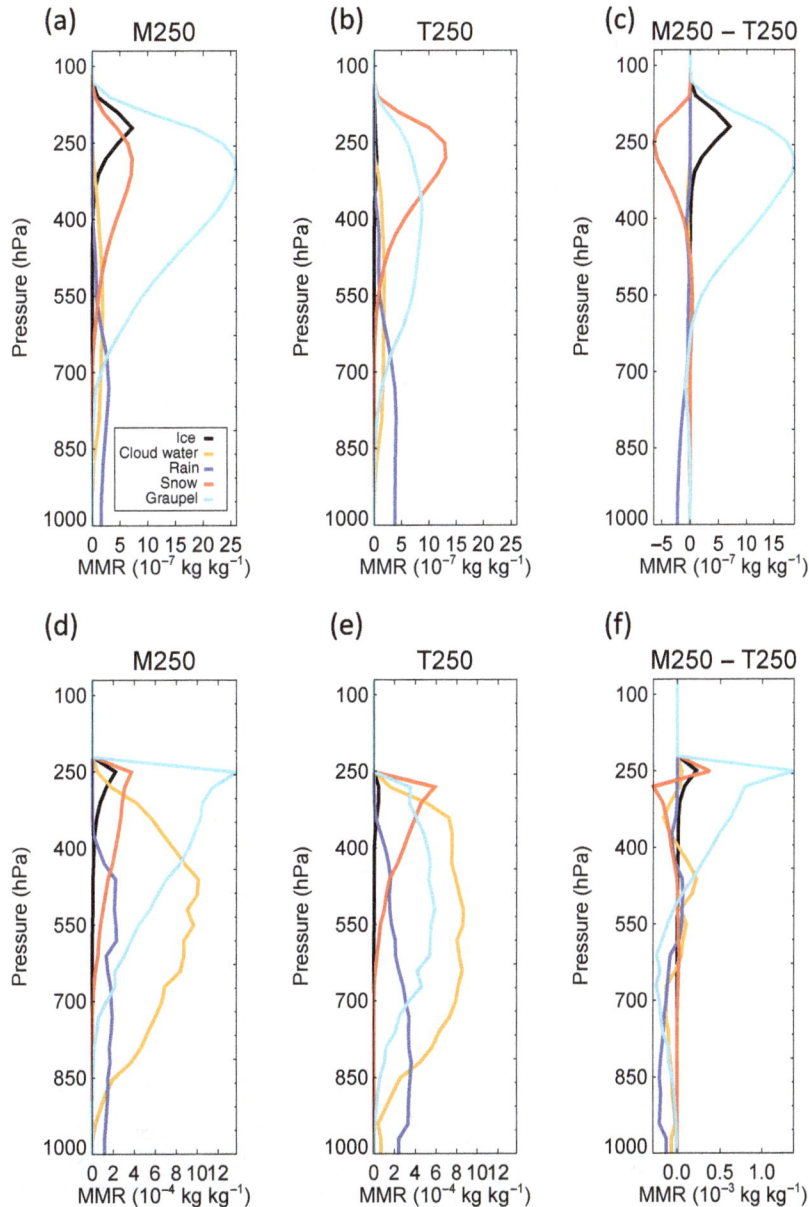

Figure 8. Idealised supercell: mean vertical profiles of hydrometeor mass mixing ratios (MMRs), as in Fig. 6, averaged over the 2 h of the supercell simulation. **(a)** SUPER-M250 domain mean, **(b)** SUPER-T250 cloudy column domain mean, **(c)** SUPER-M250 domain mean minus SUPER-T250 domain mean, **(d)** SUPER-M250 condensate mean of each hydrometeor class, **(e)** SUPER-T250 condensate mean and **(f)** SUPER-M250 condensate mean minus SUPER-T250 condensate mean.

Fan et al., 2009), we note that not only do the individual schemes respond differently to CDNC in different cases of convection (as expected), but the way the schemes differ from each other in their response to CDNC is also significantly different in the supercell case compared to the Congo case. The SUPER-MORR and SUPER-THOM cases differ qualitatively from the CONGO-MORR and CONGO-THOM cases, respectively, both in the altitudes at which the response occurs and the sign of the response of some of the hydrometeors. In the SUPER-MORR configuration, cloud water mass

increases under polluted conditions, and rain mass is suppressed at mid-levels (between 600 and 450 hPa) but shows negligible response at the surface (Fig. 9a and c). In the ice phase, cloud ice increases under polluted conditions in SUPER-MORR, while graupel and snow decrease (Fig. 9a and c). Similarly, the hydrometeor response of the SUPER-THOM case to CDNC perturbations also differs in sign and in altitude to CONGO-THOM. In SUPER-THOM, cloud water mass increases and rain mass decreases under polluted conditions (Fig. 9b and d), but unlike SUPER-MORR the

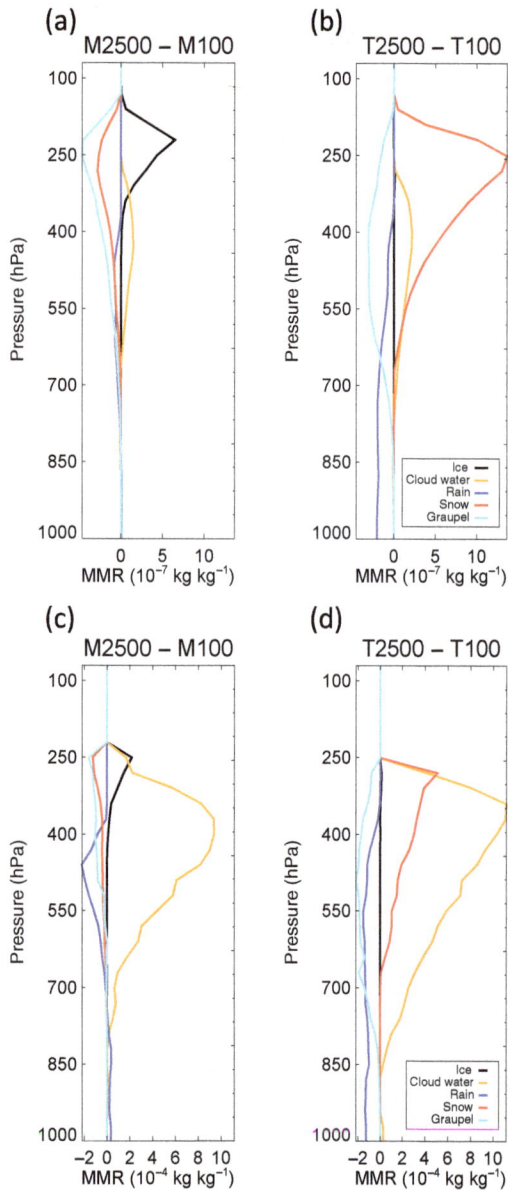

Figure 9. Idealised supercell: difference in the mean hydrometeor mixing ratio profiles under polluted and pristine conditions, as in Fig. 7, averaged over the 2 h of the supercell simulation. (a) SUPER-M2500 cloudy column domain mean minus SUPER-M100 domain mean, (b) SUPER-T2500 domain mean minus SUPER-T100 domain mean, (c) SUPER-M2500 condensate mean of each hydrometeor class minus SUPER-M100 condensate mean and (d) SUPER-T2500 condensate mean minus SUPER-T100 condensate mean.

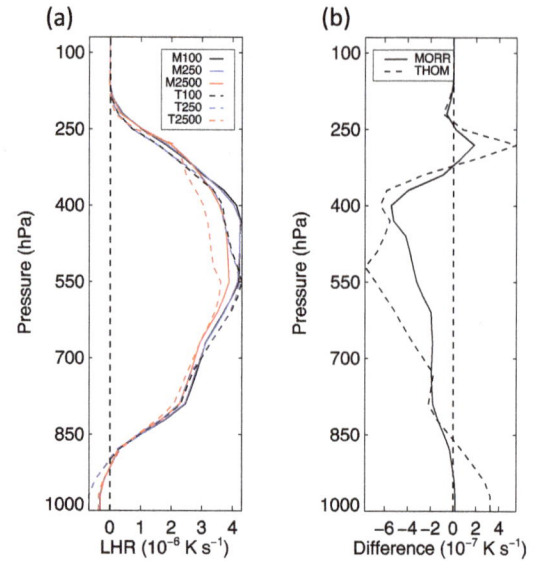

Figure 10. Idealised supercell: (a) vertical profiles of domain-mean total latent heating rate (LHR) over the 2 h of the supercell simulation for SUPER-MORR and SUPER-THOM for CDNC values of 100, 250 and 2500 cm^{-3}. (b) Difference in the total latent heating contributions over the 2 h of the supercell simulation for SUPER-M2500 minus SUPER-M100 and SUPER-T2500 minus SUPER-T100.

decrease in rain is evident at the surface. Graupel mass decreases under polluted conditions in SUPER-THOM, similarly to SUPER-MORR, but occurs over a much larger range of heights (Fig. 9b and d); this is unlike CONGO-THOM, which shows very little response to polluted conditions (Fig. 7c). Interestingly, this is in contrast to Khain and

Lynn (2009), who found an increase in graupel mass with increased CDNC in the Thompson scheme. However, their study was of 2-D idealised squall line simulations and considered CDNC values of 100, 500 and 100 drops per cm^{-3}. The dominant domain-mean response to increased CDNC perturbations in SUPER-THOM is an increase in snow mass between 550 and 150 hPa (Fig. 9b), which likely comes from lofting of an increased mass of cloud water (Fig. 9d). This is in contrast both to SUPER-MORR in which the dominant hydrometeor response occurred in the ice class (Fig. 9a), despite an almost equal increase in lofted cloud water (Fig. 9c), and to CONGO-THOM in which the dominant hydrometeor response occurred in the liquid cloud (Fig. 7b). That both schemes show an increased lofting of cloud water under polluted conditions (Fig. 7c and d), but SUPER-MORR responds by generating more cloud ice (Fig. 7a and c) while SUPER-THOM shows an increase in snow (Fig. 7b and d), suggests differences in the processes that convert cloud ice to snow. This is explored later in Sect. 3.4. We emphasise that our main result shows that the variability due to microphysics scheme dominates any aerosol impacts on microphysics. Results using the WRF-SBM in the idealised supercell case show that aerosol impacts in the bin scheme are of equal magnitude to those in the bulk schemes (Fig. S4 in the Supplement).

To further investigate the importance of the difference in microphysics representations and the difference in their response to CDNC perturbations, Fig. 10 includes the domain-mean total latent heating (sum of the latent heating from individual microphysical processes) contributions for each of the idealised supercell configurations. It can be seen that the choice of microphysics scheme can result in thermodynamic differences in the supercell system equal in magnitude to those arising from CDNC perturbations: between 500 and 250 hPa, the latent heating rate in the SUPER-M2500 configuration is almost identical to that in the SUPER-T250 configuration (solid red and dashed blue lines, Fig. 10a). Thus, the magnitude and sign of the difference in the latent heating rate between SUPER-M250 and SUPER-T250 (blue solid and dashed lines, Fig. 10a) is the same as that between SUPER-M2500 and SUPER-M250 (red and blue solid lines), and likewise the magnitude and sign of the difference in the latent heating rate between SUPER-M2500 and SUPER-T2500 (red solid and dashed lines) is the same as that between SUPER-T2500 and SUPER-T250 (red and blue dashed lines). In general, the SUPER-THOM configuration has a much stronger thermodynamic response to CDNC perturbations than the SUPER-MORR configuration, with latent heating rates consistently stronger throughout the atmosphere (Fig. 10b). Overall, there is little evidence of convective invigoration (defined here as increases in upper tropospheric heating, updraught strengths, cloud top height and surface precipitation) under increased CDNC values in either bulk microphysics scheme. Although both schemes show increased latent heating in the upper troposphere and decreased heating at mid-levels under polluted conditions (Fig. 10b), it has already been shown that there is no evidence of increased surface precipitation (Fig. 9), and the upper tropospheric peak in latent heating can be seen to correspond to an increase in ice (SUPER-M250) or snow (SUPER-T250) at these levels (Fig. 9). There is no systematic or consistent evidence of increased mean updraught velocity in the convective cores (following the method of van den Heever et al., 2006; Lebo and Seinfeld, 2011) under polluted conditions (not shown) or in increased cloud top heights of the convective cores (Fig. 2c and d). This may not be surprising, as it has been suggested that bulk microphysics schemes are unable by design to produce convective updraught invigoration effects due to limitations in their representation of nucleation, sedimentation and the way in which saturation adjustment limits diffusional growth (detailed in Khain and Lynn, 2009). Indeed, Lebo and Seinfeld (2011) found no latent heating effect of increased CCN in a bulk scheme used to simulate idealised deep convection, whereas with a bin scheme increased latent heating aloft was demonstrated. However, Lebo et al. (2012) found that saturation adjustment methods used in bulk schemes could explain differences in the response of cold pool evolution and convective dynamics between bin and bulk schemes to aerosol loading but could not explain large differences in the response of surface precipitation. Further, some simulations using bulk schemes have identified invigoration-like effects under aerosol loading. For example, Lebo and Morrison (2014) found evidence of con-

vective invigoration under increased aerosol loading in a bulk scheme under weak shear conditions (and suppressed convection under strong shear), similar to the findings of Fan et al. (2009) who found the same response in a bin scheme. Seifert and Beheng (2006a) also found higher overshooting tops and larger sizes of cumulonimbus in a weak shear environment with increased aerosol loading. Thus although our results agree with the body of the literature which does not identify a convective updraught invigoration effect when bulk microphysics schemes are used, this is not necessarily attributable to the saturation adjustment method alone and may also only hold for the particular convective environment (idealised supercell in strong shear) we consider.

3.3 WRF LES RICO

The results presented in Sect. 3.1 and 3.2 indicate that not only is the way in which the schemes differ from each other not systematic between cases of convection, but the difference between the response of the two schemes to CDNC across types of convection is also not systematic. The largest difference between the microphysics schemes in the real-data Congo basin simulations occurs in the liquid-phase hydrometeor development and response to CDNC. Making the assumption that the liquid phase is the first to respond to CDNC perturbations and the perturbation subsequently propagates to the ice phase, we consider a case of precipitating shallow cumulus convection to investigate the liquid-phase differences between the schemes. Note that the "baseline" hydrometeor profiles in Fig. 11 show data from the configurations using a prescribed CDNC value of $100\,\mathrm{cm}^{-3}$ (rather than the baseline value of $250\,\mathrm{cm}^{-3}$ used in the Congo basin and idealised supercell deep convection cases in Figs. 6 and 8), as this is more appropriate for a pristine marine environment. Even when we restrict our simulations to the liquid phase, differences in the simulated hydrometeor classes are evident. The dominant domain-mean difference between the two schemes in the RICO case is clearly in the rain profile, with RICO-T100 producing significantly more rain than RICO-M100. Very little rain is present in the RICO-M100 configuration (Fig. 11a and d), whilst the RICO-T100 configuration produces a peak domain-mean rain mass of about $10^{-6}\,\mathrm{kg\,kg}^{-1}$ (Fig. 11b). The liquid cloud profile is similar in both schemes, with RICO-M100 forming more cloud mass than RICO-T100 between 805 and 775 hPa in both the domain mean and hydrometeor-class mean sense (Fig. 11c and f).

The response of the hydrometeors to CDNC perturbations also differs between schemes in the warm-rain RICO case (Fig. 12). In the RICO-MORR configuration, domain-mean rain and cloud mass both decrease under polluted conditions, although the rain response is very weak (on the order of $10^{-8}\,\mathrm{kg\,kg}^{-1}$) and the dominant response is a reduction in liquid cloud mass (Fig. 12a). In contrast, a reduction in rain mass is the dominant hydrometeor response under

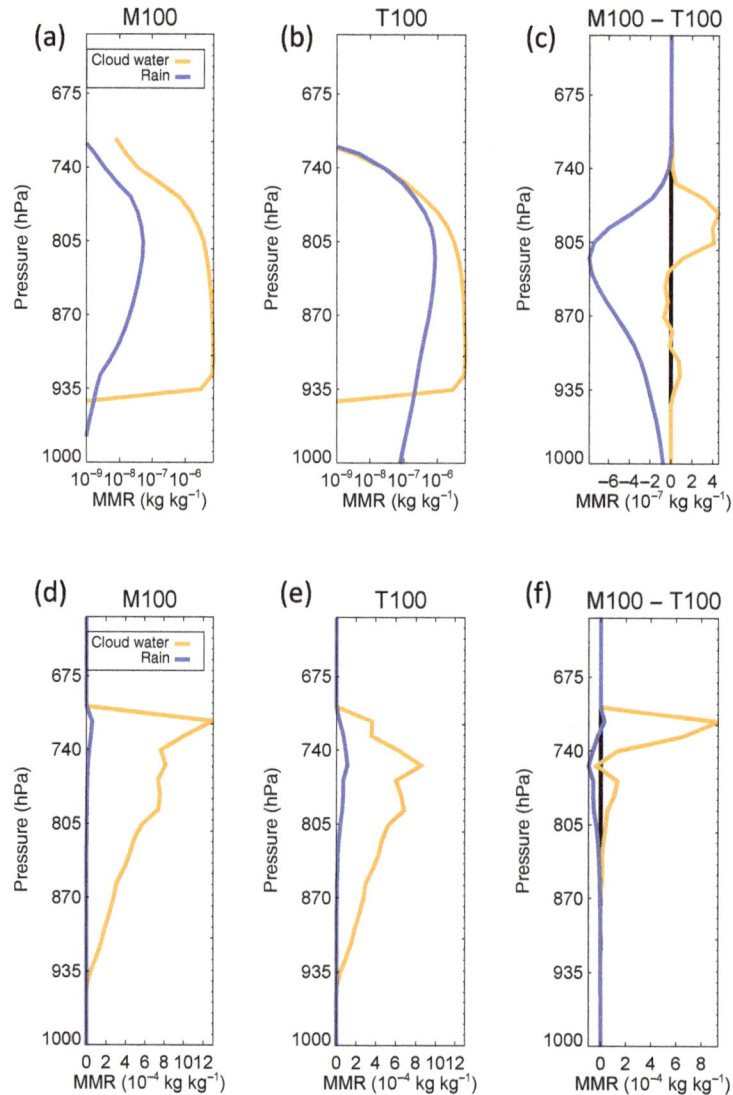

Figure 11. RICO case: mean vertical profiles of hydrometeor mass mixing ratios (MMRs), as in Fig. 6, averaged over the 24 h of the RICO simulation. **(a)** RICO-M100 cloudy column domain mean, **(b)** RICO-T100 domain mean, **(c)** RICO-M100 domain mean minus RICO-T100 domain mean, **(d)** RICO-M100 condensate mean over each hydrometeor class, **(e)** RICO-T100 condensate mean and **(f)** RICO-M100 condensate mean minus RICO-T100 condensate mean. Note that because the rain amounts are very small, especially in M100, panels **(a, b)** are shown with a logarithmic horizontal axis.

polluted conditions in the RICO-THOM configuration, and the decrease is nearly 2 orders of magnitude greater than that in RICO-MORR (Fig. 12b). The liquid cloud response to polluted conditions in RICO-THOM is weaker than the rain response but still stronger than the cloud response in RICO-MORR. Cloud mass decreases under polluted conditions between 935 and 825 hPa but increases at higher levels (Fig. 12b). Note that once again the response of the simulated hydrometeors to CDNC perturbations differs between cases: under polluted conditions, RICO-MORR exhibits a decrease in cloud and rain mass, while CONGO-MORR exhibits a decrease in rain mass with little response in the liquid cloud (Fig. 7a), and SUPER-MORR shows almost no liquid-phase

response at all (Fig. 9a). Likewise, RICO-THOM exhibits a decrease in rain and an increase in cloud mass under polluted conditions, while CONGO-THOM exhibits similar behaviour (Fig. 7b), but SUPER-THOM shows a decrease in rain mass with little response in the liquid cloud (Fig. 9b). When mean profiles of each hydrometeor class are considered, the two schemes actually show similar responses to CDNC (increased upper level cloud mass and suppressed rain; Fig. 9c and d). This indicates that the main response to CDNC in this case is not through the individual microphysical processes but through the absolute amounts of cloud and rain that are generated.

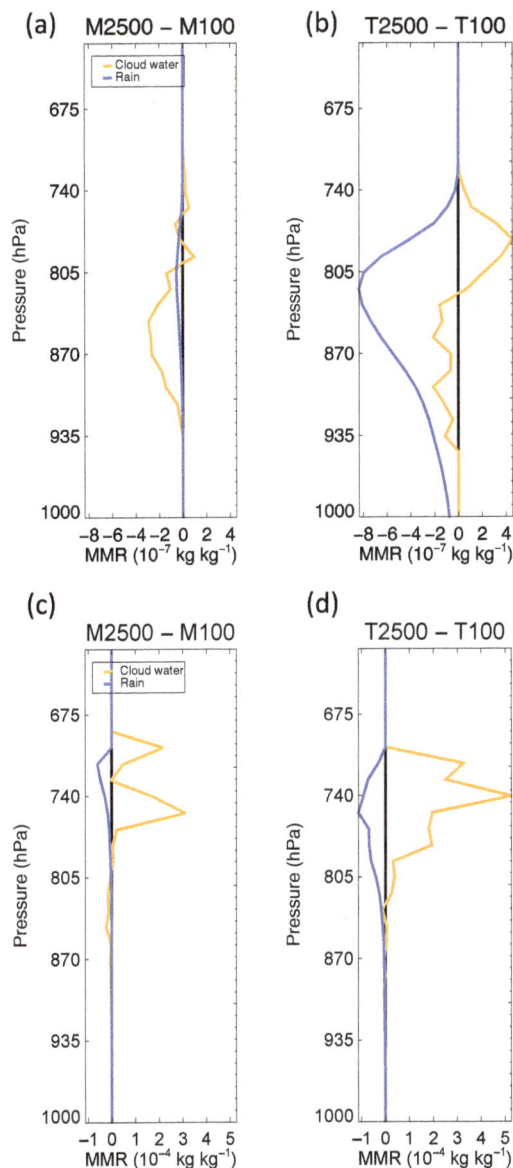

Figure 12. RICO case: difference in the mean hydrometeor mixing ratio profiles under polluted and pristine conditions in cloudy columns, as in Fig. 7, averaged over the 24 h of the RICO simulation. **(a)** RICO-M2500 cloudy column domain mean minus RICO-M100 domain mean, **(b)** RICO-T2500 domain mean minus RICO-T100 domain mean, **(c)** RICO-M2500 condensate mean over each hydrometeor class minus RICO-M100 condensate mean and **(d)** RICO-T2500 condensate mean minus RICO-T100 condensate mean.

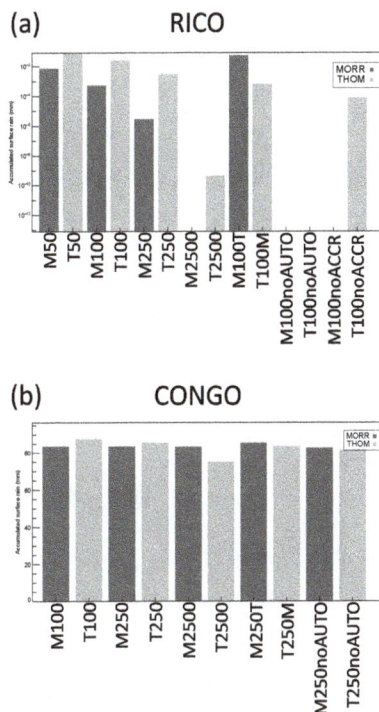

Figure 13. Total accumulated surface rain (mm) for each of the microphysics simulations, including a series of sensitivity simulations, for **(a)** the RICO case total after 24 h of simulation and **(b)** the Congo case total over the period 1–10 August 2007. Note that because the magnitude of the rain response to CDNC differs so strongly between the configurations in the RICO case, a logarithmic vertical axis is used in panel **(a)**. The horizontal dashed line in panel **(b)** indicates the total precipitation from the TRMM 2A25 product over the same period.

To illustrate the difference in the strength of response of the schemes to CDNC, total accumulated surface rain is shown for each RICO configuration in Fig. 13a along with an extra configuration using a "very pristine" CDNC value of $50 \, \mathrm{cm}^{-3}$ and a series of sensitivity tests that will be discussed later. The $50 \, \mathrm{cm}^{-3}$ CDNC configuration has been added because even at a prescribed CDNC value of $100 \, \mathrm{cm}^{-3}$ very lit-

tle rain production occurs in the RICO-MORR configuration. Warm rain formation differs strongly between schemes: very low CDNC values are required for the RICO-MORR configuration to produce any rain, whereas RICO-THOM produces significantly more rain at all CDNC values (Fig. 13a). Even under very pristine conditions, the RICO-M50 configuration produces an order of magnitude less rain than RICO-T50 (Fig. 13a). The different schemes also respond differently to CDNC perturbations. Rain production in RICO-MORR (which produces much less rain than RICO-THOM) shuts down very quickly as CDNC is increased: rain amounts are on the order of 10^2 mm at a CDNC value of $50 \, \mathrm{cm}^{-3}$, 10^1 mm at a CDNC value of $100 \, \mathrm{cm}^{-3}$ and 10^{-1} mm at a CDNC value of $250 \, \mathrm{cm}^{-3}$; rain production ceases completely at a CDNC value of $2500 \, \mathrm{cm}^{-3}$ (Fig. 13a). In contrast, rain production persists for much larger CDNC values in RICO-THOM: rain amounts are on the order of 10^3 mm at CDNC values of $50 \, \mathrm{cm}^{-3}$, 10^2 mm at CDNC values of $100 \, \mathrm{cm}^{-3}$ and 10^1 mm at CDNC values of $250 \, \mathrm{cm}^{-3}$. While rain amounts are very low at CDNC values of $2500 \, \mathrm{cm}^{-3}$ (on the order of 10^{-5} mm), rain production has not shut down completely (Fig. 13a).

3.4 Sensitivity tests

Gilmore and Straka (2008) showed that the rain rates predicted by different autoconversion formulae in bulk schemes can vary by orders of magnitude. This sensitivity of results is also well highlighted in Thompson et al. (2004); note in particular their reference to Walko et al. (1995). Autoconversion is parameterised differently in the two microphysics schemes used in the current paper. The Thompson scheme follows an adaptation of Berry and Reinhardt (1974), while the Morrison scheme follows the method of Khairoutdinov and Kogan (2000).

Thompson et al. (2004) and Thompson et al. (2008) justify their choice of an adapted version of the Berry and Reinhardt (1974) autoconversion parameterisation through favourable comparison to results from the bin scheme of Geresdi (1998). Furthermore, implementation of Berry and Reinhardt (1974) in the Thompson scheme begins the collision–coalescence production of warm rain at almost exactly 14 μm. It is known that raindrop onset begins when the mean volume radius exceeds a critical value of 13 to 14 μm (Freud and Rosenfeld, 2012; Khain et al., 2013; Rosenfeld et al., 2014). This is one of the principle reasons the Berry and Reinhardt (1974) autoconversion scheme was chosen by Thompson et al. (2004, 2008) rather than Khairoutdinov and Kogan (2000).

While the Khairoutdinov and Kogan (2000) autoconversion scheme was initially developed and applied for LES of stratocumulus, other than varying the prescribed values of cloud droplet number concentrations we run the microphysics schemes in their baseline configurations. Thus, although we do not advocate the use of Khairoutdinov and Kogan (2000) for non-stratocumulus cases, the Morrison scheme is frequently used for simulations of deep convection. Similarly, as the Khairoutdinov and Kogan (2000) autoconversion scheme was developed for LES-scale studies, the authors recognise the potential importance of subgrid cloud variability at the scales used in the present study. However, we note that we are running the model and microphysics schemes in the typical set-up for a convection-permitting model (that is, neglecting subgrid cloud variability), as one of the main aims of this study is to highlight uncertainty in commonly used model configurations which are exactly based on these schemes.

The autoconversion rates as a function of cloud water content for each of the model configurations are shown in Fig. 14. Also shown is the cloud water content (up to the mean plus 2 standard deviations) of each configuration . It is immediately clear that the threshold cloud liquid content for autoconversion in the Morrison scheme (solid lines) is significantly lower than that in the Thompson scheme (dashed lines); i.e. rain production can occur at much lower cloud

liquid water contents in Morrison. It is also clear from the mean, (mean +1 SD) and (mean +2 SD) cloud water content limits that rain production through autoconversion ought to be possible in all model configurations. However, despite the higher cloud water content threshold for autoconversion in the Thompson scheme, autoconversion rates are much greater once the threshold is reached, and liquid cloud is converted to rain much faster in Thompson than in Morrison. From Fig. 14, it appears that the threshold for autoconversion is unlikely to be reached very often in any of the T2500 cases. In the deep convective cases, rain can be generated through ice- and mixed-phase processes, but in the RICO warm-rain case this cannot occur. This explains why, compared to more pristine conditions, cloud mass increases in RICO-T2500 while rain mass decreases (Fig. 12b).

Because Fig. 14 indicates that the autoconversion threshold may be at least in part responsible for this response in the RICO-THOM case, we replace the autoconversion parameterisation in the Morrison scheme with that from the Thompson scheme and vice versa. We use the notation M100T to denote the Morrison microphysics scheme (at a CDNC value of $100 \, cm^{-3}$) with Thompson autoconversion (that of Berry and Reinhardt, 1974) and T100M to denote the Thompson scheme with Morrison autoconversion (that of Khairoutdinov and Kogan, 2000). Differences in the domain-mean hydrometeor mixing ratio profiles for each of the autoconversion swapped configurations in the RICO case are shown in Fig. 15. It is immediately clear that, in the warm-rain configuration, simply swapping the autoconversion treatment makes the hydrometeor developments of the microphysics schemes much more like each other. The differences between the RICO-M100 configuration with the Morrison and Thompson autoconversion parameterisations (Fig. 15a) is quantitatively and qualitatively very similar to the difference between the RICO-M100 and RICO-T100 configurations (Fig. 11c). Likewise, the difference between the RICO-T100 configuration with the Morrison and Thompson autoconversion parameterisations (Fig. 15b) and finally the difference between the RICO-T100 configuration with the Morrison autoconversion parameterisation and the RICO-M100 configuration with the Thompson autoconversion parameterisation (Fig. 15c) are also very similar to the difference between the RICO-M100 and RICO-T100 configurations (Fig. 11c).

Similarly, swapping the autoconversion parameterisations between the microphysics schemes in the RICO cases makes the surface rain production of the microphysics schemes much more similar. The accumulated surface rainfall in the RICO-M100T configuration looks much more similar to the surface rainfall in the RICO-T100 configuration than it does to the RICO-M100 configuration (Fig. 13a). Rain amounts are on the order of 10^2 mm in RICO-M100T and RICO-T100, whereas in RICO-M100 it is 2 orders of magnitude smaller (Fig. 13a). Likewise, the accumulated surface rainfall in the RICO-T100M configuration is on the order of 10^1 mm compared to 10^2 mm in the standard RICO-T100

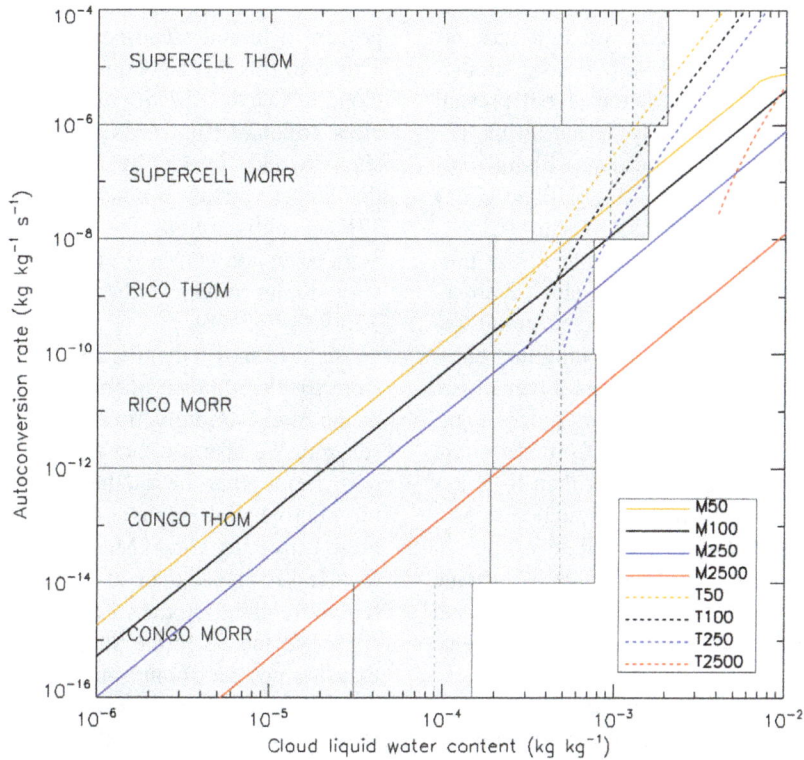

Figure 14. Autoconversion rate as a function of cloud water content for the MORR and THOM microphysics schemes (solid and dashed lines, respectively) for super pristine, pristine, moderately polluted and polluted conditions. Also shown are labelled grey bars showing the mean (solid vertical grey line) and 1 and 2 standard deviations (dashed vertical grey line and end of bar, respectively) for cloud water content averaged over all prescribed CDNC configurations for each case (note that the variability in mean cloud water content with CDNC is significantly less than the variability due to microphysics scheme).

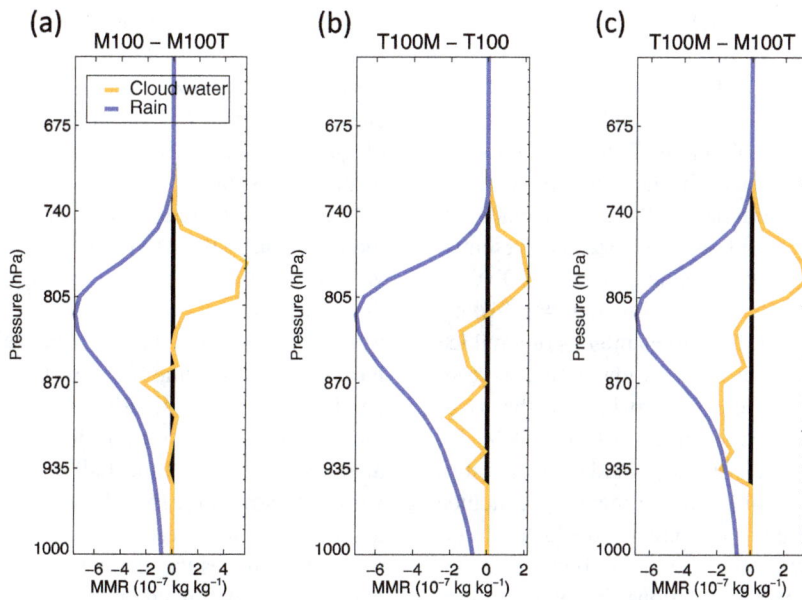

Figure 15. RICO case: difference in the cloudy column domain-mean vertical profiles of hydrometeor mass mixing ratios (MMR) between MORR and THOM, as in Fig. 11c, averaged over the 24 h of the RICO simulation for the configurations with the autoconversion treatment swapped between the microphysics schemes **(a)** M100 minus M100T, **(b)** T100M minus T100 and **(c)** T100M minus M100T.

case (Fig. 13a). To further test the importance of autoconversion in the liquid phase simulations, we first turn off autoconversion completely in the $100\,\text{cm}^{-3}$ CDNC simulations, and then allow autoconversion to occur but prevent the accretion of cloud water by rain. By design, in the absence of ice processes no precipitation occurs without autoconversion of cloud water to rain (Fig. 13a, M100noAUTO and T100noAUTO). However, in the RICO $100\,\text{cm}^{-3}$ CDNC liquid-phase configuration, the Thompson scheme can produce surface rain from autoconversion alone (albeit 2 orders of magnitude less than when rain can also accrete cloud water (Fig. 13a, T100noACCR and T100), showing that autoconversion acts almost like a "trigger" in this scheme after which accretion takes over the rain production process. (Indeed, in nearly all schemes, rain formation from accretion, once triggered, is orders of magnitude larger than from autoconversion.) In contrast, zero surface precipitation is produced in RICO M100noACCR (Fig. 13a), showing that in this (liquid-phase only) configuration the Morrison scheme requires both the autoconversion of cloud droplets to rain and the accretion of rain by cloud droplets in order to produce surface precipitation.

Despite the significant effect of autoconversion in the liquid-phase simulations, changing the autoconversion parameterisation in the idealised supercell case has very little effect on the hydrometeor development (results not shown). This is unsurprising, as ice- and mixed-phase processes will dominate this shear-driven deep convective environment. However, the Congo basin configurations show large differences between microphysics schemes in the partitioning of water into liquid and ice phases (CONGO-THOM produces much more liquid cloud; CONGO-MORR produces much more ice). In the CONGO-THOM configurations, the liquid-phase response to increased CDNC is also very similar to the RICO-THOM response (increased liquid cloud mass and decreased rain mass; Fig. 7b). When the Thompson autoconversion treatment is implemented in the Morrison scheme, rain production in the southern half of the domain ceases in CONGO-M250T, and the liquid phase is instead represented by low-level cloud with structure similar to the CONGO-T250 configuration (Fig. 16a compared to Fig. 4e). To test if radiative effects associated with large amounts of anvil ice drive or contribute to the differences in low cloud, we also set the ice extinction coefficient to zero in both the longwave and shortwave radiation schemes in CONGO-M250. However, this has no effect on the low-cloud characteristics (Fig. 16e compared to Fig. 4b), and we therefore conclude that autoconversion of cloud water to rain is the factor dominating the absence of low-level cloud in the south of the domain in the CONGO-MORR simulations. In contrast, autoconversion is a less significant process in the CONGO-T250 configuration. Implementing the Morrison autoconversion treatment in the Thompson scheme has very little effect on the hydrometeor structure in the CONGO-T250M configuration compared to the CONGO-M250 configuration (Fig. 16b

compared to Fig. 4b). As a final test, the autoconversion process is turned off in both of the microphysics schemes. This confirms that autoconversion dominates the lack of low cloud in CONGO-M250: the resulting liquid-phase hydrometeor structure (Fig. 16c) is similar to both CONGO-T250 (Fig. 7b) and CONGO-M250T (Fig. 16a). This also confirms that autoconversion is much less significant in the CONGO-THOM configurations: the bulk hydrometeor structure when autoconversion is turned off in CONGO-T250 (Fig. 16d) is very similar to both CONGO-T250 (Fig. 7b) and CONGO-T250M (Fig. 16b).

We have shown that the autoconversion process is responsible for the removal of the large cloud mass at low levels in the model configuration with the Morrison microphysics scheme. We also see that this low-level liquid-phase cloud mass forms when we run the same simulation using the WRF bin microphysics implementation (the SBM part of the Hebrew University Cloud Model; Khain et al., 2011), although to a lesser extent than in the Thompson simulations, and the warm cloud produced by the WRF-SBM produces rain (Figs. S2 and S3 in the Supplement). We therefore suggest that it is not the Thompson scheme per se which is responsible for producing the low-level cloud mass, but rather the larger-scale meteorological conditions in which these simulations are performed.

A further significant difference between the two schemes in the Congo simulations is the generation of large amounts of upper-level ice in CONGO-M250, which is not present in CONGO-T250 (Fig. 4b and e). In the Thompson scheme, the fraction of ice mass with a diameter greater than $125\,\mu\text{m}$ is instantaneously transferred into the snow category (Thompson et al., 2008). The same threshold size for cloud ice autoconversion to snow is used in the Morrison scheme, but the process is parameterised differently (Morrison et al., 2005). Because the Morrison scheme appears to produce large amounts of anvil cloudiness for the Congo case, which is not seen in the observations (Fig. 5), we perform further sensitivity tests in which we reduce the threshold size for cloud ice autoconversion in the Morrison scheme to 50 % of its original value (Fig. 16f) and 10 % of its original value (Fig. 16g). We then finally replace the autoconversion of cloud ice to snow in the Morrison scheme with the parameterisation used in the Thompson scheme (Fig. 16h). In all tests, the upper-level anvil ice is reduced significantly. Using the lowest value of the threshold size for cloud ice autoconversion reduces the anvil cloud because almost all of the ice is immediately converted to snow (Fig. 16g). However, using the Thompson ice autoconversion representation in the Morrison scheme significantly reduces the amount of cloud ice in the simulation, and all of the detrained anvil ice is removed (Fig. 16h). This suggests that for the particular Congo simulation we have investigated, the conversion of cloud ice to snow is the main factor leading to the significant difference in anvil cloudiness between the two schemes and is responsible for the difference in upper-level cloud between the CONGO-M250 simu-

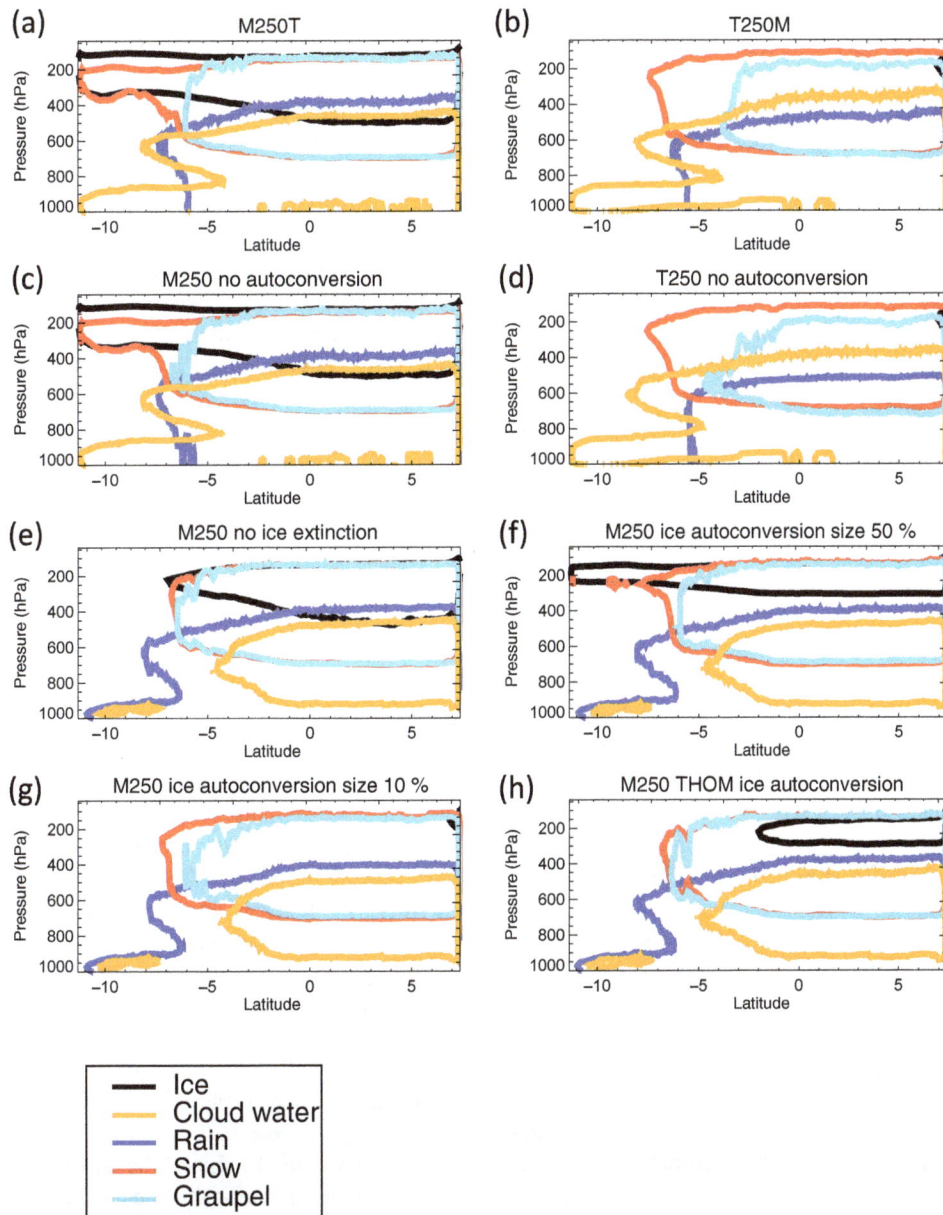

Figure 16. Congo case: zonal mean vertical sections of hydrometeor classes (colour contours) from 1 to 10 August 2007, as in Fig. 4, but for the configurations with the autoconversion treatment swapped between the microphysics schemes (**a**) CONGO-M250T and (**b**) CONGO-T250M; for the configurations with (**c**) CONGO-M250 with autoconversion turned off, (**d**) CONGO-T250 with autoconversion turned off, (**e**) CONGO-M250 with the ice extinction coefficient set to zero in the longwave and shortwave radiation schemes, (**f**) CONGO-M250 with the threshold size parameter for conversion of ice to snow reduced to 50 % of its default value, (**g**) CONGO-M250 with the threshold size parameter for conversion of ice to snow reduced to 10 % of its default value and (**h**) CONGO-M250 with the autoconversion of ice to snow replaced by that used in the Thompson microphysics scheme. Hydrometeor mass mixing ratios are contoured at 10^{-6} kg kg^{-1}.

lation and the observations (Fig. 5). Indeed, we note that in equivalent simulations performed with the WRF-SBM, the same persistent upper-level ice forms (Fig. S2 in the Supplement). This shows that differences resulting from conversion of one ice category into another is a limitation of any scheme, whether bin or bulk, which uses fixed ice categories. Our results provide further evidence that the use of discrete

ice-phase hydrometeor categories may be detrimental to the correct simulation of cloud and suggest that new schemes which do not use such partitioning may give better results (e.g. Morrison and Grabowski, 2008; Harrington et al., 2013; Morrison et al., 2015b).

Our results show little impact of aerosol on precipitation in the Congo basin (Figs. 2b and 7), which is also seen when considering total accumulated surface precipitation (Fig. 13b), although CONGO-T2500 exhibits weak precipitation suppression under polluted conditions. This may be due to the longer duration of these simulations performed over a larger domain, allowing the interaction of many cloud systems rather than considering the lifetime of a single isolated cloud. However, we also see that although the representation of autoconversion has a significant effect on the vertical hydrometeor structure in the CONGO-M250 configurations (Figs. 4b, 16a and c), it has a much weaker effect on total surface precipitation (Fig. 13b, M250, T250, M250T, T250M, M250noAUTO and T250noAUTO). This is perhaps unsurprising, as the dominant contribution to the accumulated surface precipitation over the Congo domain will be from ice processes in the convective region and not from the liquid-phase cloud. Although the lack of impact of aerosol on precipitation in the Congo simulations may be due to the use of bulk schemes in this study for the reasons detailed in Khain et al. (2015), and perhaps a different response would be seen using a bin scheme (e.g. Khain and Lynn, 2009; Lebo and Seinfeld, 2011), other studies using bulk and bin–bulk schemes have identified aerosol impacts on precipitation of up to about 15 % (e.g. van den Heever et al., 2006; Lee and Feingold, 2010, 2013; Lee, 2012; Morrison and Grabowski, 2011; Morrison, 2012; Lebo et al., 2012; Kalina et al., 2014). Indeed, even studies using bin schemes have been shown to have little impact on total precipitation, although they induce a shift in rainfall rates (Fan et al., 2013). Therefore, we note again that the choice of microphysics scheme, rather than aerosol response in either scheme, is the dominant contribution to uncertainty in the total precipitation.

4 Discussion and conclusions

This study considered the cloud and precipitation development using two double-moment bulk microphysics schemes (Morrison et al., 2009; Thompson et al., 2008) to perform cloud-system-resolving simulations of three types of convection, two of which were idealised (one deep convection case with open boundaries and one shallow cumulus case with periodic boundaries), and one real-data case of deep convection in the Congo basin using meteorological initial and boundary conditions. We tested the sensitivity of the simulated hydrometeors and precipitation to the microphysics scheme and to CDNC perturbations. The simulations were performed to explore the uncertainty in cloud and precipitation development and response to aerosol perturbations in convection-permitting models that can arise from the microphysics representation. We find that the variability among the two schemes, including the response to aerosol, differs widely between these cases. Although previous studies have found large sensitivity to the choice of microphysics schemes (e.g. Khain et al., 2015, 2016), we show this in a consistent set-up by considering different cases with the same model

and same CDNC values and constraining as many other possible sources of variability as is feasible. Our results show that for the bulk schemes used in these simulations, aerosol effects are dominated by the uncertainty in cloud and precipitation development which arises from the choice of microphysics scheme. This result was true for multiple cloud types in multiple environmental conditions.

A key finding is that the difference between the two schemes, including their response to CDNC, in different environments and cloud types is not systematic. This could perhaps be related to the nonmonotonic response to aerosol in different environments found by Kalina et al. (2014) (although their study only considered simulations of idealised supercells with a single bulk scheme and four environmental soundings). This nonmonotonic response was attributed to compensatory changes in the microphysical processes under polluted conditions.

The maximum relative difference in mass mixing ratio between each hydrometeor class in the M250 and T250 configurations for each case of convection is summarised in Table 3. Not only are the maximum differences in the domain-mean profiles of the hydrometeor classes simulated by each microphysics scheme on the order of at least tens of percent, but it is also clear that both the magnitude and sign of the difference varies between cases. In some cases, the magnitude of the difference is huge: most notably in the Congo basin case, the maximum difference in liquid cloud mass between the Morrison and Thompson schemes is on the order of $10^4\,\mathrm{kg\,kg^{-1}}$ more in Thompson (whereas in the RICO shallow cumulus case the maximum difference is on the order of $10^1\,\mathrm{kg\,kg^{-1}}$ less in Thompson). Likewise, in the RICO case the maximum difference in rain mass between the Morrison and Thompson schemes is on the order of $10^3\,\mathrm{kg\,kg^{-1}}$ more in Thompson (whereas in the Congo basin case the maximum difference is on the order of $10^1\,\mathrm{kg\,kg^{-1}}$ less in Thompson). Even for hydrometeors that have differences of the same order of magnitude, the sign of the difference can vary between cases. This result highlights the need for better observational constraints on mixed-phase and ice cloud microphysics and hydrometeors, and also perhaps the need for a shift in the development of microphysics parameterisations away from schemes which (somewhat arbitrarily) partition hydrometeors into separate categories. This is also supported by our sensitivity tests of autoconversion of cloud ice to snow in our Congo simulations.

Another key finding is that the cloud morphological difference and the difference in the hydrometeors between different schemes is significantly larger than that due to CDNC perturbations. Although we have restricted our study to the comparison of double-moment bulk microphysics schemes, this result is consistent with Khain and Lynn (2009), who found that the difference in convection between a bulk and a bin scheme was much greater than the difference within each scheme to varying aerosol concentrations. Some studies have found a significantly weaker re-

Table 3. Maximum relative difference of domain-mean hydrometeor mass mixing ratio profiles for the MORR and THOM schemes. The relative change in the hydrometeor mass mixing ratios are computed in each case for M250 minus T250.

Difference	CONGO	SUPERCELL	RICO
Liquid cloud mass	−10 900 %	−58.3 %	+17.0 %
Ice mass	+98.7 %	+96.9 %	n/a
Rain mass	+82.2 %	−138 %	−3830 %
Snow mass	+40.8 %	−99.8 %	n/a
Graupel mass	+91.6 %	+72.7 %	n/a

n/a = not applicable

sponse to aerosol when using bulk schemes compared to bin schemes; e.g. Khain and Lynn (2009); Lebo and Seinfeld (2011). In idealised simulations of continental deep convection, Lebo and Seinfeld (2011) found that increases in CCN concentrations led to increased ice mass and total condensed water mass aloft in both bin and bulk schemes but increased domain-averaged cumulative surface precipitation in the bulk scheme compared to a decrease in the bin scheme. This was because the relative increase in condensate mass aloft under polluted conditions was found to be much larger in the simulations performed with bulk microphysics as a result of increased numbers of smaller cloud particles with slower sedimentation speeds, thus resulting in reduced surface precipitation. However, in our idealised supercell simulations we find a similar magnitude of response to aerosol when using a bin scheme as in the two bulk schemes which are the focus of this study.

That cloud and precipitation development and their aerosol response differs across different cloud types in different large-scale environments is expected. Many studies have shown that aerosol effects on precipitation depend on the large-scale environment and cloud type (e.g. Khain et al., 2004; Fan et al., 2007; Lynn et al., 2005a, b; Lynn and Khain, 2007; Seifert and Beheng, 2006a; Tao et al., 2007) for reasons related to differences in different cloud types between the timescale of increased sedimentation through aerosol loading and subsequent sublimation and evaporation timescales. Further, several studies of deep convection have found that the effects of aerosol on deep convection are much weaker than those of relative humidity (e.g. van den Heever et al., 2006; Fan et al., 2007; Khain and Lynn, 2009). Fan et al. (2007) found that in idealised simulations of continental and maritime clouds using bin microphysics the magnitude and even the sign of aerosol effects on precipitation depended on relative humidity. Fan et al. (2007) found that aerosol response in idealised simulations of clouds using bin microphysics and soundings from Houston, Texas strongly depended on relative humidity with a negligible effect on cloud properties and precipitation in dry air but more significant effects in humid air. Conversely, in idealised low-

precipitation supercell simulations with bulk microphysics and dry low-level humidity performed as part of the study by Kalina et al. (2014), cold pool area decreased by 84 % and domain-averaged precipitation was reduced by 50 % under polluted conditions; however, it was insensitive to polluted conditions when a moist sounding was used. Thus, assuming that the response in our simulations would likely be more similar to the results found for bulk microphysics by Kalina et al. (2014), the magnitude of our results in the supercell case (which uses a moist sounding) may be smaller than it would be in drier environmental conditions.

In 10-day simulations of deep convection in the Congo basin in August 2007, we find that both the Morrison and Thompson schemes have a significant positive bias in cloud and surface precipitation compared to GERB and TRMM. This may be in part attributable to the positive moist bias in the Congo basin in the ERA-Interim reanalysis (used as boundary data for the Congo simulation) when compared to other reanalyses (Washington et al., 2013). Despite the positive cloud fraction bias in both schemes, we find that the Thompson scheme compares better than the Morrison scheme against observed cloud fractions, largely due to the overproduction of upper-level ice in the Morrison scheme. This is in agreement with Cintineo et al. (2014), who found that (despite the two schemes having different biases at different levels) the Thompson scheme outperformed the Morrison scheme overall against satellite observations of cloud in North America due to its more accurate upper-level cloud distribution, whereas the Morrison scheme had too much upper-level cloud through the overproduction of ice. This bias is attributable to differences in the way in which the two schemes convert cloud ice to snow. However, we also find that despite a positive surface precipitation bias in both schemes, the Morrison scheme compares better to observations in this region over this period. Morrison and Grabowski (2007) found that differences in accumulated precipitation produced by warm stratocumulus and warm cumulus clouds using different microphysics schemes were only on the order of 10 to 20 %, suggesting that accumulated rain is largely controlled by large-scale atmospheric properties. However, differences in accumulated rain in our Congo simulations can be attributed to differences in the microphysics schemes because all simulations used the same input and boundary data and therefore are under the influence of the same large-scale atmospheric conditions. That one scheme best represents cold cloud compared to observations but the other scheme better reproduces accumulated precipitation makes it difficult to conclude that one scheme outperforms another overall. It also suggests that when setting up a model configuration for research purposes, one consideration to distinguish between the use of these two particular schemes may be whether surface precipitation or radiative effects are more important to the research question.

We note here that the RRTM LW and Goddard SW radiation schemes used in these simulations are only coupled to the microphysics through the hydrometeor masses and not the numbers. This coupling therefore cannot account for changes in hydrometeor sizes, and thus some aerosol effects will be missing from these simulations. Additionally, the microphysics–radiation coupling is only through cloud water and ice and none of the other frozen species. This missing aerosol effect may have an especially important impact in our Congo simulations in which the Morrison scheme develops and retains significant amounts of upper-level ice, whereas the Thompson scheme converts nearly all the ice to snow, which the radiation scheme will not see. This could have significant radiative flux and feedback impacts (Thompson et al., 2016) which originate from the use of somewhat arbitrarily defined ice categories (e.g. if the size parameter at which cloud ice is converted to snow is changed, a bulk mass of cloud ice is removed from the radiatively coupled ice category and moved into the non-radiatively coupled snow category).

We present the new result that variability in aerosol response due to the choice of microphysics scheme differs not only between schemes, but the inter-scheme variability also differs between cases of convection. The maximum relative difference in the domain-mean hydrometeor profiles between polluted and pristine CDNC values for each of the model configurations is summarised in Table 4. It is clear that both the magnitude and the sign of the response of each hydrometeor class to CDNC differ strongly not only between microphysics schemes, but also between cases. (Note that Table 4 shows relative amounts and that the absolute difference in response to CDNC between each of the schemes and cases can also vary significantly). Whilst it is not surprising that the different cases of convection differ in their hydrometeor development and in their response to polluted conditions, it is worth noting the magnitude of and variation in the difference in response. A body of literature uses idealised model configurations to investigate storm system response to aerosol loading (e.g. Seifert and Beheng, 2006b; Khain and Lynn, 2009; Lebo and Seinfeld, 2011; Morrison, 2012) and to compare microphysics schemes (e.g. Lebo and Seinfeld, 2011). Our results highlight that the storm system response in such a model configuration may not be representative of the response over larger spatiotemporal scales, supporting similar findings of larger-scale feedbacks and life-cycle-dependent responses in idealised (Morrison and Grabowski, 2011; Lee, 2012) and real-data (van den Heever et al., 2006) studies of aerosol–convection interactions.

We note that the vertical resolution used in this study is relatively coarse and that a horizontal grid length of 4 km is at the limit of what may be considered as "convection-permitting" (Bryan et al., 2003). However, we use this grid spacing for consistency with a previous study in which 10 and 4 km grid lengths were shown to be sufficient to reproduce storm characteristics and aerosol–convection interactions in the Congo basin (Gryspeerdt et al., 2015). Previous studies have indicated sensitivity of convection to hor-

izontal grid spacing (e.g. Bryan and Morrison, 2012; Potvin and Flora, 2015) and also that the sensitivity to grid length can vary with microphysics scheme (Morrison et al., 2015b), although idealised ensemble studies of response to aerosol have shown that differences between polluted and pristine conditions were similar in simulations using horizontal grid lengths of 4, 2 and 0.5 km, respectively, and were also relatively robust to domain size (Morrison and Grabowski, 2011).

An important factor in our set-up is that we use the same values of prescribed CDNC in all of our cases. Whilst the literature also shows widely varying response to aerosol, especially between bin and bulk schemes (in which even the sign of the response may differ), Kalina et al. (2014) showed in idealised supercell simulations using 15 CCN concentrations and four environmental soundings that changes in cold pool characteristics with CCN were nonmonotonic and dependent on the environmental conditions. Therefore our use of the same CDNC values in multiple types of convection helps to minimise uncertainty due to nonmonotonic behaviour. However, considering the results of Kalina et al. (2014), we note that a caveat of the present study (and indeed of the majority of existing studies) is that the absolute values of the cloud system and precipitation response to aerosol identified here may only hold for the CDNC values used in our study.

We find that the autoconversion representation alone is sufficient to explain most of the differences between microphysics schemes in the shallow cumulus case both in terms of their representation of cloud and precipitation (consistent with Li et al., 2015) and in terms of their response to CDNC. The dominant hydrometeor difference between the microphysics schemes in the RICO simulations occurs in the rain – a different result from both the Congo basin configuration (in which the dominant difference occurs in the liquid cloud) and the idealised supercell configuration (in which the dominant difference occurs in the graupel). We also find that autoconversion of cloud droplets to rain is the mechanism that prevents the formation (or persistence) of liquid-phase cloud in the south of the domain in the Congo basin simulations using the Morrison scheme. This is in agreement with the study of Kalina et al. (2014), who found in idealised supercell simulations using the Morrison bulk microphysics scheme with a variable shape parameter for the raindrop size distribution that autoconversion rates decreased under CCN loading. The importance of autoconversion representation was shown by Gilmore and Straka (2008), who demonstrated that the rates predicted by the autoconversion formulae used in bulk schemes differ by orders of magnitude. Modelling studies and observations from RICO have found that warm-rain formation can be explained by the observed aerosol distribution (Blyth et al., 2013). In the context of our findings, this suggests that an accurate description of the autoconversion process in warm-rain regimes is fundamental not only to a realistic representation of cloud and precipitation, but also to its response to varying aerosol concentrations.

Table 4. Maximum relative difference in the response of model configurations to polluted conditions. The relative change in the domain-mean rehydrometeor mass mixing ratios are computed in each case for CDNC values of $2500\,\mathrm{cm}^{-3}$ minus $100\,\mathrm{cm}^{-3}$.

Difference	CONGO-MORR	CONGO-THOM	SUPER-MORR	SUPER-THOM	RICO-MORR	RICO-THOM
Liquid cloud mass	$-0.59\,\%$	$+32.2\,\%$	$+146\,\%$	$+169\,\%$	$-5.21\,\%$	$+44.0\,\%$
Ice mass	$+12.5\,\%$	$-5.61\,\%$	$+116\,\%$	$+29.7\,\%$	n/a	n/a
Rain mass	$-0.67\,\%$	$-62.6\,\%$	$-93.7\,\%$	$-51.6\,\%$	$-100\,\%$	$-100\,\%$
Snow mass	$+1.37\,\%$	$+13.8\,\%$	$-33.5\,\%$	$+109\,\%$	n/a	n/a
Graupel mass	$-4.60\,\%$	$-29.9\,\%$	$-19.1\,\%$	$-36.7\,\%$	n/a	n/a

n/a = not applicable

We caution that care should be taken when using autoconversion schemes in regimes other than those for which they were originally developed, such as the use of the Khairoutdinov and Kogan (2000) scheme for deep convective cases. Although not the focus of the present study, those interested in testing and improving autoconversion schemes could do so by calculating the mean volume radius and thereby the height of first raindrop formation through knowledge of the 13 to 14 µm critical radius for raindrop production (Freud and Rosenfeld, 2012; Khain et al., 2013; Rosenfeld et al., 2014). Similarly, comparison of results from bulk models to those from bin models (e.g. Thompson et al., 2004; Igel and van den Heever, 2017c) can also be a valuable tool for testing schemes. Based on the limited set of cases in our study, we would not be justified in recommending one of the autoconversion schemes over the other. Moreover, because there are so many competing processes besides autoconversion, including a number of microphysical and dynamical processes, it could be misleading to claim that one scheme is better than the other based solely on bulk comparison with observations from a few cases. For those interested in testing and evaluating the autoconversion schemes, we suggest that the best approach would be to perform offline testing based on detailed in situ observations and calculations, as was done by e.g. Wood (2005), who tested the Khairoutdinov and Kogan (2000) autoconversion scheme in such a manner.

Our results (which are shown to hold across multiple cloud types and types of simulation) have important implications not only for cloud-resolving simulations, but also for the global modelling community. Most significant, perhaps, is the radiative impact which could arise when such major differences occur in the ice phase. Our Congo simulations illustrate just how large this uncertainty may be, and our tests using a bin scheme show that this is not purely an artefact of the bulk microphysics schemes used. Further, that uncertainties due to the choice of microphysics scheme dominate any aerosol response within a given scheme has implications for global modelling studies of aerosol indirect effects (e.g. Zhang et al., 2016; Ghan et al., 2016). Once again, this highlights the continuing need of our community for tight observational constraints on cloud and precipitation processes and their response to aerosol, as well as for ongoing parameterisation development to allow these processes to be accurately represented in large domain (or global), long-term simulations.

Competing interests. The authors declare that they have no conflict of interest.

Acknowledgements. This work used the ARCHER UK National Supercomputing Service (http://www.archer.ac.uk). The research leading to these results has received funding from the European Research Council under the European Union's Seventh Framework Programme (FP7/2007–2013)/ERC grant agreement no. FP7-280025 (ACCLAIM) and grant agreement no. FP7-306284 (QUARERE). The Congo precipitation data used in this study were acquired as part of the Tropical Rainfall Measuring Mission (TRMM). The algorithms were developed by the TRMM Science Team. The data were processed by the TRMM Science Data and Information System (TSDIS) and the TRMM office; they are archived and distributed by the Goddard Distributed Active Archive Center. TRMM is an international project jointly sponsored by the Japan National Space Development Agency (NASDA) and the US National Aeronautics and Space Administration (NASA) Office of Earth Sciences. The Congo radiance data used in this study were acquired as part of the Geostationary Earth Radiation Budget Project. The CloudSat data were obtained from the CloudSat Data Processing Center. ERA-Interim data provided courtesy ECMWF. Thanks go to Laurent Labbouz (University of Oxford, UK) for helpful comments on this paper.

Edited by: Radovan Krejci

References

Albrecht, B.: Aerosols, cloud microphysics, and fractional cloudiness, Science, 245, 1227–1230, 1989.

Altaratz, O., Koren, I., Remer, L., and Hirsch, E.: Review: Cloud invigoration by aerosols Coupling between microphysics and dynamics, Atmos. Res., 140–141, 38–60, https://doi.org/10.1016/j.atmosres.2014.01.009, 2014.

Beljaars, A.: The parameterization of surface fluxes in large-scale models under free convection, Q. J. Roy. Meteor. Soc., 121, 225–270, 1994.

Berry, E. and Reinhardt, R.: An analysis of cloud drop growth by collection: Part I. Double distributions, J. Atmos. Sci., 31, 1814–1824, 1974.

Blyth, A., Lowenstein, J., Huang, Y., Cui, Z., Davies, S., and Carslaw, K.: The production of warm rain in shallowmaritime cumulus clouds, Q. J. Roy. Meteor. Soc., 139, 20–31, 2013.

Bryan, G. and Morrison, H.: Sensitivity of a simulated squall line to horizontal resolution and parameterization of microphysics, Mon. Weather Rev., 140, 202–225, https://doi.org/10.1175/MWR-D-11-00046.1, 2012.

Bryan, G., Wyngaard, J., and Fritsch, J.: Resolution requirements for the simulation of deep moist convection, Mon. Weather Rev., 131, 2394–2416, https://doi.org/10.1175/1520-0493(2003)131<2394:RRFTSO>2.0.CO;2, 2003.

Chand, D., Wood, R., Ghan, S. J., Wang, M., Ovchinnikov, M., Rasch, P. J., Miller, S., Schichtel, B., and Moore, T.: Aerosol optical depth increase in partly cloudy conditions, J. Geophys. Res.-Atmos., 117, D17207, https://doi.org/10.1029/2012JD017894, 2012.

Chou, M.-D. and Suarez, M.: An efficient thermal infrared radiation parameterization for use in general circulation models, NASA Technical Memo, 1994.

Cintineo, R., Otkin, J. A., Xue, M., and Kong, F.: Evaluating the performance of planetary boundary layer and cloud microphysical parameterization schemes in convection-permitting ensemble forecasts using synthetic GOES-13 satellite observations, Mon. Weather Rev., 142, 163–182, 2014.

Dee, D. P., Uppala, S. M., Simmons, A. J., Berrisford, P., Poli, P., Kobayashi, S., Andrae, U., Balmaseda, M. A., Balsamo, G., Bauer, P., Bechtold, P., Beljaars, A. C. M., van de Berg, L., Bidlot, J., Bormann, N., Delsol, C., Dragani, R., Fuentes, M., Geer, A. J., Haimberger, L., Healy, S. B., Hersbach, H., Hólm, E. V., Isaksen, L., Kållberg, P., Köhler, M., Matricardi, M., McNally, A. P., Monge-Sanz, B. M., Morcrette, J.-J., Park, B.-K., Peubey, C., de Rosnay, P., Tavolato, C., Thépaut, J.-N., and Vitart, F.: The ERA-Interim reanalysis: configuration and performance of the data assimilation system, Q. J. Roy. Meteor. Soc., 137, 553–597, 2011.

Dyer, A. and Hicks, B.: Flux-gradient relationships in the constant flux layer, Q. J. Roy. Meteor. Soc., 96, 715–721, 1970.

Ek, M. and Mahrt, L.: SU 1-D PBL Model User's Guide, Version 1.04, Department of Atmospheric Sciences, Department of Atmospheric Sciences, Oregon State University, 1991.

Fan, J., Zhang, R., Li, G., and Tao, W.-K.: Effects of aerosols and relative humidity on cumulus clouds, J. Geophys. Res.-Atmos., 112, D14204, https://doi.org/10.1029/2006JD008136, 2007.

Fan, J., Yuan, T., Comstock, J. M., Ghan, S., Khain, A., Leung, L. R., Li, Z., Martins, V. J., and Ovchinnikov, M.: Dominant role by vertical wind shear in regulating aerosol effects on deep convective clouds, J. Geophys. Res.-Atmos., 114, D22206, https://doi.org/10.1029/2009JD012352, 2009.

Fan, J., Leung, L. R., Li, Z., Morrison, H., Chen, H., Zhou, Y., Qian, Y., and Wang, Y.: Aerosol impacts on clouds and precipitation in eastern China: Results from bin and bulk microphysics, J. Geophys. Res.-Atmos., 117, D00K36, https://doi.org/10.1029/2011JD016537, 2012a.

Fan, J., Rosenfeld, D., Ding, Y., Leung, L. R., and Li, Z.: Potential aerosol indirect effects on atmospheric circulation and radia-tive forcing through deep convection, Geophys. Res. Lett., 39, L09806, https://doi.org/10.1029/2012GL051851, 2012b.

Fan, J., Leung, L. R., Rosenfeld, D., Chen, Q., Li, Z., Zhang, J., and Yan, H.: Microphysical effects determine macrophysical response for aerosol impacts on deep convective clouds, P. Natl. Acad. Sci. USA, 110, E4581–E4590, 2013.

Feingold, G., Stevens, B., Cotton, W., and Walko, R.: An explicit cloud microphysics/LES model designed to simulate the Twomey effect, Atmos. Res., 33, 207–233, 1994.

Freud, E. and Rosenfeld, D.: Linear relation between convective cloud drop number concentration and depth for rain initiation, J. Geophys. Res.-Atmospheres, 117, d02207, https://doi.org/10.1029/2011JD016457, 2012.

Gallus Jr., W. A. and Pfeifer, M.: Intercomparison of simulations using 5 WRF microphysical schemes with dual-Polarization data for a German squall line, Adv. Geosci., 16, 109–116, https://doi.org/10.5194/adgeo-16-109-2008, 2008.

Geresdi, I.: Idealized simulation of the Colorado hailstorm case: comparison of bulk and detailed microphysics, Atmos. Res., 45, 237–252, https://doi.org/10.1016/S0169-8095(97)00079-3, 1998.

Ghan, S., Wang, M., Zhang, S., Ferrachat, S., Gettelman, A., Griesfeller, J., Kipling, Z., Lohmann, U., Morrison, H., Neubauer, D., Partridge, D. G., Stier, P., Takemura, T., Wang, H., and Zhang, K.: Challenges in constraining anthropogenic aerosol effects on cloud radiative forcing using present-day spatiotemporal variability, P. Natl. Acad. Sci. USA, 113, 5804–5811, https://doi.org/10.1073/pnas.1514036113, 2016.

Gilmore, M. S. and Straka, J. M.: The Berry and Reinhardt autoconversion parameterization: a digest, J. Appl. Meteorol. Clim., 47, 375–396, https://doi.org/10.1175/2007JAMC1573.1, 2008.

Gryspeerdt, E., Stier, P., and Partridge, D. G.: Links between satellite-retrieved aerosol and precipitation, Atmos. Chem. Phys., 14, 9677–9694, https://doi.org/10.5194/acp-14-9677-2014, 2014.

Gryspeerdt, E., Stier, P., White, B. A., and Kipling, Z.: Wet scavenging limits the detection of aerosol effects on precipitation, Atmos. Chem. Phys., 15, 7557–7570, https://doi.org/10.5194/acp-15-7557-2015, 2015.

Harries, J., Russell, J., Hanafin, J., Brindley, H., Futyan, J., Rufus, J., Kellock, S., Matthews, G., Wrigley, R., Last, A., Mueller, J., Mossavati, R., Ashmall, J., Sawyer, E., Parker, D., Caldwell, M., Allan, P., Smith, A., Bates, M., Coan, B., Stewart, B., Lepine, D., Cornwall, L., Corney, D., Ricketts, M., Drummond, D., Smart, D., Cutler, R., Dewitte, S., Clerbaux, N., Gonzalez, L., Ipe, A., Bertrand, C., Joukoff, A., Crommelynck, D., Nelms, N., Llewellyn-Jones, D., Butcher, G., Smith, G., Szewczyk, Z., Mlynczak, P., Slingo, A., Allan, R., and Ringer, M.: The geostationary Earth Radiation Budget Project, B. Am. Meteorol. Soc., 86, 945–960, https://doi.org/10.1175/BAMS-86-7-945, 2005.

Harrington, J. Y., Sulia, K., and Morrison, H.: A method for adaptive habit prediction in bulk microphysical models. Part I: Theoretical development, J. Atmos. Sci., 70, 349–364, 2013.

Haynes, J. M., Luo, Z., Stephens, G. L., Marchand, R. T., and Bodas-Salcedo, A.: A multipurpose radar simulation package: QuickBeam, B. Am. Meteorol. Soc., 88, 1723–1727, https://doi.org/10.1175/BAMS-88-11-1723, 2007.

Hong, S.-Y., Noh, Y., and Dudhia, J.: A new vertical diffusion package with explicit treatment of entrainment processes, Mon. Weather Rev., 134, 2318–2341, 2006.

Hong, S.-Y., Lim, K.-S. S., Kim, J.-H., Lim, J.-O. J., and Dudhia, J.: Sensitivity study of cloud-resolving convective simulations with WRF using two bulk microphysical parameterizations: ice-phase microphysics versus sedimentation effects, J. Appl. Meteorol. Clim., 48, 61–76, 2009.

Huffman, G. J., Adler, R. F., Bolvin, D. T., Gu, G., Nelkin, E. J., Bowman, K. P., Hong, Y., Stocker, E. F., David, and Wolff, B.: The TRMM multi-satellite precipitation analysis: quasi-global, multi-year, combined-sensor precipitation estimates at fine scale, J. Hydrometeorol., 8, 38–55, 2007.

Igel, A., Igel, M., and van den Heever, S.: Make it a double? Sobering results from simulations using single-moment microphysics schemes, J. Atmos. Sci., 72, 910–925, https://doi.org/10.1175/JAS-D-14-0107.1, 2015.

Igel, A. L. and van den Heever, S. C.: The importance of the shape of cloud droplet size distributions in shallow cumulus clouds. Part I: Bin microphysics simulations, J. Atmos. Sci., 74, 249–258, https://doi.org/10.1175/JAS-D-15-0382.1, 2017a.

Igel, A. L. and van den Heever, S. C.: The importance of the shape of cloud droplet size distributions in shallow cumulus clouds. Part II: Bulk microphysics simulations, J. Atmos. Sci., 74, 259–273, https://doi.org/10.1175/JAS-D-15-0383.1, 2017b.

Igel, A. L. and van den Heever, S. C.: The role of the gamma function shape parameter in determining differences between condensation rates in bin and bulk microphysics schemes, Atmos. Chem. Phys., 17, 4599–4609, https://doi.org/10.5194/acp-17-4599-2017, 2017c.

Jankov, I., Grasso, L., Sengupta, M., Neiman, P., Zupanski, D., Zupanski, M., Lindsey, D., Hillger, D., Birkenheuer, D., Brummer, R., and Yuan, H.: An evaluation of five arw-wrf microphysics schemes using synthetic goes imagery for an atmospheric river event affecting the california coast, J. Hydrometeorol., 12, 618–633, https://doi.org/10.1175/2010JHM1282.1, 2011.

Jiang, H., Cotton, W., Pinto, J., Curry, J., and Weissbluth, M.: Cloud resolving simulations of mixed-phase arctic stratus observed during BASE: sensitivity to concentration of ice crystals and large-scale heat and moisture advection, J. Atmos. Sci., 57, 2105–2117, https://doi.org/10.1175/1520-0469(2000)057<2105:CRSOMP>2.0.CO;2, 2000.

Kalina, E. A., Friedrich, K., Morrison, H., and Bryan, G. H.: Aerosol effects on idealized supercell thunderstorms in different environments, J. Atmos. Sci., 71, 4558–4580, https://doi.org/10.1175/JAS-D-14-0037.1, 2014.

Kessler, E. I.: On the distribution and continuity of water substance in atmospheric circulations, Meteor. Mono., 10, 88 pp., 1969.

Khain, A. and Lynn, B.: Simulation of a supercell storm in clean and dirty atmosphere using weather research and forecast model with spectral bin microphysics, J. Geophys. Res.-Atmos., 114, D19209, https://doi.org/10.1029/2009JD011827, 2009.

Khain, A., Pokrovsky, A., Pinsky, M., Seifert, A., and Phillips, V.: Simulation of effects of atmospheric aerosols on deep turbulent convective clouds using a spectral microphysics mixed-phase cumulus cloud model. Part I: Model description and possible applications, J. Atmos. Sci., 61, 2963–2982, 2004.

Khain, A., Rosenfeld, D., Pokrovsky, A., Blahak, U., and Ryzhkov, A.: The role of CCN in precipitation and hail in a

mid-latitude storm as seen in simulations using a spectral (bin) microphysics model in a 2D dynamic frame, Atmos. Res., 99, 129–146, https://doi.org/10.1016/j.atmosres.2010.09.015, 2011.

Khain, A., Prabha, T. V., Benmoshe, N., Pandithurai, G., and Ovchinnikov, M.: The mechanism of first raindrops formation in deep convective clouds, J. Geophys. Res.-Atmos., 118, 9123–9140, https://doi.org/10.1002/jgrd.50641, 2013.

Khain, A., Lynn, B., and Shpund, J.: High resolution WRF simulations of Hurricane Irene: sensitivity to aerosols and choice of microphysical schemes, Atmos. Res., 167, 129–145, https://doi.org/10.1016/j.atmosres.2015.07.014, 2016.

Khain, A. P., Beheng, K. D., Heymsfield, A., Korolev, A., Krichak, S. O., Levin, Z., Pinsky, M., Phillips, V., Prabhakaran, T., Teller, A., van den Heever, S. C., and Yano, J.-I.: Representation of microphysical processes in cloud-resolving models: Spectral (bin) microphysics versus bulk parameterization, Rev. Geophys., 53, 247–322, https://doi.org/10.1002/2014RG000468, 2015.

Khairoutdinov, M. and Kogan, Y.: A new cloud physics parameterization in a large-eddy simulation model of marine stratocumulus, Mon. Weather Rev., 128, 229–243, 2000.

Koren, I., Kaufman, Y. J., Rosenfeld, D., Remer, L. A., and Rudich, Y.: Aerosol invigoration and restructuring of Atlantic convective clouds, Geophys. Res. Lett., 32, L14828, https://doi.org/10.1029/2005GL023187, 2005.

Kumjian, M. R. and Ryzhkov, A. V.: The impact of size sorting on the polarimetric radar variables, J. Atmos. Sci., 69, 2042–2060, https://doi.org/10.1175/JAS-D-11-0125.1, 2012.

Lebo, Z. J. and Morrison, H.: Dynamical effects of aerosol perturbations on simulated idealized squall lines, Mon. Weather Rev., 142, 991–1009, 2014.

Lebo, Z. J. and Seinfeld, J. H.: Theoretical basis for convective invigoration due to increased aerosol concentration, Atmos. Chem. Phys., 11, 5407–5429, https://doi.org/10.5194/acp-11-5407-2011, 2011.

Lebo, Z. J., Morrison, H., and Seinfeld, J. H.: Are simulated aerosol-induced effects on deep convective clouds strongly dependent on saturation adjustment?, Atmos. Chem. Phys., 12, 9941–9964, https://doi.org/10.5194/acp-12-9941-2012, 2012.

Lee, S.-S.: Effect of aerosol on circulations and precipitation in deep convective clouds, J. Atmos. Sci., 69, 1957–1974, 2012.

Lee, S.-S. and Feingold, G.: Precipitating cloud-system response to aerosol perturbations, Geophys. Res. Lett., 37, L23806, https://doi.org/10.1029/2010GL045596, 2010.

Lee, S.-S. and Feingold, G.: Aerosol effects on the cloud-field properties of tropical convective clouds, Atmos. Chem. Phys., 13, 6713–6726, https://doi.org/10.5194/acp-13-6713-2013, 2013.

Li, X., Tao, W.-K., Khain, A. P., Simpson, J., and Johnson, D. E.: Sensitivity of a cloud-resolving model to bulk and explicit bin microphysical schemes. Part I: Comparisons, J. Atmos. Sci., 66, 3–21, https://doi.org/10.1175/2008JAS2646.1, 2009a.

Li, X., Tao, W.-K., Khain, A. P., Simpson, J., and Johnson, D. E.: Sensitivity of a cloud-resolving model to bulk and explicit bin microphysical schemes. Part II: Cloud microphysics and storm dynamics interactions, J. Atmos. Sci., 66, 22–40, https://doi.org/10.1175/2008JAS2647.1, 2009b.

Li, Z., Zuidema, P., Zhu, P., and Morrison, H.: The sensitivity of simulated shallow cumulus convection and cold pools to microphysics, J. Atmos. Sci., 72, 3340–3355, 2015.

Lin, Y.-L., Farley, R., and Orville, H.: Bulk parameterization of the snow field in a cloud model, J. Climate Appl. Meteor., 22, 1065–1092, https://doi.org/10.1175/1520-0450(1983)022<1065:BPOTSF>2.0.CO;2, 1983.

Loftus, A. and Cotton, W.: Examination of CCN impacts on hail in a simulated supercell storm with triple-moment hail bulk microphysics, Atmos. Res. , 147–148, 183–204, https://doi.org/10.1016/j.atmosres.2014.04.017, 2014.

Lynn, B. and Khain, A.: Utilization of spectral bin microphysics and bulk parameterization schemes to simulate the cloud structure and precipitation in a mesoscale rain event, J. Geophys. Res.-Atmos., 112, D22205, https://doi.org/10.1029/2007JD008475, 2007.

Lynn, B. H., Khain, A. P., Dudhia, J., Rosenfeld, D., Pokrovsky, A., and Seifert, A.: Spectral (bin) microphysics coupled with a mesoscale model (MM5). Part I: Model description and first results, Mon. Weather Rev., 133, 44–58, https://doi.org/10.1175/MWR-2840.1, 2005a.

Lynn, B. H., Khain, A. P., Dudhia, J., Rosenfeld, D., Pokrovsky, A., and Seifert, A.: Spectral (bin) microphysics coupled with a mesoscale model (MM5). Part II: Simulation of a CaPE rain event with a squall line, Mon. Weather Rev., 133, 59–71, https://doi.org/10.1175/MWR-2841.1, 2005b.

Marchand, R., Mace, G. G., Ackerman, T., and Stephens, G.: Hydrometeor detection using cloudsat an earth-orbiting 94-GHz cloud radar, J. Atmos. Ocean. Tech., 25, 519–533, https://doi.org/10.1175/2007JTECHA1006.1, 2008.

Mauger, G. S. and Norris, J. R.: Meteorological bias in satellite estimates of aerosol-cloud relationships, Geophys. Res. Lett., 34, L16824, https://doi.org/10.1029/2007GL029952, 2007.

McFarquhar, G. M., Zhang, H., Heymsfield, G., Halverson, J. B., Hood, R., Dudhia, J., and Marks, F.: Factors affecting the evolution of Hurricane Erin (2001) and the distributions of hydrometeors: role of microphysical processes, J. Atmos. Sci., 63, 127–150, 2006.

Meyers, M. P., Walko, R. L., Harrington, J. Y., and Cotton, W. R.: New RAMS cloud microphysics parameterization. Part II: The two-moment scheme, Atmos. Res., 45, 3–39, https://doi.org/10.1016/S0169-8095(97)00018-5, 1997.

Milbrandt, J. A. and Yau, M. K.: A multimoment bulk microphysics parameterization. Part III: Control simulation of a hailstorm, J. Atmos. Sci., 63, 3114–3136, https://doi.org/10.1175/JAS3816.1, 2006.

Mlawer, E., Taubman, S., Brown, P., Iacono, M., and Clough, S.: Radiative transfer for inhomogeneous atmospheres: RRTM, a validated correlated k-model for the longwave, J. Geophys. Res., 102, 16663–16682, https://doi.org/10.1029/97JD00237, 1997.

Morrison, H.: On the robustness of aerosol effects on an idealized supercell storm simulated with a cloud system-resolving model, Atmos. Chem. Phys., 12, 7689–7705, https://doi.org/10.5194/acp-12-7689-2012, 2012.

Morrison, H. and Grabowski, W. W.: Comparison of bulk and bin warm-rain microphysics models using a kinematic framework, J. Atmos. Sci., 64, 2839–2861, https://doi.org/10.1175/JAS3980, 2007.

Morrison, H. and Grabowski, W. W.: A novel approach for representing ice microphysics in models: description and tests using a kinematic framework, J. Atmos. Sci., 65, 1528–1548, 2008.

Morrison, H. and Grabowski, W. W.: Cloud-system resolving model simulations of aerosol indirect effects on tropical deep convection and its thermodynamic environment, Atmos. Chem. Phys., 11, 10503–10523, https://doi.org/10.5194/acp-11-10503-2011, 2011.

Morrison, H. and Milbrandt, J.: Comparison of two-moment bulk microphysics schemes in idealized supercell thunderstorm simulations, Mon. Weather Rev., 139, 1103–1130, https://doi.org/10.1175/2010MWR3433.1, 2011.

Morrison, H. and Milbrandt, J. A.: Parameterization of cloud microphysics based on the prediction of bulk ice particle properties. Part I: Scheme description and idealized tests, J. Atmos. Sci., 72, 287–311, 2015.

Morrison, H. and Pinto, J.: Mesoscale modeling of springtime arctic mixed-phase stratiform clouds using a new two-moment bulk microphysics scheme, J. Atmos. Sci., 62, 3683–3704, https://doi.org/10.1175/JAS3564.1, 2005.

Morrison, H. and Pinto, J.: Intercomparison of bulk cloud microphysics schemes in mesoscale simulations of springtime arctic mixed-phase stratiform clouds, Mon. Weather Rev., 134, 1880–190, https://doi.org/10.1175/MWR3154.1, 2006.

Morrison, H., Curry, J., and Khvorostyanov, V.: A new double-moment microphysics parameterization for application in cloud and climate models. Part I: Description, J. Atmos. Sci., 62, 1665–1677, 2005.

Morrison, H., Thompson, G., and Tatarskii, V.: Impact of cloud microphysics on the development of trailing stratiform precipitation in a simulated squall line: comparison of one- and two-moment schemes, Mon. Weather Rev., 137, 991–1007, https://doi.org/10.1175/2008MWR2556.1, 2009.

Morrison, H., Milbrandt, J. A., Bryan, G. H., Ikeda, K., Tessendorf, S. A., and Thompson, G.: Parameterization of cloud microphysics based on the prediction of bulk ice particle properties. Part II: Case study comparisons with observations and other schemes, J. Atmos. Sci., 72, 312–339, 2015a.

Morrison, H., Morales, A., and Villanueva-Birriel, C.: Concurrent sensitivities of an idealized deep convective storm to parameterization of microphysics, horizontal grid resolution, and environmental static stability, Mon. Weather Rev., 143, 2082—2104, https://doi.org/10.1175/MWR-D-14-00271.1, 2015b.

Noppel, H., Blahak, U., Seifert, A., and Beheng, K. D.: Simulations of a hailstorm and the impact of CCN using an advanced two-moment cloud microphysical scheme, Atmos. Res., 96, 286–301, 2010.

Paulson, C.: The mathematical representation of wind speed and temperature profiles in the unstable atmospheric surface layer, J. Appl. Meteorol., 9, 857–861, 1970.

Potvin, C. and Flora, M.: Sensitivity of idealized supercell simulations to horizontal grid spacing: implications for warn-on-forecast, Mon. Weather Rev., 143, 2998–3024, https://doi.org/10.1175/MWR-D-14-00416.1, 2015.

Rajeevan, M., Kesarkar, A., Thampi, S. B., Rao, T. N., Radhakrishna, B., and Rajasekhar, M.: Sensitivity of WRF cloud microphysics to simulations of a severe thunderstorm event over Southeast India, Ann. Geophys., 28, 603–619, https://doi.org/10.5194/angeo-28-603-2010, 2010.

Rauber, R., Ochs, H. I., Di Girolamo, L., Göke, S., Snodgrass, E., Stevens, B., Knight, C., Jensen, J. B., Lenschow, D. H.,

Uncertainty from the choice of microphysics scheme in convection-permitting models significantly exceeds aerosol...

159

Rilling, R. A., Rogers, D. C., Stith, J. L., Albrecht, B. A., Zuidema, P., Blyth, A. M., Fairall, C. W., Brewer, W. A., Tucker, S., Lasher-Trapp, S. G., Mayol-Bracero, O. L., Vali, G., Geerts, B., Anderson, J. R., Baker, B. A., Lawson, R. P., Bandy, A. R., Thornton, D. C., Burnet, E., Brenguier, J.-L., Gomes, L., Brown, P. R. A., Chuang, P., Cotton, W. R., Gerber, H., Heikes, B. G., Hudson, J. G., Kollias, P., Krueger, S. K., Nuijens, L., O'Sullivan, D. W., Siebesma, A. P., and Twohy, C. H.: Rain in shallow cumulus over the ocean: the RICO campaign, B. Am. Meteorol. Soc., 88, 1912–1928, 2007.

Rosenfeld, D., Lohmann, U., Raga, G., O'Dowd, C., Kulmala, M., Fuzzi, S., Reissell, A., and M. O., A.: Flood or drought: how do aerosols affect precipitation?, Science, 321, 1309–1313, 2008.

Rosenfeld, D., Fischman, B., Zheng, Y., Goren, T., and Giguzin, D.: Combined satellite and radar retrievals of drop concentration and CCN at convective cloud base, Geophys. Res. Lett., 41, 3259–3265, https://doi.org/10.1002/2014GL059453, 2014.

Rutledge, S. and Hobbs, P.: The mesoscale and microscale structure and organisation of clouds and precipitation in midlatitude cyclones. VIII: A model for the seeder-feeder process in warm frontal rainbands, J. Atmos. Sci., 40, 1185–1206, 1983.

Saleeby, S. M. and van den Heever, S. C.: Developments in the CSU-RAMS aerosol model: emissions, nucleation, regeneration, deposition, and radiation, J. Appl. Meteorol. Clim., 52, 2601–2622, https://doi.org/10.1175/JAMC-D-12-0312.1, 2013.

Seifert, A. and Beheng, K. D.: A two-moment cloud microphysics parameterization for mixed-phase clouds. Part 2: Maritime vs. continental deep convective storms, Meteorol. Atmos. Phys., 92, 67–82, https://doi.org/10.1007/s00703-005-0113-3, 2006a.

Seifert, A. and Beheng, K. D.: A two-moment cloud microphysics parameterization for mixed-phase clouds. Part 2: Maritime vs. continental deep convective storms, Meteorol. Atmos. Phys., 92, 67–82, 2006b.

Seifert, A., Köhler, C., and Beheng, K. D.: Aerosol-cloud-precipitation effects over Germany as simulated by a convective-scale numerical weather prediction model, Atmos. Chem. Phys., 12, 709–725, https://doi.org/10.5194/acp-12-709-2012, 2012.

Shipway, B. J. and Hill, A. A.: Diagnosis of systematic differences between multiple parametrizations of warm rain microphysics using a kinematic framework, Q. J. Roy. Meteor. Soc., 138, 2196–2211, https://doi.org/10.1002/qj.1913, 2012.

Skamarock, W., Klemp, J., Dudhia, J., Gill, D., Barker, D., Duda, M., Huang, X., Wang, W., and Powers, J.: A description of the advanced research WRF version 3 NCAR Technical Note June 2008, Tech. Rep. TN-475+STR, NCAR, National Center for Atmospheric Research, Box 3000, Boulder, Colorado 80307, USA, 2008.

Stevens, B. and Feingold, G.: Untangling aerosol effects on clouds and precipitation in a buffered system, Nature, 461, 607–613, 2009.

Stevens, B., Feingold, G., Cotton, W., and Walko, R.: Elements of the microphysical structure of numerically simulated nonprecipitating stratocumulus, J. Atmos. Sci., 53, 980–1006, 1996.

Tao, W.-K., Li, X., Khain, A., Matsui, T., Lang, S., and Simpson, J.: Role of atmospheric aerosol concentration on deep convective precipitation: Cloud-resolving model simulations, J. Geophys. Res.-Atmos., 112, D24S18, https://doi.org/10.1029/2007JD008728, 2007.

Tao, W.-K., Chen, J.-P., Li, Z., Wang, C., and Zhang, C.: Impact of aerosols on convective clouds and precipitation, Rev. Geophys., 50, RG2001, https://doi.org/10.1029/2011RG000369, 2012.

Thompson, G. and Eidhammer, T.: A study of aerosol impacts on clouds and precipitation development in a large winter cyclone, J. Atmos. Sci., 71, 3636–3658, 2014.

Thompson, G., Rasmussen, R., and Manning, K.: Explicit forecasts of winter precipitation using an improved bulk microphysics scheme. Part I: Description and sensitivity analysis, Mon. Weather Rev., 132, 519–542, https://doi.org/10.1175/1520-0493(2004)132<0519:EFOWPU>2.0.CO;2, 2004.

Thompson, G., Field, P., Rasmussen, R., and Hall, W.: Explicit forecasts of winter precipitation using an improved bulk microphysics scheme. Part II: Implementation of a new snow parameterization, Mon. Weather Rev., 136, 5095–5115, 2008.

Thompson, G., Tewari, M., Ikeda, K., Tessendorf, S., Weeks, C., Otkin, J., and Kong, F.: Explicitly-coupled cloud physics and radiation parameterizations and subsequent evaluation in WRF high-resolution convective forecasts, Atmos. Res., 168, 92–104, https://doi.org/10.1016/j.atmosres.2015.09.005, 2016.

Tzivion, S., Feingold, G., and Levin, Z.: An efficient numerical solution to the stochastic collection equation, J. Atmos. Sci., 44, 3139–3149, https://doi.org/10.1175/1520-0469(1987)044<3139:AENSTT>2.0.CO;2, 1987.

van den Heever, S., Carrió, G., Cotton, W., DeMott, P., and Prenni, A.: Impacts of nucleating aerosol on florida storms. Part I: Mesoscale simulations, J. Atmos. Sci., 63, 1752–1775, 2006.

van Zanten, M. C., Stevens, B., Nuijens, L., Siebesma, A. P., Ackerman, A. S., Burnet, F., Cheng, A., Couvreux, F., Jiang, H., Khairoutdinov, M., Kogan, Y., Lewellen, D. C., Mechem, D., Nakamura, K., Noda, A., Shipway, B. J., Slawinska, J., Wang, S., and Wyszogrodzki, A.: Controls on precipitation and cloudiness in simulations of trade-wind cumulus as observed during RICO, J. Adv. Model. Earth Sy., 3, m06001, https://doi.org/10.1029/2011MS000056, 2011.

Walko, R., Cotton, W., Meyers, M., and Harrington, J.: New RAMS cloud microphysics parameterization Part I: The single-moment scheme, Atmos. Res., 38, 29–62, https://doi.org/10.1016/0169-8095(94)00087-T, 1995.

Wang, Y., Fan, J., Zhang, R., Leung, L. R., and Franklin, C.: Improving bulk microphysics parameterizations in simulations of aerosol effects, J. Geophys. Res.-Atmos., 118, 5361–5379, https://doi.org/10.1002/jgrd.50432, 2013.

Washington, R., James, R., Pearce, H., Pokam, W., and Moufouma-Okia, W.: Congo Basin rainfall climatology: can we believe the climate models?, Philos. T. R. Soc. B, 368, 20120296, https://doi.org/10.1098/rstb.2012.0296, 2013.

Webb, E.: Profile relationships: The log-linear range, and extension to strong stability, Q. J. Roy. Meteor. Soc., 96, 67–90, 1970.

Weisman, M. and Klemp, J.: The dependence of numerically simulated convective storms on vertical wind shear and buoyancy, Mon. Weather Rev., 110, 504–520, 1982.

Weisman, M. and Klemp, J.: The structure and classification of numerically simulated convective storms in directionally varying wind shears, Mon, Weather Rev., 112, 2479–2498, 1984.

Weisman, M. and Rotunno, R.: The use of vertical wind shear versus helicity in interpreting supercell dynamics, J. Atmos. Sci., 57, 1452–1472, 2000.

Weverberg, K. V., Vogelmann, A. M., Lin, W., Luke, E. P., Cialella, A., Minnis, P., Khaiyer, M., Boer, E. R., and Jensen, M. P.: The role of cloud microphysics parameterization in the simulation of mesoscale convective system clouds and precipitation in the Tropical Western Pacific, J. Atmos. Sci., 70, 1104–1128, 2013.

Weverberg, K. V., Goudenhoofdt, E., Blahak, U., Brisson, E., Demuzere, M., Marbaix, P., and van Ypersele, J.-P.: Comparison of one-moment and two-moment bulk microphysics for high-resolution climate simulations of intense precipitation, Atmos. Res., 147–148, 145–161, 2014.

Wood, R.: Drizzle in stratiform boundary layer clouds. Part II: Microphysical aspects, J. Atmos. Sci., 62, 3034–3050, https://doi.org/10.1175/JAS3530.1, 2005.

Wu, L. and Petty, G. W.: Intercomparison of bulk microphysics schemes in model simulations of polar lows, Mon. Weather Rev., 138, 2211–2228, https://doi.org/10.1175/2010MWR3122.1, 2010.

Yamaguchi, T. and Feingold, G.: Technical note: Large-eddy simulation of cloudy boundary layer with the Advanced Research WRF model, J. Adv. Model. Earth Sy., 4, m09003, https://doi.org/10.1029/2012MS000164, 2012.

Zhang, J., Reid, J. S., and Holben, B. N.: An analysis of potential cloud artifacts in MODIS over ocean aerosol optical thickness products, Geophys. Res. Lett., 32, L15803, https://doi.org/10.1029/2005GL023254, 2005.

Zhang, S., Wang, M., Ghan, S. J., Ding, A., Wang, H., Zhang, K., Neubauer, D., Lohmann, U., Ferrachat, S., Takeamura, T., Gettelman, A., Morrison, H., Lee, Y., Shindell, D. T., Partridge, D. G., Stier, P., Kipling, Z., and Fu, C.: On the characteristics of aerosol indirect effect based on dynamic regimes in global climate models, Atmos. Chem. Phys., 16, 2765–2783, https://doi.org/10.5194/acp-16-2765-2016, 2016.

9

Technical note: The US Dobson station network data record prior to 2015, re-evaluation of NDACC and WOUDC archived records with WinDobson processing software

Robert D. Evans[2], Irina Petropavlovskikh[1], Audra McClure-Begley[1], Glen McConville[1], Dorothy Quincy[2], and Koji Miyagawa[3]

[1]Cooperative Institute for Research in Environmental Sciences, University of Colorado, Boulder, CO 80309, USA
[2]Retired from NOAA/ESRL, Global Monitoring Division, Boulder, CO 80305, USA
[3]Visitor with NOAA/ESRL, Global Monitoring Division, Boulder, CO 80305, USA

Correspondence to: Robert D. Evans (robert.d.evans@noaa.gov)

Abstract. The United States government has operated Dobson ozone spectrophotometers at various sites, starting during the International Geophysical Year (1 July 1957 to 31 December 1958). A network of stations for long-term monitoring of the total column content (thickness of the ozone layer) of the atmosphere was established in the early 1960s and eventually grew to 16 stations, 14 of which are still operational and submit data to the United States of America's National Oceanic and Atmospheric Administration (NOAA). Seven of these sites are also part of the Network for the Detection of Atmospheric Composition Change (NDACC), an organization that maintains its own data archive. Due to recent changes in data processing software the entire dataset was re-evaluated for possible changes. To evaluate and minimize potential changes caused by the new processing software, the reprocessed data record was compared to the original data record archived in the World Ozone and UV Data Center (WOUDC) in Toronto, Canada. The history of the observations at the individual stations, the instruments used for the NOAA network monitoring at the station, the method for reducing zenith-sky observations to total ozone, and calibration procedures were re-evaluated using data quality control tools built into the new software. At the completion of the evaluation, the new datasets are to be published as an update to the WOUDC and NDACC archives, and the entire dataset is to be made available to the scientific community. The procedure for reprocessing Dobson data and the results of the reanalysis on the archived record are presented in this paper. A summary of historical changes to 14 station records is also provided.

1 Background

The Dobson ozone spectrophotometer was designed in the 1920s and is still in use today. The instrument is fully described elsewhere (Dobson, 1931, 1968) but, briefly, it measures the relative intensity of solar radiation between selected wavelength pairs in the range of 300–350 nm. These pairs are named A (305.5 and 325.4 nm), C (311.5 and 334.4 nm), C′ (334.4 and 453.6 nm) and D (317.5 and 339.8 nm) and are combined in the measurement process as either A and D (AD), C and C′ (CC′) or C and D (CD). Observations on the direct solar beam have DS (direct sun) attached to the wavelength identifier (for example, ADDS is an observation using the A and D pairs, on the direct solar beam). Observations on the zenith sky have ZB for zenith clear, ZC for zenith cloudy, and ZS for general zenith-sky observations. The optical arrangement of the instrument is presented in Fig. 1. Measurements on either direct solar light or light scattered from the zenith can be used to calculate the amount of ozone between the instrument and the top of the atmosphere (total ozone column or TOC). Approximately 90 % of this TOC resides in the region between 15 and 30 km above the Earth's surface, which is defined as the ozone layer.

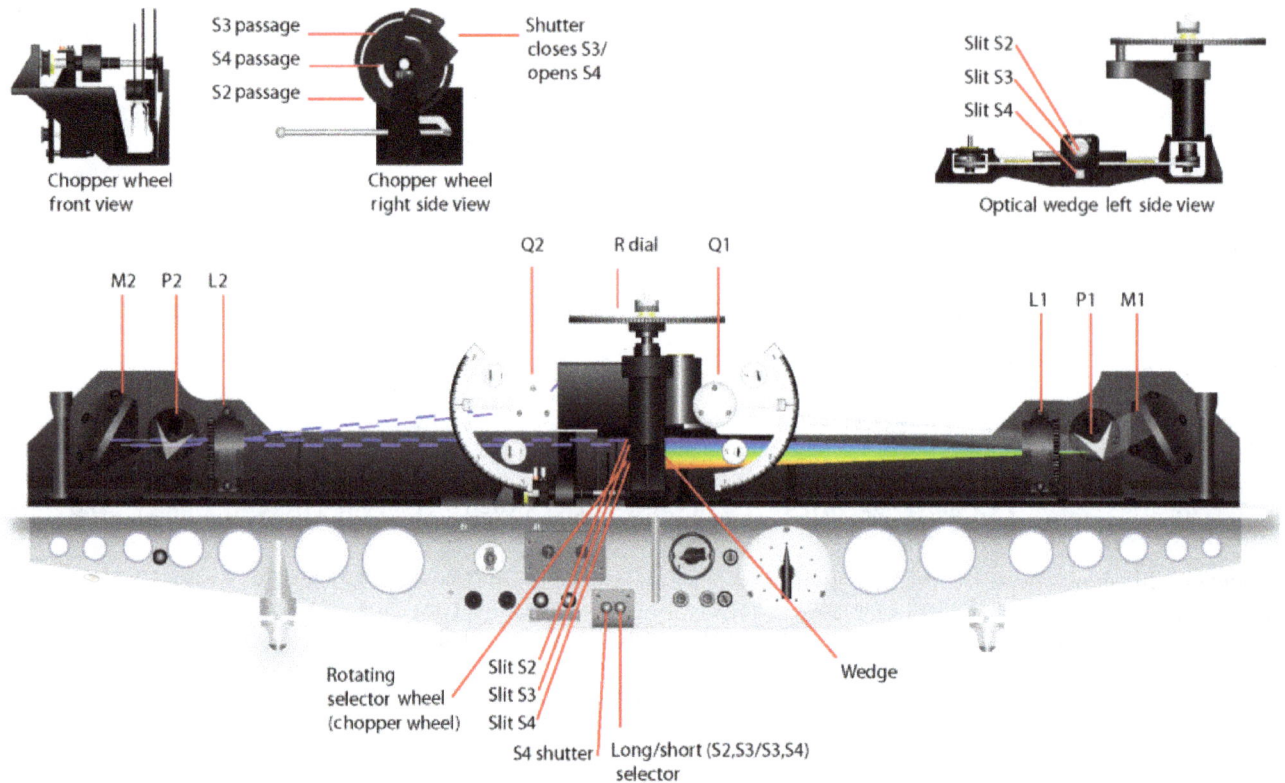

Figure 1. Diagram of the Dobson instrument, with cover omitted from view (some components shown are actually mounted in the cover).

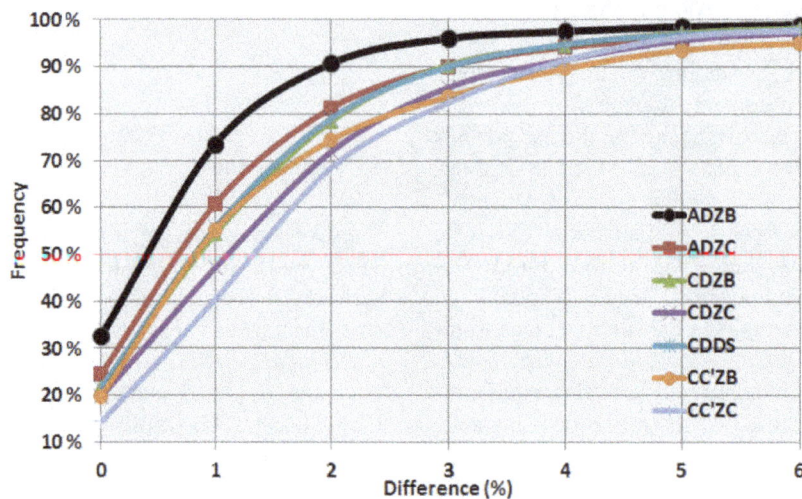

Figure 2. Distribution of cumulative differences between results from direct sun (ADDS) compared to zenith measurements on the same day. The frequency of compared zenith and ADDS total ozone (y axis) is accumulated between 0 and 6 % (x axis). Results are shown for other types of zenith-sky measurements denoted by colors in the legend. Results are the average of 12 stations in the US network except for the CC' results, which are based on the SPO data record.

The relative intensity of wavelength pairs measured if instrument was operated outside the Earth's atmosphere is referred to as the extraterrestrial constant (ETC). The instrument's ETC is determined either through a Langley plot method (Langley, 1884) or by direct comparison with a stan-

dard Dobson instrument (Komhyr and Evans, 2008). The concept of measurement of the Dobson spectrophotometer exploits the change in the ratio of solar light intensities measured at the respective wavelength pairs as caused by the passage of UV light through the ozone layer. The light enters the instrument (Fig. 1), and the right side of the optics produces a spectrum projected on a slit arrangement containing slits S_2, S_3 and S_4 (only used for C' measurements). The left side of the optics combines the images of the slits into the photomultiplier tube. The measurement of the relative intensities is performed by moving a neutral density filter (the optical "wedge") across the light path of the wavelength less absorbed by ozone, specifically the light passing through slit S_3. The wedge is moved to reduce the light intensity of S_3 to equal the intensity of S_2, the wavelength more absorbed by ozone, as seen by the photomultiplier tube. The instrument's output, R value, indicates the position of the neutral density filter as indicated on an engraved plate, which is the primary measurement and is specific to the Dobson instrument optical properties.

The N value represents the attenuation by ozone and by other scattering and absorption during the UV light's passage through the atmosphere between the sun and the instrument. The relationship between R values and N values is both wavelength and instrument specific. R values are converted to N values using tables called R to N tables (N tables). These tables change during the instruments' lifespans due to repairs, updates and aging; thus each set has a limited period of application. The applicability of the N table is monitored by means of intercomparisons with standard Dobson instruments and with the use of instrument-specific reference lamps. The calculation of ozone from observations made on the DS light (or reflected light from the moon) is with a defined algorithm based on Beer's law. The resolution of the measurement is 1 DU, and the precision (uncertainty) is considered to be ± 1 % (Grant, 1989). Accuracy is another issue. The accuracy is dependent on knowledge of the ozone and temperature profile at the time of the measurement to correctly calculate the ozone absorption cross section. As this information is not available for individual observations, some assumptions must be made. A standard algorithm for the reduction of DS observations (Komhyr et al., 1993) is used by all organizations reporting daily values to the WOUDC or NDACC archives. The accuracy is also dependent on the knowledge of the ozone cross-section datasets used to determine the absorption coefficients in the reduction algorithm (Redondas et al., 2014). The reduction of measurements on the ZS is more complicated, as it is based on statistical analysis of DS and ZS observations close in time. The precision (uncertainty) of the ZS empirical model is found to be 2–5 % in this work and is dependent on the wavelength pairs used and the sky conditions. An accepted method of statistical analysis is not defined in the standard operating procedures; different organizations using the instrument employ different methods to build the empirical relation between direct-sun and zenith-sky measurements (Josefsson and Löfvenius, 2008).

The Dobson instrument has limitations in the accuracy of measurements at certain observing conditions (Basher, 1982). Internal stray light is one such limitation. Moreover,

each Dobson instrument has unique optical components that result in an instrument-specific level of the stray light. The quality and aging stability of the individual wedge construction has improved over time, especially for instruments within the NOAA network, which had optical components replaced with those of a more robust design during instrument rebuilding in the 1980s.

Data reduction algorithms are fully discussed in Sect. 7 of the Operations Handbook (Komhyr and Evans, 2008; http://www.wmo.int/pages/prog/arep/gaw/documents/GAW183-Dobson-WEB.pdf).

2 Station history

There are measurements of TOC in the USA prior to 1960 made by university and federal organizations (Brönnimann, 2003), but the development of a coherent network of observing sites within the US Weather Service started in the 1960s under the guidance of Walter Komhyr. The network was transferred to NOAA's Global Monitoring for Climate Change (GMCC) in the early 1970s and is currently operated by NOAA's Earth System Research Laboratory's Global Monitoring Division (ESRL/GMD). As many as 16 stations comprised the network since its establishment. One station was closed; another was transferred to another parent authority. Table 1 displays the stations reporting at end of 2015. Originally, observations using the Dobson instruments were recorded with pen or pencil on forms designed to assist manual calculations (https://youtu.be/w1rV_96UChk). As computer power increased, the data were transcribed to punched cards for processing and then to direct entry by keyboard. By the mid-1990s, the NOAA instruments were equipped with computers and encoders, and the data were recorded in a "dayfile" at the time of the measurement. Six stations were equipped with fully automated instruments in the 1980s.

2.1 Data processing

TOC is normally archived as a single representative value of TOC selected for each day. This not an average value, but the result of the "best" observation during the day. As the exact instrumentation and observational scheduling vary from station to station, the number of observations made daily also vary. The full record of observations is available per request from NOAA Dobson network personnel listed at https://www.esrl.noaa.gov/gmd/ozwv/dobson/contact.html. In this publication, the term "select value" means the daily value produced in the NOAA processing stream. An earlier reprocessing of the stations' data was done in the 1990s (Komhyr et al., 1995); the report also details much of the early history of measurements in the US system of stations.

To convert measurements to TOC values, calibration N tables and information (reference lamp adjustment) from monthly instrument tests using reference lamps are required.

Table 1. Current stations in the NOAA network using Dobson ozone spectrophotometers.

Station Name	NOAA station code and WMO station number	Station Dobson record started	Responsible organizations (archives)	Current automation status
Mauna Loa GMD Observatory	MLO 31	1963	NOAA (NDACC and WOUDC)	WinDobson full
South Pole	SPO 111	1963	NOAA (NDACC and WOUDC)	NOAA semi-auto
Bismarck, North Dakota	BIS 19	1962	NOAA	NOAA semi-auto
Caribou, Maine	CAR 20	1962	NOAA	NOAA semi-auto
Nashville, Tennessee	BNA 106	1962	NOAA	NOAA semi-auto
Fairbanks, Alaska	FBK; POK 105; 217	1965	NOAA; University of Alaska	WinDobson full
Boulder, Colorado	BDR 67	1966	NOAA (NDACC and WOUDC)	WinDobson full
Wallops Is., Virginia	WAI 107	1967	NOAA; NASA (NDACC and WOUDC)	NOAA semi-auto
Barrow GMD Observatory	BRW 199	1973	NOAA	NOAA semi-auto
American Samoa, GMD Observatory	SMO 191	1976	NOAA (NDACC and WOUDC)	NOAA semi-auto
Haute Provence, France	OHP 40	1983	NOAA; Centre National de la Recherche Scientifique (NDACC and WOUDC)	WinDobson full
Fresno and Hanford, California	FAT; HNX 244; 341	1983	NOAA	NOAA semi-auto
Perth, Australia	PTH 159	1984	NOAA; Australian Bureau Meteorology	NOAA full
Lauder, New Zealand	LDR 256	1987	NOAA; National Institute for Water and Atmosphere (NDACC and WOUDC)	WinDobson full

The N tables are defined by comparison of the station instrument to a reference standard. These referencings are normally done on a 4- to 6-year schedule. The calibration of the wedge is normally measured at the same time. The reference lamp tests are an indication of the instrument's aging but are only a single point test in the instrument measurement range. The comparison process measures instrument performance over a typical mu range of 1.15 to 3.85 but this is often adjusted with consideration to circumstances such as an instrument's location. Mu is the normalized optical path length through the ozone layer and is calculated from the solar zenith angle (SZA). The calibration N tables are changed when the difference between the station and reference instrument is greater than the equivalent of 1 % in TOC. When the calibration N tables are changed due to a drift (determined from an inspection of the past calibrations, instrument operational history and, if possible, comparison with other instrumental records), the existing dataset from the last calibration change to the new calibration was reprocessed and republished in the archives.

The set of computer programs used for the NOAA processing were written in the FORTRAN language and, by the 2010s, were difficult to use and maintain due to changes in computer hardware and personnel. The decision was made to convert the NOAA processing to processing using the WinDobson software package, as the fully automated instruments were updated to a modern system based on this software. Developed by personnel of the Japan Meteorological Agency

(Miyagawa, 1996), WinDobson is a software package for operations, data analysis and quality assurance of Dobson spectrophotometer observations. The algorithm for the reduction of ozone from DS observations with the Dobson is the standard method used by the NOAA software, but the ZS observations are reduced with a method described later in this paper. For the NOAA application, new components were developed. These new components are available from NOAA to other users of WinDobson. It is applicable for both TOC and Umkehr (ozone vertical profile) measurements. As this software has a different statistical method for the reduction of the zenith measurements, and set of rules (see Sect. 2.4) for determining the representative value of total ozone for each day with observations, the entire data record of each operational station was reprocessed in the WinDobson system to minimize the effect of the change when future data are placed in the archive. In the development of the data files and calibration information for WinDobson processing, the entire record of observations, repair and calibration checks of each station was investigated and re-evaluated. This investigation allows for correction of past errors.

2.2 Data format conversion and initial comparison of datasets

The NOAA processed data were converted to "long line format" (LLF) files. These files are actually the image of the information sent to printers in the 1990s version of the data

Table 2. Statistics of the overall differences between WOUDC and NDACC records and WinDobson record (WinDobson–WOUDC, NDACC).

Station code	Offset WinDobson–WOUDC	Linear trend WinDobson–WOUDC per year	Offset WinDobson–NDACC	Linear trend WinDobson–NDACC per year
MLO	$-0.1 \pm 1.6\%$	$+0.014 \pm 0.001\%$	$-0.1 \pm 1.8\%$	$+0.015 \pm 0.001\%$
SPO	$-0.0 \pm 4.0\%$	$-0.016 \pm 0.003\%$	$-0.5 \pm 6.9\%$	$-0.026 \pm 0.006\%$
BIS	$+0.1 \pm 2.2\%$	$-0.004 \pm 0.001\%$	N/A	N/A
CAR	$+0.2 \pm 3.2\%$	$+0.022 \pm 0.002\%$	N/A	N/A
BNA	$+0.6 \pm 2.7\%$	$+0.002 \pm 0.001\%$	N/A	N/A
FBK	$-0.4 \pm 2.8\%$	$+0.033 \pm 0.003\%$	N/A	N/A
BDR	$+0.3 \pm 1.7\%$	$+0.007 \pm 0.001\%$	$+0.3 \pm 1.5\%$	$-0.001 \pm 0.001\%$
WAI	$-0.1 \pm 3.3\%$	$+0.024 \pm 0.003\%$	$+0.0 \pm 1.6\%$	$+0.032 \pm 0.006\%$
BRW	$+0.7 \pm 2.8\%$	$+0.011 \pm 0.004\%$	N/A	N/A
SMO	$-0.1 \pm 1.7\%$	$+0.042 \pm 0.002\%$	$-0.1 \pm 2.3\%$	$+0.042 \pm 0.002\%$
OHP	$-0.1 \pm 1.8\%$	$-0.002 \pm 0.003\%$	$-0.1 \pm 1.7\%$	$-0.004 \pm 0.003\%$
FAT/HNX	$+0.0 \pm 1.5\%$	$-0.003 \pm 0.002\%$	N/A	N/A
PTH	$+0.3 \pm 1.6\%$	$+0.022 \pm 0.002\%$	N/A	N/A
LDR	$-0.1 \pm 1.8\%$	$-0.022 \pm 0.003\%$	$-0.3 \pm 1.3\%$	$-0.066 \pm 0.002\%$

stream. The select values for the WOUDC and NDACC archives were originally produced from these files, using a process of both machine and inspection by personnel. Programs were developed to convert the LLF and dayfiles into formats compatible with the WinDobson data stream. Files with instrument, station and calibration information (parafiles) were also developed to complete the structure of the WinDobson system. Connections to other sources of TOC information (satellite data records, for example) were developed so that comparisons with these values could be performed using tools internal to WinDobson. Reference lamp values were extracted from the LLF records for time periods prior to 1995 and from the dayfiles afterwards. By the end of 2015, all operational stations' data were being processed in WinDobson.

Initially, the datasets of only ADDS (fundamental wavelength pairs) observations from the two processing streams were compared with the expectation that the results should agree within ± 1 DU. The ADDS observations are considered the most reliable (fundamental), as the equation derived for conversion to ozone minimizes the Rayleigh scattering term, and the aerosol term can be considered to be zero. Time periods with differences greater than this ± 1 DU were investigated to determine the source of the problem, and correct any differences. When the ADDS differences were reconciled, the ZS observations were compared to the DS observations to define a polynomial method within the WinDobson system for converting the ZS observations to TOC. Separate polynomials were defined for various time periods related to instrument repairs and calibration changes. The change in the methods of reduction of ZS measurements often produced large changes in reported TOC values. The improvement in the ZS results with respect to the ADDS results is

displayed in Fig. 2 and in Table 3. The new method has resulted in $\sim 91\%$ of ZS-derived total ozone (ADZB) within 2% of the coincident DS ozone column (ADDS). This is an improvement over the 78% value reported in the Operations Handbook (Komhyr and Evans, 2008). Results of observations made on the direct sun using the CD wavelength pairs differ from those made on AD pairs. The differences come primarily from imperfect knowledge of the ozone cross sections used to determine the absorption coefficients used in the algorithm and of the optical characteristics of the instrument (Redondas et al., 2014). The differences in observational results within a specific SZA range were analyzed, and a multiplying factor was established to bring the average of the CD results to that of the AD results with in the WinDobson system.

2.3 Comparison of WinDobson representative values with archived daily values

The individual station records are archived as daily values in the World Ozone and Ultraviolet Radiation Data Centre (WOUDC) in Canada (http://woudc.org/home.php). The format of reporting is a single value for the local day, but in UTC (universal time coordinated) time with a resolution of an hour. The NDACC (http://www.ndsc.ncep.noaa.gov/data/) archive has the same TOC values for a subset of the WOUDC stations but in a different format, with date and time in UTC. The reprocessed datasets will be archived in WOUDC and NDACC. For each station, tools in WinDobson were used to make a dataset of daily representative TOC values. These datasets were compared to the datasets of select values downloaded from the WOUDC and NDACC, and the differences were investigated. The history of the instrument calibrations was again reviewed, and changes in the N tables and the pe-

Table 3. Displayed is the cumulative agreement in percent for specific ZS and CDDS results compared to ADDS results on the same day. For example, an agreement of 2 % occurs in 91 % of the cases for ADZB observations. Displayed are the average of 12 stations in the NOAA network (Barrow, Fairbanks, Caribou, Bismarck, Haute Provence, Boulder, Wallops Island, Mauna Loa, Tutuila, Perth, Lauder and South Pole). Definitions of all abbreviations are in Sect. 1.

% difference from ADDS	ADZB (%)	ADZC (%)	CDZB (%)	CDZC (%)	CDDS (%)	CC′ZB (%)	CC′ZC (%)
0	33	25	22	20	22	20	14
1	74	61	54	47	56	55	41
2	91	81	78	72	79	74	68
3	96	90	90	86	90	84	82
4	98	94	95	92	95	90	91
5	99	96	97	96	97	94	96
6	99	97	98	97	98	95	98
7	99	98	99	98	98	96	98
8	100	99	100	98	99	96	98
Frequency	ADZB	ADZC	CDZB	CDZC	CDDS	CC′ZB	CC′ZC
85	1.5	2.3	2.5	3.0	2.5	3.3	3.3
90	1.9	3.0	3.0	3.6	3.0	4.0	3.8

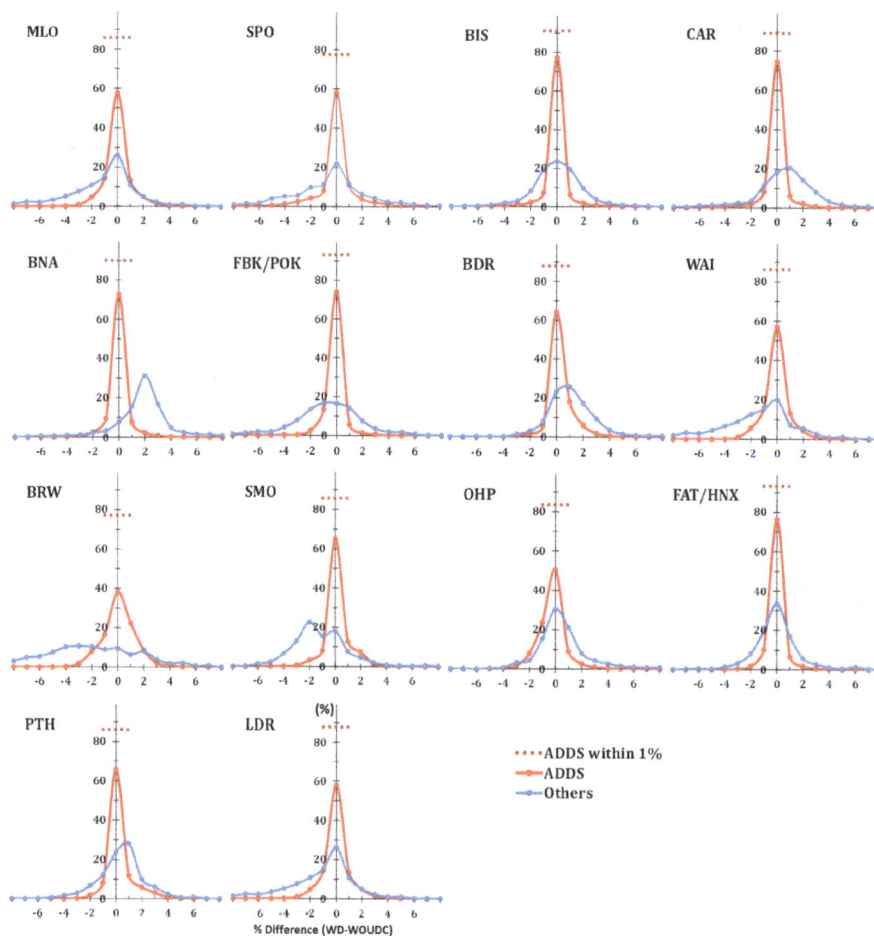

Figure 3. The probability of a daily value changing by a particular percentage for each station. The red line is for ADDS-type observations; the blue for all other types; the horizontal line is the percent of ADDS in the range ±1 % differences.

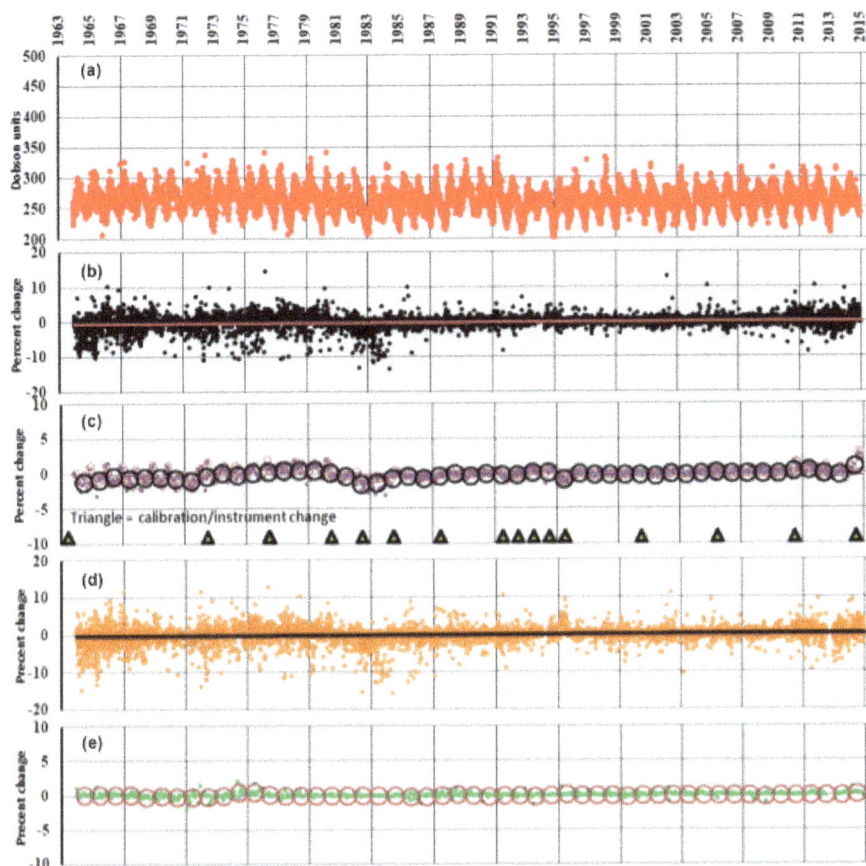

Figure 4. Graphic representation of the changes in the Mauna Loa Observatory, Hawai'i, USA (19° N, 156° W; NDACC station), record after the conversion into WinDobson processing. **(a)** The time record of total ozone measured at the station from the start of observations through 2015 (or until the station was converted to WinDobson processing. **(b)** Percent difference between daily WinDobson total ozone records compared to the WOUDC record (WinDobson–WOUDC). The red line is a linear fit. **(c)** The same as **(b)** but for monthly and yearly averages (based on all the values in the month in each dataset). The small white circles are averages made from DS observations only; the red symbols represent averages using all Dobson total ozone records; the large black open circles are yearly averages of all observations, based on monthly averages. Large triangle symbols indicate major calibration or instrument changes that lead to creating the new N tables; however, not all calibration checks of the station record are shown. For NDACC stations only: panel **(d)** is the same as **(b)** but for comparisons with the data archived at NDACC center (WinDobson–NDACC). The black line is a linear fit. Panel **(e)** is the same as **(c)** for comparisons with the NDACC archived monthly and yearly averages. NDACC values are not recorded as observation type.

riods of the use of N tables within the WinDobson system were made as needed. The differences stem from a number of reasons.

– There are data in the WOUDC dataset for some stations that were reported by earlier organizations. The processing and selection rules for these data are unknown.

– The older (1995) processing included time periods of special processing to attempt to account for specific problems in the older optics of specific instruments. This was accomplished by a modification to the reference lamp correction used in the data processing. The lamp corrections for the pre-1995 processing were extracted from the LLF format and applied to the WinDobson data process to introduce the correction applied in the earlier processing. In some cases, the full correction was not possible to reproduce, so special N tables were reconstructed from the information in the LLF format and applied on an annual basis. These periods are displayed graphically and discussed in the individual station reports. The problems that the special processing were attempting to correct were as follows:

– The so-called mu dependency (Komhyr et al., 1995), where DS results are lower at low sun angles. As this effect is dependent on the intensity of the input solar beam, and thus on the TOC, no attempt was made to account for this effect in WinDobson processing. This problem is related to the internal scattered light in the instrument, which is difficult to evaluate.

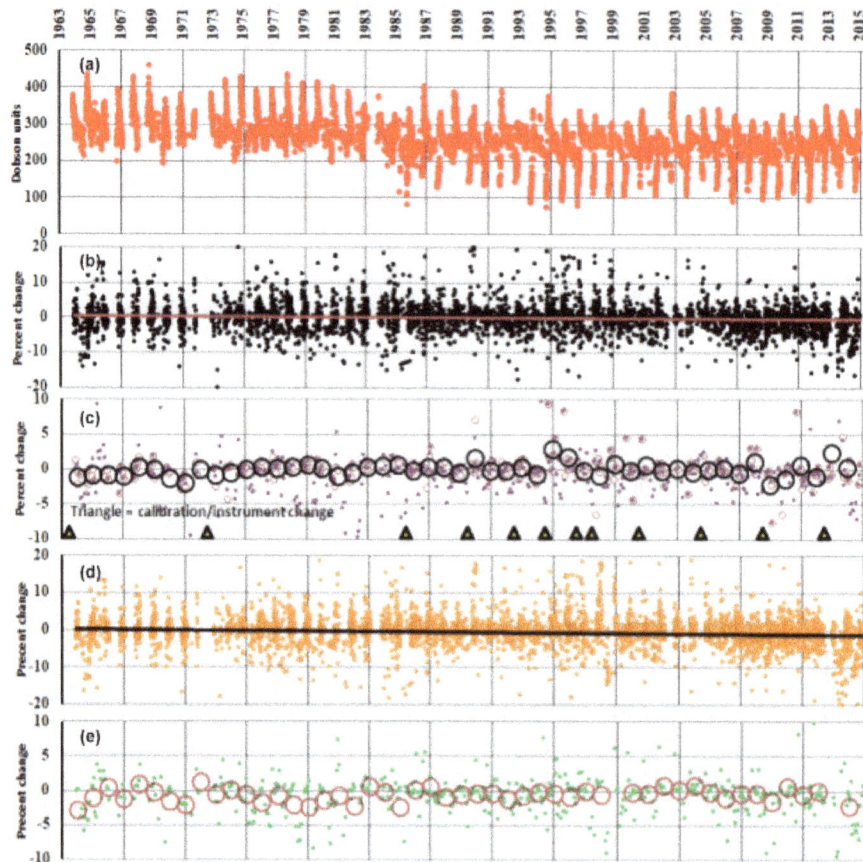

Figure 5. Graphic representation of the changes in the South Pole, Antarctica (90° S, 59° E; NDACC station), record with the conversion into WinDobson processing.

– Drifts in the shape of "wedge" calibration. It is unclear how the drift correction was actually performed in earlier processing; no attempt was made to account for this effect in WinDobson processing. Newer construction of the optical wedges used in the instrument have proved to have a much more stable calibration.

– Drifts in the "extraterrestrial constant" as part of the calibration. This was done in the WinDobson processing of later data, but with a different scheme – multiple N tables with shorter time periods of applicability.

– There was a weakness in the NOAA processing in choosing a select value for each day. During the original review of observations, certain observations were rejected for selection; this rejection was not recorded in the LLF files, and thus rejected observations appeared in the WinDobson dataset. We scrutinized the record for these discrepancies and amended the results.

– The results of the zenith measurements changed due to updates to the reduction method, and these types of

changes affect all stations – some of the changes are large and are discussed in the individual station reports.

– For some stations, it is common for observations to be made throughout the local day but with later observations being on the next consecutive UTC day. This occurs at Lauder (LDR), Samoa (SMO) and the South Pole (SPO), where UTC dates change during normal observing period. For SPO, observations on a local day can differ by 22 h; thus the choice of the selected/representative ozone in the change from NOAA to WinDobson processing and selection may differ by 22 h. At certain times of the year, the TOC can change appreciatively during this time period at SPO.

– Data archives sometimes failed to be updated after a calibration drift was detected during an intercomparison with a standard. This is not necessarily a failure of the internal WOUDC archiving process. NDACC appears to capture these periods more correctly.

– The rules for choosing the NOAA selected value for the day were similar to that of WinDobson but were not consistent throughout the record or across stations,

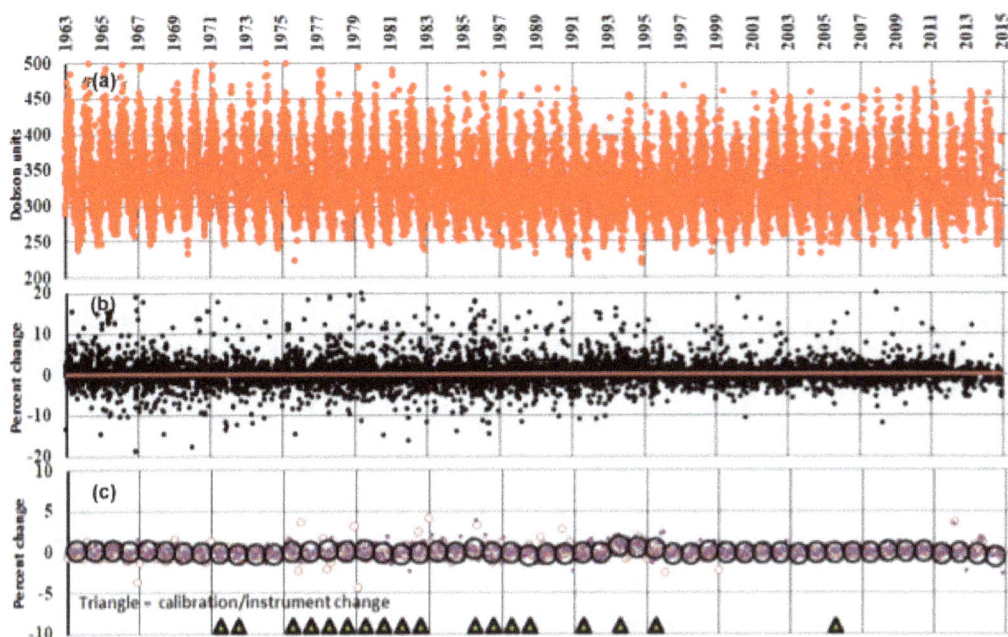

Figure 6. Graphic representation of the changes in the Bismarck, North Dakota, USA (47° N, 101° W), record with the conversion into WinDobson processing.

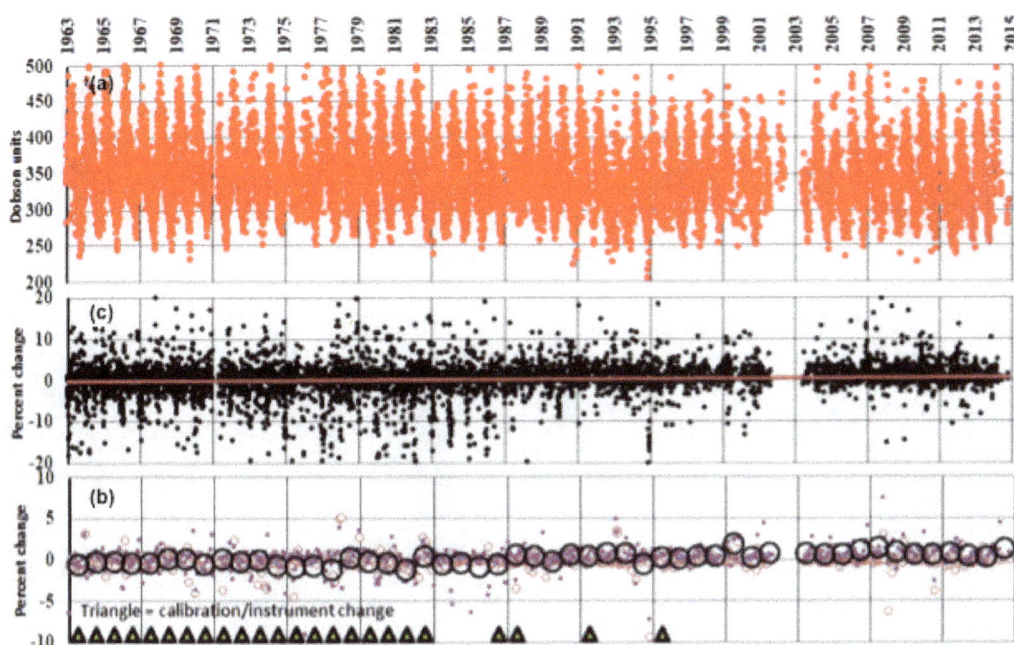

Figure 7. Graphic representation of the changes in the Caribou, Maine, USA (47° N, 68° W), record with the conversion into WinDobson processing.

and the documentation of these rules is incomplete. For the WinDobson processing, the same rules are applied throughout the record and stations and are described in Sect. 2.4.

– Our investigations of the station and instrument operation history revealed several periods for which different N tables were used in the archived records as compared to the historical record of NOAA N tables. Also, when a station instrument is compared to a standard instrument, and the results are within the uncer-

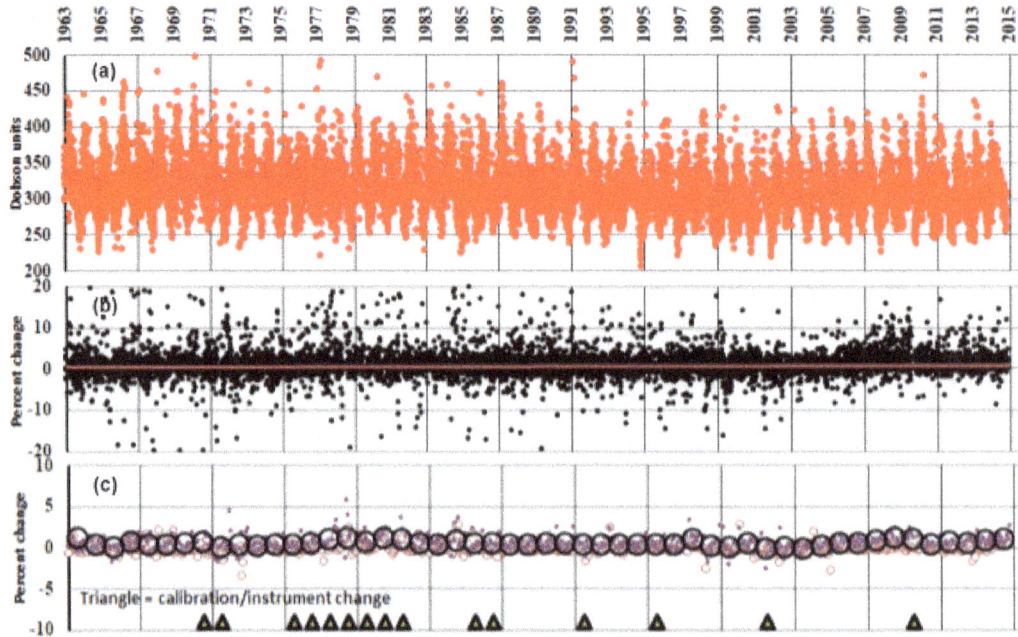

Figure 8. Graphic representation of the changes in the Nashville, Tennessee, USA (36° N, 87° W), record with the conversion into WinDobson processing.

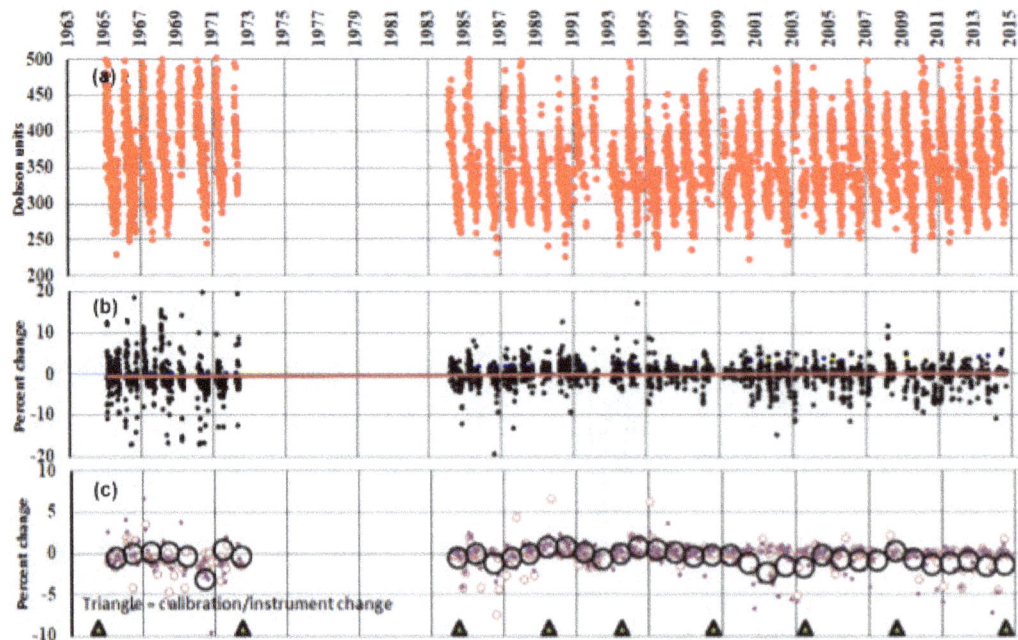

Figure 9. Graphic representation of the changes in the Fairbanks, Alaska, USA (65° N, 148° W), record with the conversion into WinDobson processing.

tainty of the measurements (±1 %), the station instrument's calibration is considered to be stable and thus is not changed. Otherwise, the instrument's calibration is changed and the existing data record starting from the time of the last comparison against a standard is reprocessed with the assumption that instrument's calibra-

tion has changed in a linear manner. Using the tools in WinDobson, our studies of the stations' records allowed comparisons with long-term records indicating TOC. These comparisons showed that at certain stations, the calibration change was not linear. Further investigation of stations' history revealed damage to the instrument

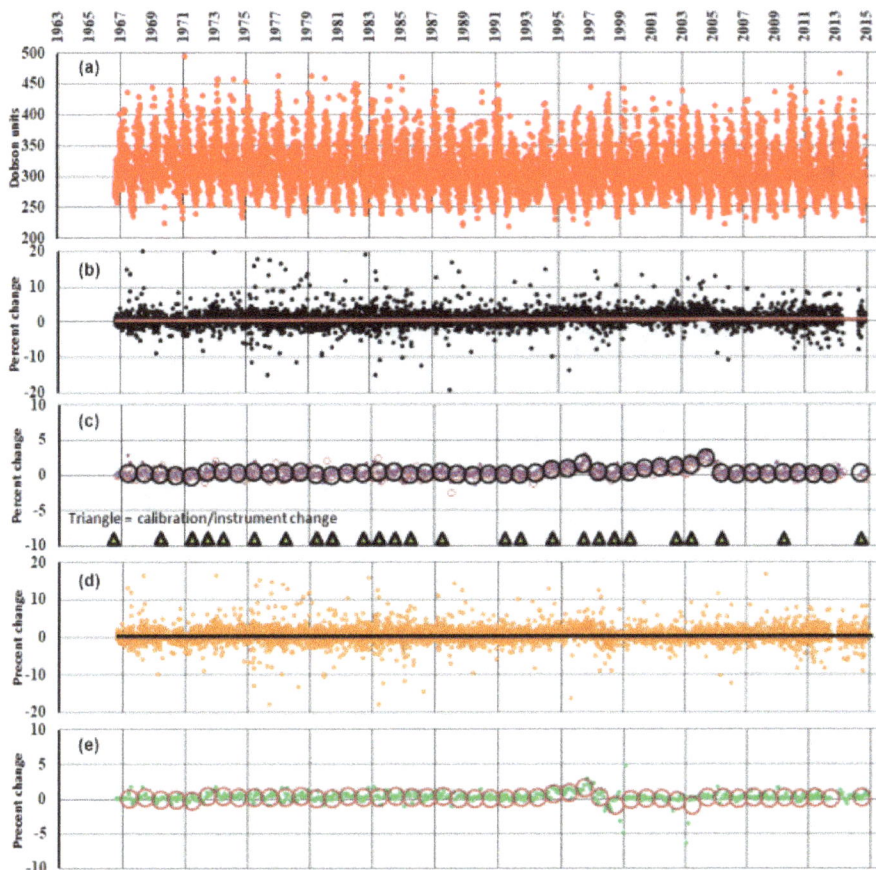

Figure 10. Graphic representation of the changes in the Boulder, Colorado, USA (40° N, 105° W, NDACC station), record with the conversion into WinDobson processing.

2.4 WinDobson selection rules

Often there are multiple observations on an individual day. The observations are given an internal numeric code in Win-Dobson, based on the observation type and operator input about the observation. The representative value is chosen by the software with the priority groups given below, highest to lowest. These groups are based on Table 2 in the Operations Handbook (Komhyr and Evans, 2008). If there are multiple observations of the highest priority on that day, the observation closest in time to local noon is chosen. After the automatic selection, the daily representative values are reviewed by human inspection with possible intervention to select a different value. The WinDobson software also has quality control routines that rates individual observations as good, questionable (flagged yellow) and likely bad (flagged red), based on internal consistencies of the measurements. If

an observation is rejected by the human inspector, the observation is not removed from the data record but flagged as "not included".

Priority groups are listed here; operator inputs as to sky quality are included in determining priority:

1. Direct-sun observations using the AD pair combination with or without ground quartz plate (diffuser) in the instrument's inlet window. Observations with diffuser have higher priority.

2. Zenith-sky observations using the AD pair combination. Observations on the clear zenith have higher priority over those on cloudy conditions.

3. Direct-sun observations using the CD pair combination with ground quartz plate (diffuser) in the instrument's inlet window. Observations without diffuser have lower priority.

4. Zenith-sky observations using the CD pair combination. Observations on the clear zenith have higher priority over those on cloudy conditions.

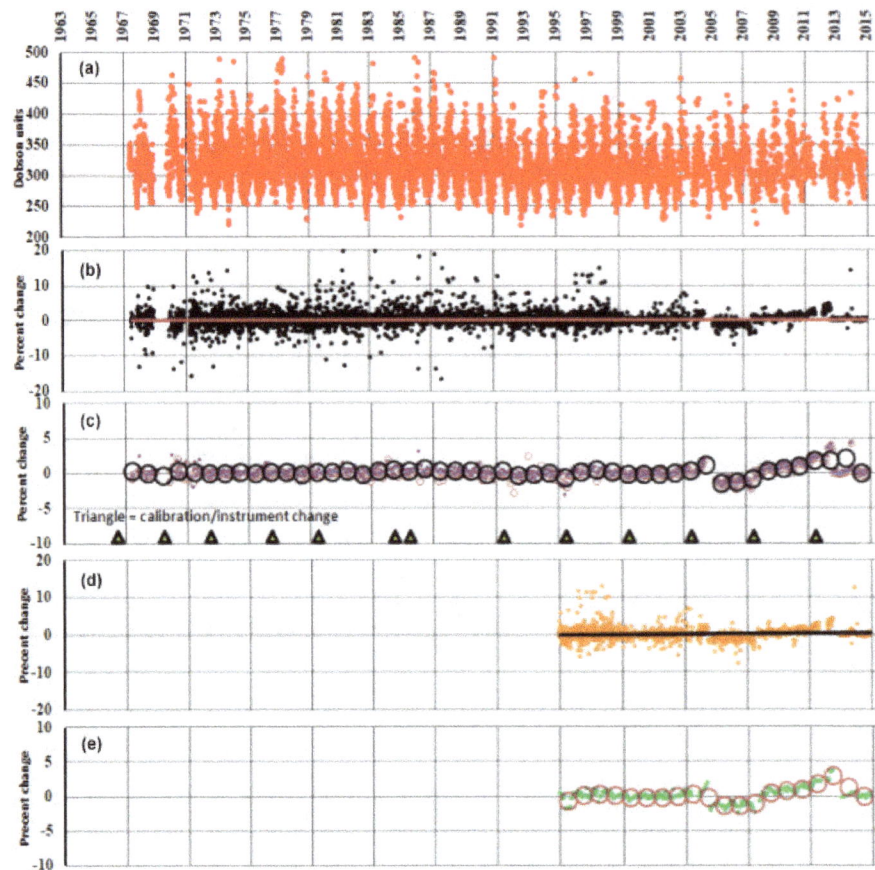

Figure 11. Graphic representation of the changes in the Wallops Island Flight Center, Virginia, USA (38° N, 76° W; NDACC station), record with the conversion into WinDobson processing.

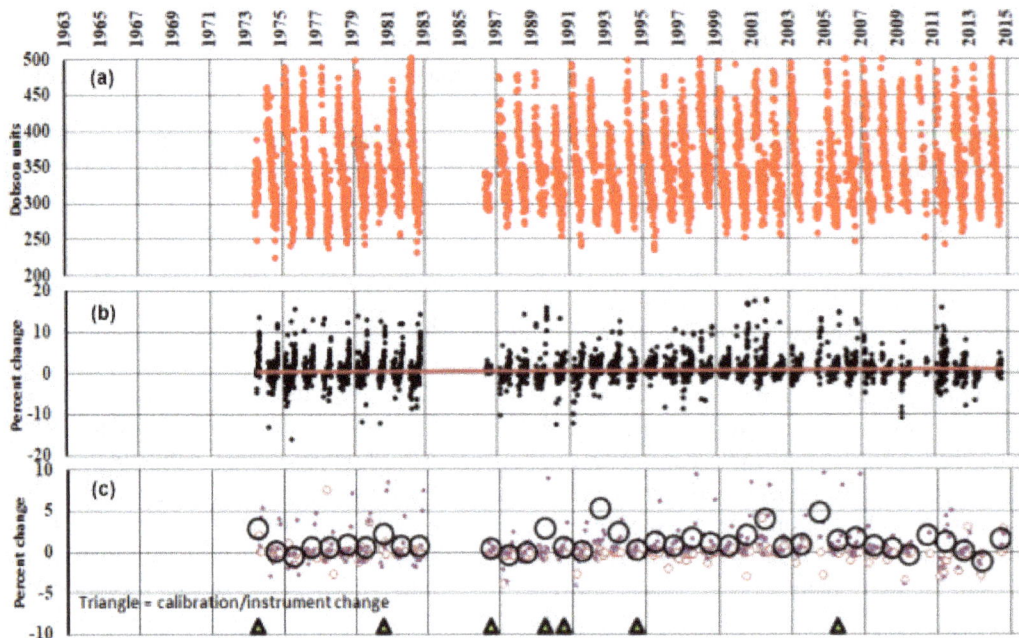

Figure 12. Graphic representation of the changes in the NOAA/ESRL/GMD Observatory, Barrow, Alaska, USA (71° N, 157° W), record with the conversion into WinDobson processing.

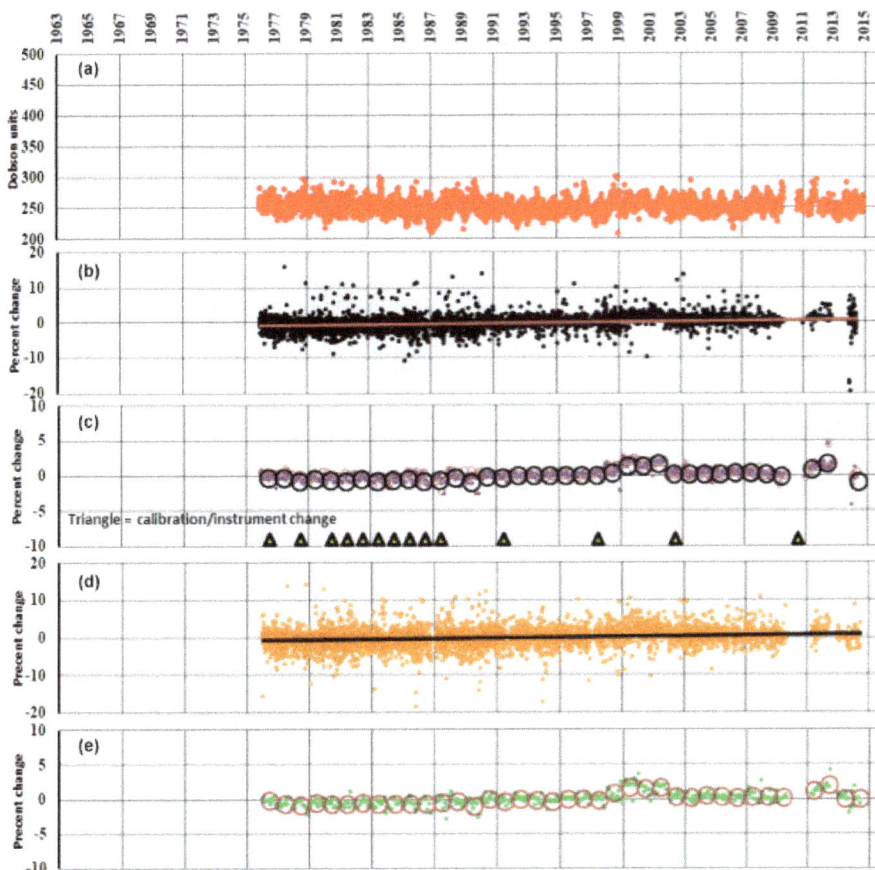

Figure 13. Graphic representation of the changes in the NOAA/ESRL/GMD Observatory, American Samoa (14° S, 171° W; NDACC station), record with the conversion into WinDobson processing.

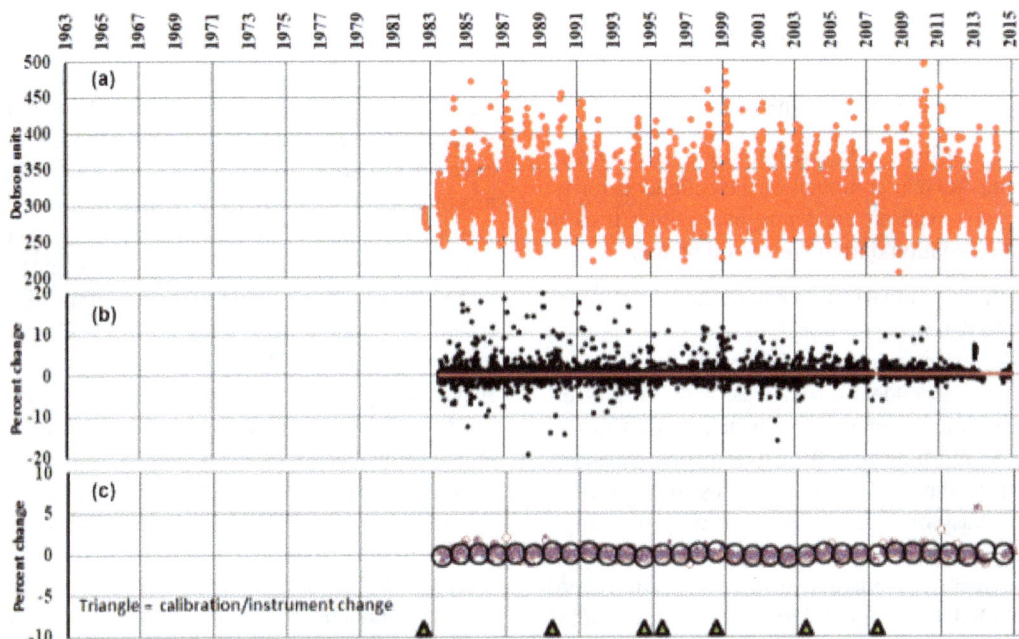

Figure 14. Graphic representation of the changes in the Fresno and Hanford, California, USA (36° N, 120° W), record with the conversion into WinDobson processing.

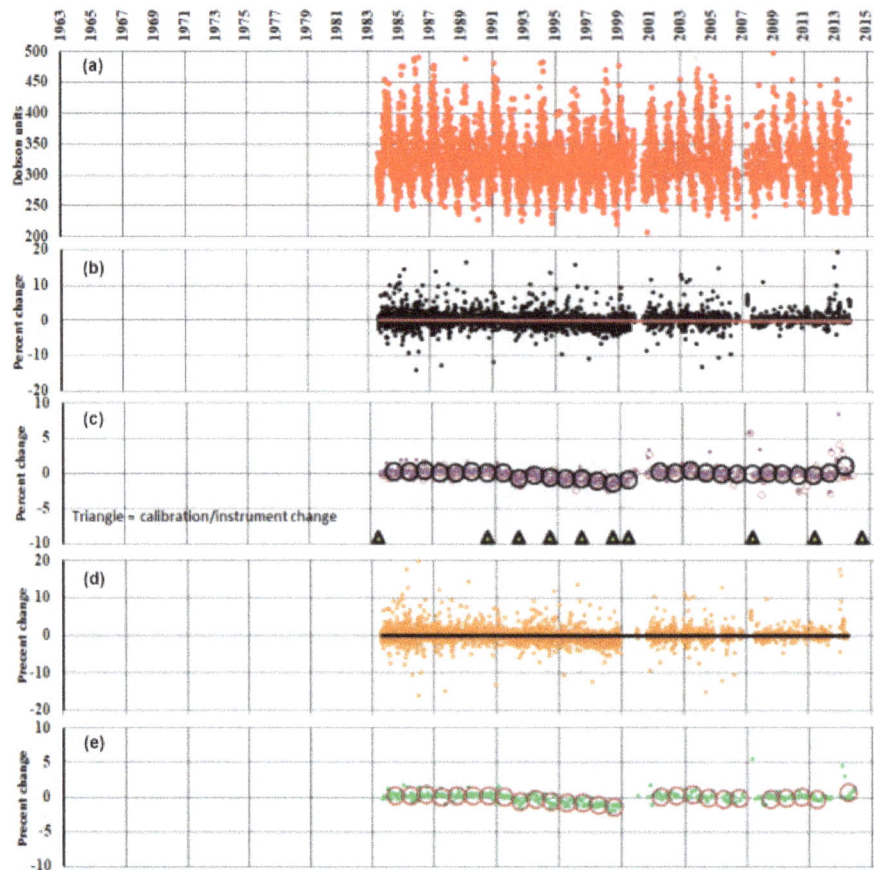

Figure 15. Graphic representation of the changes in the Observatoire de Haute-Provence, France (44° N, 6° E; NDACC station), record with the conversion into WinDobson processing.

5. Zenith-sky observations using the CC′ pair combination. Observations on the clear zenith have higher priority over those on cloudy conditions.

6. Observations on light reflected from the moon. Observations using AD pair combination have higher priority. Note these observations are rarely made other than at the South Pole station during the austral winter.

A discussion of the individual station records and the changes is presented in the following section.

The station discussions are accompanied by a referenced graphic of the time-dependent differences, consisting of either three panels (all stations) or five panels (NDACC stations). Panel A is the time record of total ozone measured at the station from the start of observations through 2014 (or until station was converted to WinDobson processing). Panel B is the percent difference between daily WinDobson total ozone records compared to the WOUDC record (WinDobson–WOUDC). The red line is a linear fit. Panel C is the same as the second but for monthly and yearly averages (based on all the values in the month in each dataset). The small white circles are averages made from DS observations only, the red symbols represent averages using all Dob-

son total ozone records and the large black open circles are yearly averages of all observations, based on monthly averages. Large triangle symbols indicate major calibration or instrument changes that lead to creating the new N tables; however, not all calibration checks of the station record are shown as not all calibration checks revealed problems. For NDACC stations only, panel D is the same as panel B but for comparisons with the data archived at NDACC center (WinDobson–NDACC). The black line is a linear fit. Panel E is the same as panel C for comparisons with the NDACC archived monthly and yearly averages. NDACC values are not recorded as observation type. Table 2 displays standard statistics of the differences between WinDobson and WOUDC and between WinDobson and NDACC records.

Assessment of changes in the WinDobson representative dataset relative to WOUDC record is analyzed in the form of probability distributions, where percent differences in TOC are plotted (Fig. 3) as a function of likely change when the archive is updated. The datasets analyses are separated into ADDS and other type of measurements. The ADDS curves are symmetric and indicate that the vast majority of ADDS values will be unchanged. The "other" curves are less symmetric and are driven by the updated ZS reduction polynomi-

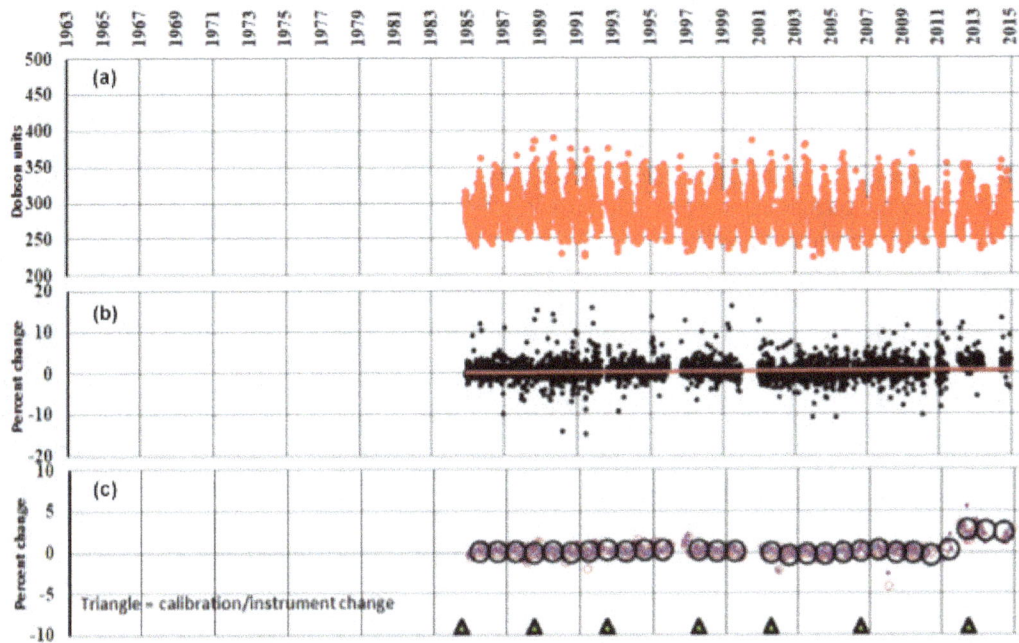

Figure 16. Graphic representation of the changes in the Perth Airport, Western Australia, Australia (32° S, 116° E), record with the conversion into WinDobson processing.

als. As the overall record average offsets are small ($< 1.0\%$), this is an indication of the number of ADDS observations vs. other observation types.

2.4.1 Mauna Loa Observatory, Hawai'i, USA (19° N, 156° W; NDACC station)

Observations at MLO were started in December 1957. The instrument was damaged in 1961, and thus the calibration is unknown prior to 1963. Before 1984, the primary instrument was D063, with short periods with other instruments. The data in the archive prior to 1984 were not processed in the standard method in an attempt to account for instrument calibration drifts and other instrument problems, which causes larger variation in the comparison of original to the WinDobson record prior to 1984. The automated instrument D076 was installed at the station in 1984 after rebuilding in Boulder. A mirror deteriorated, so the calibration in the period 1990–1995 (indicated by the yearly N table triangles) is based on comparisons with World Standard Dobson D083 while it was on station for Langley plot campaigns. (The Langley plot method is used to establish an extraterrestrial constant for an instrument; Langley, 1884.) This new calibration is not reflected in the WOUDC or NDACC archives. The instrument was rebuilt and the WinDobson automation installed in June 2010. Data from 2010 to 2014 were processed in the NOAA system after converting WinDobson data files to a format compatible with the NOAA system. The NDACC archive appears to have updates not reflected in the WOUDC Archive, but there are periods with data missing from the NDACC archive (July–December 2012). The difference between the WOUDC and NDACC archive records processed in the NOAA system and WinDobson system is presented graphically in Fig. 4.

2.4.2 South Pole, Antarctica (90° S, 59° E; NDACC station)

The South Pole station was established in 1957. The first Dobson instrument failed due to the extreme cold. Observations started again in 1961 and these results are in the NOAA archive, but the calibration record dates from 1963. The normal routine established in 1985 was to change the instrument every 4 years for calibration checks, but this was not always achieved. This station has the possibility of large changes in reported daily values in the WinDobson, primarily due to the extended daily observation period, and high variation in total ozone during certain periods of the year. The station local day is the same as that of Christchurch, New Zealand, for ease of logistics, but the Dobson observations are reported in the WOUDC in UTC date and hour. The date and time combination often is misleading (for example, in the WOUDC archive, 14 November 1994 has a time of 28 h UTC, which matches the WinDobson and NDACC 15 November 1994 values). The calculation of the astronomical parameters used in the algorithm for reducing reflected moon observations was incorrect in the NOAA program throughout the period of record. Changes in the method of deriving total ozone from ZS observations improved the average with respect to DS averages but creates differences between the old and new archives. There are several periods missing from the WOUDC and NDACC archives (for example, July through December 2002). The difference between the WOUDC and NDACC archive records processed in the NOAA system and WinDobson system is presented graphically in Fig. 5. The exclusion of low TOC values in early October in the archived data (small white circles are outside of the plot range) in some years also produces large percentage differences in the averages (see large deviations in open circles seen in some years in the panels c and e). An example is October 1994, where there are 25 reported days in the WinDobson record but only 18 reported in the WOUDC and only 10 in the NDACC archive. These inconsistencies can produce large percentage differences, especially during low ozone conditions.

The rules for selection and inclusion of days in the archives appear have been inconsistent in earlier (NOAA) processing and archiving. The NDACC archive prior to 1999 has TOC expressed as vertical column density (molecules cm^{-2}). These numbers appear to have been calculated from DU, as this archive is derived from the WOUDC archive. There are periods where this calculation was done incorrectly (for example, October 1998, where the NDACC values differ by more than 100 DU when converted back to DU). While the NDACC archive is supposed to be derived from the same internal NOAA archive as WOUDC, there are random differences (for example, February 1981 is missing from the NDACC archive). The change in the yearly cycle of TOC (panel A) is evident in the austral spring due the depletion related to chlorofluorocarbon release (Farman etal, 1985). Station and observing schedules were changed to accommodate research needs after 1985.

2.4.3   Bismarck, North Dakota, USA (47° N, 101° W)

The instrument is operated by the US National Weather Service office at Bismarck Airport. There are observations in the archive from the late 1950s, but the documented record starts in December 1962. The difference between the WOUDC and NDACC archive records processed in the NOAA system and WinDobson system is presented graphically in Fig. 6. The periods where the N tables were reconstructed from the results of the special processing in 1995 are indicated by the yearly N table triangles. The instrument's calibration has been quite stable since 1995.

2.4.4   Caribou, Maine, USA (47° N, 68° W)

The instrument is operated at the National Weather Service office at the Caribou Airport. There are observations in the archive from the late 1950s, but the documented record starts in August 1962. The Weather Service office was rebuilt in the early 2000s, with data gaps during that period of the record. The difference between the WOUDC and NDACC archive records processed in the NOAA system and WinDobson sys-

tem is presented graphically in Fig. 7. The periods where the N tables were reconstructed from the results of the special processing in 1995 are indicated by the yearly N table triangles. The instrument's calibration has been quite stable since 1995.

2.4.5   Nashville, Tennessee, USA (36° N, 87° W)

The instrument is operated at the National Weather Service office near Old Hickory, Tennessee. There are observations in the archive from the late 1950s, but the documented record starts in July 1962. This station record shows a larger offset (+0.6 %) between the WOUDC and WinDobson datasets due to the change to the zenith observations results. The difference between the WOUDC and NDACC archive records processed in the NOAA system and WinDobson system is presented graphically in Fig. 8. The periods where the N tables were reconstructed from the results of the special processing in 1995 are indicated by the yearly N table triangles.

2.4.6   Fairbanks, Alaska, USA (65° N, 148° W)

Observations were started at the Fairbanks airport in 1964 using instrument D076 but ceased in 1972. The values in the WOUDC archive in the 1964–1972 period do not correspond to the values in the older NOAA internal archive for reasons not determined. Observations were restarted at the Poker Flat Research Range (65° N, 147° W) in 1985. The mission of the range changed in 1993 and the Dobson shelter was moved to the roof of the Geophysical Institute at the University of Fairbanks. Operations restarted in April 1994. This station is at 65° N, with observations on low sun with high ozone amounts common, especially in March and April. Researchers are advised that this instrument shows patterns in the comparison with other instrumentation that imply an under estimation of ozone on the ADDS wavelength under conditions of low sun and high ozone. The older NOAA processing and selection of observations was different from other stations, as CD pair combinations were often selected over AD pair combinations, while WinDobson uses the same rules for all stations. This change in selection is reflected in the variability in the comparison with WOUDC archive. The difference between the WOUDC archive records processed in the NOAA system and WinDobson system is presented graphically in Fig. 9.

2.4.7   Boulder, Colorado, USA (40° N, 105° W, NDACC station)

Dobson observations were started at the University of Colorado east campus in 1966. Earlier observations were made either at the National Center for Atmospheric Research or at the Table Mountain facility north of Boulder. The station was moved to the David Skaggs Research Center in 1999. Multiple instruments have been used here in the record, especially prior to the automation of Dobson instrument D061 in 1980.

The observations made after 1980 automation do not include CC′ zenith observations. The instrument was rebuilt with the WinDobson automation, but the data were processed in the NOAA system until the beginning of 2015. There is data in the WOUDC archive prior to 1966 but not connected to a calibration. The data for July 2013 to July 2014 are missing from the WOUDC and NDACC archives. The periods 1992–1996 and 1998–2005 were not processed or archived using the correct calibration information. The difference between the WOUDC and NDACC archive records processed in the NOAA system and WinDobson system is presented graphically in Fig. 10. The instrument's calibration is tracked more closely than at other stations, as the World Standard Dobson D083 is kept in Boulder.

2.4.8   Wallops Island Flight Center, Virginia, USA (38° N, 76° W; NDACC station)

Dobson observations were started at WIFC in 1967 as support for balloon- and rocket-borne experiments. The station has moved several times to different sites within the facility. Since 1995 only ADDS observations are made to support ozonesonde flights. There are periods in the WOUDC and NDACC archives with either missing data or archived with incorrect calibration. The difference between the WOUDC and NDACC archive records processed in the NOAA system and WinDobson system is presented graphically in Fig. 11.

2.4.9   NOAA/ESRL/GMD Observatory, Barrow, Alaska, USA (71° N, 157° W)

Dobson observations at the NOAA observatory began in 1973. The instrument was out of operation between 1983 and 1986 due to lack of funding. The difference between the WOUDC archive processed in the NOAA system and WinDobson system is presented graphically in Fig. 12. The station's weather is far cloudier than at other stations, with the station reporting 58 % ZS observations. The change in the method for retrieving TOC from these observations is evident in the variability in the differences between the archives.

2.4.10   NOAA/ESRL/GMD Observatory, American Samoa (14° S, 171° W; NDACC station)

Dobson observations were started at the NOAA observatory in 1976. The station is in a warm, humid marine environment, which caused instrument degradation in the early part of the record. The original processing pre-1995 was not standard and not repeatable. The periods where the N tables were reconstructed from the results of the special processing in 1995 are indicated by the yearly N table triangles. An earthquake and tsunami on 29 September 2009 damaged the station and instrument and observations were interrupted for several years. The difference between the WOUDC and NDACC archive records processed in the NOAA system and WinDobson system is presented graphically in Fig. 13. The period

1999–2001 was not processed or archived using the correct calibration information. The WOUDC and NDACC were not completely updated after observations were restarted due to perceived instrument problems which since have been resolved.

2.4.11   Fresno and Hanford, California, USA (36° N, 120° W)

Dobson observations were started at the Fresno Weather Service Office, California (37° N, 120° W), in 1982, with observations starting the next year. The Weather Service Office was moved to Hanford in March 1995. The difference between the WOUDC and NDACC archive records processed in the NOAA system and WinDobson system is presented graphically in Fig. 14. There are very few issues with the Fresno and Hanford records.

2.4.12   Observatoire de Haute-Provence, France (44° N, 6° E; NDACC station)

Dobson observations were started at the Observatoire de Haute-Provence (Station Géophysique Gérard Mégie) in 1983 with an automated instrument. This instrument was updated to the WinDobson automation and data processing in 2014. The station and instrument are operated by the Centre National de la Recherché Scientifique (CNRS). The period of 1990 to 1999 was reprocessed to account for calibration drift but has not yet been updated in WOUDC and NDACC. The difference between the WOUDC and NDACC archive records processed in the NOAA system and WinDobson system is presented graphically in Fig. 15. The instrument was damaged several time in its history; inspection within WinDobson resulted in the removal of days from inclusion in the record. This produced several months of higher differences (February 2013, for example).

2.4.13   Perth Airport, Western Australia, Australia (32° S, 116° E)

Dobson observations were started originally in 1969 at Perth Airport weather radar, Western Australia, then the NOAA automated instrument D081 was installed in 1984. The instrument is operated by the Australian Bureau of Meteorology (BoM). In the late 1990s, the station was moved to the newly constructed weather station. There are periods of missing data in the WOUDC archive. The period after 2012 in the WOUDC archive does not have correct calibration information, as the BoM recalibrated the instrument, and this information was not included in NOAA's database of calibrations. The difference between the WOUDC and NDACC archive records processed in the NOAA system and WinDobson system is presented graphically in Fig. 16.

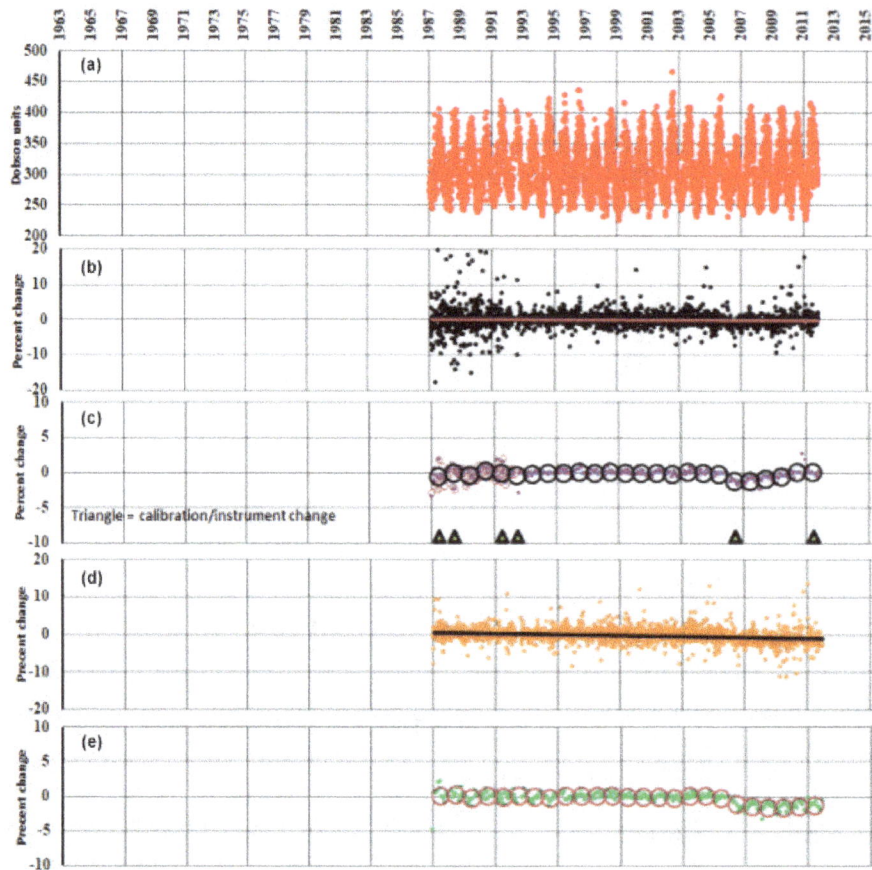

Figure 17. Graphic representation of the changes in the Lauder, Central Otago, New Zealand (45° S, 170° E; NDACC station), record with the conversion into WinDobson processing.

2.4.14   Lauder, Central Otago, New Zealand (45° S, 170° E; NDACC station)

Dobson observations began in early 1987 at the research station in Central Otago, South Island, New Zealand. The instrument is operated by New Zealand's National Institute of Water and Atmospheric Research (NIWA). The station's time zone is UTC + 12, which means the UTC day changes at noon local standard time. The calculation of the local day and UTC day for reporting the selected value was incorrect prior to 1992, indicated by the higher scatter in the comparison of the old and new archives in that time period. Also, a selected value could be from the afternoon of one local day and the representative value from the morning of the following local day while still being in the same UTC day. When inspected during the WinDobson processing, the instrument record between 2006 and 2011 revealed rain damage following reinstallation shortly after the 2006 intercomparison in Melbourne. The 2012 calibration information determined before the rebuilding of the instrument was used to process the data in WinDobson during 2006 through 2012. The inspection also determined that the calibration was stable from 1992 to 2006, while the instrument calibration was checked

in 1997 and 2001. The WOUDC and NDACC records are not yet updated. The instrument was rebuilt at the beginning of 2012 and has been operated with the data reduction in WinDobson since that time. The difference between the WOUDC and NDACC archive records processed in the NOAA system and WinDobson system is presented graphically in Fig. 17.

3   Conclusions

NOAA has submitted nearly a half-century's data into the WOUDC and NDACC archives. Personnel and data processing protocols changed many times throughout that period, and knowledge of early techniques was slowly lost. Furthermore NOAA personnel tended to use a larger and more comprehensive database when performing research, so the accuracy of data within the WOUDC and NDACC archives was seldom questioned. Our experiences in the investigation of the long-term archived NOAA Dobson data records should alert other Dobson data producers of the importance of regular review and intercomparisons of the archived station's records residing at multiple archives. The Dobson station data processing procedures and software tend to change over time after the new knowledge or technology becomes available. Although the reprocessing of historical datasets is extremely difficult due to lost documentation or even raw data, the benefit of the investigation is record's homogenization and adjustment to conform to the WMO/GAW operating procedure guidance (Komhyr and Evans, 2008). With the advent of WinDobson software and its newer technique for calculating TOC from zenith observations and selecting representative observations, we felt it was prudent to reprocess all previous measurements for the sake of homogeneity. It also seemed logical to compare and replace data within the WOUDC and NDACC archives with the newly reprocessed data. The overall changes are small (~ 0.1 % offset), but several individual stations have a larger offset (maximum 0.7 %) driven by the changes in the ZC reduction polynomials. During comparisons between WinDobson dataset and the existing NDACC and WOUDC archives, we were able identify periods with either missing data or incorrectly processed data. The differences between the historical and the new version of Dobson data have overall small offsets and trends (Table 2), but within the long-term record there are periods with greater differences of which researchers should be aware (see Figs. 4–16 and description of the individual station histories). The paper includes a section that describes individual station histories, which provides information on specific-to-station updates and their effects on the total ozone record. The offsets and trends for differences between the old and the new version of the data are not the same for WOUDC and NDACC archives, as the NDACC set of data is not a perfect match to the one available from the WOUDC archive. For example, the Wallops Island NDACC record is 1995–2014, while the WOUDC record is 1967–2014. When the NDACC and WOUDC archives are updated, these archived datasets will be complete and homogenized. Moreover, after all calibrations and the applicable periods were reviewed, the history method of applying calibrations to all of the instruments in the networks has been standardized. The new, complete (all observations) WinDobson operational database, available to researchers on request, will allow investigators to improve the accuracy and consistency of the Dobson retrieval algorithms and stations records. The expectation is that in the future the WOUDC and NDACC archives will consist of all observation results.

Competing interests. The authors declare that they have no conflict of interest.

Special issue statement. This article is part of the special issue "Twenty-five years of operations of the Network for the Detection of Atmospheric Composition Change (NDACC) (AMT/ACP/ESSD inter-journal SI)". It is not associated with a conference.

Acknowledgements. The authors would like to acknowledge the work done in the past by such people as Walter Komhyr, Robert Grass and Kent Leonard in establishing the US Dobson ozone network.

The Dobson observations at Lauder are supported through NIWA's core research funded by the NZ Ministry of Business, Innovation and Employment; at Perth by the Bureau of Meteorology, an executive agency of the Australian government; and at l'Observatoire du Haute-Provence by the National Center for Scientific Research (CNRS), under the responsibility of the French Ministry of Education and Research. Support for updating of the automation at several NDACC sites was provided by the NOAA Joint Polar Satellite System (JPSS) Calibration/Validation program.

Edited by: Hal Maring

References

Basher, R. E.: Review of the Dobson Spectrophotometer and its Accuracy, Global Ozone Research and Monitoring Project, Report No. 13, available at: http://www.esrl.noaa.gov/gmd/ozwv/dobson/papers/report13/report13.html (last access: 10 June 2017), 1982.

Brönnimann, S., Staehelin, J., Farmer, S. F. G., Cain, J. C., Svendby, T., and Svenøe, T.: Total ozone observations prior to the IGY. I: A history, Q. J. Roy. Meteor. Soc., 129, 2797–2817, https://doi.org/10.1256/qj.02.118, 2003.

Dobson, G. M. B.: A photoelectric spectrophotometer for measuring the amount of atmospheric ozone, P. Phys. Soc., 43, 324–339, https://doi.org/10.1088/0959-5309/43/3/308, 1931.

Dobson, G. M. B.: Forty years' research on atmospheric ozone at Oxford: a history, Appl. Opt., 7, 387–405, https://doi.org/10.1364/AO.7.000387, 1968.

Farman, J. C., Gardiner, B. G., and Shanklin, J. D. Large losses of total ozone in Antarctica reveal seasonal ClO_x/NO_x interaction, Nature, 315, 207–210, https://doi.org/10.1038/315207a0, 1985.

Grant, W. B.: Ozone Measuring Instruments for the Stratosphere, Optical Society of America, Washington, D. C., 1989.

Josefsson, W. and Löfvenius, M. O.: Total Ozone from Zenith Radiance Measurements – an Empirical Model Approach, SHMI Report No. 130, available at: https://www.smhi.se/polopoly_fs/1.1735!/meteorologi_130%5B1%5D.pdf (last access: 10 June 2017), 2008.

Komhyr, W. D. and Evans, R. D.: Operations Handbook – Ozone Observations with a Dobson Spectrophotometer, Revised 2008 World Meteorological Organization GAW Report no. 183, WMO, Geneva Switzerland, available at: https://www.wmo.int/pages/prog/arep/gaw/documents/GAW183-Dobson-WEB.pdf (last access: 10 June 2017), 2008.

Komhyr, W. D., Mateer, C. L., and Hudson, R. D.: Effective Bass–Paur 1985 ozone absorption coefficients for use with Dobson ozone spectrophotometers, J. Geophys. Res.-Atmos., 98, 20451–20465, https://doi.org/10.1029/93JD00602, 1993.

Komhyr, W. D., Quincy, D. M., Grass, R. D., and Koenig, G. L.: Re-evaluation of total and Umkehr ozone data from NOAA-CMDL Dobson spectrophotometer observatories. Final report, UNT Digital Library, United States, available at: http://digital.

library.unt.edu/ark:/67531/metadc693696/, last access: 7 October 2016, 1995.

Langley, S. P.: Determination of the solar constant, in: Researches on Solar Heat and its Absorption by the Earth's Atmosphere: a Report on the Mount Whitney Expedition, Professional Papers of the Signal Service XV, Chapt. 10, 124–128, Government Printing Office, Washington, DC, USA, available at: https://ia802706.us.archive.org/0/items/researchesonsola00lang/researchesonsola00lang.pdf (last access: 10 June 2017), 1884.

Miyagawa, K.: Development of automated measuring system for Dobson ozone spectrophotometer, in: Atmospheric Ozone: Proceedings of the XVIII Quadrennial Ozone Symposium L'Aquila, Italy, 12–21 September 1996, Parco Scientifico e Tecnologici d'Abruzzo, L'Aquila, Italy, vol. 2, 951–954, 1996.

Redondas, A., Evans, R., Stuebi, R., Köhler, U., and Weber, M.: Evaluation of the use of five laboratory-determined ozone absorption cross sections in Brewer and Dobson retrieval algorithms, Atmos. Chem. Phys., 14, 1635–1648, https://doi.org/10.5194/acp-14-1635-2014, 2014.

10

Regional modelling of polycyclic aromatic hydrocarbons: WRF-Chem-PAH model development and East Asia case studies

Qing Mu[1], Gerhard Lammel[1,2], Christian N. Gencarelli[3], Ian M. Hedgecock[3], Ying Chen[1,4], Petra Přibylová[2],
Monique Teich[4], Yuxuan Zhang[1], Guangjie Zheng[5], Dominik van Pinxteren[4], Qiang Zhang[6], Hartmut Herrmann[4],
Manabu Shiraiwa[1,7], Peter Spichtinger[8], Hang Su[1,9], Ulrich Pöschl[1], and Yafang Cheng[1,9]

[1]Multiphase Chemistry Department, Max Planck Institute for Chemistry, Mainz, Germany
[2]Research Centre for Toxic Compounds in the Environment, Masaryk University, Brno, Czech Republic
[3]CNR-Institute of Atmospheric Pollution Research, Division of Rende, Rende, Italy
[4]Leibniz Institute for Tropospheric Research, Leipzig, Germany
[5]Brookhaven National Laboratory, Brookhaven, USA
[6]Ministry of Education Key Laboratory for Earth System Modeling, Department of Earth System Science,
Tsinghua University, Beijing, China
[7]Department of Chemistry, University of California, Irvine, USA
[8]Institute for Atmospheric Physics, Johannes Gutenberg University, Mainz, Germany
[9]Institute for Environmental and Climate Research, Jinan University, Guangzhou, China

Correspondence to: Yafang Cheng (yafang.cheng@mpic.de) and Gerhard Lammel (g.lammel@mpic.de)

Abstract. Polycyclic aromatic hydrocarbons (PAHs) are hazardous pollutants, with increasing emissions in pace with economic development in East Asia, but their distribution and fate in the atmosphere are not yet well understood. We extended the regional atmospheric chemistry model WRF-Chem (Weather Research Forecast model with Chemistry module) to comprehensively study the atmospheric distribution and the fate of low-concentration, slowly degrading semivolatile compounds. The WRF-Chem-PAH model reflects the state-of-the-art understanding of current PAHs studies with several new or updated features. It was applied for PAHs covering a wide range of volatility and hydrophobicity, i.e. phenanthrene, chrysene and benzo[*a*]pyrene, in East Asia. Temporally highly resolved PAH concentrations and particulate mass fractions were evaluated against observations. The WRF-Chem-PAH model is able to reasonably well simulate the concentration levels and particulate mass fractions of PAHs near the sources and at a remote outflow region of East Asia, in high spatial and temporal resolutions. Sensitivity study shows that the heterogeneous reaction with ozone and the homogeneous reaction with the nitrate radical significantly influence the fate and distributions

of PAHs. The methods to implement new species and to correct the transport problems can be applied to other newly implemented species in WRF-Chem.

1 Introduction

Polycyclic aromatic hydrocarbons (PAHs), released into the atmosphere as by product of all kinds of combustion processes, are harmful for human health via inhalation as well as ingestion pathways (WHO, 2003; Lv et al., 2016) and for ecosystems (Hylland, 2006). In the atmospheric environment PAHs are partly readily degradable, partly undergo long-range transport and reach remote areas (Keyte et al., 2013). PAHs have been included in the United Nations Economic Commission for Europe Convention on Long-range Transboundary Air Pollution and Protocol on Persistent Organic Pollutants. Hazardous substances, mostly benzo[*a*]pyrene (BaP), are criteria pollutants in many countries, including the European Union, USA and Japan. The United States Environmental Protection Agency (USEPA) prioritised 16 PAHs in the 1970s, which have been mostly targeted in the environment since then, but this selection is questionable considering toxicity and occurrence of PAHs (Andersson and Achten, 2015).

As PAHs are mainly generated from incomplete combustion of carbonaceous bio- and fossil fuels, their emissions increase dramatically in Asia due to rapid economic development and energy consumption. According to Zhang and Tao (2009), the total emission of the 16 PAH compounds listed in the USEPA priority control list was about $290\,\mathrm{Gg\,yr^{-1}}$ in Asia in the year 2004, accounting for more than half of total global emissions. High emissions of PAHs in Asia pose a hazard to the ecosystems and human health on an intercontinental or even global scale (Hung et al., 2005).

In an attempt to elucidate the spatiotemporal distributions of PAH ambient concentrations and processes governing their atmospheric fate, several numerical modelling studies have been published. Lagrangian frameworks have been used for Europe (van Jaarsveld et al., 1997; Halsall et al., 2001) and China (Lang et al., 2007, 2008). Other studies focused on the multicompartmental behaviour using box models (Yaffe et al., 2001; Prevedouros et al., 2004, 2008). Eulerian chemical transport models have been developed and applied for regions, i.e. Europe (Aulinger et al., 2007; Matthias et al., 2009; Bieser et al., 2012; San Jose et al., 2013; Efstathiou et al., 2016), North America (Galarneau et al., 2014) and East Asia (Zhang et al., 2009, 2011a, b; Inomata et al., 2012, 2013), or on the global scale (Sehili and Lammel, 2007; Lammel et al., 2009; Friedman and Selin, 2012; Friedman et al., 2014a, b; Shen et al., 2014; Shrivastava et al., 2017). The aforementioned studies differ in many respects relating to the PAH species examined, the temporal variability of their emissions, the spatial and process resolutions of the models. However, the up-to-date representations of complex PAH processes (e.g. gas–particle partitioning, heterogeneous chemistry and re-volatilisation) are not reflected in the previous modelling studies, nor have these processes been evaluated against limited monitoring data. In addition, a high temporal resolution like the diurnal cycle is not seen.

Since its initial release in 2002, Weather Research Forecast model with Chemistry module (WRF-Chem) has been widely applied and verified for regional air quality (Zhang et al., 2010, 2013) and climate (Liao et al., 2014; Yahya et al., 2016) study with high temporal and spatial resolutions. Compared to the previous PAH modelling studies, this work is unique in four aspects: (1) it includes all relevant state-of-the-art processes of PAH into WRF-Chem which are important for its cycling in the atmospheric environment over land (i.e. new heterogeneous degradation scheme, several oxidants in homogeneous degradation processes, and re-volatilisation from soil, among others), (2) predicts and validates against observed gas- and particulate-phase PAH concentrations separately, (3) validates atmospheric concentrations and particulate mass fraction against diurnal variable PAH observations, and (4) explores the significance of PAH

heterogeneous reaction with ozone (O_3) and gas-phase reaction with nitrate radical (NO_3).

2 Model development

WRF-Chem-PAH is based on the open-source community model WRF-Chem (version 3.6.1), which is a fully coupled, "online" regional model with integrated meteorological, gas-phase chemistry, and aerosol components (Grell et al., 2005). It is built on the Advanced Research WRF core, which handles the dynamics, physics, and transport processes. Gas-phase chemistry and aerosol schemes are integrated over the same time step as transport processes, allowing for full coupling between the schemes. The short chemistry time step also made the model ideal for studying short-lived PAH species with high levels of spatial heterogeneity.

By considering the compatibility with the WRF-Chem Kinetic PreProcessor (KPP) (Sandu et al., 2003; Sandu and Sander, 2006) and similarities of PAH chemistry to volatile organic aerosol formations, we have made the following choices. The Regional Atmospheric Chemistry Mechanism (RACM) (Stockwell et al., 1997) is used for homogeneous gas-phase reactions. The aerosol module includes the Modal Aerosol Dynamics Model for Europe (MADE) (Ackermann et al., 1998) for the inorganic fraction, and the Secondary Organic Aerosol Model (SORGAM) (Schell et al., 2001) for the secondary organic aerosols. MADE/SORGAM in WRF-Chem uses the modal approach with three log-normally distributed modes (nuclei, accumulation and coarse mode). All pollutant species normally simulated in the standard RACM/SORGAM mechanism are also simulated in WRF-Chem-PAH. In order to include PAHs (and organics in general) in air–soil gas exchange processes, the Noah soil scheme (Chen and Dudhia, 2001) is utilised.

2.1 Framework

Figure 1 shows the framework of PAH extensions in WRF-Chem, where modules/subroutines that have been modified in terms of embedding PAH extensions are listed. All the new variables related to emissions and concentration fields of PAHs and those intermediate variables used in different chemical and physical processes, such as air–soil gas exchange, gas-phase/heterogeneous reactions, cloud scavenging, dry/wet deposition, advective transport and cumulus convection, are first defined in registry.chem and then included in respective sub-modules/routines.

The subroutine chem_driver is the main driver for handling chemistry-related tasks on a particular time step, including emissions, photolysis, gas- and particulate-phase reactions, convective transport, cloud chemistry, and dry/wet depositions. Based on the existing structure, all the chemical reactions involving gaseous and particulate PAHs mentioned in Sect. 2.2.1 and 2.2.2 have been added to

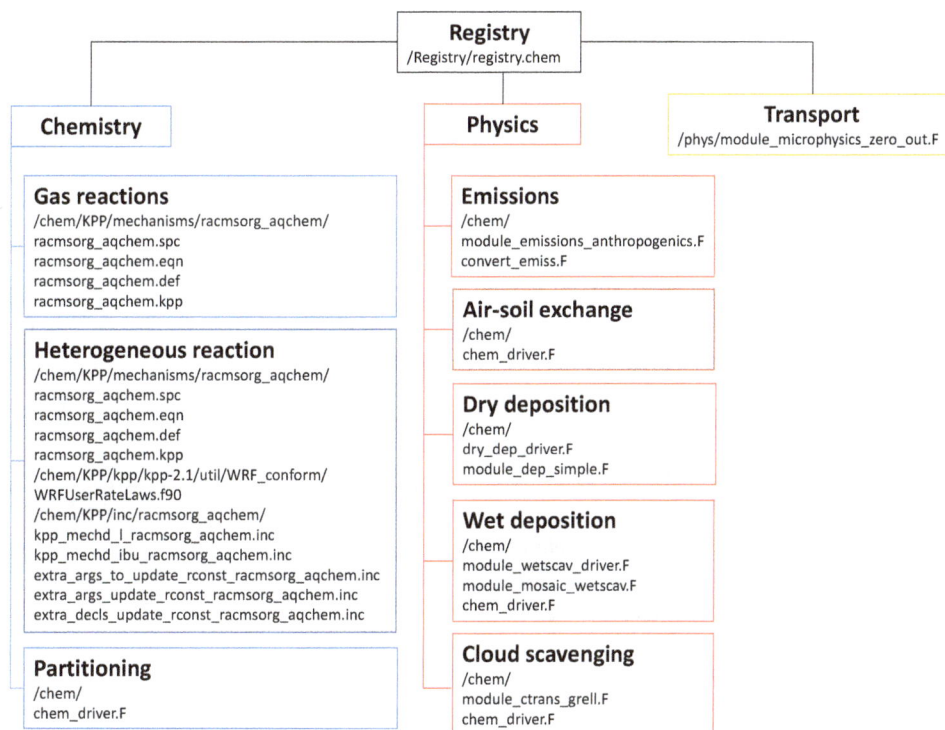

Figure 1. Framework of PAH extensions in WRF-Chem.

the RACMSORG_AQCHEM chemical mechanism by using KPP and the WRF-Chem KPP Coupler (Salzmann and Lawrence, 2006). Fixed rate coefficients are used for gas-phase reactions, while for heterogeneous reactions oxidant/temperature/humidity-dependent functions are formulated. Gas–particle partitioning of semivolatile PAHs (and organic compounds in general) is implemented based on substance-specific empiric equilibrium relationships in addition to the MADE/SORGAM module. Dry deposition, wet deposition and wet scavenging from cumulus convection are calculated in the respective subroutines, using the dry deposition velocity and the fraction of gaseous species dissolved in cloud water calculated in module_dep_simple and module_mosaic_wetscav, respectively. The air–soil gas exchange for organic compounds (secondary emissions) is implemented in addition to the primary emission module.

The simulated PAHs in our current WRF-Chem-PAH model include phenanthrene (PHE), chrysene (CHR) and BaP, representing volatile, semivolatile and non-volatile PAH compounds, respectively. The WRF-Chem-PAH framework is not limited to these three species, and in general, all semivolatile compounds can be similarly implemented following our practices.

2.2   Atmospheric processes of PAHs

2.2.1   Gas-phase reactions

Reactions of gas-phase PAHs with hydroxyl radical (OH), NO_3 and O_3 are considered in this model. While NO_3 reactions appear to be less significant than OH reactions as the main PAH degradation process, the observed considerably high nitro-PAH yields suggest that nighttime reactions of PAHs with NO_3 may be a significant contributor of these compounds in the atmosphere (Keyte et al., 2013). The significance of NO_3 oxidation is further discussed in Sect. 6. Here, the PAH oxidative loss is calculated as a second-order process using the model-predicted OH and NO_3 concentration (Table 1). At this stage, PAH reaction products are not tracked in the model.

2.2.2   Heterogeneous degradation of particulate BaP

In this study, we have developed and applied a more elaborate parameterisation of the heterogeneous reaction kinetics of BaP degradation by ozone as a function of temperature and relative humidity based on the experimental data of Zhou et al. (2013) and the kinetic multi-layer model KM-SUB (Shiraiwa et al., 2010). Our model approach and parameterisation build on a Langmuir–Hinshelwood reaction mechanism involving the decomposition of ozone and formation of long-lived reactive oxygen intermediates (Shiraiwa et al., 2011; Berkemeier et al., 2016). Table S2 in the Supplement lists the parameterisations of the reaction rates. Text S1 in the Supplement reviews previous study of the heterogeneous degradation of BaP.

2.2.3 Gas–particle partitioning

The unmodified (conventional) configuration of WRF-Chem uses a thermodynamic equilibrium scheme (SORGAM) to simulate gas-to-particle mass distribution of condensable and water-soluble species (Binkowski and Roselle, 2003). This approach is inappropriate for semivolatile and hydrophobic substances. Instead, empiric equilibrium partitioning expressions for PAHs are applied. We consider the absorption processes and chemically specific adsorption processes, both of which were found to be significant contributors to PAH gas–particle partitioning (Dachs and Eisenreich, 2000; Lohmann and Lammel, 2004).

We use a second equilibrium partitioning expression, which accounts for two contributions, absorption into organic matter and adsorption onto black carbon (BC) (Dachs and Eisenreich, 2000):

$$K_p = 10^{-12} \left(\frac{1.5 f_{oc}}{\rho_{oct} K_{OA}} + f_{bc} K_{SA} \right) = \frac{\sum c/c_{TSP}}{c_g}, \qquad (1)$$

$$\theta = \left(1 + \frac{1}{K_p c_{TSP}} \right)^{-1}, \qquad (2)$$

where ρ_{oct} is the bulk density of octanol ($0.82\,\mathrm{kg\,L^{-1}}$), f_{oc} is the organic carbon (OC) fraction of the particulate matter (the 1.5 multiplier converts OC to organic matter, which is assumed to be well represented by octanol), K_{OA} is the octanol–air partition coefficient (dimensionless, temperature dependent; Odabasi et al., 2006), f_{bc} is the BC fraction of the particulate matter, K_{SA} is the soot–air partition coefficient ($\mathrm{L\,kg^{-1}}$), $\sum c_p$ is the particulate PAH concentration across all the size bins ($\mathrm{ng\,m^{-3}}$), c_{TSP} is the total particulate matter concentration ($\mathrm{\mu g\,m^{-3}}$), and c_g is the gas-phase concentration ($\mathrm{ng\,m^{-3}}$).

Since direct K_{SA} measurements are not available for PAHs, soot–air partitioning coefficients ($K_{SA}\,\mathrm{L\,kg^{-1}}$) are estimated as the ratios of soot–water (K_{SW}; Jonker and Koelmans, 2002) and the air–water (K_{AW}) partitioning coefficients (dimensionless, temperature dependent; Bamford et al., 1999). Values of K_{SW} vary substantially (up to a factor of 47 for the PAHs considered here) among relevant soot. The representative values for atmospheric BC are determined by weighting the reported K_{SW} values by the contribution of their related combustion processes to the total emitted fine particulate matter used in the inventory of Galarneau et al. (2007).

2.2.4 Air–soil gas exchange

Semivolatile PAHs are subject to revolatilisation (Lammel et al., 2009; Galarneau et al., 2014). An air–soil gas exchange module is therefore included in WRF-Chem-PAH.

Table 1. Physical and chemical properties of PAHs and soil.

Parameter	Symbol	Unit	PHE	CHR	BaP	Reference
Molecular weight	MW	g mol⁻¹	178.2	228.3	252.3	
Gas-phase OH reaction rate constant	k_{OH}	cm³ molec⁻¹ s⁻¹	[a]3.1 × 10⁻¹¹	[b]5.0 × 10⁻¹¹	[b]1.5 × 10⁻¹⁰	[a] Atkinson et al. (1989); [b]Klöpffer et al. (2008)
Gas-phase NO₃ reaction rate constant	k_{NO_3}	cm³ molec⁻¹ s⁻¹	[a]1.2 × 10⁻¹³	4 × 10⁻¹²	[b]5.4 × 10⁻¹¹	[a]Kwok et al. (1994); [b]Klöpffer et al. (2008)
Gas-phase O₃ reaction rate constant	k_{O_3}	cm³ molec⁻¹ s⁻¹	[a]4.0 × 10⁻¹⁹	[b]4.0 × 10⁻¹⁹	[b]2.6 × 10⁻¹⁷	[a]Kwok et al. (1994); [b]Klöpffer et al. (2008)
Soot–water partitioning coefficient	K_{SW}	L kg⁻¹	4.34 × 10⁵	2.82 × 10⁷	9.59 × 10⁷	Jonker and Koelmans (2002)
Octanol–air partitioning coefficient	K_{OA}-m	dimensionless	3293	4754	5382	Odabasi et al. (2006)
	K_{OA}-b	dimensionless	-3.37	-5.65	-6.5	log $K_{OA} = m/T(K) + b$
Air-water partitioning coefficient	K_{AW}-m	dimensionless	-5689.2	-12136.16	-4437.1	Bamford et al. (1999)
	K_{AW}-b	dimensionless	12.75	32.235	3.9881	ln $K_{AW} = m/T(K) + b$
Soil degradation rate	k_{soil}	s⁻¹	1.00 × 10⁻⁸	1.00 × 10⁻⁸	1.00 × 10⁻⁸	Mackay and Paterson (1991)
Water content of soil	l	dimensionless	0.3	0.3	0.3	Jury et al. (1983)
Air content of soil	a	dimensionless	0.2	0.2	0.2	Jury et al. (1983)
Soil depth	z_s	m	0.15	0.15	0.15	Jury et al. (1983)
Bulk density of soil	ρ_s	kg m⁻³	1350	1350	1350	Jury et al. (1983)
Organic carbon fraction of soil	f_{oc}	dimensionless	0.0125	0.0125	0.0125	Jury et al. (1983)
Air diffusion coefficient of soil	D_{air}	m² s⁻¹	5.00 × 10⁻⁶	5.00 × 10⁻⁶	5.00 × 10⁻⁶	Jury et al. (1983)
Water diffusion coefficient of soil	D_{water}	m² s⁻¹	5.00 × 10⁻¹⁰	5.00 × 10⁻¹⁰	5.00 × 10⁻¹⁰	Jury et al. (1983)

Air–soil gas exchange is parameterised following Strand and Hov (1996), which is based on Jury et al. (1983).

Model soil is a 0.15 m thick layer consisting of fixed volumes of soil organic matter, air and water. The soil layer is assumed to have the standard properties suggested by Jury et al. (1983) (Table 1). The change in PAH concentrations in soil/air, c_s/c_a, with time is expressed by

$$\frac{\partial c_s}{\partial t} = \frac{1}{z_s}\left(F_{exc,soil} + F_{wet}\right) - k_{soil}c_s, \tag{3}$$

$$\frac{\partial c_a}{\partial t} = -\frac{1}{z_a}F_{exc,soil}, \tag{4}$$

where z_s and z_a are the soil and atmospheric layer depths (m), respectively, $F_{exc,soil}$ is the air–soil gas exchange flux, F_{wet} is the wet deposition flux, and k_{soil} is the degradation rate in soil. The air–soil gas exchange flux is given by

$$F_{exc,soil} = v_s\left(c_a - \frac{c_s}{K_{soil-air}}\right), \tag{5}$$

where v_s is the exchange velocity, c_a is the PAH concentrations in air, and $K_{soil-air}$ is the partitioning coefficient between soil and air. The exchange velocity is given by

$$v_s = \frac{D_{air}a^{10/3}(1-l-a)^{-2} + D_{water}l^{10/3}K_{WA}(1-l-a)^{-2}}{z_s/2}, \tag{6}$$

where D_{air} is the air diffusion coefficient, D_{water} is the liquid diffusion coefficient, K_{WA} is the water-air partitioning coefficient depending on the soil temperature and equals the inverse of K_{AW}, and l and a are the water and air fractions in soil, respectively. Partitioning between soil and air is given by Karickhoff (1981):

$$K_{soil-air} = 4.11 \times 10^{-4} \times \rho_s f_{oc} K_{OA}, \tag{7}$$

where ρ_s is the soil density, f_{oc} is the soil OC fraction and 4.11×10^{-4} is a constant with units of $m^3\,kg^{-1}$. PAHs are subject to biodegradation in soil, processes which are actually not well quantified. The degradation rate in soil k_{soil} is estimated to be $10^{-8}\,s^{-1}$ based on a laboratory model ecosystems study (Lu et al., 1977), following a global PAH model (Friedman et al., 2014a, b).

Due to a lack of monitoring data, the PAH concentrations in soil are initialised by the global multicompartmental model output: a pseudo-steady state of anthracene (ANT), fluoranthene (FLT) and BaP concentrations in the soil compartment had been safely reached by a global simulation over 10 years with $2.8° \times 2.8°$ horizontal resolution (Lammel et al., 2009). PHE/CHR concentrations in soil are scaled from ANT/FLT according to the ratio upon primary emission. Figure S1 in the Supplement shows the air–soil gas exchange flux at a receptor site based on the above air–soil exchange scheme.

2.2.5 Wet and dry depositions

Dry deposition of gas-phase species in WRF-Chem is treated using the standard resistance approach (Wesely, 1989). The original WRF-Chem routines have been adapted to include the deposition of gas-phase PAH compounds, and the deposition flux is calculated from the product of the deposition velocity and gaseous PAH concentration in the lowermost model level. We consider the particulate-phase PAH species to be bound to the atmospheric particulate matter in the accumulation mode, whose dry deposition flux is calculated using WRF-Chem particulate deposition parameterisations.

The model accounts for wet deposition of PAH species through the schemes for gas and particulate convective transport, in-cloud and below-cloud scavenging of PAH species (sub-grid resolution, following the UCI (University of California, Irvine) chemistry transport model; Neu and Prather, 2012).

3 Modification of transport scheme for low-concentration species

The transport of BaP seems stopped (Fig. 2a) when we follow the WRF-Chem manual's suggestion by running with monotonic advection (chem, moist, scalar_adv_opt = 2) while the transport of other chemical species behaves normally, e.g. BC, as shown in Fig. 2d. This is because the atmospheric concentration of BaP is too low, down to 10^{-9} to 10^{-12} ppmv. Other species are dealt with similarly, when artificially brought to extremely low concentrations, as confirmed for BC: the species does not undergo transport when dividing the BC concentration by 10^{10} before and then multiplying by 10^{10} after the advection subroutine. Figure 2c and d show the clear differences between the transportation of BC in these two cases. Furthermore, the near-source concentration of BaP is too low compared with the observation, because in the unmodified (conventional) transport non-zero BaP concentrations in air are limited to the immediate vicinities to strong sources and undergo fast degradation.

One of the important features of how the chemical transport model (Chem) couples with WRF in the WRF-Chem model is that the transport of chemical species is done by WRF. In WRF, monotonic advection is not a positive definite option, so that mp_zero_out = 2 is usually set to make sure that the transport tendency of all the moisture variables will not grow below zero. To this end, the mp_zero_out_thresh is set as suggested to a small value of 10^{-8}, and the transport tendency of moisture variables will be mapped to 0 when concentrations are smaller than mp_zero_out_thresh. In the coupled WRF-Chem model, when dealing with chemical transport, WRF advection module treats all the chemical species as if they are moisture variables following the same criterion of exceedance of mp_zero_out_thresh. This is usually not a problem, because the concentrations of species

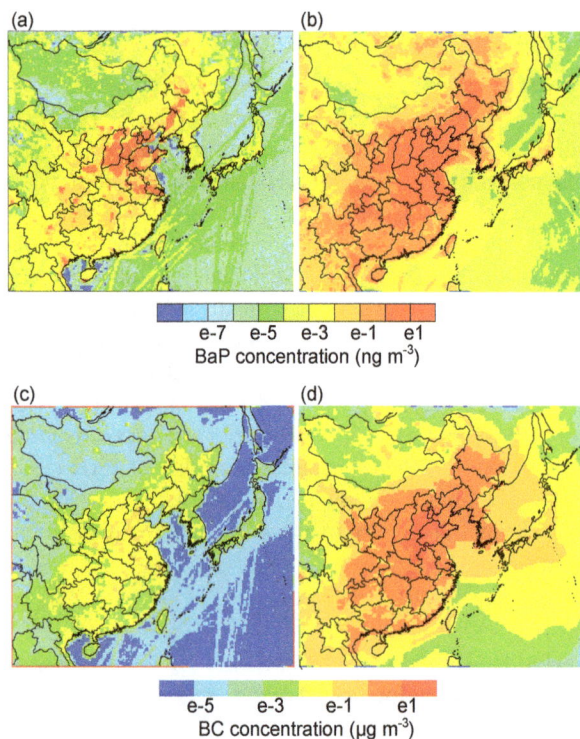

Figure 2. Simulated near-ground concentrations of BaP with **(a)** conventional transport scheme and **(b)** modified transport scheme for low-concentration species, and BC with **(c)** scaled low concentration and **(d)** normal concentration with conventional transport scheme averaged on 14 February 2003.

transported are in general higher than this threshold. However, it is not the case for BaP, and the threshold truncates the BaP concentration and its transport tendency is forced to 0 and thus no transport occurs. To cope with this, we set mp_zero_out_thresh $= 10^{-22}$ for PAHs species but leave it to 10^{-8} to moisture variables and all other chemical species. After this modification in advective transportation, BaP adequately undergoes transport in the model and the near-source concentration of BaP is elevated too (Fig. 2b). This solution can be applied to all newly implemented low-concentration species in WRF-Chem.

4   Case study in East Asia

4.1   Model configuration

To apply the WRF-Chem-PAH model, we configured a domain that covers East China and Japan (15–55° N, 95–155° E) with a horizontal resolution of 27 km by 27 km and 39 vertical layers up to 0.01 hPa. The spatial coverage of the domain is shown in Fig. 2.

The physics options applied in this study are summarised as follows (also see Table S1): the Purdue–Lin scheme (Lin et al., 1983) is used for microphysics, which includes six

classes of hydrometeors (water vapour, cloud water, rain, cloud ice, snow and graupel). The planetary boundary layer is parameterised by the Mellor–Yamada–Janjić scheme (Janjic, 1994). It describes vertical sub-grid-scale fluxes due to eddy transport in the whole atmospheric column, while the horizontal eddy diffusivity is calculated with a Smagorinsky first-order closure. The surface layer parameterisation employed is the Eta Similarity surface layer scheme (Janjic, 1994). The land surface model to describe interactions between the soil and atmosphere is Noah Land Surface Model (Chen and Dudhia, 2001). The Grell 3D ensemble scheme (Grell and Devenyi, 2002) is used for cumulus parameterisation. The long- and shortwave radiation is calculated online with rapid radiative transfer model (Mlawer et al., 1997) and the Goddard schemes (Chou and Suarez, 1994), respectively. Photolysis rates are calculated using the Fast-J photolysis scheme (Wild et al., 2000) and updated every 60 min.

For simulation of standard aerosol precursors and aerosol species in the WRF-Chem model, anthropogenic emissions for NO_x, CO, non-methane volatile organic compounds, SO_2, NH_3, BC, and OC are taken from the EDGAR-HTAP global monthly inventory (http://edgar.jrc.ec.europa.eu/national_reported_data/htap.php) in the year 2010. The emissions of BC and OC in 2010 are further extrapolated to simulation year based on annual scaling factors taken from Lu et al. (2011), while no annual changes have been applied to emissions of other species. The EDGAR-HTAP inventory has a horizontal resolution of 0.1°. Hereby, biomass burning emissions are from the monthly Quick Fire Emissions Dataset (QFED) (Darmenov and Silva, 2013). Biogenic volatile organic compounds emissions are calculated from the Model of Emissions of Gases and Aerosols from Nature (MEGAN) (Guenther et al., 2006). Anthropogenic PAH emissions are re-gridded from a 0.1° × 0.1° global annual PAH emission inventory for the year 2008, with 69 detailed source types (Shen et al., 2013). For specific simulation period, inter-annual scaling factors in the simulated domain are taken from Shen et al. (2013), based on historical fuel consumption data and IPCC SRES A1 scenario supposing a future world of rapid economic growth. Monthly scaling factors are taken from Zhang and Tao (2008). A diurnal cycle of the PAH emissions are applied with two maxima, around 08:00 and 19:00 local time, following that of BC. Biogenic contributions to PAH emission have been neglected. Figure S2 shows the average distributions of PHE, CHR and BaP emissions in July 2013. Figure S3 shows the inter-annual, monthly and hourly scaling factors of PAH.

Meteorological initial and boundary conditions are based on the National Center for Environmental Prediction Final Analysis' (NCEP-FNL) reanalysis data. Meteorology (temperature, horizontal wind, and moisture) is nudged at all vertical levels. Chemical initial and boundary conditions of standard species are from the global Model for Ozone and Related Chemical Tracers (MOZART-4) (Emmons et al., 2010), simulations performed using 1.9° × 2.5° horizontal resolutions. Initial PAH concentrations at all lateral boundaries are set to zero because China is the dominant emission country. To reach a steady-state equilibrium concentration of PAHs in air, a spin-up time of 48 h is used.

4.2 Model evaluation

The WRF-Chem-PAH model is developed to capture the PAH transport episode in higher temporal and spatial resolution, i.e. in diurnal to daily scales and in both concentration level and particulate mass fraction. To this end, two sets of continuous PAH field campaign data with at least daily resolution and in both gaseous and particulate phases are chosen.

The first dataset provides both daytime and nighttime samples. As part of the Program of Campaigns of Air Quality Research in Beijing and Surrounding Region (CAREBeijing) 2013 campaign, measurement was made at the Xianghe Atmospheric Observatory (39.80° N, 116.96° E). The Xianghe site is located in a PAH near-source area, 45 km southeast of Beijing and 70 km northwest of Tianjin (Fig. S2). The site is surrounded by residential suburban areas and distanced some 5 km from the local town centre. Particulate- and gas-phase samples were collected twice a day (daytime samples 08:00–18:00 LT; nighttime samples 20:00–06:00 LT) during 11–22 July 2013. EC and OC samples of PM_{10} were also collected separately twice a day (daytime samples: 06:00–18:00 LT; nighttime samples: 18:00–06:00 LT on the following morning) at the same site, which are important to evaluate gas/particle partition scheme of semivolatile PAHs. Details of the sampling methods and data quality control are described in Text S2.

Another PAH observation dataset is taken at the Gosan station (33.28° N, 126.17° E; 72 m above sea level) on Jeju Island, in the northern part of the East China Sea, about 100 km south of the Korean Peninsula (Fig. S2). Gosan is a representative background site and an ideal location for studying long-range transport of air pollutants in East Asia (Han et al., 2006). Although the Gosan observation covers a long period from November 2001 to August 2003 (Kim et al., 2007, 2012), due to the high computational cost of WRF-Chem, we focus on an intensive measurement period (14–25 February 2003) with continuous gas- and particulate-phase PAHs (daily samples of 08:00–08:00 LT in the following morning) to represent a polluted continental outflow in East Asia in wintertime. However, to further demonstrate the general model performance in a seasonal scale, we make additional simulation for a continuous summer period 6–17 June 2003. Details of sampling and analysis methods are given in Kim et al. (2012).

4.2.1 Evaluation at the near-source areas

PAH diurnal variabilities are well captured for both gas- and particulate-phase species at the Xianghe site, with correlation coefficients of 0.42–0.69 (Fig. 3, Table S3) compared

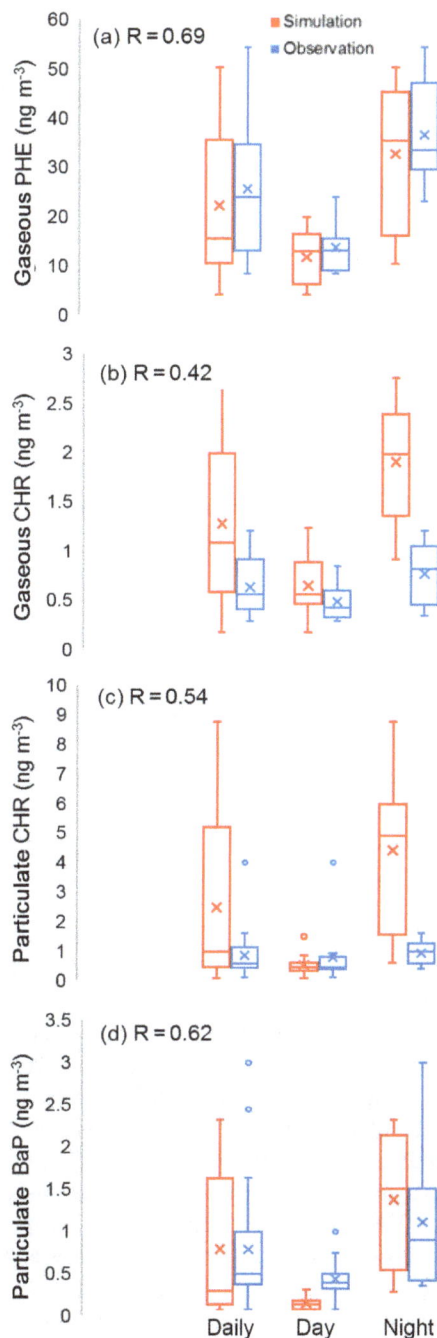

Figure 3. Simulated (red) and observed (blue) concentrations of **(a)** gaseous PHE, **(b)** gaseous CHR, **(c)** particulate CHR and **(d)** particulate BaP at the Xianghe site averaged over 11–22 July 2013. The line and "X" in each box are the median and mean, while the boxes represent the 25th and 75th percentiles. Upper whisker is quartile 3 (Q3) + 1.5 × interquartile range (IQR) or maximum value, whichever is smaller; lower whisker is quartile 1 (Q1) − 1.5 × IQR or minimum value, whichever is larger. The Spearman's rank correlation coefficients R use combined daytime and nighttime data sets.

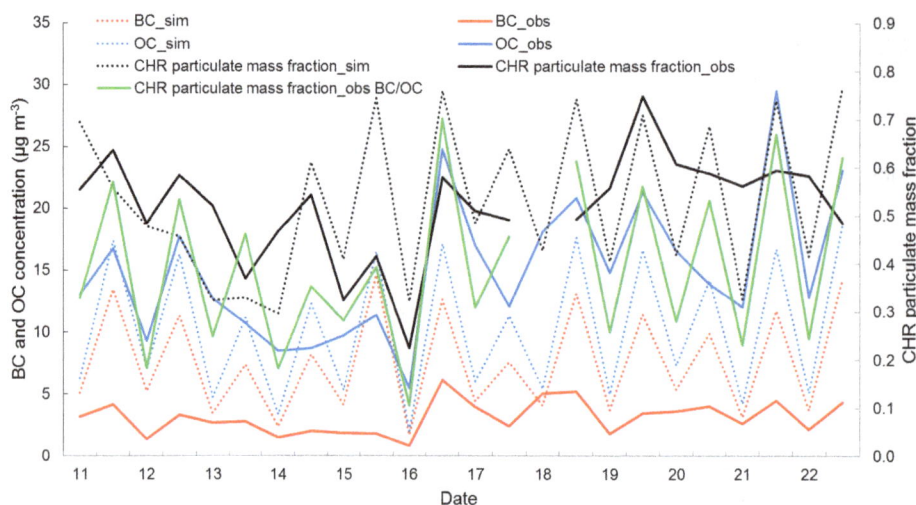

Figure 4. Simulated and observed BC, OC and particulate mass fraction of CHR at the Xianghe site during 11–22 July 2013 (10 h means). Calculated particulate mass fraction of CHR based on observed BC and OC is shown at the green solid line. Calculation explained in the text.

with 0.30–0.58 in Beijing (Xianghe is a semi-urban town in the Beijing metropolitan area) by Inomata et al. (2012). This demonstrates the model's good capability in predicting the vertical and horizontal transport of PAH. The simulated gas-phase PHE has the best correlation rate of 0.69 and also the best predicted average concentrations among the other PAH species (Fig. 3). The model well catches the observed daily average concentration of particulate BaP (observed 0.78 ng m^{-3}, simulated 0.78 ng m^{-3}), while Inomata et al. (2012) underestimated daily concentration of BaP in Beijing by about a factor of 2. The night and day mean levels of particulate BaP, 1.10 and 0.43 ng m^{-3}, respectively, are predicted as 1.37 and 0.14 ng m^{-3}, respectively. Both predicted gas- and particulate-phase CHR are overestimated (Fig. 3b–c). Further diurnal comparisons reveal that such overestimate of daily CHR concentrations mainly comes from nighttime rather than daytime (Table S3). One reason may be that the same hourly, monthly and annual scaling factors of CHR emissions are applied all over the domain. Furthermore, as partitioning strongly influences atmospheric lifetime, the bias in predicted particulate mass fraction (θ) can lead to the bias in predicted concentrations of CHR.

Although the simulated absolute values of θ are close to the measured values (black solid and dotted lines in Fig. 4): the observed (simulated) average particulate mass fraction, is 0.52 (0.53) for 24 h, 0.49 (0.42) for daytime and 0.54 (0.65) for nighttime. The prediction might shield the fact of the overestimated OC and compensating underestimated BC concentrations (blue and red lines in Fig. 4). When applying the gas–particle partitioning parameterisations to the measured OC and BC concentrations, slightly lower than observed is predicted, but the correlation of simulated and observed θ is significantly improved, from $r = 0.31$ to $r = 0.74$

(green line in Fig. 4) when the simulated bias in OC/BC concentration is eliminated. It also implies that adsorption to BC is more important than absorption by OC in determining partitioning, and that the partitioning scheme used in this model is suitable for this East Asian source area.

Overall, the model is found to predict the diurnal variations of PAH concentrations and particulate mass fractions reasonably well at the suburban site in the source region.

4.2.2   Evaluation of the Asian outflow

PAH predictions at remote sites are more challenging as the uncertainties in chemistry and gas–particle partitioning propagate. Model validation so far had been limited to seasonal features (Zhang et al., 2011a, b), while higher temporal features had not been addressed yet. For example, discrepancies of a factor of 16–476 between predicted and observed average PAH (BaP, CHR, BbF, BkF, IcdP, DahA, BghiP) concentrations at the Waliguan site, a continental background site for ambient air monitoring in western China, were found much larger than at urban or suburban sites (Zhang et al., 2009). In our study, the predicted concentration levels in the Gosan winter case agree well with observations: the observed (simulated) average concentrations of PAHs are 0.020 (0.022) ng m^{-3} for particulate BaP, 0.81 (1.73) ng m^{-3} for gaseous PHE, 0.029 (0.029) ng m^{-3} for gaseous CHR and 0.45 (0.24) ng m^{-3} for particulate CHR (Fig. 5, Table S4). Compared with previous studies, our simulated average concentrations of BaP agreed well with the observation (deviation < 10 %), while Zhang et al. (2011a) underestimated BaP by about 50 %. For the Gosan summer case, our simulated average BaP concentration is 0.006 ng m^{-3} (Fig. S6), much closer to the observed value of 0.012 ng m^{-3} than the simulated BaP concentration of ≈ 0.001 ng m^{-3} by Zhang

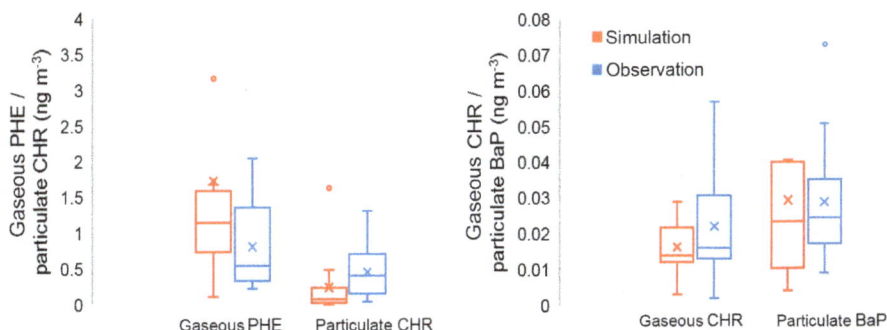

Figure 5. Simulated (red) and observed (blue) concentrations of gaseous PHE, particulate CHR, gaseous CHR and particulate BaP at the Gosan site averaged over 14–25 February 2003. The line and "X" in each box are the median and mean, while the boxes represent the 25th and 75th percentiles. Upper whisker is quartile 3 (Q3) + 1.5 × interquartile range (IQR) or maximum value, whichever is smaller; lower whisker is quartile 1 (Q1) − 1.5 × IQR or minimum value, whichever is larger.

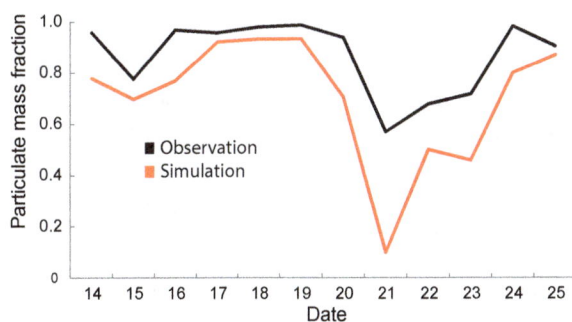

Figure 6. Simulated and observed daily averaged particulate mass fraction of CHR at the Gosan site during 14–25 February 2003.

et al. (2011a). In general, WRF-Chem-PAH model shows good/reasonable agreement with observations in both winter and summer seasons. However, it is worth noting that although the daily average concentration levels of PAHs are reasonably well simulated, the diurnal variations are not well captured at the remote back ground site Gosan (Fig. S5).

The correlation of observed and simulated daily average CHR particulate mass fractions in the Gosan winter case is high ($r = 0.73$; see Fig. 6). The correlation is significantly lower at the Xianghe site ($r = 0.37$), which may due to the proximity to sources. The phase equilibrium of CHR may not be established shortly after emission and the model may not resolve its spatial gradients. An underestimation of simulated particulate mass fraction can be seen in the both winter and summer cases (Figs. 6 and S6). Such underestimation may be caused by the combined effects of uncertainties, such as the emission, degradation, dry/wet deposition and the long-range transport of OC/BC.

These results suggest that our newly developed WRF-Chem-PAH is reasonably accurate in simulating the concentration levels and particulate mass fractions of PAHs for the Asian outflow.

4.3 Distributions of PAHs in East Asia

To illustrate the PAHs distribution in East Asia in both summer and wintertime, Fig. 7 shows the surface concentrations of three representative PAHs averaged over summer 11–22 July 2013 and winter 14–25 February 2003 simulation periods. Another simulation period 6–17 June 2003 agrees with the period 11–22 July 2013 in reflecting summer time distribution characteristics. The lifetimes of PAH over eastern China (20–42° N, 107–122° E, mainland China) are 1.5–9 h for PHE, 2–11 h for CHR, 2 h–3 days for BaP in summer and 9.5 h–3.5 days for PHE, 11 h–4.5 days for CHR, 1.5–6.5 days for BaP in winter, respectively. Due to the relatively short lifetime of PAH species, the spatial distribution of PAH concentrations in the atmosphere is dominated largely by local emissions (Hafner et al., 2005) and the concentration of PAH decreases rapidly away from the source regions. There is a major eastward transport and outflow pathway that plumes with high levels of PAHs from the eastern part of China are swept to the East China Sea and further to the western Pacific Ocean. The simulated average concentration of $0.006\,\mathrm{ng\,m^{-3}}$ for particulate BaP during 14–25 February 2003 at a monitoring background site (26.19° N, 127.75° E) in Okinawa, Japan, which is located on the outflow pathway from China, is close to the monthly-averaged observation value $0.013\,\mathrm{ng\,m^{-3}}$ (http://tenbou.nies.go.jp/gis/monitor/). The concentrations of PAHs are higher in winter than summer, mainly due to higher emissions and slower degradation rates.

The model-calculated mean particulate mass fractions are also shown in Fig. 7. The particulate mass fractions are higher in winter than in summer and in North China than in other regions. This is largely due to the distribution of OC/BC concentrations (high in North China), and seasonal and latitudinal variation of temperature. PHE has extremely small particulate mass fractions, while $\theta\,\mathrm{BaP} \approx 1$ over most of China in winter but only 95 % over North China in summer. On the other hand, CHR shows the largest spatial and

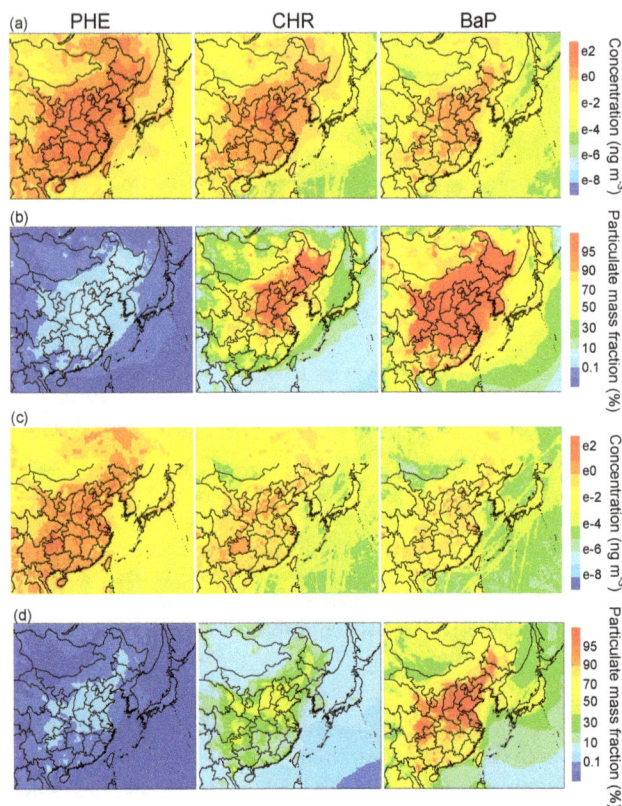

Figure 7. Simulated **(a)** surface concentrations and **(b)** particulate mass fractions of PHE, CHR and BaP averaged over 14–25 February 2003. Simulated **(c)** surface concentrations and **(d)** particulate mass fractions of PHE, CHR and BaP averaged over 11–22 July 2013.

seasonal variations among these three compounds. Our predicted PAH distributions and particulate mass fractions generally agree with the previous studies in East Asia (Zhang et al., 2011b; Inomata et al., 2012).

5 Significance of PAH heterogeneous reaction with ozone and gas-phase reaction with nitrate radical

We test the impact of heterogeneous and homogeneous reaction of BaP, as well as NO_3 gas reactions of CHR and PHE. In other model studies of PAHs, these processes were usually neglected (Zhang et al., 2011a, b).

As shown in Fig. 8, the simulation without any BaP reaction significantly overestimates the average concentration of BaP. Compared with the kinetic model scheme, the model-calculated BaP concentration increased from 0.14 to 1.32 $ng\,m^{-3}$ for daytime, 1.37 to 2.86 $ng\,m^{-3}$ for nighttime and 0.78 to 2.09 $ng\,m^{-3}$ for the whole day, which move further away from the observed daily average of 0.78 $ng\,m^{-3}$ (Fig. 8). Furthermore, the homogeneous gas-phase reaction of BaP is of negligible efficiency as compared to the het-

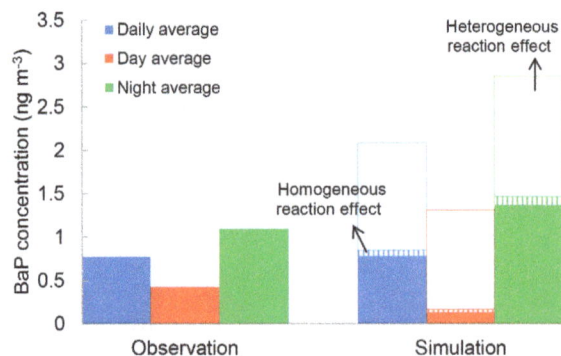

Figure 8. Simulated concentrations of BaP compared with observation at the Xianghe site during 11–22 July 2013. The contributions of heterogeneous and homogeneous reaction are shown are shown as blank areas and as small vertical bars, respectively, in the simulation bars.

erogeneous reaction. This confirms that BaP heterogeneous degradation is indispensable. A companion manuscript discussing the comparison of different heterogeneous degradation schemes is now in preparation.

Figure 9 shows the simulated concentrations of gas-phase PHE and gas- and particulate-phase CHR with and without NO_3 gas-phase reactions compared with observations at the Xianghe site. It is found that during nights with high NO_3 (48.5 and 43.0 $ng\,m^{-3}$, or ≈ 18 and ≈ 16 pptv as the 10 h mean, 12–14 July) the NO_3 reaction causes a significant nighttime drop of PAH levels, i.e. PHE and CHR by $\approx -50\%$ and -50 to -75%, respectively. This is surprisingly drastic for PHE regarding the rate coefficient, $k_{NO_3} = 1.2 \times 10^{-13}\,cm^3\,molec^{-1}\,s^{-1}$ (Table 1), corresponding to a lifetime of $\tau_{NO_3} \approx 5\,h$, but $\tau_{NO_3} \approx 10\,min$ for CHR ($k_{NO_3} = 4.0 \times 10^{-12}\,cm^3\,molec^{-1}\,s^{-1}$, Table 1). The impact on the concentration of particulate-phase CHR is as significant as gas-phase CHR at night.

6 Conclusions and discussion

We have developed the WRF-Chem-PAH model based on the WRF-Chem model to simulate the atmospheric fate of volatile, semivolatile and non-volatile PAH compounds. The implemented state-of-the-art processes for PAHs are gas–particle partitioning, air–soil gas exchange, homogeneous gas-phase and heterogeneous reactions, cloud scavenging, dry and wet deposition, advective transport and cumulus convection. The simulated PAHs in our current WRF-Chem-PAH model include PHE, CHR and BaP, representing volatile, semivolatile and non-volatile PAH compounds, respectively. Also, the model can be applied for any similar semivolatile trace organic compound.

The model has been applied for East Asia. The model predicts observations (both atmospheric concentration of PAHs and the particulate mass fraction of semivolatile CHR) at

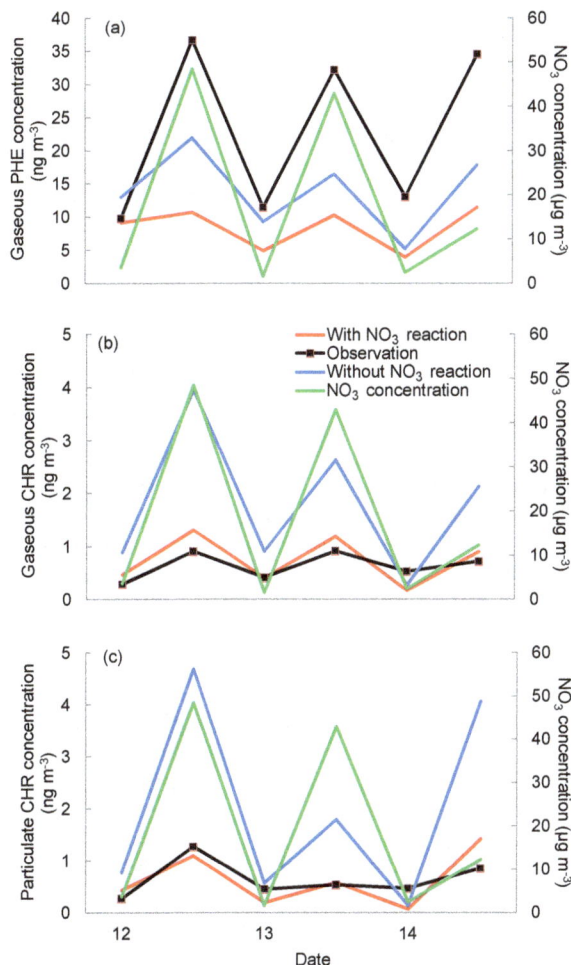

Figure 9. Simulated concentrations of **(a)** gaseous PHE, **(b)** gaseous CHR and **(c)** particulate CHR with and without reactions of NO$_3$ compared with observation and simulated concentrations of NO$_3$ during 12–14 July 2013.

both a near-source and remote site in the continental outflow reasonably well considering big uncertainties in our current knowledge, most notably with regard to the emission, gas–particle partitioning and atmospheric chemistry.

Both the testing of heterogeneous O$_3$ reaction and the homogeneous NO$_3$ reaction emphasise the importance of these reactions for the fate and distributions of the selected PAHs in the polluted atmospheric environment. PAH modelling should include these reactions to better assess the fate and the distributions in polluted and remote environments. However, chemical kinetic measurement and understanding of pathways are limited, in particular for semivolatile species and for heterogeneous chemistry in general (Keyte et al., 2013). Intensive laboratory studies covering semivolatile PAHs, various aerosol matrices, and scenarios of particle mixing and aging are needed to improve PAH modelling.

The model accounts for secondary PAH emissions, i.e. re-volatilisation from soil (semivolatile PAHs only). These emissions add to the pollutant distributions. However, as a consequence of prevailing westerly winds in combination with emissions being concentrated in eastern China, unlike in South Asia (very little emissions in areas east of northeastern India and Bangladesh) and other continents (Lammel et al., 2009; Galarneau et al., 2014), there is a large geographical overlap between secondary sources and primary sources over East Asia. Secondary emissions (re-volatilisation) from the sea surface of the Yellow Sea and the adjacent shelf areas (East and South China seas) also influence the regional PAH distributions over the mainland. This was not addressed in this study (process not included) and should be investigated in order to better assess trans-Pacific transport of PAHs and their even more toxic metabolites, nitro-PAHs (Zhang et al., 2011a).

Author contributions. YFC, GL, UP, and HS and conceived the study. QM did model development, case simulation, data processing and visualisation. MS, YFC, HS and UP developed the new heterogeneous degradation scheme for model implementation. CNG and IMH provided the code of WRF-Chem-Hg (Gencarelli et al., 2014) and helped with model development. YC contributed to data processing. PP, MT, YXZ, GJZ, DvP, QZ and HH provided the observation data at the Xianghe site. YFC, QM, GL, HS, UP and PS discussed the results. QM, YFC, GL and UP wrote the manuscript.

Competing interests. The authors declare that they have no conflict of interest.

Acknowledgements. This work is supported by the Max Planck Society. The work of Yafang Cheng and Hang Su is also supported by the National Natural Science Foundation of China (41330635). We thank Young-Sung Ghim for providing the Gosan measurement data. We thank Georg Grell and Bill Skamarock for the explanation of WRF-Chem transport scheme. The coding work is supported by Yvlu Qiu, Feng Wang, Mega Octaviani, Tabish Ansari, Chao Wei and Stephan Nordmann. The description of the Xianghe sampling is supported by Pourya Shahpoury. We also thank Huizhong Shen, Rong Wang, Ye Huang, Fumo Yang and Pasquale Sellitto for valuable comments.

Edited by: Ronald Cohen

References

Ackermann, I. J., Hass, H., Memmesheimer, M., Ebel, A., Binkowski, F. S., and Shankar, U.: Modal aerosol dynamics model for Europe: Development and first applications, Atmos. Environ., 32, 2981–2999, 10.1016/S1352-2310(98)00006-5, 1998.

Andersson, J. T. and Achten, C.: Time to Say Goodbye to the 16 EPA PAHs? Toward an Up-to-Date Use of PACs for Environmental Purposes, Polycycl. Aromat. Comp., 35, 330–354, https://doi.org/10.1080/10406638.2014.991042, 2015.

Atkinson, R., Baulch, D. L., Cox, R. A., Hampson, R. F., Kerr, J. A., and Troe, J.: Evaluated Kinetic and Photochemical Data for Atmospheric Chemistry – Supplement-Iii, Int. J. Chem. Kinet., 21, 115–150, https://doi.org/10.1002/kin.550210205, 1989.

Aulinger, A., Matthias, V., and Quante, M.: Introducing a partitioning mechanism for PAHs into the Community Multiscale Air Quality modeling system and its application to simulating the transport of benzo(a) pyrene over Europe, J. Appl. Meteorol. Clim., 46, 1718–1730, https://doi.org/10.1175/2007jamc1395.1, 2007.

Bamford, H. A., Poster, D. L., and Baker, J. E.: Temperature dependence of Henry's law constants of thirteen polycyclic aromatic hydrocarbons between 4 °C and 31 °C, Environ. Toxicol. Chem., 18, 1905–1912, https://doi.org/10.1002/etc.5620180906, 1999.

Berkemeier, T., Steimer, S. S., Krieger, U. K., Peter, T., Pöschl, U., Ammann, M., and Shiraiwa, M.: Ozone uptake on glassy, semi-solid and liquid organic matter and the role of reactive oxygen intermediates in atmospheric aerosol chemistry, Phys. Chem. Chem. Phys., 18, 12662–12674, https://doi.org/10.1039/c6cp00634e, 2016.

Bieser, J., Aulinger, A., Matthias, V., and Quante, M.: Impact of Emission Reductions between 1980 and 2020 on Atmospheric Benzo[a]pyrene Concentrations over Europe, Water Air Soil Poll., 223, 1393–1414, https://doi.org/10.1007/s11270-011-0953-z, 2012.

Binkowski, F. S. and Roselle, S. J.: Models-3 community multiscale air quality (CMAQ) model aerosol component – 1. Model description, J. Geophys. Res., 108, 4183, https://doi.org/10.1029/2001jd001409, 2003.

Chen, F. and Dudhia, J.: Coupling an advanced land surface-hydrology model with the Penn State-NCAR MM5 modeling system. Part I: Model implementation and sensitivity, Mon. Weather. Rev., 129, 569–585, https://doi.org/10.1175/1520-0493(2001)129<0569:Caalsh>2.0.Co;2, 2001.

Chou, M. and Suarez, M.: An efficient thermal infrared radiation parameterization for use in general circulation models, NASA Tech. Memo, 104606, 85 pp., 1994.

Dachs, J. and Eisenreich, S. J.: Adsorption onto aerosol soot carbon dominates gas-particle partitioning of polycyclic aromatic hydrocarbons, Environ. Sci. Technol., 34, 3690–3697, https://doi.org/10.1021/Es991201+, 2000.

Darmenov, A. S. and Silva, A. D.: The Quick Fire Emissions Dataset (QFED) – Documentation of versions 2.1, 2.2 and 2.4, NASA Technical Report Series on Global Modeling and Data Assimilation, 32, 1–183, 2013.

Efstathiou, C. I., Matejovicová, J., Bieser, J., and Lammel, G.: Evaluation of gas-particle partitioning in a regional air quality model for organic pollutants, Atmos. Chem. Phys., 16, 15327–15345, https://doi.org/10.5194/acp-16-15327-2016, 2016.

Emmons, L. K., Walters, S., Hess, P. G., Lamarque, J.-F., Pfister, G. G., Fillmore, D., Granier, C., Guenther, A., Kinnison, D., Laepple, T., Orlando, J., Tie, X., Tyndall, G., Wiedinmyer, C., Baughcum, S. L., and Kloster, S.: Description and evaluation of the Model for Ozone and Related chemical Tracers, version 4 (MOZART-4), Geosci. Model Dev., 3, 43–67, https://doi.org/10.5194/gmd-3-43-2010, 2010.

Friedman, C. L. and Selin, N. E.: Long-Range Atmospheric Transport of Polycyclic Aromatic Hydrocarbons: A Global 3-D Model Analysis Including Evaluation of Arctic Sources, Environ. Sci.

Technol., 46, 9501–9510, https://doi.org/10.1021/Es301904d, 2012.

Friedman, C. L., Pierce, J. R., and Selin, N. E.: Assessing the Influence of Secondary Organic versus Primary Carbonaceous Aerosols on Long-Range Atmospheric Polycyclic Aromatic Hydrocarbon Transport, Environ. Sci. Technol., 48, 3293–3302, https://doi.org/10.1021/Es405219r, 2014a.

Friedman, C. L., Zhang, Y. X., and Selin, N. E.: Climate Change and Emissions Impacts on Atmospheric PAH Transport to the Arctic, Environ. Sci. Technol., 48, 429–437, https://doi.org/10.1021/Es403098w, 2014b.

Galarneau, E., Makar, P. A., Sassi, M., and Diamond, M. L.: Estimation of atmospheric emissions of six semivolatile polycyclic aromatic hydrocarbons in southern Canada and the United States by use of an emissions processing system, Environ. Sci. Technol., 41, 4205–4213, https://doi.org/10.1021/Es062303k, 2007.

Galarneau, E., Makar, P. A., Zheng, Q., Narayan, J., Zhang, J., Moran, M. D., Bari, M. A., Pathela, S., Chen, A., and Chlumsky, R.: PAH concentrations simulated with the AURAMS-PAH chemical transport model over Canada and the USA, Atmos. Chem. Phys., 14, 4065–4077, https://doi.org/10.5194/acp-14-4065-2014, 2014.

Gencarelli, C. N., De Simone, F., Hedgecock, I. M., Sprovieri, F., and Pirrone, N.: Development and application of a regional-scale atmospheric mercury model based on WRF/Chem: a Mediterranean area investigation, Environ. Sci. Pollut. R., 21, 4095–4109, https://doi.org/10.1007/s11356-013-2162-3, 2014.

Grell, G. A. and Devenyi, D.: A generalized approach to parameterizing convection combining ensemble and data assimilation techniques, Geophys. Res. Lett., 29, 1693, https://doi.org/10.1029/2002gl015311, 2002.

Grell, G. A., Peckham, S. E., Schmitz, R., McKeen, S. A., Frost, G., Skamarock, W. C., and Eder, B.: Fully coupled "online" chemistry within the WRF model, Atmos. Environ., 39, 6957–6975, https://doi.org/10.1016/j.atmosenv.2005.04.027, 2005.

Guenther, A., Karl, T., Harley, P., Wiedinmyer, C., Palmer, P. I., and Geron, C.: Estimates of global terrestrial isoprene emissions using MEGAN (Model of Emissions of Gases and Aerosols from Nature), Atmos. Chem. Phys., 6, 3181–3210, https://doi.org/10.5194/acp-6-3181-2006, 2006.

Hafner, W. D., Carlson, D. L., and Hites, R. A.: Influence of local human population on atmospheric polycyclic aromatic hydrocarbon concentrations, Environ. Sci. Technol., 39, 7374–7379, https://doi.org/10.1021/es0508673, 2005.

Halsall, C. J., Sweetman, A. J., Barrie, L. A., and Jones, K. C.: Modelling the behaviour of PAHs during atmospheric transport from the UK to the Arctic, Atmos. Environ., 35, 255–267, https://doi.org/10.1016/S1352-2310(00)00195-3, 2001.

Han, J. S., Moon, K. J., Lee, S. J., Kim, Y. J., Ryu, S. Y., Cliff, S. S., and Yi, S. M.: Size-resolved source apportionment of ambient particles by positive matrix factorization at Gosan background site in East Asia, Atmos. Chem. Phys., 6, 211–223, https://doi.org/10.5194/acp-6-211-2006, 2006.

Hung, H., Blanchard, P., Halsall, C. J., Bidleman, T. F., Stern, G. A., Fellin, P., Muir, D. C. G., Barrie, L. A., Jantunen, L. M., Helm, P. A., Ma, J., and Konoplev, A.: Temporal and spatial variabilities of atmospheric polychlorinated biphenyls (PCBs), organochlorine (OC) pesticides and polycyclic aromatic hydrocarbons (PAHs) in the Canadian Arctic: Results from a decade of monitoring, Sci. Total Environ., 342, 119–144, https://doi.org/10.1016/j.scitotenv.2004.12.058, 2005.

Hylland, K.: Polycyclic aromatic hydrocarbon (PAH) ecotoxicology in marine ecosystems, J. Toxicol. Env. Heal. A., 69, 109–123, https://doi.org/10.1080/15287390500259327, 2006.

Inomata, Y., Kajino, M., Sato, K., Ohara, T., Kurokawa, J. I., Ueda, H., Tang, N., Hayakawa, K., Ohizumi, T., and Akimoto, H.: Emission and Atmospheric Transport of Particulate PAHs in Northeast Asia, Environ. Sci. Technol., 46, 4941–4949, https://doi.org/10.1021/Es300391w, 2012.

Inomata, Y., Kajino, M., Sato, K., Ohara, T., Kurokawa, J., Ueda, H., Tang, N., Hayakawa, K., Ohizumi, T., and Akimoto, H.: Source contribution analysis of surface particulate polycyclic aromatic hydrocarbon concentrations in northeastern Asia by source-receptor relationships, Environ. Pollut., 182, 324–334, https://doi.org/10.1016/j.envpol.2013.07.020, 2013.

Janjic, Z. I.: The Step-Mountain Eta Coordinate Model – Further Developments of the Convection, Viscous Sublayer, and Turbulence Closure Schemes, Mon. Weather. Rev., 122, 927–945, https://doi.org/10.1175/1520-0493(1994)122<0927:Tsmecm>2.0.Co;2, 1994.

Jonker, M. T. O. and Koelmans, A. A.: Sorption of Polycyclic Aromatic Hydrocarbons and Polychlorinated Biphenyls to Soot and Soot-like Materials in the Aqueous Environment: Mechanistic Considerations, Environ. Sci. Technol., 36, 3725–3734, https://doi.org/10.1021/es020019x, 2002.

Jury, W. A., Spencer, W. F., and Farmer, W. J.: Behavior Assessment Model for Trace Organics in Soil .1. Model Description, J. Environ. Qual., 12, 558–564, 1983.

Karickhoff, S. W.: Semi-empirical estimation of sorption of hydrophobic pollutants on natural sediments and soils, Chemosphere, 10, 833–846, 1981.

Keyte, I. J., Harrison, R. M., and Lammel, G.: Chemical reactivity and long-range transport potential of polycyclic aromatic hydrocarbons – a review, Chem. Soc. Rev., 42, 9333–9391, https://doi.org/10.1039/C3cs60147a, 2013.

Kim, J. Y., Ghim, Y. S., Song, C. H., Yoon, S. C., and Han, J. S.: Seasonal characteristics of air masses arriving at Gosan, Korea, using fine particle measurements between November 2001 and August 2003, J. Geophys. Res.-Atmos., 112, D07202, https://doi.org/10.1029/2005jd006946, 2007.

Kim, J. Y., Lee, J. Y., Choi, S. D., Kim, Y. P., and Ghim, Y. S.: Gaseous and particulate polycyclic aromatic hydrocarbons at the Gosan background site in East Asia, Atmos. Environ., 49, 311–319, https://doi.org/10.1016/j.atmosenv.2011.11.029, 2012.

Klöpffer, W., Wagner, B. O., and Steinhäuser, K. G.: Atmospheric Degradation of Organic Substances: Persistence, Transport Potential, Spatial Range, Wiley, 2008.

Kwok, E. S. C., Harger, W. P., Arey, J., and Atkinson, R.: Reactions of Gas-Phase Phenanthrene under Simulated Atmospheric Conditions, Environ. Sci. Technol., 28, 521–527, https://doi.org/10.1021/es00052a027, 1994.

Lammel, G., Sehili, A. M., Bond, T. C., Feichter, J., and Grassl, H.: Gas/particle partitioning and global distribution of polycyclic aromatic hydrocarbons – A modelling approach, Chemosphere, 76, 98–106, https://doi.org/10.1016/j.chemosphere.2009.02.017, 2009.

Lang, C., Tao, S., Zhang, G., Fu, J., and Simonich, S.: Outflow of polycyclic aromatic hydrocarbons from Guangdong, Southern China, Environ. Sci. Technol., 41, 8370–8375, https://doi.org/10.1021/es071853v, 2007.

Lang, C., Tao, S., Liu, W. X., Zhang, Y. X., and Simonich, S.: Atmospheric transport and outflow of polycyclic aromatic hydrocarbons from China, Environ. Sci. Technol., 42, 5196–5201, https://doi.org/10.1021/Es800453n, 2008.

Liao, J., Wang, T., Wang, X., Xie, M., Jiang, Z., Huang, X., and Zhu, J.: Impacts of different urban canopy schemes in WRF/Chem on regional climate and air quality in Yangtze River Delta, China, Atmos. Res., 145–146, 226–243, https://doi.org/10.1016/j.atmosres.2014.04.005, 2014.

Lin, Y.-L., Farley, R. D., and Orville, H. D.: Bulk Parameterization of the Snow Field in a Cloud Model, J. Clim. Appl. Meteorol., 22, 1065–1092, https://doi.org/10.1175/1520-0450(1983)022<1065:BPOTSF>2.0.CO;2, 1983.

Lohmann, R. and Lammel, G.: Adsorptive and absorptive contributions to the gas-particle partitioning of polycyclic aromatic hydrocarbons: State of knowledge and recommended parametrization for modeling, Environ. Sci. Technol., 38, 3793–3803, https://doi.org/10.1021/Es035337q, 2004.

Lu, P. Y., Metcalf, R. L., Plummer, N., and Mandel, D.: The environmental fate of three carcinogens: Benzo-(α)-pyrene, benzidine, and vinyl chloride evaluated in laboratory model ecosystems, Arch. Environ. Con. Tox., 6, 129–142, https://doi.org/10.1007/BF02097756, 1977.

Lu, Z., Zhang, Q., and Streets, D. G.: Sulfur dioxide and primary carbonaceous aerosol emissions in China and India, 1996–2010, Atmos. Chem. Phys., 11, 9839–9864, https://doi.org/10.5194/acp-11-9839-2011, 2011.

Lv, Y., Li, X., Xu, T. T., Cheng, T. T., Yang, X., Chen, J. M., Iinuma, Y., and Herrmann, H.: Size distributions of polycyclic aromatic hydrocarbons in urban atmosphere: sorption mechanism and source contributions to respiratory deposition, Atmos. Chem. Phys., 16, 2971–2983, https://doi.org/10.5194/acp-16-2971-2016, 2016.

Mackay, D. and Paterson, S.: Evaluating the multimedia fate of organic chemicals: a level III fugacity model, Environ. Sci. Technol., 25, 427–436, https://doi.org/10.1021/es00015a008, 1991.

Matthias, V., Aulinger, A., and Quante, M.: CMAQ simulations of the benzo(a)pyrene distribution over Europe for 2000 and 2001, Atmos. Environ., 43, 4078–4086, https://doi.org/10.1016/j.atmosenv.2009.04.058, 2009.

Mlawer, E. J., Taubman, S. J., Brown, P. D., Iacono, M. J., and Clough, S. A.: Radiative transfer for inhomogeneous atmospheres: RRTM, a validated correlated-k model for the longwave, J. Geophys. Res., 102, 16663–16682, https://doi.org/10.1029/97jd00237, 1997.

Neu, J. L. and Prather, M. J.: Toward a more physical representation of precipitation scavenging in global chemistry models: cloud overlap and ice physics and their impact on tropospheric ozone, Atmos. Chem. Phys., 12, 3289–3310, https://doi.org/10.5194/acp-12-3289-2012, 2012.

Odabasi, M., Cetin, E., and Sofuoglu, A.: Determination of octanol-air partition coefficients and supercooled liquid vapor pressures of PAHs as a function of temperature: Application to gas–particle partitioning in an urban atmosphere, Atmos. Environ., 40, 6615–6625, https://doi.org/10.1016/j.atmosenv.2006.05.051, 2006.

Prevedouros, K., Jones, K. C., and Sweetman, A. J.: Modelling the atmospheric fate and seasonality of polycyclic aromatic hydrocarbons in the UK, Chemosphere, 56, 195–208, https://doi.org/10.1016/j.chemosphere.2004.02.032, 2004.

Prevedouros, K., Palm-Cousins, A., Gustafsson, O., and Cousins, I. T.: Development of a black carbon-inclusive multi-media model: Application for PAHs in Stockholm, Chemosphere, 70, 607–615, https://doi.org/10.1016/j.chemosphere.2007.07.002, 2008.

Salzmann, M. and Lawrence, M.: Automatic coding of chemistry solvers in WRF-Chem using KPP, 7th WRF Users Workshop, Boulder, Colorado, USA, 2006.

San Jose, R., Perez, J. L., Callen, M. S., Lopez, J. M., and Mastral, A.: BaP (PAH) air quality modelling exercise over Zaragoza (Spain) using an adapted version of WRF-CMAQ model, Environ. Pollut., 183, 151–158, https://doi.org/10.1016/j.envpol.2013.02.025, 2013.

Sandu, A. and Sander, R.: Technical note: Simulating chemical systems in Fortran90 and Matlab with the Kinetic PreProcessor KPP-2.1, Atmos. Chem. Phys., 6, 187–195, https://doi.org/10.5194/acp-6-187-2006, 2006.

Sandu, A., Daescu, D. N., and Carmichael, G. R.: Direct and adjoint sensitivity analysis of chemical kinetic systems with KPP: Part I – theory and software tools, Atmos. Environ., 37, 5083–5096, https://doi.org/10.1016/j.atmosenv.2003.08.019, 2003.

Schell, B., Ackermann, I. J., Hass, H., Binkowski, F. S., and Ebel, A.: Modeling the formation of secondary organic aerosol within a comprehensive air quality model system, J. Geophys. Res., 106, 28275–28293, https://doi.org/10.1029/2001jd000384, 2001.

Sehili, A. M. and Lammel, G.: Global fate and distribution of polycyclic aromatic hydrocarbons emitted from Europe and Russia, Atmos. Environ., 41, 8301–8315, https://doi.org/10.1016/j.atmosenv.2007.06.050, 2007.

Shen, H. Z., Huang, Y., Wang, R., Zhu, D., Li, W., Shen, G. F., Wang, B., Zhang, Y. Y., Chen, Y. C., Lu, Y., Chen, H., Li, T. C., Sun, K., Li, B. G., Liu, W. X., Liu, J. F., and Tao, S.: Global Atmospheric Emissions of Polycyclic Aromatic Hydrocarbons from 1960 to 2008 and Future Predictions, Environ. Sci. Technol., 47, 6415–6424, https://doi.org/10.1021/Es400857z, 2013.

Shen, H. Z., Tao, S., Liu, J. F., Huang, Y., Chen, H., Li, W., Zhang, Y. Y., Chen, Y. C., Su, S., Lin, N., Xu, Y. Y., Li, B. G., Wang, X. L., and Liu, W. X.: Global lung cancer risk from PAH exposure highly depends on emission sources and individual susceptibility, Sci. Rep., 4, 6561, https://doi.org/10.1038/Srep06561, 2014.

Shiraiwa, M., Pfrang, C., and Pöschl, U.: Kinetic multi-layer model of aerosol surface and bulk chemistry (KM-SUB): the influence of interfacial transport and bulk diffusion on the oxidation of oleic acid by ozone, Atmos. Chem. Phys., 10, 3673–3691, https://doi.org/10.5194/acp-10-3673-2010, 2010.

Shiraiwa, M., Sosedova, Y., Rouviere, A., Yang, H., Zhang, Y. Y., Abbatt, J. P. D., Ammann, M., and Pöschl, U.: The role of long-lived reactive oxygen intermediates in the reaction of ozone with aerosol particles, Nat. Chem., 3, 291–295, https://doi.org/10.1038/Nchem.988, 2011.

Shrivastava, M., Lou, S., Zelenyuk, A., Easter, R. C., Corley, R. A., Thrall, B. D., Rasch, P. J., Fast, J. D., Massey Simonich, S. L., Shen, H., and Tao, S.: Global long-range transport and lung cancer risk from polycyclic aromatic hydrocarbons shielded by coatings of organic aerosol, P. Natl. Acad. Sci. USA, 114, 1246–1251, https://doi.org/10.1073/pnas.1618475114, 2017.

Stockwell, W. R., Kirchner, F., Kuhn, M., and Seefeld, S.: A new mechanism for regional atmospheric chemistry modeling, J. Geophys. Res., 102, 25847–25879, https://doi.org/10.1029/97jd00849, 1997.

Strand, A. and Hov, O.: A model strategy for the simulation of chlorinated hydrocarbon distributions in the global environment, Water Air Soil Poll., 86, 283–316, https://doi.org/10.1007/Bf00279163, 1996.

van Jaarsveld, J. A., VanPul, W. A. J., and DeLeeuw, F. A. A. M.: Modelling transport and deposition of persistent organic pollutants in the European region, Atmos. Environ., 31, 1011–1024, https://doi.org/10.1016/S1352-2310(96)00251-8, 1997.

Wesely, M. L.: Parameterization of Surface Resistances to Gaseous Dry Deposition in Regional-Scale Numerical-Models, Atmos. Environ., 23, 1293–1304, https://doi.org/10.1016/0004-6981(89)90153-4, 1989.

WHO (World Health Organization): Polynuclear aromatic hydrocarbons in Drinking-water, Background document for development of WHO Guidelines for Drinking-water Quality, Geneva, 2003.

Wild, O., Zhu, X., and Prather, M. J.: Fast-j: Accurate simulation of in- and below-cloud photolysis in tropospheric chemical models, J. Atmos. Chem., 37, 245–282, https://doi.org/10.1023/A:1006415919030, 2000.

Yaffe, D., Cohen, Y., Arey, J., and Grosovsky, A. J.: Multimedia analysis of PAHs and nitro-PAH daughter products in the Los Angeles basin, Risk Anal., 21, 275–294, https://doi.org/10.1111/0272-4332.212111, 2001.

Yahya, K., Wang, K., Campbell, P., Glotfelty, T., He, J., and Zhang, Y.: Decadal evaluation of regional climate, air quality, and their interactions over the continental US and their interactions using WRF/Chem version 3.6.1, Geosci. Model Dev., 9, 671–695, https://doi.org/10.5194/gmd-9-671-2016, 2016.

Zhang, Y. X. and Tao, S.: Seasonal variation of polycyclic aromatic hydrocarbons (PAHs) emissions in China, Environ. Pollut., 156, 657–663, https://doi.org/10.1016/j.envpol.2008.06.017, 2008.

Zhang, Y. and Tao, S.: Global atmospheric emission inventory of polycyclic aromatic hydrocarbons (PAHs) for 2004, Atmos. Environ., 43, 812–819, 2009.

Zhang, Y. X., Tao, S., Shen, H. Z., and Ma, J. M.: Inhalation exposure to ambient polycyclic aromatic hydrocarbons and lung cancer risk of Chinese population, P. Natl. Acad. Sci. USA, 106, 21063–21067, https://doi.org/10.1073/pnas.0905756106, 2009.

Zhang, Y., Pan, Y., Wang, K., Fast, J. D., and Grell, G. A.: WRF/Chem-MADRID: Incorporation of an aerosol module into WRF/Chem and its initial application to the TexAQS2000 episode, J. Geophys. Res., 115, D18202, https://doi.org/10.1029/2009jd013443, 2010.

Zhang, Y., Tao, S., Ma, J., and Simonich, S.: Transpacific transport of benzo[a]pyrene emitted from Asia, Atmos. Chem. Phys., 11, 11993–12006, https://doi.org/10.5194/acp-11-11993-2011, 2011a.

Zhang, Y. X., Shen, H. Z., Tao, S., and Ma, J. M.: Modeling the atmospheric transport and outflow of polycyclic aromatic hydrocarbons emitted from China, Atmos. Environ., 45, 2820–2827, https://doi.org/10.1016/j.atmosenv.2011.03.006, 2011b.

Zhou, S. M., Shiraiwa, M., McWhinney, R. D., Pöschl, U., and Abbatt, J. P. D.: Kinetic limitations in gas-particle reactions arising from slow diffusion in secondary organic aerosol, Faraday Discuss., 165, 391–406, https://doi.org/10.1039/C3fd00030c, 2013.

11

Aerosol surface area concentration: a governing factor in new particle formation in Beijing

Runlong Cai[1,*], **Dongsen Yang**[2,*], **Yueyun Fu**[1], **Xing Wang**[2], **Xiaoxiao Li**[1], **Yan Ma**[2], **Jiming Hao**[1], **Jun Zheng**[2], and **Jingkun Jiang**[1]

[1]State Key Joint Laboratory of Environment Simulation and Pollution Control, School of Environment, Tsinghua University, Beijing, 100084, China
[2]Collaborative Innovation Center of Atmospheric Environment and Equipment Technology, Nanjing University of Information Science & Technology, Nanjing 210044, China
*These authors contributed equally to this work.

Correspondence to: Jingkun Jiang (jiangjk@tsinghua.edu.cn) and Jun Zheng (zheng.jun@nuist.edu.cn)

Abstract. The predominating role of aerosol Fuchs surface area, A_{Fuchs}, in determining the occurrence of new particle formation (NPF) events in Beijing was elucidated in this study. The analysis was based on a field campaign from 12 March to 6 April 2016 in Beijing, during which aerosol size distributions down to ~ 1 nm and sulfuric acid concentrations were simultaneously monitored. The 26 days were classified into 11 typical NPF days, 2 undefined days, and 13 non-event days. A dimensionless factor, L_Γ, characterized by the relative ratio of the coagulation scavenging rate over the condensational growth rate (Kuang et al., 2010), was applied in this work to reveal the governing factors for NPF events in Beijing. The three parameters determining L_Γ are sulfuric acid concentration, the growth enhancement factor characterized by contribution of other gaseous precursors to particle growth, Γ, and A_{Fuchs}. Different from other atmospheric environments, such as in Boulder and Hyytiälä, the daily-maximum sulfuric acid concentration and Γ in Beijing varied in a narrow range with geometric standard deviations of 1.40 and 1.31, respectively. A positive correlation between the estimated new particle formation rate, $J_{1.5}$, and sulfuric acid concentration was found with a mean fitted exponent of 2.4. However, the maximum sulfuric acid concentrations on NPF days were not significantly higher (even lower, sometimes) than those on non-event days, indicating that the abundance of sulfuric acid in Beijing was high enough to initiate nucleation, but may not necessarily lead to NPF events. Instead, A_{Fuchs} in Beijing varied greatly among days with a geometric standard deviation of 2.56, whereas the variabilities of A_{Fuchs} in Tecamac, Atlanta, and Boulder were reported to be much smaller. In addition, there was a good correlation between A_{Fuchs} and L_Γ in Beijing ($R^2 = 0.88$). Therefore, it was A_{Fuchs} that fundamentally determined the occurrence of NPF events. Among 11 observed NPF events, 10 events occurred when A_{Fuchs} was smaller than $200\,\mu\text{m}^2\,\text{cm}^{-3}$. NPF events were suppressed due to the coagulation scavenging when A_{Fuchs} was greater than $200\,\mu\text{m}^2\,\text{cm}^{-3}$. Measured A_{Fuchs} in Beijing had a good correlation with its PM$_{2.5}$ mass concentration ($R^2 = 0.85$) since A_{Fuchs} in Beijing was mainly determined by particles in the size range of 50–500 nm that also contribute to the PM$_{2.5}$ mass concentration.

1 Introduction

New particle formation (NPF) is closely related to atmospheric environment. It is a common atmospheric phenomenon, which has been observed all over the world (Kulmala et al., 2004). High concentrations of ultrafine particles are formed intensively during NPF events. It has been illustrated through both theoretical modeling and field observations that these ultrafine particles can grow and serve as cloud condensation nuclei (Kuang et al., 2009; Spracklen et al., 2008) and thus affect climate (IPCC, 2013). The increased number concentration of ultrafine particles also raises concerns about human health (HEI, 2013).

New particles are formed by nucleation from gaseous precursors, such as sulfuric acid, ammonia, and organics. Newly formed particles either grow by condensation or are lost by coagulation with other particles (McMurry, 1983). Aerosol Fuchs surface area, A_{Fuchs}, is a parameter that describes the coagulation scavenging effect quantitatively. In addition to gaseous precursors participating in nucleation and subsequent condensational growth, there has been a consensus that the occurrence of a NPF event is also limited by A_{Fuchs}, because the survival possibility of nucleated particles is suppressed when the coagulation scavenging effect is significant (Weber et al., 1997; Kerminen et al., 2001; Kuang et al., 2012). Reported average A_{Fuchs} (or in the form of a condensation sink) on NPF days was found to be lower than that on non-event days at several locations (Dal Maso et al., 2005; Gong et al., 2010; Qi et al., 2015).

A dimensionless criterion, L_{Γ}, was proposed to characterize the ratio of particle scavenging loss rate over condensational growth rate, and to predict the occurrence of NPF events in diverse atmospheric environments (Kuang et al., 2010). By definition, L_{Γ} is determined by three factors, i.e., the sulfuric acid concentration, the growth enhancement factor representing contributions of other gaseous precursors in addition to the sulfuric acid concentration, Γ, and A_{Fuchs}. The diurnal sulfuric acid concentration can vary drastically due to the substantial change in radiation (e.g., from several thousand to $\sim 1.5 \times 10^6 \, \#\,cm^{-3}$ in this campaign) and the increase in sulfuric acid concentration after the sunrise can potentially lead to nucleation. The values of A_{Fuchs}, however, were usually reported within a narrow range at locations, such as Tecamac, Atlanta, and Boulder (Kuang et al., 2010). The sulfuric acid concentration in Atlanta and Hyytiälä can differ significantly among days (Eisele et al., 2006; Petäjä et al., 2009). Therefore, the sulfuric acid concentration often governs nucleation and subsequent growth in the sulfur-rich atmosphere, such as in Atlanta (McMurry et al., 2005). The growth enhancement factor, Γ, at Hyytiälä varied in a wide range, while those at Tecamac and Boulder were found in a relatively narrow range.

Aerosol concentrations in Beijing are usually much higher than those in clean environments. The annual average $PM_{2.5}$ mass concentration in 2016 was $73 \, \mu g \, m^{-3}$ (reported by the Beijing Municipal Environmental Protection Bureau), and the average A_{Fuchs} measured in Beijing by this campaign was $381.5 \, \mu m^2 \, cm^{-3}$, which is approximately a magnitude higher than those measured in clean environments, such as in Hyytiälä (Dal Maso et al., 2002). Differently from the comparatively slow accumulation and depletion process of aerosol concentrations in clean environments, A_{Fuchs} in Beijing may change rapidly because of changes in air mass origins (Wehner et al., 2008) or accumulation of pollutants.

The sulfuric acid concentration is needed to estimate L_{Γ} and direct measurement of particle size distribution down to $\sim 1 \, nm$ will help to better quantify NPF events. Although sulfuric acid has been measured around the world (Erupe

et al., 2010) and analyses based on sub-3 nm size distributions have been conducted sporadically since the development of diethylene glycol scanning mobility particle spectrometers (DEG-SMPS, Jiang et al., 2011a, b; Kuang et al., 2012) and particle size magnifiers (PSMs, Vanhanen et al., 2011; Kulmala et al., 2013), there are limited data on atmospheric sulfuric acid concentrations and directly measured sub-3 nm particle size distributions in China. A campaign in Beijing during the 2008 Olympic Games (Yue et al., 2010; Zheng et al., 2011) characterized atmospheric sulfuric acid concentration and its correlation with the new particle formation rate. The exponent in the correlation of the formation rate, J_3, with the sulfuric acid concentration was found to be 2.3. The exponent for correlating derived $J_{1.5}$ with the sulfuric acid concentration was 2.7 (Wang et al., 2011). They were different from the exponents between 1 and 2 often reported in other places around the world (Riipinen et al., 2007; Sihto et al., 2006; Kuang et al., 2008). The same instrument used in the Beijing campaign was also deployed in Kaiping to measure the sulfuric acid concentration during a 1-month campaign in 2008 (Wang et al., 2013a). Sub-3 nm particle size distributions have not been reported previously in China, except for the 1–3 nm particle number concentration in Shanghai in the winter of 2013 inferred by a PSM (Xiao et al., 2015). Due to the limitation of observation data, although a good correlation between the new particle formation rate and the sulfuric acid concentration in Beijing was found and the ratio of the sulfuric acid concentration over A_{Fuchs} was reported to positively correlate with the number concentration of 3–6 nm particles (Wang et al., 2011), the roles of the sulfuric acid concentration and A_{Fuchs} in determining the occurrence of NPF events have not been quantitatively illustrated.

In this study, we aimed to examine the roles of A_{Fuchs} and the sulfuric acid concentration in determining whether a NPF event will occur on a particular day in Beijing. The data analysis was based on simultaneous measurement of particle size distributions down to $\sim 1 \, nm$ and sulfuric acid. The correlation between particle formation rate, $J_{1.5}$, and the sulfuric acid concentration was examined. L_{Γ} was used to predict the occurrence of NPF events. Daily variations of the three parameters determining L_{Γ}, i.e., the sulfuric acid concentration, Γ, and A_{Fuchs}, were compared. A nominal value of A_{Fuchs} was suggested to predict the occurrence of NPF events in Beijing. The relationship between the $PM_{2.5}$ mass concentration and NPF events was also examined.

2 Experiments

A field campaign studying NPF in Beijing was carried out from 7 March 2016 to 7 April 2016. The campaign site was located on the campus of Tsinghua University. Details of this site can be found elsewhere (Cai and Jiang, 2017; He et al., 2001). A home-made DEG SMPS was used to measure sub-5 nm particle size distributions and a particle size distribution system (including a TSI aerodynamic particle sizer and two parallel SMPSs, equipped with a TSI nanoDMA and a TSI long DMA, respectively) was used to measure size distributions of particles from 3 nm (in electrical mobility diameter) to 10 µm (in aerodynamic diameter, Liu et al., 2016). A specially designed miniature cylindrical differential mobility analyzer (mini-cyDMA) for effective classification of sub-3 nm aerosol was equipped with the DEG-SMPS (Cai et al., 2017). A cyclone was used at the sampling inlet to remove particles larger than 10 µm. The sampled aerosol was subsequently dried by a silica-gel diffusion drier. The diameter change due to drying was neglected when calculating A_{Fuchs} since the mean daytime relative humidity during the campaign period was $\sim 25\%$. Diffusion losses, charging efficiency, penetration efficiencies through the DMAs, detection efficiencies of particle counters, and multi-charging effect were considered during data inversion. The particle density was assumed to be $1.6\,\mathrm{g\,cm^{-3}}$ according to local observation results (Hu et al., 2012).

Sulfuric acid was measured by a modified high-resolution time-of-flight chemical ionization mass spectrometer (HR-ToF-CIMS, Aerodyne Research Inc.). Instead of using a radioactive ion source, a home-made corona discharge (CD) ion source was utilized with the HR-TOF-CIMS. The CD ion source was designed to be able to operate from a few Torr up to near atmospheric pressure and has been successfully implemented in measuring ambient amine (Zheng et al., 2015a) and formaldehyde (Ma et al., 2016). In this study, nitrate reagent ions were used to measure gaseous sulfuric acid (Zheng et al., 2010). The detailed ion chemistry to generate nitrate ions and the calibration procedure for sulfuric acid measurement have been reported in Zheng et al. (2015b). Ambient sulfuric acid concentration in Beijing has been reported only once in a field campaign conducted in 2008 (Zheng et al., 2011; Wang et al., 2011). Compared to that campaign, the sulfuric acid concentration measured in this study displayed similar diurnal variations, but with lower daily-maximum values. This might be caused by the relatively weak solar radiation intensity encountered in this springtime observation compared with the previous summertime campaign. To verify the precision of sulfuric acid measurement, the instrument was calibrated daily at night and background checks were performed for ~ 3 min each hour during daytime.

A meteorological station (Davis 6250) measuring temperature, relative humidity, wind speed, wind direction, and precipitation was located ~ 10 m away from the sampling inlet. The PM$_{2.5}$ mass concentration measured in the nearest national monitoring station (Wanliu station, ~ 5 km away to the southwest of our campaign site) was also used for analysis. Backward trajectories were obtained from the online HYSPLIT server of the National Oceanic and Atmospheric Administration (NOAA).

3 Theory

Nucleation is only the first step of new particle formation. The random collisions of gaseous precursor molecules can form clusters together by Van der Waals forces and/or chemical bonds. These clusters become particles if they are more likely to grow by condensation rather than evaporate. However, particles formed by nucleation may be scavenged through coagulation with larger particles before they grow large enough to be detected (McMurry, 1983; Zhang et al., 2012). Nucleation only refers to the process where stable molecular clusters formed spontaneously from gaseous precursors. New particle formation also requires subsequent condensational growth of freshly nucleated particles. That is, the occurrence of nucleation is mainly determined by gaseous precursors (e.g., sulfuric acid and organics) in atmospheric environments, while new particle formation is also influenced by the coagulation scavenging effect of pre-existing aerosols. A possibility exists that nucleation occurs while NPF events are not observed because of the short lifetime of nucleated particles due to a strong coagulation scavenging (Kerminen et al., 2001). In fact, nucleation can also be suppressed when the aerosol concentration is high since vapors and clusters may also be scavenged by aerosol surfaces.

Aerosol Fuchs surface area, A_{Fuchs}, is a representative parameter of coagulation scavenging based on kinetic theory. It is corrected for particles whose size falls in the transition regime (Davis et al., 1980; McMurry, 1983). The formula assuming a unity mass accommodation coefficient (sticking probability) is shown in Eq. (1),

$$A_{Fuchs} = \frac{4\pi}{3} \int_{d_{min}}^{\infty} d_p^2 \times \left(\frac{Kn + Kn^2}{1 + 1.71\,Kn + 1.33\,Kn^2} \right) \tag{1}$$
$$\times n \times \mathrm{d}d_p,$$

where d_p is the particle diameter, d_{min} is the smallest particle diameter in theory and the smallest detected one in practice, Kn is the Knudsen number, and n is the particle size distribution function, $\mathrm{d}N/\mathrm{d}d_p$. The condensation sink and coagulation sink can also describe how rapidly gaseous precursors and particles are scavenged by pre-existing aerosols, respectively (Kerminen et al., 2001; Kulmala et al., 2001). Since the condensation sink is proportional to A_{Fuchs} (McMurry et al., 2005) and the coagulation sink can be approximately converted to the condensation sink using a simple formula (Lehtinen et al., 2007), only A_{Fuchs} is used in this study to describe the coagulation scavenging effect. Condensation sink values reported in previous studies are referred to in the form of A_{Fuchs}. The diffusion coefficient of sulfuric acid was assumed to be $0.117\,\mathrm{cm^{-2}\,s^{-1}}$ (Gong et al., 2010) when converting the condensation sink into A_{Fuchs}.

A dimensionless criterion, L_Γ, was proposed to predict the occurrence of NPF events (Kuang et al., 2010). It is defined as

$$L_\Gamma = \frac{\bar{c} \times A_{\text{Fuchs}}}{4\beta_{11}N_1} \times \frac{1}{\Gamma}, \qquad (2)$$

where $\bar{c}$ is the mean thermal speed of sulfuric acid that can be calculated from molecular kinetic theory; β_{11} is the coagulation coefficient between sulfuric acid monomers that can be calculated using Eq. (13.56) in Seinfeld and Pandis (2006); N_1 is the number concentration of sulfuric acid; Γ is a growth enhancement factor and is defined as

$$\Gamma = \frac{2\,\text{GR}}{v_1 N_m \bar{c}}, \qquad (3)$$

where GR is the observed mean growth rate; v_1 is the corresponding volume of sulfuric acid monomer and was estimated to be 1.7×10^{-28} m^3 (the volume of a hydrated sulfuric acid molecule, Kuang et al., 2010); and N_m is the maximum number of sulfuric acid concentration during a whole NPF event period. Since other gaseous precursors in addition to sulfuric acid might also contribute to the condensational growth of particles formed by nucleation (O'Dowd et al., 2002; Ristovski et al., 2010) and only sulfuric acid concentration is used in Eq. (2), the ratio of measured growth rate over the sulfuric acid condensational growth rate (Weber et al., 1997), i.e., Γ, was used for correction. It should be clarified that L_Γ in Eq. (2) is defined similarly to that in McMurry et al. (2005) but slightly differently from that in Kuang et al. (2010), since L_Γ in this study presents time-resolved values rather than event-specific ones. Theoretically, Γ can also be time- and size-resolved when using time- and size-resolved GR and time-resolved sulfuric acid (Kuang et al., 2012). However, Γ during each NPF event is assumed to be constant in Eq. (3) because further evaluations are needed for this time- and size-resolved model. Note that in Eq. (2) the absolute sulfuric acid concentrations were effectively normalized by the corresponding daily-maximum sulfuric acid concentrations and thus have no influence on L_Γ values and conclusions based on L_Γ reported in this study.

A new balance formula to estimate the new particle formation rate was proposed recently (Cai and Jiang, 2017) and is given below:

$$J_k = \frac{dN_{[d_k,d_u)}}{dt} + \sum_{d_g=d_k}^{d_{u-1}} \sum_{d_i=d_{\min}}^{+\infty} \beta_{(i,g)} N_{[d_i,d_{i+1})} N_{[d_g,d_{g+1})}$$
$$- \frac{1}{2} \sum_{d_g=d_{\min}}^{d_{u-1}} \sum_{d_i=\max(d_{\min}^3,d_k^3-d_{\min3})}^{d_{i+1}^3+d_{g+1}^3 \leq d_u^3} \beta_{(i,j)} N_{[d_i,d_{i+1})} N_{[d_g,d_{g+1})}$$
$$+ n_u \times \text{GR}_u, \qquad (4)$$

where J_k is the formation rate of particles at the size of d_k, $N_{[d_k,d_u)}$ is the total number concentration of particles from d_k to d_u (not included), d_u is the upper bound of the size range for calculation (25 nm in this study), d_{u-1} is the lower bound

of the last size bin, and $d_{\min}$ is the size of the smallest cluster in theory and the smallest detected size in practice (1.3 nm in this study). The second and third terms on the right-hand side of Eq. (4) are the coagulation sink term (*CoagSnk*) and the coagulation source term (*CoagSrc*), respectively. The difference between *CoagSnk* and *CoagSrc* is the net *CoagSnk* representing the net rate of particles from d_k to d_u, i.e., lost by coagulation scavenging. The last term is often negligible according to the determination criteria for d_u. dN/dt is the balance result of J_k and net *CoagSnk*.

4 Results and discussion

A total of 26 days from 12 March to 6 April was classified by the occurrence of a daytime NPF event. A typical NPF day is featured with distinct and persisting increases in the sub-3 nm particle number concentration and subsequent growth of these nucleated particles. A non-event day means that neither of these two features was observed. As shown in Fig. 1, there are 11 typical NPF days and 13 non-event days. The other 2 days, i.e., 19 and 30 March, were classified as undefined days. On these days, the increase in the sub-3 nm particle number concentration and subsequent growth were both observed. However, the sub-3 nm particle number concentration was relatively low and the evolution of particle size distributions was not continuous. NPF events mainly occurred when wind came from northwest of Beijing and non-event days were associated with air masses from the southwest (as summarized in Table 1). Air masses coming from the north usually experience less influence from urban pollution (Wehner et al., 2008; Wang et al., 2013b); i.e., the A_{Fuchs} values on days dominated by the northerly wind are usually lower than those on days dominated by the southwesterly wind (Wu et al., 2007).

The occurrence of NPF events on most days can be predicted by L_Γ if unity was empirically chosen as the threshold value. Greater L_Γ indicates higher possibilities of nucleated particles being scavenged by coagulation before they can continue to grow. Growth rates on non-event days were assumed to be 2.4 nm h^{-1}, the mean value of observed growth rates on NPF days (the range is 1.2 to 3.3 nm h^{-1}). A threshold value of L_Γ can not be theoretically predicted but can be empirically estimated; 0.7 was suggested as the threshold value by Kuang et al. (2010). However, unity suggested by McMurry et al. (2005) appeared to work better for results from this campaign in Beijing. As shown in Table 1, the median and mean values of L_Γ on NPF days observed in this campaign were 0.55 and 0.71 (with a standard deviation of 0.40), respectively, compared to 3.05 and 3.45 on non-event days (with a standard deviation of 1.79), respectively. However, some exceptions were also observed. On the 2 undefined days, L_Γ were 1.40 and 0.64, respectively, and weak nucleation was observed. Although the estimated L_Γ value on 18 March was 1.75, a comparatively weak but still distinct

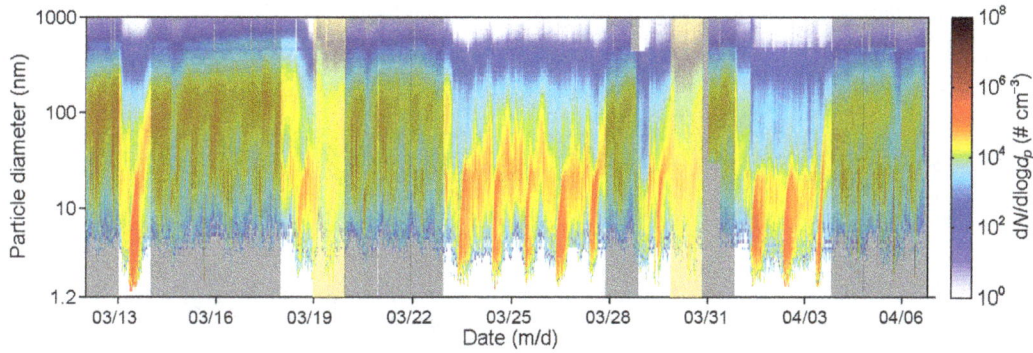

Figure 1. Contour of measured particle size distributions during 12 March to 6 April. The identified 13 non-event days and 2 undefined days are shadowed by grey and yellow background, respectively.

Table 1. Characteristics of each campaign day.

Date (mm/dd)	Classification	Max $J_{1.5}$ ($cm^{-3} s^{-1}$)	N_{1-3} (No. cm^{-3})	A_{Fuchs} ($\mu m^2 cm^{-3}$)	L_Γ	Wind direction[a]
03/12	Non-event	–	0	919.5	3.63	SW
03/13	NPF	156.0	26 347.5	119.7	0.71	NW
03/14	Non-event	–	0	632.7	3.05	NW
03/15	Non-event	–	0	733.9	3.73	SW
03/16	Non-event	–	0	796.2	4.15	WSW
03/17	Non-event	–	0	1140.1	9.04	WSW
03/18	NPF	33.8	741.2	329.0	1.75	WNW
03/19	Undefined	Weak[b]	1643.7	240.8	1.40	SE
03/20	Non-event	–	137.9	348.8	1.74	NNW
03/21	Non-event	–	0	512.0	2.76	SSW
03/22	Non-event	–	0	457.6	2.58	E
03/23	NPF	30.1	3846.3	76.1	0.57	NNW
03/24	NPF	46.8	5576.7	145.2	0.76	NNW
03/25	NPF	57.0	4637.7	126.7	0.52	NNE
03/26	NPF	41.5	9640.9	100.4	0.71	N
03/27	NPF	31.2	2806.2	90.6	0.44	NW
03/28	Non-event	–	0	508.1	2.86	W
03/29	NPF	32.3	2449.8	121.0	0.69	NW
03/30	Undefined	17.7	2885.7	88.8	0.64	NW
03/31	Non-event	–	0	767.0	4.21	SW
04/01	NPF	50.9	5477	51.7	0.22	WNW
04/02	NPF	46.9	10 002	63.1	0.31	NW
04/03	NPF	21.6	10 962.9	105.7	0.24	NW
04/04	Non-event	–	442	398.2	3.09	SW
04/05	Non-event	–	185	391.2	2.33	NW
04/06	Non-event	–	0	365.5	1.71	SW

[a] Indicated by 12 h backward trajectory (starting at noon, 500 m in altitude).
[b] Difficult to estimate.

NPF event was observed. Despite these few exceptions, L_Γ works well on most days in this campaign and was verified in other places (Kuang et al., 2010). The following discussion is focused on the contribution of different factors, i.e., the sulfuric acid concentration, Γ, and A_{Fuchs}.

4.1 The role of gaseous precursors

There was a positive correlation between the estimated new particle formation rate, $J_{1.5}$, and the sulfuric acid concentration during most NPF periods (typically 08:00–16:00 when the estimated $J_{1.5}$ was greater than zero). On NPF days, an increase in the sub-3 nm particle number concentration was often accompanied by an increase in the sulfuric acid con-

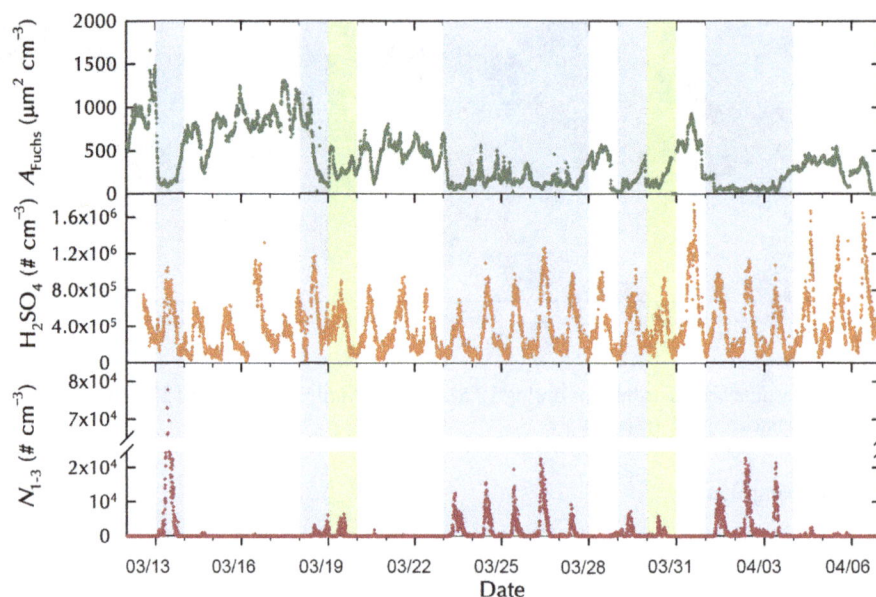

Figure 2. Time series for Fuchs surface area (A_{Fuchs}), the sulfuric acid concentration, and number concentration of 1–3 nm particles. Typical NPF days and undefined days are shadowed by light blue and light green background, respectively.

centration (as shown in Fig. 2). Considering the possible sensitivity of the fitted parameters to the fitting time period (Kuang et al., 2008), the correlation between $J_{1.5}$ and the sulfuric acid concentration was only examined for NPF periods. We found that the mean coefficient of determination (R^2) in this campaign was 0.53. The exponents for correlating the $J_{1.5}$ and the sulfuric acid concentration ranged from 1.5 to 4.0 in the 10 days, with a mean value of 2.4 (29 March was not included because of insignificant correlation). This is in agreement with the previously reported mean exponent of 2.3 using J_3 in Beijing (Wang et al., 2011). However, the exponent is quite different from the exponents no greater than 2 observed in North America and Europe (Kuang et al., 2008; Riipinen et al., 2007; Sihto et al., 2006), indicating that activation or kinetic nucleation alone can not explain all NPF events observed in this campaign.

Although the correlation between the sulfuric acid concentration and the particle formation rate was significant, sulfuric acid appeared not to be the determining factor for whether a NPF event would occur in Beijing. As illustrated by the temporal trend of the sulfuric acid concentration in Fig. 2, a significant diurnal variation was observed every day. However, the differences among the daily-maximum sulfuric acid concentrations were small. The variations of daily-maximum sulfuric acid concentration were significantly less than those of A_{Fuchs}. The geometrical standard deviation and relative standard deviation of maximum sulfuric acid concentration on each day were 1.40 and 0.34, respectively, while those of the daily-averaged A_{Fuchs} values were 2.56 and 0.82, respectively. The sulfuric acid concentrations during NPF periods were not significantly higher than those between 08:00 and

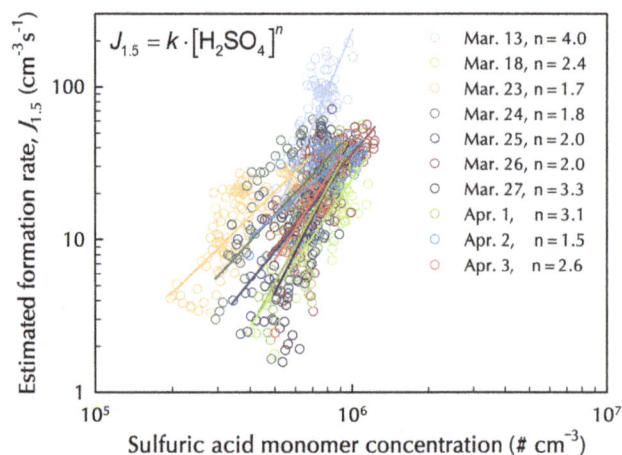

Figure 3. The correlations between the estimated new particle formation rate, $J_{1.5}$, and the sulfuric acid concentration during the NPF event period on each NPF day. The regression line of $J_{1.5}$ versus the sulfuric acid concentration was exponentially fitted. n is the exponent. Data on 29 March were not included because the correlation was not significant ($p = 0.34$).

16:00 on non-event days (significant value, $p = 1$). In addition, comparatively high sulfuric acid concentrations, e.g., on 4–6 April, did not necessarily lead to NPF events.

The influence of the growth enhancement factor, Γ, on the occurrence of NPF events also needs to be addressed because sulfuric acid alone may not explain the observed growth rates. The estimated Γ value for each event was normalized by the geometric mean Γ value for the whole campaign to make it comparable with those obtained from pre-

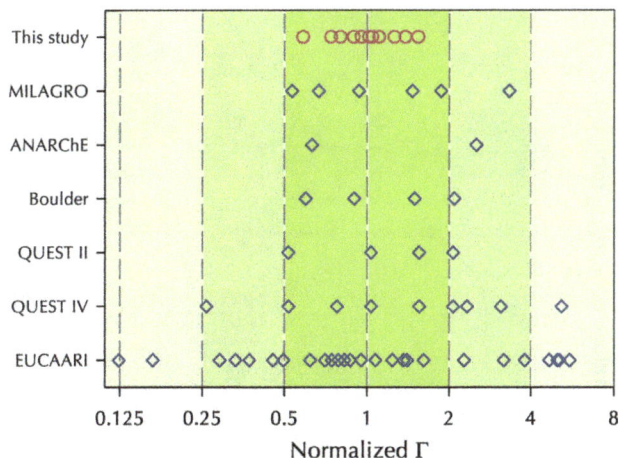

Figure 4. Normalized growth enhancement factor, Γ, in this campaign in comparison to those reported for other campaigns. Γ was normalized by the geometric mean value in each campaign.

vious studies (Kuang et al., 2010): MILAGRO in Tecamac (Iida et al., 2008); ANARChE (McMurry et al., 2005) in Atlanta; Boulder (Iida et al., 2006); and QUEST II (Sihto et al., 2006), QUEST IV (Riipinen, et al., 2007), and EUCAARI (Manninen et al., 2009) at the SMEAR II station in Hyytiälä. It should be clarified that the relative value of Γ can improve the comparability by overcoming some uncertainties in the measured sulfuric acid concentrations in different studies. Figure 4 indicates that Γ values observed in this study are distributed in a relatively narrow range, similar to those observed in Tecamac, Atlanta, and Boulder, while being different from the widely spreading characteristics of Γ values in Hyytiälä. Geometric standard deviations of Γ values were 1.31, 1.75, 2.23, 1.87, 1.62, 2.77, and 2.87 in this campaign, MILAGRO, ANARChE, Boulder, QUEST II, QUEST IV, and EUCAARI, respectively. The daily variations of Γ values in Beijing were less than those observed in other places. They were also less than the daily variations of A_{Fuchs} values measured in this campaign. Considering the small daily variations of both the sulfuric acid concentration and Γ values, it is reasonable to conclude that the abundance of gaseous precursors, such as sulfuric acid, in Beijing during the campaign period was sufficiently high for nucleation to occur, but the occurrence of NPF events appeared to be governed by A_{Fuchs}.

4.2 Relationship between A_{Fuchs} and NPF events

Comparatively lower A_{Fuchs} values were found during most of the NPF days, whereas the sulfuric acid concentrations on NPF days were not significantly higher than those on non-event days. NPF events mainly occurred when A_{Fuchs} was smaller than $200\,\mu m^2\,cm^{-3}$ (the corresponding condensation sink is $0.027\,s^{-1}$). Non-event days mainly corresponded to a real-time A_{Fuchs} value greater than $200\,\mu m^2\,cm^{-3}$ and an average A_{Fuchs} value greater than $350\,\mu m^2\,cm^{-3}$ (Fig. 5). The

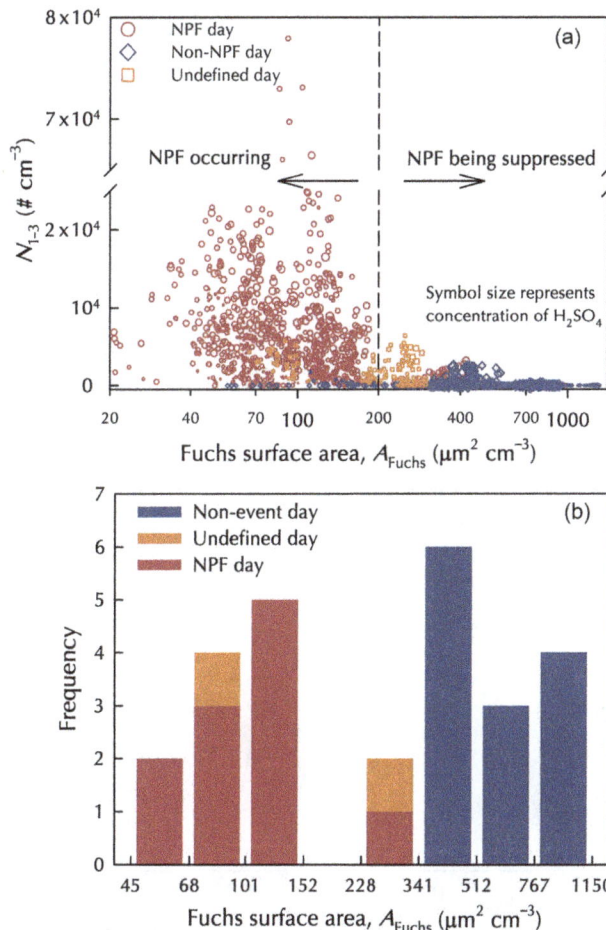

Figure 5. (a) The relationship between Fuchs surface area and number concentration of 1–3 nm particles, N_{1-3}. The relative concentration of measured sulfuric acid is represented by symbol size; i.e., the higher the relative concentration, the bigger the symbol size. Data points are 5 min resolved. **(b)** Frequencies of observed NPF days, undefined days, and non-event days in comparison to the daily-average A_{Fuchs}. On typical NPF days and undefined days, A_{Fuchs} was averaged during NPF event periods. On non-event days, it was averaged between 08:00 and 16:00. A_{Fuchs} values were binned in a logarithmic scale ranging from 45 to 1150.

value of $200\,\mu m^2\,cm^{-3}$ appeared to be an empirical division between NPF days and non-event days. If A_{Fuchs} was lower than this value, a NPF event tended to occur. Otherwise, the occurrence of NPF events was suppressed because of the predominant coagulation scavenging effect. A similar threshold (the condensation sink of $0.02\,s^{-1}$) was found in Budapest, Hungary (Salma et al., 2017).

The variation of L_{Γ} in Beijing was governed by A_{Fuchs}. The measured L_{Γ} and A_{Fuchs} values had a good correlation with the coefficient of determination (R^2) of 0.88. The mean relative error of fitted L_{Γ} using A_{Fuchs} was 11.4 % compared to the measured ones (Fig. 6a). It should be clarified that GR on non-event days in this campaign was assumed to be the

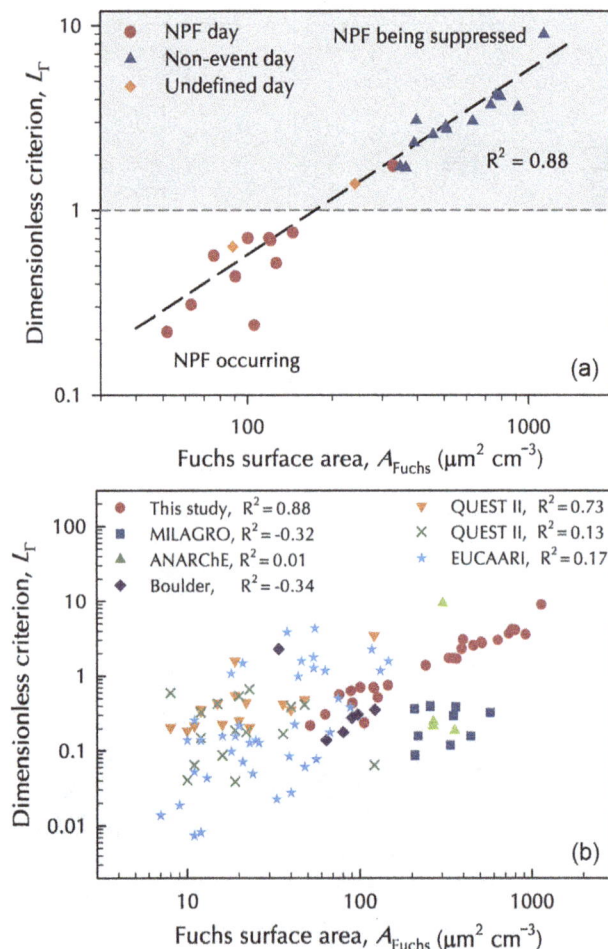

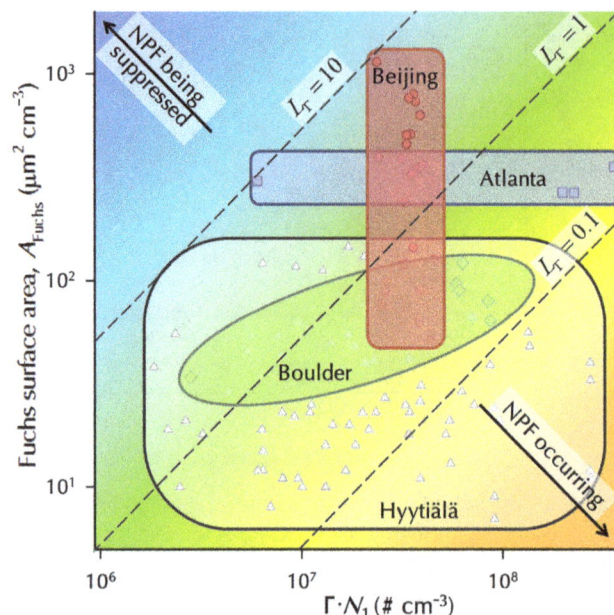

Figure 7. The schematic of governing factors for L_Γ at different locations. Concentration of growth relevant gaseous precursors is represented by $\Gamma \times N_1$, where Γ is the growth enhancement factor and N_1 is the sulfuric acid number concentration. Background color represents the magnitude of L_Γ. Data for each location are shown as different symbols (circle: Beijing; square: Atlanta; diamond: Boulder; triangle: Hyytiälä). The ellipse and the boxes were artificially drawn to illustrate the variations. Tecamac was not included due to the lack of data on non-event days. Both axes are in log scale.

Figure 6. (a) The correlation between L_Γ and A_{Fuchs} (data from Table 1) in this campaign. NPF days, non-event days, and undefined days are shown as different symbols. The regression was based on all campaign days. **(b)** The correlation between L_Γ and A_{Fuchs} estimated for this study in comparison to other campaigns.

same ($2.4 \, \text{nm} \, \text{h}^{-1}$, an average of the fitted values on NPF days). The correlation between L_Γ and A_{Fuchs} on NPF days alone had an R^2 of 0.89. The A_{Fuchs} of $200 \, \mu\text{m}^2 \, \text{cm}^{-3}$ corresponds to an L_Γ of approximately unity in this campaign. Since L_Γ has been verified as a proper nucleation criterion in diverse atmospheric environments, it is reasonable to conclude that A_{Fuchs} was the governing factor of the occurrence of NPF events observed in this campaign.

The characteristics of A_{Fuchs} dominated NPF events in Beijing are different from those at other locations. As shown in Fig. 6b, L_Γ and A_{Fuchs} in most other places do not correlate well, indicating that A_{Fuchs} alone can not predict the occurrence of NPF events at these locations. The variations of these parameters at various locations are illustrated in Fig. 7. In Atlanta and Boulder, A_{Fuchs} values fluctuated within relatively narrow ranges, while the concentrations of gaseous precursors participating in nucleation differed significantly. The variations of L_Γ at these locations were mainly caused

by the relatively large variations in the concentrations of gaseous precursors. However, the contribution of gaseous precursors to L_Γ in Beijing was relatively stable and the variations of L_Γ were mainly caused by the variations in A_{Fuchs} values.

The predominant role of A_{Fuchs} in Beijing can also be explained using the balance formula shown as Eq. (4). It is dN / dt rather than the formation rate, J, that directly reflects whether a NPF event has occurred or not. dN / dt is the balanced result of the formation rate and the net $CoagSnk$. Differently from L_Γ, that is, the ratio of the particle loss rate over the growth rate, the ratio of the net $CoagSnk$ over J represents how many nucleated particles are lost due to the coagulation scavenging. The surviving particles are accounted for by the increment in the number concentration of particles in the nucleation mode (1–25 nm). The nucleation mode was used in this study to estimate dN / dt caused by nucleation because newly formed particles seldom grew beyond 25 nm in the evaluated time period. Surviving possibilities of nucleated particles can also be inferred using the growth rate and A_{Fuchs} (Weber et al., 1997; Kerminen and Kulmala, 2002; Kuang et al., 2012). However, the ratio of the net $CoagSnk$ over J was used because it is based on measured particle size distributions. Note that theoretically the ratio of the net

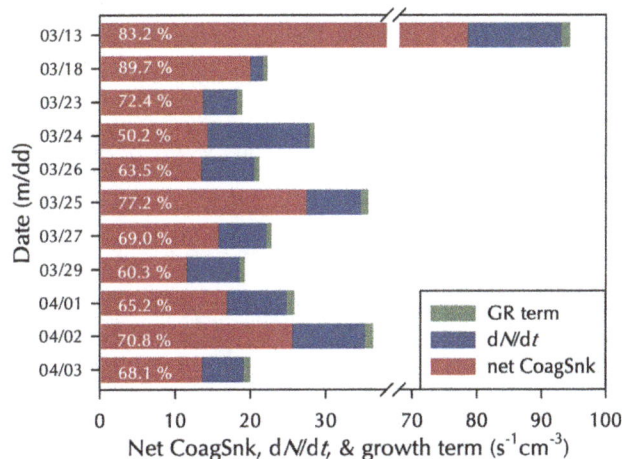

Figure 8. Average contribution of the net CoagSnk, dN/dt, and the condensational growth term (GR term) to the estimated new particle formation rate, $J_{1.5}$, on identified typical NPF days. The percentage presented in each column is the relative ratio of the net CoagSnk compared to $J_{1.5}$ of that NPF event. Note that only the time period when dN/dt was positive during a NPF event was taken into account when calculating the average contribution.

CoagSnk over J can be greater than unity. This would correspond to a negative dN/dt value. For a better description of the occurrence of NPF events rather than the whole process including termination, only NPF periods when dN/dt was positive were considered here. On average, 70 % of particles formed by nucleation were lost due to coagulation scavenging on NPF days (as shown in Fig. 8), indicating high coagulation losses in Beijing even on NPF days. When the A_{Fuchs} value was much greater, most nucleated particles were lost due to the coagulation scavenging rather than were grown into larger sizes, such that NPF events were less likely to be observed.

It should be clarified that although with much less possibility, NPF events may also occur in Beijing when A_{Fuchs} was greater than $200\,\mu m^2\,cm^{-3}$. In this campaign, a distinct NPF event was observed with a comparatively high A_{Fuchs} value of $329\,\mu m^2\,cm^{-3}$ (on 18 March). It was significantly higher than the suggested threshold value of $200\,\mu m^2\,cm^{-3}$. As indicated by Table 1, this exception was caused by the failure of L_Γ rather than A_{Fuchs} alone; i.e., NPF events occurred when estimated L_Γ was greater than unity (the empirical threshold value). The comparatively low number concentration of sub-3 nm particles together with the moderate particle formation rate indicated that the NPF event was suppressed. In addition, previous studies in Beijing also observed some NPF events when A_{Fuchs} values were relatively high (Wu et al., 2007; Wang et al., 2013c, 2017), e.g., an A_{Fuchs} value of $\sim 555\,\mu m^2\,cm^{-3}$ (Kulmala et al., 2016). These reported A_{Fuchs} values might be overestimated since the daily-average value rather than the average only over NPF event periods was used. A_{Fuchs} in Beijing during non-event periods can be

significantly higher. Nevertheless, A_{Fuchs} can be considered the major determining factor of the occurrence of NPF events in Beijing while admitting that exceptions can occasionally occur at a medium L_Γ value greater than unity (corresponding to the A_{Fuchs} value of $200\,\mu m^2\,cm^{-3}$).

4.3 A case study of 3 days

Three continuous days, including 2 NPF days and 1 non-event day, are shown in Fig. 9 to further illustrate the roles of A_{Fuchs} and sulfuric acid (together with other gaseous precursors) in affecting the occurrence of NPF events in Beijing. On 2 April, A_{Fuchs} remained at a relatively low level. A NPF event occurred after sunrise (together with an increase in the sulfuric acid concentration) and ended in the afternoon when the sulfuric acid concentration decreased to a low level. The whole NPF event began at approximately 07:30 and ended at approximately 14:30, which was also the typical time period for other NPF events observed in this campaign. However, when wind direction changed from northwest to southwest at noon on 3 April, the sulfuric acid concentration decreased and A_{Fuchs} increased rapidly because of particles transported from the south. This led to an increase in L_Γ. The ongoing NPF event was interrupted and no newly nucleated particles were detected even when the sulfuric acid concentration increased again later. On 4 April, A_{Fuchs} stayed at a high level. L_Γ was always greater than unity. The maximum sulfuric acid concentrations on 4 April were even higher than those on 2 and 3 April. However, no NPF event was observed. It supports the argument that the abundance of gaseous precursors in Beijing is often high enough for nucleation to happen; however, whether or not a NPF event occurs is mainly governed by A_{Fuchs}.

4.4 Predicting NPF days using PM$_{2.5}$ mass concentration

The PM$_{2.5}$ mass concentration in Beijing serves as a rough but simple parameter to predict whether a NPF event can happen. The value of A_{Fuchs} is affected by particle size distributions. Accumulation mode particles ranging from 50 to 500 nm in Beijing were the major contribution to A_{Fuchs}. Normalized size distributions of accumulation mode particles were relatively stable at various A_{Fuchs} levels (as shown in Fig. 10). On NPF days when A_{Fuchs} was relatively low, particles smaller than 30 nm in diameter formed by nucleation and subsequent growth also contributed to A_{Fuchs}, although A_{Fuchs} was still governed by accumulation mode particles. Thus, A_{Fuchs} should show better correlation with the particle mass concentration rather than the particle number concentration. Figure 11 indicates that there was a good correlation between A_{Fuchs} and the PM$_{2.5}$ mass concentration in Beijing, with R^2 of 0.85, although the correlation at a high A_{Fuchs} level was generally better than that at a low A_{Fuchs} level because particles formed by nucleation significantly

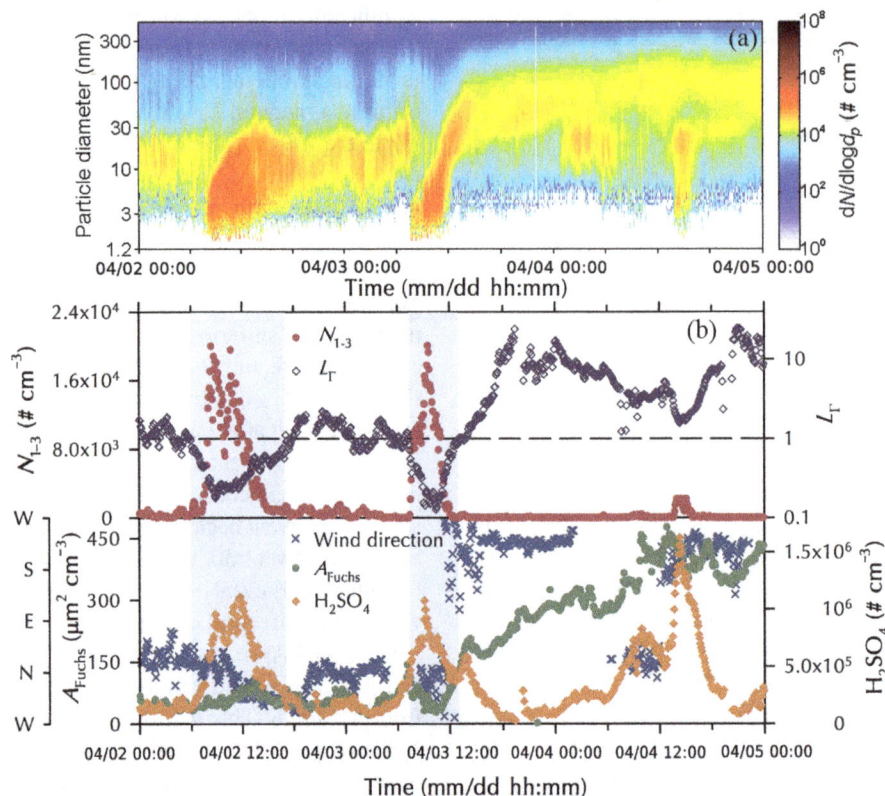

Figure 9. (a) Contour of measured particle size distributions on 2, 3, and 4 April. **(b)** Representative parameters on these 3 NPF days. Time periods when L_Γ was lower than 1.0 are shadowed by light blue background. When wind speed was close to zero, the corresponding wind direction data were not included in the plot.

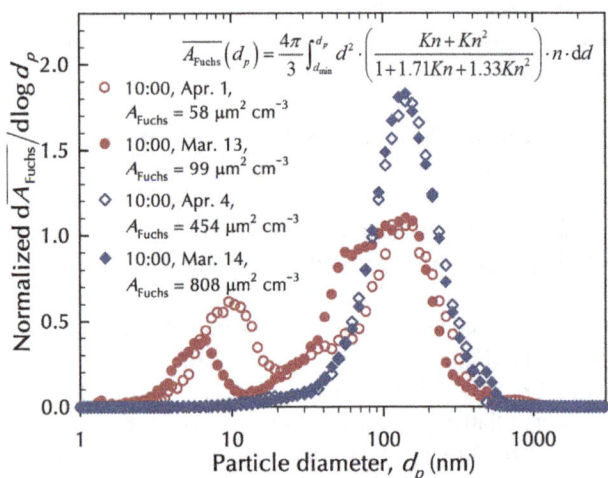

Figure 10. Normalized distribution of cumulative Fuchs surface area, $\overline{A_{Fuchs}}$, as a function of the particle diameter, d_p, on 2 NPF days (red circle) and 2 non-event days (blue diamond). $\overline{A_{Fuchs}}$ is equal to A_{Fuchs} when d_p approaches positive infinity. $d\overline{A_{Fuchs}}/d\log d_p$ is normalized by A_{Fuchs}.

changed the shape of particle size distribution functions on NPF days. Measured PM$_{2.5}$ mass concentrations in the 26 days ranged from 3 to 420 µg m^{-3}, wide enough to represent both relatively clean days and severely polluted days in Beijing. The PM$_{2.5}$ mass concentrations during NPF event periods were mostly lower than 30 µg m^{-3}, except for the event on 18 March. On non-event days, the PM$_{2.5}$ mass concentrations between 08:00 and 16:00 were typically greater than 30 µg m^{-3}. Note that this threshold PM$_{2.5}$ value of 30 µg m^{-3} may not be valid for the whole year. This campaign was in March and early April. Emissions and radiation intensity are different in different seasons, such that the concentrations of gaseous precursors can vary with seasons as well.

The criterion of PM$_{2.5}$ mass concentration was applied to predict NPF events measured at the same site in Beijing in April and May 2014. Among 38 days in that campaign, 11 typical NPF events were identified. For 9 NPF events, average PM$_{2.5}$ mass concentrations during event periods were lower than 30 µg m^{-3}. For the other 2 events, it was 49.8 and 40.5 µg m^{-3}, respectively. In another campaign in Beijing during January 2016 (Jayaratne et al., 2017), 14 NPF events were observed. Among them, 12 events occurred when the daily-average PM$_{2.5}$ mass concentration was lower than 30 µg m^{-3}. The daily-average PM$_{2.5}$ mass concentrations on 16 non-event days were all greater than 40 µg m^{-3}.

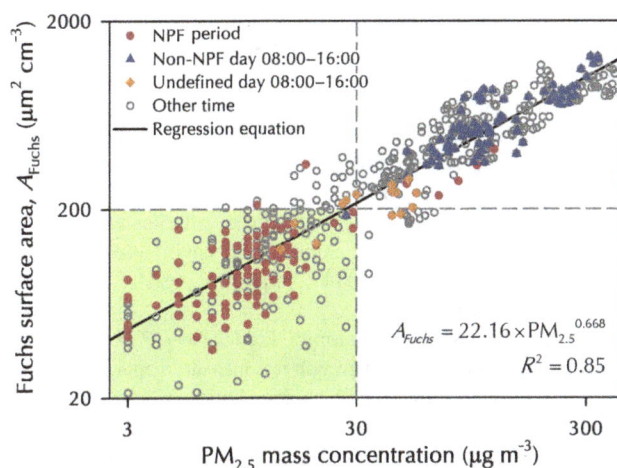

Figure 11. Relationship between hourly averaged A_{Fuchs} and the PM$_{2.5}$ mass concentration in Beijing. Data when A_{Fuchs} changed rapidly were not included to avoid potential influence caused by the distance between Wanliu station and our campaign site. NPF period, daytime (08:00–16:00) on non-event days and undefined days, and other time are shown as different symbols. The regression of A_{Fuchs} versus the PM$_{2.5}$ mass concentration was based on all the data. The proposed criterion for the occurrence of NPF events, i.e., A_{Fuchs} is lower than 200 µm^2 cm^{-3} (the PM$_{2.5}$ mass concentration is lower than 30 µg cm^3), is shadowed by light green background.

5 Conclusions

Factors governing the occurrence of NPF events in Beijing were examined using data from a field campaign during 12 March 2016 to 6 April 2016. In these 26 days, 11 typical NPF events were observed. The rest were 2 undefined days and 13 non-event days. The new particle formation rate, $J_{1.5}$, had a positive correlation with the sulfuric acid concentration, with a fitted mean exponent of 2.4. However, the sulfuric acid concentrations on NPF days were not significantly higher than those on non-event days. A dimensionless criterion proposed by Kuang et al. (2010), L_Γ, was found to be applicable to predict NPF events in most days. Theoretically, L_Γ is determined by the sulfuric acid concentration, the enhancement factor, Γ, and the aerosol Fuchs surface area, A_{Fuchs}, together. In Beijing, however, A_{Fuchs} alone was found to be in a good correlation with L_Γ ($R^2 = 0.88$). Differently from NPF events observed at other locations, such as Hyytiälä, the daily-maximum sulfuric acid concentration and the enhancement factor in Beijing only varied in a narrow range with geometric standard deviations of 1.40 and 1.31, respectively, while A_{Fuchs} varied significantly among days with a geometric standard deviation of 2.56. It was inferred that the concentrations of gaseous precursors, such as sulfuric acid, in Beijing were high enough to initiate nucleation, while it was A_{Fuchs} that determined whether a NPF event would occur or not. An A_{Fuchs} value of 200 µm^2 cm^{-3} was proposed as the empirical threshold in Beijing below

which NPF events are highly likely to occur. NPF events will be suppressed when A_{Fuchs} is higher than this threshold value. The A_{Fuchs} dominated characteristics in Beijing are different from those at other locations, such as Atlanta, Boulder, and Hyytiälä. Since A_{Fuchs} in Beijing was mainly governed by accumulation mode particles (50 to 500 nm) and the normalized $d\overline{A_{Fuchs}}/d\log d_p$ in this size range was relatively stable at different A_{Fuchs} levels in Beijing, measured A_{Fuchs} had a good correlation with the PM$_{2.5}$ mass concentration ($R^2 = 0.85$). Accordingly, the PM$_{2.5}$ mass concentration may also serve as a rough and simple parameter to predict the occurrence of NPF events in Beijing. An empirical PM$_{2.5}$ threshold value of 30 µg m^{-3} was proposed based on data from this field campaign and was found to also work well for other field campaigns in Beijing.

Competing interests. The authors declare that they have no conflict of interest.

Acknowledgements. Financial supports from the National Science Foundation of China (21422703, 41227805, 21521064, 21377059 and 41575122) and the National Key R&D Program of China (2014BAC22B00, 2016YFC0200102 and 2016YFC0202402) are acknowledged.

Edited by: Veli-Matti Kerminen

References

Cai, R., Chen, D.-R., Hao, J., and Jiang, J.: A Miniature Cylindrical Differential Mobility Analyzer for sub-3 nm Particle Sizing, J. Aerosol. Sci., 106, 111–119, https://doi.org/10.1016/j.jaerosci.2017.01.004, 2017.

Cai, R. and Jiang, J.: A new balance formula to estimate new particle formation rate: reevaluating the effect of coagulation scavenging, Atmos. Chem. Phys. Discuss., https://doi.org/10.5194/acp-2017-199, in review, 2017.

Dal Maso, M., Kulmala, M., Lehtinen, K. E., Mäkelä, J. M., Aalto, J., and O'Dowd, C. D.: Condensation and coagulation sinks and formation of nucleation mode particles in coastal and boreal forest boundary layers, J. Geophys. Res., 107, 1–10, https://doi.org/10.1029/2001jd001053, 2002.

Dal Maso, M., Kulmala, M., Riipinen, I., Wagner, R., Hussein, T., Aalto, P., and Lehtinen, K. E.: Formation and growth of fresh atmospheric aerosols: eight years of aerosol size distribution data from SMEAR II, Hyytiälä, Finland, Boreal Environ. Res., 10, 323–336, 2005.

Davis, E. J., Ravindran, P., and Ray, A. K.: A Review of Theory and Experiments on Diffusion from Submicroscopic Particles, Chem. Eng. Commun., 5, 251–268, 1980.

Eisele, F. L., Lovejoy, E. R., Kosciuch, E., Moore, K. F., Mauldin, R. L., Smith, J. N., McMurry, P. H., and Iida, K.: Negative atmospheric ions and their potential role in ion-induced nucleation, J. Geophys. Res., 111, D04305, https://doi.org/10.1029/2005jd006568, 2006.

Erupe, M. E., Benson, D. R., Li, J., Young, L.-H., Verheggen, B., Al-Refai, M., Tahboub, O., Cunningham, V., Frimpong, F., Viggiano, A. A., and Lee, S.-H.: Correlation of aerosol nucleation rate with sulfuric acid and ammonia in Kent, Ohio, An atmospheric observation, J. Geophys. Res., 115, D23216, https://doi.org/10.1029/2010jd013942, 2010.

Gong, Y., Hu, M., Cheng, Y., Su, H., Yue, D., Liu, F., Wiedensohler, A., Wang, Z., Kalesse, H., and Liu, S.: Competition of coagulation sink and source rate: New particle formation in the Pearl River Delta of China, Atmos. Environ., 44, 3278–3285, https://doi.org/10.1016/j.atmosenv.2010.05.049, 2010.

He, K., Yang, F., Ma, Y., Zhang, Q., Yao, X., Chan, C. K., Cadel, S., Chan, T., and Mulawa, P.: The characteristics of $PM_{2.5}$ in Beijing, China, Atmos. Environ., 35, 4959–4970, 2001.

HEI Review Panel on Ultrafine Particles: Understanding the Health Effects of Ambient Ultrafine Particles, HEI Perspectives 3, Health Effects Institute, Boston, MA, 2013.

Hu, M., Peng, J., Sun, K., Yue, D., Guo, S., Wiedensohler, A., and Wu, Z.: Estimation of size-resolved ambient particle density based on the measurement of aerosol number, mass, and chemical size distributions in the winter in Beijing, Environ. Sci. Technol., 46, 9941–9947, https://doi.org/10.1021/es204073t, 2012.

Iida, K., Stolzenburg, M., McMurry, P., Dunn, M. J., Smith, J. N., Eisele, F., and Keady, P.: Contribution of ion-induced nucleation to new particle formation, Methodology and its application to atmospheric observations in Boulder, Colorado, J. Geophys. Res.-Atmos., 111, D23201, https://doi.org/10.1029/2006jd007167, 2006.

Iida, K., Stolzenburg, M. R., McMurry, P. H., and Smith, J. N.: Estimating nanoparticle growth rates from size-dependent charged fractions: Analysis of new particle formation events in Mexico City, J. Geophys. Res.-Atmos., 113, D05207, https://doi.org/10.1029/2007jd009260, 2008.

IPCC.: Climate Change 2013, IPCC Fifth Assessment Report (AR5), Cambridge University Press, 2013.

Jayaratne, R., Pushpawela, B., He, C., Li, H., Gao, J., Chai, F., and Morawska, L.: Observations of particles at their formation sizes in Beijing, China, Atmos. Chem. Phys., 17, 8825–8835, https://doi.org/10.5194/acp-17-8825-2017, 2017.

Jiang, J., Chen, M., Kuang, C., Attoui, M., and McMurry, P. H.: Electrical Mobility Spectrometer Using a Diethylene Glycol Condensation Particle Counter for Measurement of Aerosol Size Distributions Down to 1 nm, Aerosol Sci. Tech., 45, 510–521, https://doi.org/10.1080/02786826.2010.547538, 2011a.

Jiang, J., Zhao, J., Chen, M., Eisele, F. L., Scheckman, J., Williams, B. J., Kuang, C., and McMurry, P. H.: First Measurements of Neutral Atmospheric Cluster and 1–2 nm Particle Number Size Distributions During Nucleation Events, Aerosol Sci. Tech., 45, D04305, https://doi.org/10.1080/02786826.2010.546817, 2011b.

Kerminen, V.-M., Pirjola, L., and Kulmala, M.: How significantly does coagulational scavenging limit atmospheric particle production? J. Geophys. Res.-Atmos., 106, 24119–24125, https://doi.org/10.1029/2001jd000322, 2001.

Kerminen, V. M. and Kulmala, M.: Analytical formulae connecting the "real" and the "apparent" nucleation rate and the nuclei number concentration for atmospheric nucleation events, J. Aerosol Sci., 33, 609–622, 2002.

Kuang, C., McMurry, P. H., McCormick, A. V., and Eisele, F. L.: Dependence of nucleation rates on sulfuric acid vapor concentration in diverse atmospheric locations, J. Geophys. Res., 113, D10209, https://doi.org/10.1029/2007jd009253, 2008.

Kuang, C., McMurry, P. H., and McCormick, A. V.: Determination of cloud condensation nuclei production from measured new particle formation events, Geophys. Res. Lett., 36, L09822, https://doi.org/10.1029/2009gl037584, 2009.

Kuang, C., Riipinen, I., Sihto, S.-L., Kulmala, M., McCormick, A. V., and McMurry, P. H.: An improved criterion for new particle formation in diverse atmospheric environments, Atmos. Chem. Phys., 10, 8469–8480, https://doi.org/10.5194/acp-10-8469-2010, 2010.

Kuang, C., Chen, M., Zhao, J., Smith, J., McMurry, P. H., and Wang, J.: Size and time-resolved growth rate measurements of 1 to 5 nm freshly formed atmospheric nuclei, Atmos. Chem. Phys., 12, 3573–3589, https://doi.org/10.5194/acp-12-3573-2012, 2012.

Kulmala, M., Dal Maso, M., Mäkelä, J. M., Pirjola, L., Väkevä, M., Aalto, P., Miikkulainen, P., Hämeri, K., and O'Dowd, C. D.: On the formation, growth and composition of nucleation mode particles, Tellus, 53, 479–490, 2001.

Kulmala, M., Vehkamäki, H., Petäjä, T., Dal Maso, M., Lauri, A., Kerminen, V. M., Birmili, W., and McMurry, P. H.: Formation and growth rates of ultrafine atmospheric particles: a review of observations, J. Aerosol Sci., 35, 143–176, https://doi.org/10.1016/j.jaerosci.2003.10.003, 2004.

Kulmala, M., Kontkanen, J., Junninen, H., Lehtipalo, K., Manninen, H. E., Nieminen, T., Petäjä, T., Sipilä, M., Schobesberger, S., Rantala, P., Franchin, A., Jokinen, T., Jarvinen, E., Äijälä, M., Kangasluoma, J., Hakala, J., Aalto, P. P., Paasonen, P., Mikkilä, J., Vanhanen, J., Aalto, J., Hakola, H., Makkonen, U., Ruuskanen, T., Mauldin III, R. L., Duplissy, J., Vehkamäki, H., Bäck, J., Kortelainen, A., Riipinen, I., Kurtén, T., Johnston, M. V., Smith, J. N., Ehn, M., Mentel, T. F., Lehtinen, K. E., Laaksonen, A., Kerminen, V.-M., and Worsnop, D. R.: Direct observations of atmospheric aerosol nucleation, Science, 339, 943–946, https://doi.org/10.1126/science.1227385, 2013.

Kulmala, M., Petäjä, T., Kerminen, V.-M., Kujansuu, J., Ruuskanen, T., Ding, A., Nie, W., Hu, M., Wang, Z., Wu, Z., Wang, L., and Worsnop, D. R.: On secondary new particle formation in China, Front. Environ. Sci. En., 10, 8, https://doi.org/10.1007/s11783-016-0850-1, 2016.

Lehtinen, K. E. J., Dal Maso, M., Kulmala, M., and Kerminen, V.-M.: Estimating nucleation rates from apparent particle formation rates and vice versa, Revised formulation of the Kerminen–Kulmala equation, J. Aerosol Sci., 38, 988–994, https://doi.org/10.1016/j.jaerosci.2007.06.009, 2007.

Liu, J., Jiang, J., Zhang, Q., Deng, J., and Hao, J.: A spectrometer for measuring particle size distributions in the range of 3 nm to 10 µm, Front. Environ. Sci. En., 10, 63–72, https://doi.org/10.1007/s11783-014-0754-x, 2016.

Ma, Y., Diao, Y., Zhang, B., Wang, W., Ren, X., Yang, D., Wang, M., Shi, X., and Zheng, J.: Detection of formaldehyde emissions from an industrial zone in the Yangtze River Delta region of China using a proton transfer reaction ion-drift chemical ionization mass spectrometer, Atmos. Meas. Tech., 9, 6101–6116, https://doi.org/10.5194/amt-9-6101-2016, 2016.

Manninen, E. H., Petäjä, T., Asmi, E., Riipinen, I., Nieminen, T., Mikkilä, J., Hõrrak, U., Mirme, A., Mirme, S., Laakso, L., Kerminen, V.-M., and Kulmala, M.: Long-term field measurements of charged and neutral clusters using Neutral cluster and Air Ion

Spectrometer (NAIS), Boreal Environ. Res., 14, 591–605, 2009.

McMurry, P. H.: New particle formation in the presence of an aerosol: rates, time scales, and sub-0.01 µm size distributions, J. Colloid Interf. Sci., 95, 72–80, 1983.

McMurry, P. H., Fink, M., Sakurai, H., Stolzenburg, M. R., Mauldin, R. L., Smith, J., Eisele, F., Moore, K., Sjostedt, S., Tanner, D., Huey, L. G., Nowak, J. B., Edgerton, E., and Voisin, D.: A criterion for new particle formation in the sulfur-rich Atlanta atmosphere, J. Geophys. Res., 110, D22S02, https://doi.org/10.1029/2005jd005901, 2005.

O'Dowd, C. D., Aalto, P., Hmeri, K., Kulmala, M., and Hoffmann, T.: Aerosol formation: atmospheric particles from organic vapours, Nature, 416, 497–498, https://doi.org/10.1038/416497a, 2002.

Petäjä, T., Mauldin, R. L., Kosciuch, E., McGrath, J., Nieminen, T., Paasonen, P., Boy, M., Adamov, A., Kotiaho, T., and Kulmala, M.: Sulfuric acid and OH concentrations in a boreal forest site, J. Aerosol. Sci., 9, 7435–7448, 2009.

Qi, X. M., Ding, A. J., Nie, W., Petäjä, T., Kerminen, V.-M., Herrmann, E., Xie, Y. N., Zheng, L. F., Manninen, H., Aalto, P., Sun, J. N., Xu, Z. N., Chi, X. G., Huang, X., Boy, M., Virkkula, A., Yang, X.-Q., Fu, C. B., and Kulmala, M.: Aerosol size distribution and new particle formation in the western Yangtze River Delta of China: 2 years of measurements at the SORPES station, Atmos. Chem. Phys., 15, 12445–12464, https://doi.org/10.5194/acp-15-12445-2015, 2015.

Riipinen, I., Sihto, S.-L., Kulmala, M., Arnold, F., Dal Maso, M., Birmili, W., Saarnio, K., Teinilä, K., Kerminen, V.-M., Laaksonen, A., and Lehtinen, K. E. J.: Connections between atmospheric sulphuric acid and new particle formation during QUEST III-IV campaigns in Heidelberg and Hyytiälä, Atmos. Chem. Phys., 7, 1899–1914, https://doi.org/10.5194/acp-7-1899-2007, 2007.

Ristovski, Z. D., Suni, T., Kulmala, M., Boy, M., Meyer, N. K., Duplissy, J., Turnipseed, A., Morawska, L., and Baltensperger, U.: The role of sulphates and organic vapours in growth of newly formed particles in a eucalypt forest, Atmos. Chem. Phys., 10, 2919–2926, https://doi.org/10.5194/acp-10-2919-2010, 2010.

Salma, I., Németh, Z., Kerminen, V.-M., Aalto, P., Nieminen, T., Weidinger, T., Moln'ar, Á., Imre, K., and Kulmala, M.: Regional effect on urban atmospheric nucleation, Atmos. Chem. Phys., 16, 8715–8728, https://doi.org/10.5194/acp-16-8715-2016, 2016.

Seinfeld, J. H. and Pandis, S. N.: Atmospheric Chemistry and Physics (2nd Ed), John Wiley and Sons, Inc., New Jersey, 2006.

Sihto, S.-L., Kulmala, M., Kerminen, V.-M., Dal Maso, M., Petäjä, T., Riipinen, I., Korhonen, H., Arnold, F., Janson, R., Boy, M., Laaksonen, A., and Lehtinen, K. E. J.: Atmospheric sulphuric acid and aerosol formation: implications from atmospheric measurements for nucleation and early growth mechanisms, Atmos. Chem. Phys., 6, 4079–4091, https://doi.org/10.5194/acp-6-4079-2006, 2006.

Spracklen, D. V., Carslaw, K. S., Kulmala, M., Kerminen, V.-M., Sihto, S.-L., Riipinen, I., Merikanto, J., Mann, G. W., Chipperfield, M. P., Wiedensohler, A., Birmili, W., and Lihavainen, H.: Contribution of particle formation to global cloud condensation nuclei concentrations, Geophys. Res. Lett., 35, L06808, https://doi.org/10.1029/2007gl033038, 2008.

Vanhanen, J., Mikkilä, J., Lehtipalo, K., Sipilä, M., Manninen, H. E., Siivola, E., Petäjä, T., and Kulmala, M.: Particle Size Magni-

fier for Nano-CN Detection, Aerosol Sci. Technol., 45, 533–542, https://doi.org/10.1080/02786826.2010.547889, 2011.

Wehner, B., Birmili, W., Ditas, F., Wu, Z., Hu, M., Liu, X., Mao, J., Sugimoto, N., and Wiedensohler, A.: Relationships between submicrometer particulate air pollution and air mass history in Beijing, China, 2004–2006, Atmos. Chem. Phys., 8, 6155–6168, https://doi.org/10.5194/acp-8-6155-2008,2008.

Wang, Z. B., Hu, M., Yue, D. L., Zheng, J., Zhang, R. Y., Wiedensohler, A., Wu, Z. J., Nieminen, T., and Boy, M.: Evaluation on the role of sulfuric acid in the mechanisms of new particle formation for Beijing case, Atmos. Chem. Phys., 11, 12663–12671, https://doi.org/10.5194/acp-11-12663-2011, 2011.

Wang, Z. B., Hu, M., Yue, D. L., He, L. Y., Huang, X. F., Yang, Q., Zheng, J., Zhang, R. Y., and Zhang, Y. H.: New particle formation in the presence of a strong biomass burning episode at a downwind rural site in PRD, China, Tellus B, 65, https://doi.org/10.3402/tellusb.v65i0.19965, 2013a.

Wang, Z. B., Hu, M., Wu, Z. J., Yue, D. L., He, L. Y., Huang, X. F., Liu, X. G., and Wiedensohler, A.: Long-term measurements of particle number size distributions and the relationships with air mass history and source apportionment in the summer of Beijing, Atmos. Chem. Phys., 13, 10159—10170, https://doi.org/10.5194/acp-13-10159-2013, 2013b.

Wang, Z. B., Hu, M., Sun, J. Y., Wu, Z. J., Yue, D. L., Shen, X. J., Zhang, Y. M., Pei, X. Y., Cheng, Y. F., and Wiedensohler, A.: Characteristics of regional new particle formation in urban and regional background environments in the North China Plain, Atmos. Chem. Phys., 13, 12495–12506, https://doi.org/10.5194/acp-13-12495-2013, 2013c.

Wang, Z., Wu, Z., Yue, D., Shang, D., Guo, S., Sun, J., Ding, A., Wang, L., Jiang, J., Guo, H., Gao, J., Cheung, H. C., Morawska, L., Keywood, M., and Hu, M.: New particle formation in China: Current knowledge and further directions, Sci. Total Environ., 577, 258–266, https://doi.org/10.1016/j.scitotenv.2016.10.177, 2017.

Weber, R. J., Marti, J. J., McMurry, P. H., Eisele, F. L., Tanner, D. J., and Jefferson, A.: Measurements of new particle formation and ultrafine particle growth rates at a clean continental site, J. Geophys. Res.-Atmos., 102, 4375–4385, https://doi.org/10.1029/96jd03656, 1997.

Wu, Z., Hu, M., Liu, S., Wehner, B., Bauer, S., Maßling, A., Wiedensohler, A., Petäjä, T., Dal Maso, M., and Kulmala, M.: New particle formation in Beijing, China, Statistical analysis of a 1-year data set, J. Geophys. Res., 112, D09209, https://doi.org/10.1029/2006jd007406, 2007. v

Xiao, S., Wang, M. Y., Yao, L., Kulmala, M., Zhou, B., Yang, X., Chen, J. M., Wang, D. F., Fu, Q. Y., Worsnop, D. R., and Wang, L.: Strong atmospheric new particle formation in winter in urban Shanghai, China, Atmos. Chem. Phys., 15, 1769–1781, https://doi.org/10.5194/acp-15-1769-2015, 2015.

Yue, D. L., Hu, M., Zhang, R. Y., Wang, Z. B., Zheng, J., Wu, Z. J., Wiedensohler, A., He, L. Y., Huang, X. F., and Zhu, T.: The roles of sulfuric acid in new particle formation and growth in the mega-city of Beijing, Atmos. Chem. Phys., 10, 4953–4960, https://doi.org/10.5194/acp-10-4953-2010, 2010.

Zhang, R., Khalizov, A., Wang, L., Hu, M., and Xu, W.: Nucleation and Growth of Nanoparticles in the Atmosphere, Chem. Rev., 112, 1957–2011, https://doi.org/10.1021/cr2001756, 2012.

Zheng, J., Khalizov, A., Wang, L., and Zhang, R.: Atmospheric pressure-ion drift chemical ionization mass spectrometry for detection of trace gas species, Anal. Chem., 82, 7302–7308, https://doi.org/10.1016/j.atmosenv.2014.05.024, 2010.

Zheng, J., Hu, M., Zhang, R., Yue, D., Wang, Z., Guo, S., Li, X., Bohn, B., Shao, M., He, L., Huang, X., Wiedensohler, A., and Zhu, T.: Measurements of gaseous H2SO4 by AP-ID-CIMS during CAREBeijing 2008 Campaign, Atmos. Chem. Phys., 11, 7755–7765, https://doi.org/10.5194/acp-11-7755-2011, 2011.

Zheng, J., Ma, Y., Chen, M., Zhang, Q., Wang, L., Khalizov, A. F., Yao, L., Wang, Z., Wang, X., and Chen, L.: Measurement of atmospheric amines and ammonia using the high resolution time-of-flight chemical ionization mass spectrometry, Atmos. Environ., 102, 249–259, https://doi.org/10.1016/j.atmosenv.2014.12.002, 2015a.

Zheng, J., Yang, D., Ma, Y., Chen, M., Cheng, J., Li, S., and Wang, M.: Development of a new corona discharge based ion source for high resolution time-of-flight chemical ionization mass spectrometer to measure gaseous H_2SO_4 and aerosol sulfate, Atmos. Environ., 119, 167–173, https://doi.org/10.1016/j.atmosenv.2015.08.028, 2015b.

Permissions

List of Contributors

Yu Wang, HaoWang, Hai Guo and Xiaopu Lyu
Air Quality Studies, Department of Civil and Environmental Engineering, the Hong Kong Polytechnic University, Hong Kong SAR, China

Hairong Cheng
Department of Environmental Engineering, School of Resource and Environmental Sciences, Wuhan University, Wuhan 430079, China

Zhenhao Ling
School of Atmospheric Sciences, Sun Yat-sen University, Guangzhou, China

Peter K. K. Louie
Air Group, Hong Kong Environmental Protection Department, Hong Kong SAR, China

Isobel J. Simpson, Simone Meinardi and Donald R. Blake
Department of Chemistry, University of California, Irvine, CA, USA

Matthieu Pommier
LATMOS/IPSL, UPMC Univ. Paris 06 Sorbonne Universités, UVSQ, CNRS, Paris, France
Norwegian Meteorological Institute, Oslo, Norway

Pierre-Francois Coheur
Spectroscopie de l'Atmosphère, Chimie Quantique et Photophysique, Université Libre de Bruxelles (ULB), Brussels, Belgium

Cathy Clerbaux
LATMOS/IPSL, UPMC Univ. Paris 06 Sorbonne Universités, UVSQ, CNRS, Paris, France
Spectroscopie de l'Atmosphère, Chimie Quantique et Photophysique, Université Libre de Bruxelles (ULB), Brussels, Belgium

Adam Kristensson, Kristina Eriksson Stenström and Erik Swietlicki
Division of Nuclear Physics, Lund University, P.O. Box 118, 22100, Lund, Sweden

Guillaume Monteil
Department of Physical Geography, Lund University, Lund, P.O. Box 118, 22100, Lund, Sweden

Moa K. Sporre
Department of Geosciences, University of Oslo, P.O. Box 1022, Blindern, 0315, Oslo, Norway

Anne Maria Kaldal Hansen and Marianne Glasius
Department of Chemistry and iNANO, Aarhus University, Langelandsgade 140, 8000, Aarhus C, Denmark

Johan Martinsson
Division of Nuclear Physics, Lund University, P.O. Box 118, 22100, Lund, Sweden
Centre for Environmental and Climate Research, Lund University, Ecology Building, 22362, Lund, Sweden

Stefanie Falk and Björn-Martin Sinnhuber
Institute of Meteorology and Climate Research, Karlsruhe Institute of Technology, Karlsruhe, Germany

Gisèle Krysztofiak
LPC2E, Université d'Orléans, CNRS, UMR7328, Orléans, France

Patrick Jöckel and Phoebe Graf
Deutsches Zentrum für Luft- und Raumfahrt e.V., Oberpfaffenhofen, Germany

Sinikka T. Lennartz
Geomar, Helmholtz Centre for Ocean Research Kiel, Kiel, Germany

Hannah Meusel, Uwe Kuhn, Kathrin Reinmuth-Selzle, Guo Li, Xiaoxiang Wang and Ulrich Pöschl
Max Planck Institute for Chemistry, Multiphase Chemistry Department, Mainz, Germany

Thorsten Bartels-Rausch and Markus Ammann
Paul Scherrer Institute, Villigen, Switzerland

Christopher J. Kampf
Johannes Gutenberg University of Mainz, Institute for Organic Chemistry, Mainz, Germany

Jos Lelieveld
Max Planck Institute for Chemistry, Atmospheric Chemistry Department, Mainz, Germany

Thorsten Hoffmann
Johannes Gutenberg University of Mainz, Institute for Inorganic and Analytical Chemistry, Mainz, Germany

Yafang Cheng and Hang Su
Max Planck Institute for Chemistry, Multiphase Chemistry Department, Mainz, Germany
Institute for Environmental and Climate Research, Jinan University, Guangzhou, China

Yasin Elshorbany
NASA Goddard Space Flight Center, Greenbelt, Maryland, USA
Earth System Science Interdisciplinary Center, University of Maryland, College Park, Maryland, USA

Zhe Peng and Jose L. Jimenez
Cooperative Institute for Research in Environmental Sciences and Department of Chemistry, University of Colorado, Boulder, CO80309, USA

Aldenor Gomes Santos, Jailson Bittencourt de Andrade and Gisele Olímpio da Rocha
Institute of Chemistry, University of São Paulo, São Paulo – SP, 05508-000, Brazil

Kimmo Teinilä and Risto Hillamo
Finnish Meteorological Institute, P.O. Box 503, 00101 Helsinki, Finland

Célia A. Alves
CESAM & Department of Environment, University of Aveiro, Aveiro, 3810-193, Portugal

Huang Xian and Rajasekhar Balasubramanian
Department of Civil and Environmental Engineering, National University of Singapore, E1A 07-03, 117576, Singapore

Maria de Fátima Andrade
Institute of Astronomy, Geophysics and Atmospheric Sciences, University of São Paulo, São Paulo – SP, 05508-090, Brazil

Guilherme Martins Pereira and Pérola de Castro Vasconcellos
Institute of Chemistry, University of São Paulo, São Paulo – SP, 05508-000, Brazil
INCT for Energy and Environment, Federal University of Bahia, Salvador – BA, 40170-115, Brazil

Danilo Custódio
Institute of Chemistry, University of São Paulo, São Paulo – SP, 05508-000, Brazil
CESAM & Department of Environment, University of Aveiro, Aveiro, 3810-193, Portugal

Prashant Kumar
Global Centre for Clean Air Research (GCARE), Department of Civil and Environmental Engineering, Faculty of Engineering and Physical Sciences, University of Surrey, Guildford GU2 7XH, UK
Environmental Flow Research Centre, Faculty of Engineering and Physical Sciences,
University of Surrey, Guildford GU2 7XH, UK

Bethan White and Philip Stier
Atmospheric, Oceanic and Planetary Physics, University of Oxford, Oxford, UK

Edward Gryspeerdt
Institute for Meteorology, Universität Leipzig, Leipzig, Germany

Hugh Morrison and Gregory Thompson
National Center for Atmospheric Research, Boulder, Colorado, USA

Zak Kipling
European Centre for Medium-Range Weather Forecasts, Shinfield Park, Reading, UK

Irina Petropavlovskikh, Audra McClure-Begley and Glen McConville
Cooperative Institute for Research in Environmental Sciences, University of Colorado, Boulder, CO 80309, USA

Robert D. Evans and Dorothy Quincy
Retired from NOAA/ESRL, Global Monitoring Division, Boulder, CO 80305, USA

Koji Miyagawa
Visitor with NOAA/ESRL, Global Monitoring Division, Boulder, CO 80305, USA

Qing Mu, Yuxuan Zhang and Ulrich Pöschl
Multiphase Chemistry Department, Max Planck Institute for Chemistry, Mainz, Germany

Petra Přibylová
Research Centre for Toxic Compounds in the Environment, Masaryk University, Brno, Czech Republic

Christian N. Gencarelli and Ian M. Hedgecock
CNR-Institute of Atmospheric Pollution Research, Division of Rende, Rende, Italy

Monique Teich, Hartmut Herrmann and Dominik van Pinxteren
Leibniz Institute for Tropospheric Research, Leipzig, Germany

Guangjie Zheng
Brookhaven National Laboratory, Brookhaven, USA

Qiang Zhang
Ministry of Education Key Laboratory for Earth System Modeling, Department of Earth System Science, Tsinghua University, Beijing, China

Peter Spichtinger
Institute for Atmospheric Physics, Johannes Gutenberg University, Mainz, Germany

Gerhard Lammel
Multiphase Chemistry Department, Max Planck Institute for Chemistry, Mainz, Germany
Research Centre for Toxic Compounds in the Environment, Masaryk University, Brno, Czech Republic

Ying Chen
Multiphase Chemistry Department, Max Planck Institute for Chemistry, Mainz, Germany
Leibniz Institute for Tropospheric Research, Leipzig, Germany

Manabu Shiraiwa
Multiphase Chemistry Department, Max Planck Institute for Chemistry, Mainz, Germany
Department of Chemistry, University of California, Irvine, USA

Hang Su and Yafang Cheng
Multiphase Chemistry Department, Max Planck Institute for Chemistry, Mainz, Germany
Institute for Environmental and Climate Research, Jinan University, Guangzhou, China

Runlong Cai, Yueyun Fu, Xiaoxiao Li, Jiming Hao and Jingkun Jiang
State Key Joint Laboratory of Environment Simulation and Pollution Control, School of Environment, Tsinghua University, Beijing, 100084, China

Dongsen Yang, Xing Wang, Yan Ma and Jun Zheng
Collaborative Innovation Center of Atmospheric Environment and Equipment Technology, Nanjing University of Information Science & Technology, Nanjing 210044, China

Index

www.ingramcontent.com/pod-product-compliance
Lightning Source LLC
Chambersburg PA
CBHW080636200326
41458CB00013B/4645